COMPOSITE FERMIONS

When electrons are confined to two dimensions, cooled to near absolute zero temperature, and subjected to a strong magnetic field, they form an exotic new collective state of matter, which rivals superfluidity and superconductivity in both its scope and the elegance of the phenomena associated with it. Investigations into this state began in the 1980s with the observations of integral and fractional quantum Hall effects, which are among the most important discoveries in condensed matter physics. The fractional quantum Hall effect and a stream of other unexpected findings are explained by a new class of particles: composite fermions.

A self-contained and pedagogical introduction to the physics and experimental manifestations of composite fermions, this textbook is ideal for graduate students and academic researchers in this rapidly developing field. The topics covered include the integral and fractional quantum Hall effects, the composite fermion Fermi sea, geometric observations of composite fermions, various kinds of excitations, the role of spin, edge state transport, electron solid, and bilayer physics. The author also discusses fractional braiding statistics and fractional local charge. This textbook contains numerous exercises to reinforce the concepts presented in the book.

JAINENDRA JAIN is Erwin W. Mueller Professor of Physics at the Pennsylvania State University. He is a fellow of the John Simon Guggenheim Memorial Foundation, the Alfred P. Sloan Foundation, and the American Physical Society. Professor Jain was co-recipient of the Oliver E. Buckley Prize of the American Physical Society in 2002.

Pre-publication praise for Composite Fermions:
"Everything you always wanted to know about composite fermions by its primary architect and champion. Much gorgeous theory, of course, but also an excellent collection of the relevant experimental data. For the initiated, an illuminating account of the relationship between the composite fermion model and other models on stage. For the novice, a lucid presentation and dozens of valuable exercises."
Horst Stormer, Columbia University, NY and Lucent Technologies. Winner of the Nobel Prize in Physics in 1998 for discovery of a new form of quantum fluid with fractionally charged excitations.

COMPOSITE FERMIONS

Jainendra K. Jain
The Pennsylvania State University

CAMBRIDGE UNIVERSITY PRESS
Cambridge, New York, Melbourne, Madrid, Cape Town,
Singapore, São Paulo, Delhi, Tokyo, Mexico City

Cambridge University Press
The Edinburgh Building, Cambridge CB2 8RU, UK

Published in the United States of America by Cambridge University Press, New York

www.cambridge.org
Information on this title: www.cambridge.org/9781107404250

First published 2007
First paperback edition 2011

A catalogue record for this publication is available from the British Library

ISBN 978-0-521-86232-5 Hardback
ISBN 978-1-107-40425-0 Paperback

To Manju, Sunil, and Saloni

Contents

Preface

Odd how the creative power at once brings the whole universe to order.
Virginia Woolf

When electrons are confined to two dimensions, cooled to near absolute zero temperature, and subjected to a strong magnetic field, they form a quantum fluid that exhibits unexpected behavior, for example, the marvelous phenomenon known as the fractional quantum Hall effect. These properties result from the formation of a new class of particles, called "composite fermions," which are bound states of electrons and quantized microscopic vortices. The composite fermion quantum fluid joins superconductivity and Bose–Einstein condensation in providing a new paradigm for collective behavior.

This book attempts to present the theory and the experimental manifestations of composite fermions in a simple, economical, and logically coherent manner. One of the gratifying aspects of the theory of composite fermions is that its conceptual foundations, while profoundly nontrivial, can be appreciated by anyone trained in elementary quantum mechanics. At the most fundamental level, the composite fermion theory deals directly with the solution of the Schrödinger equation, its physical interpretation, and its connection to the observed phenomenology. The basics of the composite fermion (CF) theory are introduced in Chapter 5. The subsequent chapters, with the exception of Chapter 12, are an application of the CF theory in explaining and predicting phenomena. Detailed derivations are given for many essential facts. Formulations of composite fermions using more sophisticated methods are also introduced, for example, the topological Chern–Simons field theory. To keep the book within a manageable length, many developments are mentioned only briefly, but my hope is that this book will at least serve as a useful first resource for any reader interested in the field. It can be used as a textbook for a graduate level special topics course, or selected portions from it can be used in the standard graduate course on condensed matter physics or many-body theory. Many simple exercises have been included to provide useful breaks.

Disclosure: Personally, the most difficult aspect that I have faced while writing this book has been my own long and intimate involvement with composite fermions, which, one might hope, would enhance the probability that the exposition is sometimes well thought out, but makes it difficult to ensure the kind of objectivity that can come only with distance, both in space and in time. Fortunately, much is known that is indisputable, which allows one to distinguish opinions and speculations from facts. The selection of topics, the emphasis, and the logic of presentation reflect my views on what is firmly established, what is important, and how it should be taught. For the theoretical part, my preference has been for concepts and formulations that directly relate to laboratory and/or computer experiments. If work in which I have participated appears more often than it deserves, it is because that is what I know and understand best. My sincere apologies are extended to those who might feel their work is not adequately represented or, worse, misrepresented. I have made an effort to supply the original references to the best of my knowledge and ability, but the list is surely incomplete. I have collected, over the years, many nuggets of knowledge by osmosis, and my suspicion is that some of them may have crept into the book without proper attribution. The book is not intended, and ought not to be taken, as a historical account.

This book is an account of the collective contributions of too many scientists to name individually. It is a pleasure to acknowledge my profound debt to many colleagues whose wisdom and collaboration have benefited me over the years. These include, but are not limited to, Alexei Abrikosov, Phil Allen, Phil Anderson, Jayanth Banavar, G. Baskaran, Lotfi Belkhir, Nick Bonesteel, Moses Chan, Albert Chang, Chiachen Chang, Sankar Das Sarma, Goutam Dev, Rui Du, Jim Eisenstein, Herb Fertig, Eduardo Fradkin, Steve Girvin, Gabriele Giuliani, Fred Goldhaber, Vladimir Goldman, Ken Graham, Devrim Güçlü, Duncan Haldane, Bert Halperin, Hans Hansson, Jason Ho, Gun Sang Jeon, Shivakumar Jolad, Thierry Jolicoeur, Rajiv Kamilla, Woowon Kang, Anders Karlhede, Tetsuo Kawamura, Steve Kivelson, Klaus von Klitzing, Paul Lammert, Bob Laughlin, Patrick Lee, Seung Yeop Lee, Jon Magne Leinaas, Mike Ma, Allan MacDonald, Jerry Mahan, Sudhansu Mandal, Noureddine Meskini, Alexander Mirlin, Ganpathy Murthy, Wei Pan, Kwon Park, Vittorio Pellegrini, Mike Peterson, Aron Pinczuk, John Quinn, T. V. Ramakrishnan, Sumathi Rao, Nick Read, Nicolas Regnault, Ed Rezayi, Tarek Sbeouelji, Vito Scarola, R. Shankar, Mansour Shayegan, Chuntai Shi, Boris Shklovskii, Steve Simon, Shivaji Sondhi, Doug Stone, Horst Stormer, Aron Szafer, David Thouless, Csaba Tőke, Nandini Trivedi, Dan Tsui, Cyrus Umrigar, Susanne Viefers, Giovanni Vignale, Xiao Gang Wu, Xincheng Xie, Frank Yang, Jinwu Ye, Fuchun Zhang, and Lizeng Zhang.

I am indebted to Jayanth Banavar, Vin Crespi, Herb Fertig, Fred Goldhaber, Ken Graham, Devrim Güçlü, Gun Sang Jeon, Shivakumar Jolad, Paul Lammert, Ganpathy Murthy, Mike Peterson, Csaba Tőke, Giovanni Vignale, and Dave Weiss for a careful and critical reading of parts of the manuscript. Thanks are also due to Wei Pan for making available the trace used for the cover page, to Gabriele Giuliani for bringing to my attention the quotation at the beginning of the Preface, and to the National Science Foundation for financially supporting my research on composite fermions.

I am tremendously grateful to Gun Sang Jeon for his help with numerous figures.

Finally, I express my deepest gratitude to my family, Manju, Sunil, and Saloni, who patiently put up with my preoccupation during the writing of this book. Little had I realized at the beginning how major an undertaking it would be. But the experience has been instructive and also greatly rewarding. I hope that the reader will find the book useful.

Symbols and abbreviations

b	Chern–Simons magnetic field
B	external magnetic field
B^*	effective magnetic field experienced by composite fermions
	also denoted B_{eff}, B_{CF} or ΔB in the literature
CF	composite fermion
^{2p}CF	composite fermion carrying $2p$ vortices
CFCS	composite fermion Chern–Simons
CF-LL	Λ level
CS	Chern–Simons
e^*	local charge of an excitation
e_{CS}	Chern–Simons electric field
ϵ	dielectric function of the background material ($\epsilon \approx 13$ for GaAs)
FQHE	fractional quantum Hall effect
$\hbar\omega_{\text{c}}$	cyclotron energy
$\hbar\omega_{\text{c}}^*$	cyclotron energy of composite fermion
IQHE	integral quantum Hall effect
k	wave vector
ℓ	magnetic length ($\ell = \sqrt{\hbar c / eB}$)
L	total orbital angular momentum in spherical geometry, or
	total z component of the angular momentum in the disk geometry
ΛL	Λ level; Landau-like level of composite fermions
LL	Landau level
LLL	lowest Landau level
m_{a}^*	activation mass of composite fermion
m_{b}	electron band mass ($m_{\text{b}} = 0.067 m_{\text{e}}$ in GaAs)
m_{e}	electron mass in vacuum
m_{p}^*	polarization mass of composite fermion
n	integral filling factor; LL or Λ-level index
N	number of electrons/composite fermions
ν	filling factor of electrons
ν^*	filling factor of composite fermions

ϕ_0	flux quantum ($\phi_0 = hc/e$)
Φ_{ν^*}	wave function of noninteracting electrons at ν^*
$\mathcal{P}_{\text{LLL}}$	lowest Landau level projection operator
Ψ_ν	wave function of interacting electrons at ν
Q	monopole strength
Q^*	effective monopole strength
QHE	quantum Hall effect
R_{H}, ρ_{xy}	Hall resistance
R_{K}	the von Klitzing constant ($R_{\text{K}} = h/e^2$)
R_L, ρ_{xx}	longitudinal resistance
RPA	random phase approximation
ρ	two-dimensional density
SMA	single mode approximation
TL	Tomonaga–Luttinger
TSG	Tsui–Stormer–Gossard
2DES	two-dimensional electron system
V_{C}	Coulomb energy scale ($V_{\text{C}} \equiv \frac{e^2}{\epsilon\ell} \approx 50\sqrt{B[\text{T}]}$ K for GaAs)
V_m	interaction pseudopotentials
V_m^{CF}	interaction pseudopotentials for composite fermions
z	position in 2D ($z \equiv x - \text{i}y$)

1
Overview

The past two and a half decades have witnessed the birth and evolution of a new quantum fluid that has produced some of the most profoundly beautiful structures in physics. This fluid is formed when electrons are confined to two dimensions, cooled to near absolute zero temperature, and subjected to a strong magnetic field. In this overview chapter, results and ideas are stated without explanation; they reappear later in the book with greater elaboration.

1.1 Integral quantum Hall effect

The field began in 1980 with the discovery of the integral quantum Hall effect (IQHE), when von Klitzing observed (Fig. 1.1) plateaus in the plot of the Hall resistance as a function of the magnetic field. The Hall resistance on these plateaus is precisely quantized at

$$R_{\mathrm{H}} = \frac{h}{ne^2},$$

(1.1)

where n is an integer, h is the Planck constant, and e is the electron charge. Quantizations in physics are as old as quantum mechanics itself. What is remarkable about the quantization of the Hall resistance is that it is a universal property of a complex, macroscopic system, independent of materials details, sample type or geometry; it is also robust to variation in temperature and disorder, provided they are sufficiently small. Furthermore, concurrent with the quantized plateaus is a "superflow," i.e., a dissipationless current flow in the limit of zero temperature.

It has been known since the early days of quantum mechanics that a magnetic field quantizes the kinetic energy of an electron confined in two dimensions. These quantized kinetic energy levels are called Landau levels (LLs), which are separated by a "cyclotron energy" gap. The number of filled Landau levels, which depends on the electron density (ρ) and the magnetic field (B), is called the filling factor (ν), given by

$$\nu = \frac{\rho h c}{eB}.$$

(1.2)

The $R_{\mathrm{H}} = h/ne^2$ plateau occurs in the magnetic field range where $\nu \approx n$.

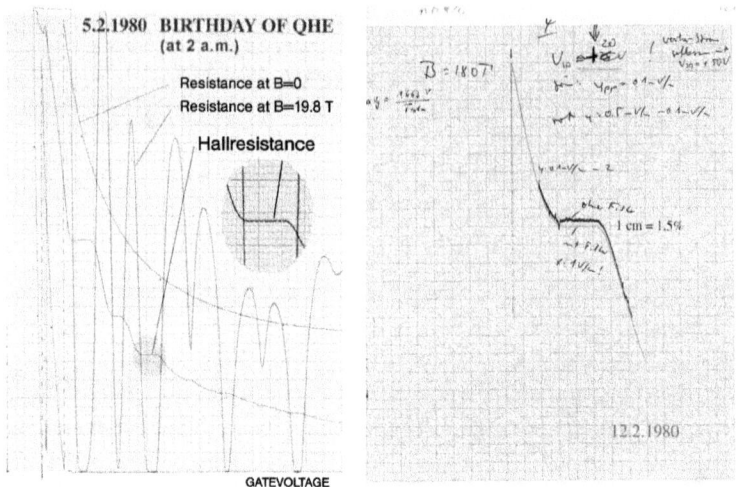

Fig. 1.1. Left: Discovery of the integral quantum Hall effect by K. von Klitzing (February 5, 1980). Right: The first high precision measurement (February 12, 1980). Courtesy: K. von Klitzing. (Reprinted with permission.)

The IQHE is a consequence of the LL quantization of the electron's kinetic energy in a magnetic field, and can be understood in terms of free electrons. When the filling factor is an integer ($\nu = n$), the ground state is especially simple: N electrons occupy the N single particle orbitals of the lowest n Landau levels, described by a wave function that is a single Slater determinant, denoted Φ_n. The state is "incompressible," i.e., its excitations cost a nonzero energy. This incompressibility at integral fillings, combined with disorder-induced Anderson localization, leads to an explanation of the IQHE [368].

1.2 Fractional quantum Hall effect

The 1982 discovery by Tsui, Stormer and Gossard of the 1/3 effect, i.e., of a quantized plateau at

$$R_{\mathrm{H}} = \frac{h}{\frac{1}{3}e^2},\tag{1.3}$$

(Fig. 1.2) raised the field to a whole new level. This plateau is seen when the lowest Landau level is approximately 1/3 full (i.e., $\nu \approx 1/3$). Laughlin [369] stressed the correlated nature of the "1/3 state," and wrote an elegant wave function for the ground state at $\nu = 1/m$:

$$\Psi_{1/m} = \prod_{j<k}(z_j - z_k)^m\, e^{-\frac{1}{4}\sum_i |z_i|^2}.\tag{1.4}$$

Here, $z_j = x_j - iy_j$ denotes the two-dimensional coordinates of the jth electron as a complex number, and the magnetic length $\ell = \sqrt{\hbar c/eB}$ has been chosen as the unit of length. The

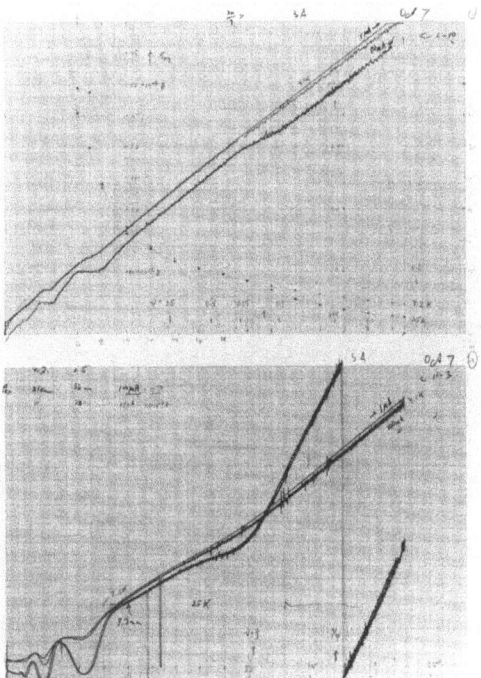

Fig. 1.2. Discovery of the fractional quantum Hall effect (FQHE) by D. C. Tsui, H. L. Stormer and A. C. Gossard. October 7, 1981. Source: H. L. Stormer, *Rev. Mod. Phys.* **71** 875–889 (1999). (Reprinted with permission.)

exponent m must be an odd integer to ensure antisymmetry. Laughlin argued that this state is incompressible, and that the excitations have a fractional charge of magnitude e/m.

The 1/3 plateau had offered only the first glimpse into an extraordinary new collective state of matter. Subsequent experimentation revealed new phenomena whose breathtaking beauty and richness would have been impossible to anticipate, by any stretch of the imagination, at the time of the first observation of 1/3. Many more quantized plateaus,

$$R_{\mathrm{H}} = \frac{h}{fe^2},$$
(1.5)

labeled by different fractions f, were observed. The number of fractions rapidly mushroomed over the years with refinements in experimental conditions, and new fractions are being discovered even now. Figure 1.3 shows a more recent magnetoresistance trace. An explanation of the physics of this new collective state of matter, which rivals superconductivity and Bose–Einstein superfluidity in both scope and elegance of the phenomena associated with it, has been an exciting challenge for, and accomplishment of, the modern many-body condensed matter theory.

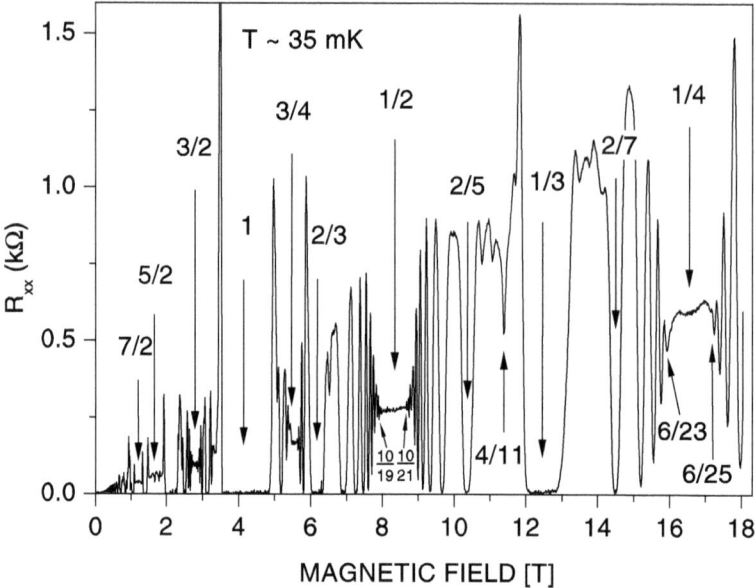

Fig. 1.3. The magnificent FQHE skyline. Diagonal resistance as a function of the magnetic field for a two-dimensional electron system with a mobility of 10 million cm²/V s. A FQHE or an IQHE state is associated with each minimum. Many arrows only indicate the positions of filling factors (for example, 1/2, 1/4, etc.) and have no FQHE associated with them. Source: W. Pan, H. L. Stormer, D. C. Tsui, L. N. Pfeiffer, K. W. Baldwin, and K. W. West, *Phys. Rev. Lett.* **88**, 176802 (2002). (Reprinted with permission.)

1.3 Strongly correlated state

At high magnetic fields all electrons occupy the lowest Landau level. Their kinetic energy is then constant, hence irrelevant. Interacting electrons in a high magnetic field are mathematically described by the Hamiltonian

$$H = \sum_{j<k} \frac{1}{r_{jk}} \quad \text{(lowest Landau level),} \tag{1.6}$$

which is to be solved in the lowest Landau level (LLL) subspace. (The quantity $r_{jk} = |z_j - z_k|$ is the distance between the electrons j and k.) The "pure" FQHE problem has no parameters. The Hamiltonian looks simple until one makes the following (related) observations:

(i) **The no-small-parameter problem** The theory contains no parameters. (The Coulomb interaction merely sets the overall energy scale.)

(ii) **The degeneracy problem** The number of degenerate ground states in the absence of interaction (with $H = 0$) is astronomically large.

(iii) **The no-normal-state problem** In some instances, a nontrivial collective phenomenon can be understood as an instability of a "normal state," which is the state that would be obtained if the interaction could be switched off. For the FQHE problem, switching off the interaction does not produce a meaningful state. The FQHE has no normal state. The physics of the FQHE state is "strongly" non-perturbative, in the sense that no weak coupling limit exists in which the solution is close to a known solution. We do not have the luxury of calculating small deviations from a normal state, but must determine "full" answers.

We wish to identify the principle responsible for the dramatic physics revealed by experiment, but do not even know where to begin. Yet this has turned out to be a theorist's dream problem, in which, thanks to an abundance of experimental clues and an incredible amount of luck, a precise and secure understanding has been achieved. A new language has been developed to describe the concepts governing the behavior of this state. Many facets of the FQHE physics, as well as numerous related phenomena, follow directly from a single unifying principle: the formation of "composite fermions."

1.4 Composite fermions

With the observation of many fractions, an analogy with the IQHE could be identified, which has led to an explanation of the FQHE (and more). The eigenfunctions and eigenenergies for the ground and (low-energy) excited states of strongly interacting electrons at an arbitrary LLL filling ν are expressed in terms of the known solutions of the noninteracting electron problem at the LL filling ν^* as follows:

$$\Psi_\nu = \mathcal{P}_{\mathrm{LLL}} \prod_{j<k} (z_j - z_k)^{2p} \Phi_{\nu^*}, \tag{1.7}$$

and

$$E_\nu = \frac{\langle \Psi_\nu | \sum_{j<k} \frac{1}{r_{jk}} | \Psi_\nu \rangle}{\langle \Psi_\nu | \Psi_\nu \rangle} + V_{\mathrm{el-bg}} + V_{\mathrm{bg-bg}}, \tag{1.8}$$

where

$$\nu = \frac{\nu^*}{2p\nu^* \pm 1}, \tag{1.9}$$

$$B^* = B - 2p\rho\phi_0. \tag{1.10}$$

Here, Φ_{ν^*} are the eigenfunctions of noninteracting electrons at ν^*, $\mathcal{P}_{\mathrm{LLL}}$ projects the wave function to its right into the lowest Landau level, p is an integer, B^* is an "effective" magnetic field, ρ is the two-dimensional density, $\phi_0 = hc/e$ is called the flux quantum, and $V_{\mathrm{el-bg}}$ and $V_{\mathrm{bg-bg}}$ are electron–background and background–background interaction energies (assuming a uniform positively charged neutralizing background). These equations define a one-to-one correspondence between the ground and excited states at filling factor ν (or magnetic field B) and ν^* (magnetic field B^*). They have the following physical interpretation (which is more generally valid than Eqs. (1.7) and (1.8)):

- The Jastrow factor $\prod_{j<k} (z_j - z_k)^{2p}$ attaches $2p$ quantized vortices to each electron in Φ_{ν^*}. The bound state of an electron and $2p$ vortices is interpreted as a particle called the "composite fermion."

It is sometimes pictured and modeled as the bound state of an electron and an even number of magnetic flux quanta, although that picture should not be taken literally. Electrons capture vortices to turn into composite fermions because that is how they minimize their interaction energy.

- As composite fermions move about, the vortices bound to them produce Berry phases, which cancel part of the Aharonov–Bohm phases originating from the external magnetic field. Composite fermions thus sense a magnetic field B^* that is much smaller than the applied magnetic field, and can even be zero. This property of composite fermions distinguishes them from electrons, and lies at the root of most of the dramatic phenomenology. Composite fermions form Landau-like levels (called Λ levels) in the reduced magnetic field, and occupy ν^* of them.
- The right hand side of the expression for Ψ_ν (Eq. 1.7) is interpreted as the wave function of composite fermions at filling factor ν^*.

Detailed quantitative calculations have shown Eqs. (1.7) and (1.8) to be accurate, and experiments have confirmed their interpretation in terms of composite fermions. Composite fermions have been directly observed in many experiments and their numerous consequences have been verified in repeated tests over the last decade and a half. Composite fermions embody the nonperturbative reorganization that takes place when a collection of two-dimensional electrons is subjected to a strong magnetic field.

Equations (1.7) and (1.8) represent the "quantum mechanical wave function" formulation of composite fermions. The physics of composite fermions has also been implemented through other calculational frameworks, most notably a topological field theory known as the composite fermion Chern–Simons (CFCS) theory. The Laughlin wave function is recovered as a special case of Eq. (1.7) with $\nu^* = 1$, $2p = m - 1$, and with Φ_1 taken as the ground state at filling factor one.

Composite fermions are bizarre particles in many respects. They represent a new class of particles realized in nature. Previously known fermions were either elementary fermions or their bound states. A composite fermion, on the other hand, is the bound state of an electron and an even number of quantized vortices. The vortex is a collective, topological, quantum object. It is not a degree of freedom in the Hamiltonian but an emergent state of *all* electrons; it has quantum mechanical phases associated with it; and it is a topological entity because the quantum mechanical phase associated with a closed loop around a vortex is exactly 2π, independent of the shape and the size of the loop. (The topological character of vortices is implicit in the fact that we *count* them.) As a result, composite fermions are collective, topological, quantum particles. We note: (a) Even a single composite fermion is a collective bound state of all electrons. It is surprising that composite fermions behave as almost free, ordinary particles to a great extent. (b) The quantum mechanical phases associated with the vortex give the composite fermion an inherently quantum mechanical character. While quantum mechanics describes all particles, it is responsible for the very creation of composite fermions. A purely classical world would have no composite fermions. (c) All fluids of composite fermions are topological quantum fluids. The topological quantization of the vorticity of composite fermions is responsible for Hall quantization, the effective magnetic field, and numerous other phenomena. The emergence of such a complex particle is a testament to the genuinely collective character of this quantum fluid.

1.5 Origin of the FQHE

The formation of composite fermions leads to an immediate understanding of the enigmatic FQHE of electrons as a manifestation of the IQHE of composite fermions. The latter occurs when the CF filling factor is an integer, i.e., $\nu^* = n$. The integral fillings of composite fermions correspond to electron filling factors given by

$$\nu = \frac{n}{2pn \pm 1},\tag{1.11}$$

which nicely match the prominently observed fractions. This explanation clarifies:

- the origin of gaps at certain fractional fillings (composite fermions fill n Landau-like levels);
- appearance of sequences of fractions (fractions are derived from the sequence of integers);
- their order of stability;
- the abundance of odd-denominator fractions;
- exactness of the Hall quantization (the right hand side of Eq. (1.11) is made up of whole numbers, hence invariant under weak perturbations; the integral value of p has a topological origin).

At relatively low magnetic fields, when the spin degree of freedom is not frozen, several FQHE states with different spin polarizations become possible at each of the fractions in Eq. (1.11). These are explained in terms of composite fermions carrying spin.

1.6 The composite fermion quantum fluid

In addition to explaining the FQHE at the fractions in Eq. (1.11), composite fermions open the door into a new world where many other phenomena can be studied experimentally and explained or predicted theoretically. Some notable states of composite fermions are (also including their IQHE for completeness):[1]

- IQHE of composite fermions [$\nu = n/(2pn \pm 1)$]
- composite fermion Fermi sea ($\nu = 1/2$)
- Bardeen–Cooper–Schrieffer-like p-wave paired state ($\nu = 5/2$)
- FQHE of composite fermions (e.g., $\nu = 4/11$)
- quantum crystal of composite fermions*
- composite fermion stripes and bubble crystals*

Many phenomena of composite fermions (other than the IQHE and the FQHE) have been measured:

- Shubnikov–de Haas oscillations
- commensurability oscillations
- surface acoustic wave absorption

---

[1] The items marked by an asterisk have not been observed experimentally yet.

- thermopower
- magnetic focusing
- bilayer drag
- compressibility

These have enabled a determination of the quantum numbers and other parameters of composite fermions:

- charge
- spin
- exchange statistics
- effective mass
- magnetic moment
- Fermi wave vector
- cyclotron orbits

Composite fermions have been directly observed in geometric experiments, where they can be manipulated like billiard balls, injected at one point and collected at another. Such experiments have measured trajectories of the current-carrying objects and found them to be consistent with the effective magnetic field, confirming that the current carriers are not electrons but composite fermions.

Additionally, many excitations of composite fermions have been observed:

- charged excitations
- excitons
- rotons
- bi-rotons
- skyrmions
- spin-flip excitations
- cyclotron resonance
- flavor-altering excitations

Excited composite fermions are called CF-quasiparticles. The topological character of composite fermions results in the following unusual properties for the CF-quasiparticles:

- Fractional "local" charge: The charge excess associated with a CF-quasiparticle is a precise fraction of an electron charge.
- Fractional "braiding" statistics: CF-quasiparticles are believed to acquire, theoretically, nontrivial phases under braiding around one another. They are currently the most promising candidate for a physical realization of "anyons," i.e., particles obeying fractional braiding statistics.

Experimental evidence exists for fractional charge. The braiding property has not yet been verified experimentally.

1.7 An "ideal" theory

The composite fermion theory possesses many qualities desirable in a theory.

Unification   It explains very many experimental facts starting from a single principle. It obtains all fractions in an equivalent manner, unifies the fractional and the integral quantum Hall effects, and describes states that do not exhibit FQHE (for example, the compressible states at even denominator fractions). It gives a unified theory of quasiparticles, quasiholes, and various other excitations of FQHE states.

Simplicity   It provides simple intuitive explanations for the basic phenomenology of the fractional quantum Hall effect, e.g., for the appearance of certain sequences of odd-denominator fractions, and the lack of FQHE at even-denominator fractions (with one exception).

Uniqueness   An aim of theoretical physics is to reduce the number of adjustable parameters to the maximum extent possible. The composite fermion principle is "super-constraining" in that it provides parameter-free answers for many quantities of interest. In particular, it uniquely identifies wave functions for the incompressible ground states and their low-energy excitations.

Accuracy   Rigorous, unbiased and detailed tests against exact "computer experiments" on finite systems have shown that the CF theory gives a faithful counting of the low-energy eigenstates, and that the wave functions of Eq. (1.7) are practically exact: They typically have close to 100% overlap with the exact eigenstates, and their energies are accurate to better than 0.1%. This level of accuracy is especially noteworthy in view of the absence of adjustable parameters in the theory.

Microscopic theory   Condensed matter theories aspire to explain macroscopic phenomena by solving the microscopic Hamiltonian describing a collection of interacting electrons and ions. In practice, however, the path from the microscopic Hamiltonian to the macroscopic phenomenology often passes through one or more layers of approximation that can be justified only with the help of detailed testing against experiment. The enormously successful BCS theory, for example, is built upon the crucial assumption that the normal state can be modeled in terms of weakly interacting Landau quasiparticles, which are the variables in terms of which the BCS wave function is expressed.[2] In contrast, the wave functions of the CF theory are the actual wave functions that occur in the laboratory. They are written in terms of the real electron coordinates, and represent the solutions of the Hamiltonian of Eq. (1.6) in a quantum chemistry sense.

Falsifiability   Many sharp and nontrivial consequences and predictions stemming from the CF theory have been verified during the past 1.5 decades. Predictions play a more

---

[2] A lack of such simplification lies at the root of the difficulties encountered in the explanation of high-temperature superconductivity.

important role in the establishment of a theory than explanations of known facts. Many nontrivial predictions of the CF theory, of both qualitative and quantitative natures, have been confirmed.

New particles   Emergent physics often expresses itself through new particles. In spite of their rather complex and unconventional nature, composite fermions behave as legitimate particles with mass, charge, spin, statistics, and other properties that we associate with particles. Composite fermions have numerous other particle-like manifestations; they exhibit many phenomena that electrons do.[3]

1.8 Miscellaneous remarks

The overview concludes with some other interesting facts about the CF liquid.

Macroscopic quantum phenomena without BEC   We instinctively associate one or another kind of Bose–Einstein condensation (BEC) with the phrase "macroscopic quantum phenomenon," be it a ^{4}He superfluid or a BCS superconductor (the latter being, crudely, a BEC of Cooper pairs). These systems are characterized by the formation of a macroscopic quantum state with phase rigidity. The CF fluid constitutes a distinct paradigm for macroscopic quantum behavior. It *is* a quantum fluid, because quantum mechanical phases play a central role in determining its macroscopic behavior. These phases enter, however, in the form of quantized vortices, which are captured by electrons to produce new physics.

Order without order parameter   The order in the CF fluid is characterized by the formation of composite fermions. There is no off-diagonal long-range order in the CF fluid, nor an order parameter in the sense defined by Landau. The CF fluid is a topological quantum fluid. The topological nature of the state is evidenced most directly through the effective magnetic field, which is responsible for the FQHE, the CF Fermi sea, and numerous other phenomena.

"Meissner effect" without induced currents   In a "type-I" superconductor, an applied magnetic field is fully screened, known as the Meissner effect. Macroscopic currents are induced that produce a magnetic field of their own that exactly cancels the external field. The partial "cancellation" of the external magnetic field by the CF fluid is not caused by an induced current, but by induced Berry phases, and has a purely quantum mechanical origin. The cancellation is totally "internal" to composite fermions. An external magnetometer will assess the full applied magnetic field, but when composite fermions themselves are used to measure the magnetic field, the reduced effective magnetic field is detected.

---

[3] "They're as real as Cooper pairs." H. L. Stormer [623]

Mass generation without Anderson–Higgs mechanism The Anderson–Higgs mechanism plays a central role in generating mass in the standard model of particle theory. The CF physics illustrates a new mechanism for mass generation without any spontaneously broken symmetry. The Hamiltonian in Eq. (1.6) has no kinetic energy or mass term in it, because the LLL limit is equivalent to setting the electron mass equal to zero. Nonetheless, through the nonperturbative act of vortex assimilation, massless electrons transform into massive composite fermions. The CF mass, which originates entirely from interelectron interactions, manifests itself through quantized "kinetic energy" levels of composite fermions, their Fermi sea, and various other phenomena. (Some similarity to the Anderson–Higgs mechanism does exist, in that the "massless" electrons of the lowest Landau level "eat up" part of the vector potential to turn into massive composite fermions.)

History has taught us that nature has the economical habit of recycling any given principle in many different contexts, although often in such delightfully concealed guises that it takes deep insights to reveal the connection. Many concepts discovered in condensed matter physics, e.g., spontaneous symmetry breaking and the Anderson–Higgs mechanism, have found crucial applications in other areas of physics. It is irresistible to wonder what other important physics problems await resolution with the help of the exotic new structures and paradigms realized in the FQHE and the CF fluid. Could particles like composite fermions or anyons emerge in other contexts? Could they provide an underlying model for some known particles? Are there deeper conceptual implications of the composite fermion theory that we have not yet fully grasped?

2

Quantum Hall effect

Transport anomalies have given the first inkling for many new states of matter, for example superconductivity. That is also the case for the quantum fluid considered in this book, whose first clue came from the discovery of the "quantum Hall effect" (QHE), which is the topic of this chapter. An understanding of the effect has led to numerous other phenomena, which are discussed in later chapters.

We begin with the status of our understanding of the Hall effect prior to 1980, to prepare us for fully appreciating the surprise caused by the discovery of the QHE. The discussion of the classical Hall effect is brief, as it is a part of every textbook on solid state physics.

2.1 The Hall effect

Ohm's law, $I = V/R$, implies that the current through a resistor is proportional to the applied voltage. In most situations, current flows in the same direction as the applied electric field. The local, and more fundamental form of the law is

$$J = \sigma E \tag{2.1}$$

where σ is the conductivity, and $J = q\rho v$ is the current density for particles of charge q and density ρ moving with a velocity v.

In 1879, E. H. Hall [227] discovered that, in the presence of a magnetic field, current flows in a direction *perpendicular* to the direction of the applied field. Alternatively, passage of a current induces a voltage perpendicular to the direction of the current flow. This is known as the Hall effect. See Fig. 2.1. (In the presence of disorder, the induced voltage also has a component in the direction of the current flow.)

The phenomenon has a classical origin. The Lorentz equation of electrodynamics

$$F = q\left(E + \frac{1}{c}v \times B\right) \tag{2.2}$$

gives the force on a particle of charge q, moving with a velocity v, in the presence of electric and magnetic fields. A consequence of this equation is that for crossed electric and magnetic fields, say $E = E\hat{y}$ and $B = B\hat{z}$, the charged particle drifts in a direction perpendicular to the plane containing the two fields, with a velocity $v = c(E/B)\hat{x}$. (Because we are interested

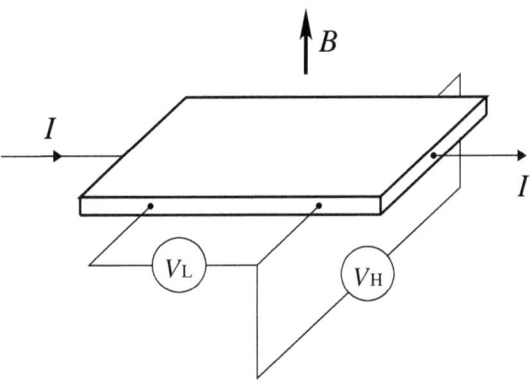

Fig. 2.1. Schematics of magnetotransport measurements. I, V_L, and V_H are the current, longitudinal voltage, and the Hall voltage, respectively. The longitudinal and Hall resistances are defined as $R_L \equiv V_L/I$ and $R_H \equiv V_H/I$.

in linear response, E is taken to be arbitrarily small.) This can be derived by integrating Newton's equations of motion. This can also be seen by "boosting" to a frame of reference moving with the velocity $v = c(E/B)\hat{x}$. From the relativistic transformation law for the electromagnetic field, the electric field in the moving frame is given by

$$E'_x = E_x,$$ (2.3)

$$E'_y = \gamma \left(E_y - \frac{v}{c} B_z \right),$$ (2.4)

$$E'_z = \gamma \left(E_z + \frac{v}{c} B_y \right),$$ (2.5)

where $\gamma = 1/\sqrt{1 - v^2/c^2}$. The electric field has been completely eliminated in the frame moving with our choice of v. Depending on initial conditions, a particle in the moving frame can execute various trajectories. A particle at rest in the moving frame remains at rest, which can also be seen in the laboratory frame by noting that no net force acts on a particle moving with velocity $v = c(E/B)\hat{x}$. In general, the projection of the particle motion in the x–y plane is a circle (in the moving frame), called the cyclotron orbit, the center of which does not move. (The center can move in the $\hat{z}$ direction, but that is of no relevance to the considerations here: no net current flows in that direction, because, for each particle moving in the $+\hat{z}$ direction, another is moving in the $-\hat{z}$ direction.) In the laboratory frame, the motion appears complicated, but the center of the cyclotron orbit moves with a drift velocity $v = c(E/B)\hat{x}$.

Current density is given by $J = q\rho v$, where ρ is the (three-dimensional) density of particles. That produces the Hall resistivity

$$\rho_H = \frac{E_y}{J_x} = \frac{B}{\rho q c}.$$ (2.6)

The proportionality of the Hall resistivity to B has been fully confirmed. The Hall measurement is used routinely to measure the density ρ of the mobile charges, as well as the sign of the charge carriers (i.e., whether they are electrons or holes).

The Hall resistance is conceptually simpler to understand than the ordinary resistance of Ohm's law. In the absence of a magnetic field, a charged particle in a pure system will keep accelerating under the influence of an electric field, producing an ever-increasing current; scattering from impurities is essential to a theoretical derivation of Ohm's law via a Drude-type approach. The Hall current, in contrast, remains finite even in the absence of disorder. The name "Hall resistance" is somewhat of a misnomer: even though it has the dimensions of resistance, no energy dissipation is associated with the Hall current. No work is needed to maintain the Hall current, because no force acts on the particle. (These statements refer an idealized situation. Real experimental systems with disorder and phonons also have an ordinary resistance, called the longitudinal resistance, which requires a power input to maintain a steady current.)

In two dimensions, the observed behavior is dramatically different from that described above. We begin with a brief discussion of what the phrase "two-dimensional electron system" (2DES) means and how a 2DES is produced in the laboratory.

2.2 Two-dimensional electron system

In physics literature, the phrase "two-dimensional electron system" sometimes refers to electrons in a hypothetical universe that has no third (spatial) dimension. In that case, Gauss's law $\nabla^2 V(r) = -4\pi\rho$, where ρ is the two-dimensional charge density, implies a logarithmic interaction between the electrons. That is *not* what "2DES" means in this book, which is concerned with electrons confined to two dimensions in a three-dimensional universe. The interaction between electrons has the standard Coulomb form

$$V(r) = \frac{e^2}{\epsilon r},\tag{2.7}$$

where ϵ is the dielectric constant of the host material. The electric field of an electron in the 2D plane extends in all three dimensions, and the electron interacts with charges off the plane.

The Fourier transform of the interaction in three dimensions is $4\pi e^2/\epsilon q^2$. For electrons confined to two dimensions, we have

$$V(q) = \int d^2r\, V(r) e^{-i k \cdot r}$$

$$= \int_0^\infty dr\, rV(r) \int_0^{2\pi} d\theta \cos(qr\cos\theta)$$

$$= 2\pi \int_0^\infty dr \; rV(r)J_0(qr)$$

$$= \frac{2\pi e^2}{\epsilon q}, \tag{2.8}$$

where the last step assumes $V(r) = e^2/\epsilon r$.

A 2DES is achieved in artificial structures, created by molecular beam epitaxy (MBE), which allows a controllable, layer-by-layer growth, in which one type of semiconductor, say GaAs, can be grown on top of another, approximately lattice-matched semiconductor, say $Al_xGa_{1-x}As$, to produce an atomically sharp interface. Of importance is the fact that the band gaps of the two semiconductors are different, for example, 2.2 eV for AlAs and 1.5 eV for GaAs. We consider here two common geometries:

(i) **Semiconductor quantum well** For the AlGaAs–GaAs–AlGaAs geometry, the band structure is shown in Fig. 2.2 in the absence of doping. (A theoretical understanding of the relative alignment of the bands, or the determination of the "band offset" is a complicated problem.) This band structure provides a realization of the familiar quantum well, studied in undergraduate quantum mechanics as a toy problem. The energy levels in the z direction are quantized into what are known as subbands. The depth of the quantum well (~ 0.4 eV) is often much larger than the other energy scales, and can be taken to be infinite in the simplest approximation. Then, the kinetic energy of a particle in the nth subband is given by

$$E_{n,k} = \frac{\hbar^2 k^2}{2m_b} + \frac{n^2 \pi^2 \hbar^2}{2mw^2}, \tag{2.9}$$

where k is the 2D wave vector in the $x-y$ plane, w is the width of the well (typically 15–50 nm), and $n = 1, 2, \ldots$ is the subband index. Upon doping the AlGaAs region, electrons fall into the GaAs region and occupy, depending on the density of electrons, one or more subbands in the quantum well.

(ii) **Semiconductor heterostructure** A 2DES also results in the heterostructure geometry, shown in Fig. 2.2, when the AlAs is doped with donor atoms. To simplify matters, let us consider a situation in which the donors reside on a plane at $z = -d$, as can be achieved in MBE. (This is called delta doping.) Again, because of the band off-set, the donated electrons fall into the GaAs side to lower their energy. Once on the GaAs side, they are attracted toward the positive charge of the donor ions left behind. Unable to overcome the energy barrier, the best compromise is for them to sit right at the interface, thus forming a 2DES at the interface. A simple model describing this situation is considered in Exercise 2.1.

The potential shown in Fig. 2.2 is for two undoped semiconductors. The actual confinement potential seen by electrons also receives contributions from the fields produced by the donor ions and the free electrons. This contribution was neglected in the above discussion of the quantum well potential. A more accurate description of the transverse

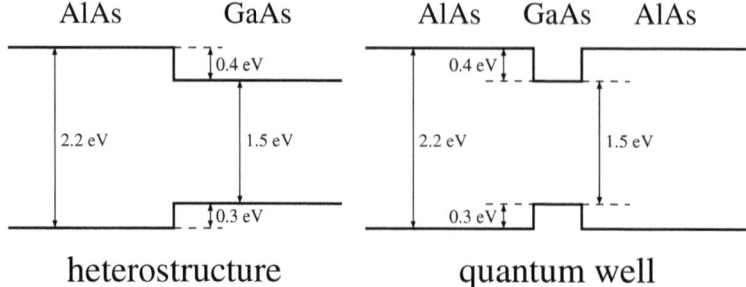

Fig. 2.2. The band diagrams of (undoped) structures containing AlAs and GaAs. The growth direction is the horizontal direction on the page. A heterostructure (left) contains a single, sharp interface with AlAs and GaAs on the two sides. A quantum well (right) contains two nearby interfaces, producing a square well potential in both the conduction and the valence bands. Upon doping the AlAs region with n-type impurities, electrons fall into the GaAs region and form a two-dimensional electron system, as explained in the text.

confinement, subbands, and the effective interaction (below) can be achieved by determining self-consistent solutions. References to relevant literature can be found in Ref. [10].

Given that the electron wave function in the third direction is not a delta function, one may ask: "When and in what sense is the system two dimensional?" The system is two dimensional when only the lowest band of the confinement potential is occupied. Because no degree of freedom is associated with the third dimension, the dynamics of electrons is formally equivalent to that of electrons confined strictly in two dimensions.

The nonzero transverse extent of the electron wave function does have a quantitative effect. The effective two-dimensional interaction is given by

$$V^{\text{eff}}(r) = \frac{e^2}{\epsilon} \int dz_1 \int dz_2 \frac{|\xi(z_1)|^2 |\xi(z_2)|^2}{[r^2 + (z_1 - z_2)^2]^{1/2}}, \tag{2.10}$$

where z_1 and z_2 denote the coordinates normal to the plane containing the electrons, $r^2 = (x_1 - x_2)^2 + (y_1 - y_2)^2$, and $\xi(z)$ is the wave function in the transverse direction. The effective interaction, $V^{\text{eff}}(r)$, is indistinguishable from $e^2/\epsilon r$ at large distances ($r \gg w$, w being the "width" of the wave function), but is less repulsive than Coulomb at short distances. Very roughly, electrons can be viewed as disks rather than point particles. Using the relation

$$\int d^2r \frac{1}{\sqrt{r^2 + d^2}} e^{-i k \cdot r} = 2\pi \int_0^\infty dr \frac{r}{\sqrt{r^2 + d^2}} J_0(qr)$$

$$= \frac{2\pi}{q} e^{-qd}, \tag{2.11}$$

the Fourier transform of the effective interaction is seen to be

$$V^{\text{eff}}(q) = f(q) \frac{2\pi e^2}{\epsilon q}, \tag{2.12}$$

where the form factor is given by

$$f(q) = \int dz_1 \int dz_2 |\xi(z_1)|^2 |\xi(z_2)|^2 \, e^{-q|z_1-z_2|}. \tag{2.13}$$

At the simplest level, the transverse wave function ξ can be taken to be the wave function of an electron in an infinite square quantum well. More sophisticated treatments determine ξ by self-consistently solving the Schrödinger and Poisson equations, taking into account the interaction effects through the local density approximation (LDA) including the exchange correlation potential (Ortalano, He, and Das Sarma [488]; Stern and Das Sarma [612]), which yields a density and geometry dependent wave function.

2.3 The von Klitzing discovery

The "integral quantum Hall effect" (IQHE), also known as the "von Klitzing effect," was discovered by von Klitzing and collaborators[1] [659,660] in 1980 (Figs. 2.3 and 2.4), almost exactly one hundred years after the discovery of the Hall effect, in their study of the Hall effect of two-dimensional electrons in silicon MOSFET (metal oxide-semiconductor field-effect transistor). In two dimensions, the Hall resistance is defined as

$$R_{\mathrm{H}} = \frac{V_{\mathrm{H}}}{I}, \tag{2.14}$$

which, from classical electrodynamics, is expected to be proportional to the magnetic field B. That is indeed what is observed in small magnetic fields. At sufficiently large B, however, quantum mechanical effects appear in a rather dramatic manner. The difference between the behaviors in three and two dimensions is shown schematically in Fig. 2.5.

The two essential observations are:

(i) **Hall quantization** When plotted as a function of the magnetic field B, the Hall resistance exhibits numerous plateaus. On any given plateau, R_{H} is precisely quantized at values given by

$$R_{\mathrm{H}} = \frac{h}{ne^2}, \tag{2.15}$$

where n is an integer. This is referred to as the "integral quantum Hall effect" or the "von Klitzing effect." The $R_{\mathrm{H}} = h/ne^2$ plateau occurs in the vicinity of $\nu \equiv Be/\rho hc = n$, where ν is the "filling factor," i.e., the number of filled Landau levels. (More details are given in the next chapter.)

(ii) **Superflow** In the plateau region, the longitudinal resistance exhibits an Arrhenius behavior:

$$R_{xx} \sim \exp\left(-\frac{\Delta}{2k_{\mathrm{B}}T}\right). \tag{2.16}$$

This gives an energy scale Δ, which is interpreted as a gap in the excitation spectrum. (The gap depends on the filling factor ν.) R_{xx} vanishes in the limit $T \to 0$, indicating dissipationless transport.

---

[1] The history of events leading to the discovery has been recounted by Landwehr [367].

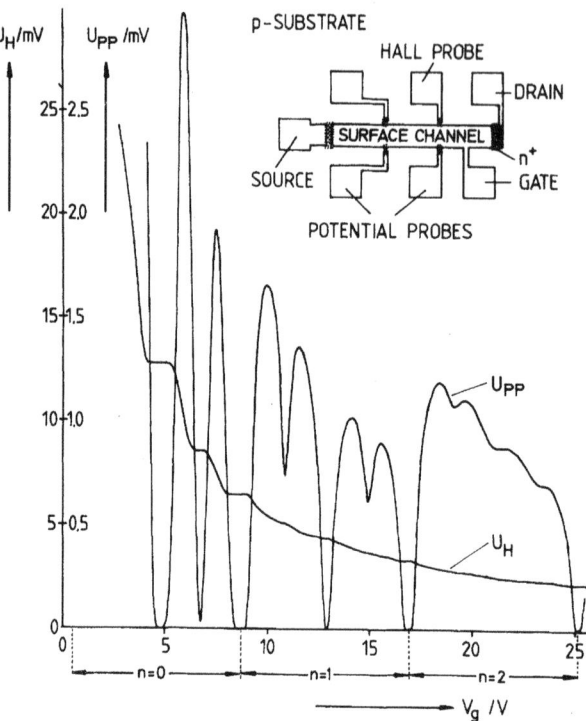

Fig. 2.3. The Hall voltage U_H as a function of density for a fixed magnetic field $B = 18$ T and a fixed current of 1 μA. The density is controlled by the gate voltage V_g. U_{PP} is the voltage drop along the direction of the current. Source: K. von Klitzing, G. Dorda, and M. Pepper, *Phys. Rev. Lett.* **45**, 494 (1980). (Reprinted with permission.)

The following aspects are noteworthy.

- The combination h/e^2 has the unit of resistance, a fact that was little appreciated prior to the discovery of the IQHE.
- The inherently quantum mechanical origin of the effect is evident from the facts that: (i) the Hall resistance is quantized; (ii) the Planck constant makes its appearance prominently in the quantized value of the Hall resistance; and (iii) the results are inconsistent with the classical theory.
- The resis*tance* is quantized, not the resis*tivity*. They both have the same units in two dimensions.
- The Hall quantization is universal. It is independent of the sample type, geometry, and various materials parameters, such as the band mass of the electron or the dielectric constant of the semiconductor. (Figure 2.4 shows the trace for two different samples.) The quantization is also not affected by the unavoidable presence of disorder in condensed matter systems. The effective interaction between electrons can be changed by using materials with different dielectric functions, by working with different transverse confinements, or by placing a conducting plane nearby. All of these also do not produce any correction to the value of the Hall resistance. We do not intend to imply that the Hall resistance is always quantized. The quantization can surely be destroyed, for example, by increasing the disorder or raising the temperature by a substantial amount. But there is a broad range of parameters for which universally quantized Hall plateaus are observed.

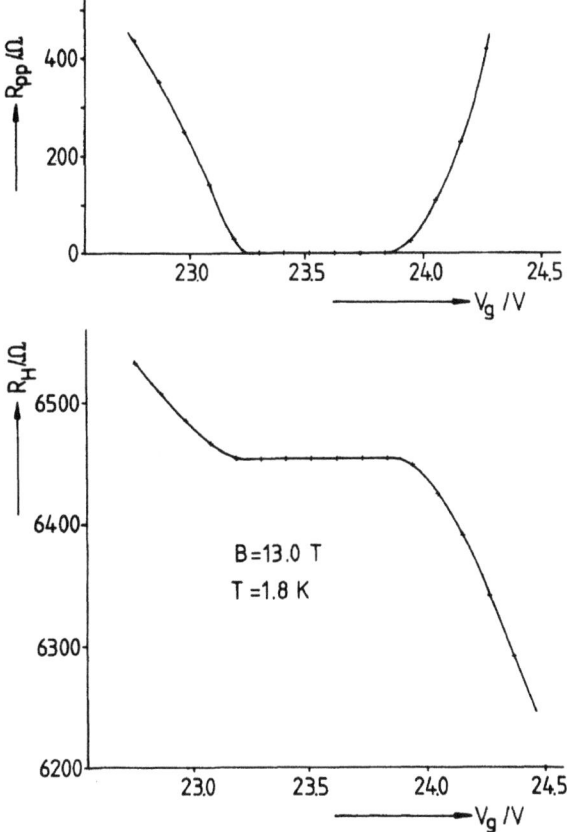

Fig. 2.4. The Hall resistance trace for two different samples for certain parameter ranges. In the plateau region, the resistance is $6453.3 \pm 0.1\ \Omega$. The straight horizontal line depicts the position of $h/4e^2 = 6453.204\ \Omega$. Source: K. von Klitzing, G. Dorda, and M. Pepper, *Phys. Rev. Lett.* **45**, 494 (1980). (Reprinted with permission.)

- A total lack of resistance at $T = 0$ is nontrivial. For ordinary metals, the resistance at $T \to 0$, called the residual resistance, is nonzero and proportional to disorder.
- The Arrhenius behavior for R_{xx} implies an absence of a phase transition as a function of temperature.

2.4 The von Klitzing constant

An important application of the quantum Hall effect has been in metrology. The quantized Hall resistance on the plateau characterized by the integer one has the experimental value

$$25\,813.807\ldots\ \Omega,\qquad\qquad (2.17)$$

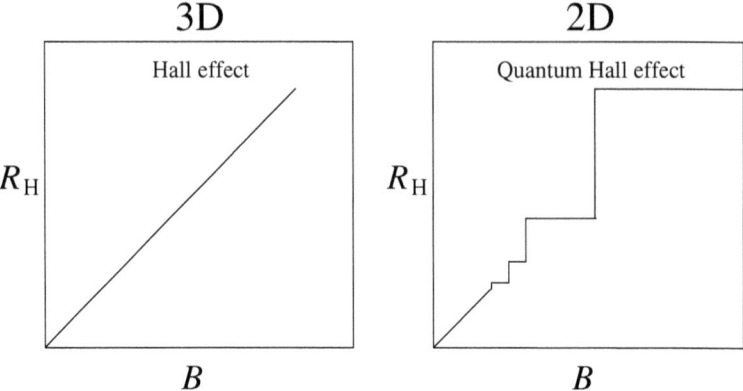

Fig. 2.5. Schematic difference between the Hall effect in three and two dimensions.

which is indistinguishable, within measurement errors, from h/e^2. This number is accepted to be exactly equal to h/e^2, called the von Klitzing constant:

$$R_{\rm K} = \frac{h}{e^2}. \tag{2.18}$$

It was measured by von Klitzing in his discovery article [659] with an accuracy of approximately two parts in ten thousand. The state-of-the-art on the measurement of the von Klitzing constant has been summarized in a recent review (von Klitzing [661]). The uncertainty in the experimental value is dominated by the uncertainty in the realization of the ohm in SI units. The absolute value of the von Klitzing constant is known at the level of 2 parts in 10^7, but its stability and reproducibility in different samples has been confirmed (Bachmair *et al.* [17]; Delahaye *et al.* [111, 112]; Jeckelmann, Inglis and Jeanneret [290]; Jeffery *et al.* [291]) to better than 2 parts in 10^9. The quality of the plateaus is diminished with increasing temperature or decreasing sample size, but no *intrinsic* correction to the quantization (i.e., a correction that would survive in very large samples in the limit of $T \to 0$) is known at present.

The combination h/e^2 also occurs in the definition of the fine structure constant:

$$\alpha = \frac{e^2}{\hbar c}, \tag{2.19}$$

the value of which is approximately 1/137. (Substituting $e = 4.803\,24 \times 10^{-10}$ esu, $\hbar = 1.054\,59 \times 10^{-27}$ erg s, and $c = 2.997\,925 \times 10^{10}$ cm/s gives $\alpha^{-1} = 137.0363$.) Because the speed of light is known much more precisely than the other fundamental constants involved here, the Hall effect measurements in dirty, solid state systems provide an accurate value for α.[2] Because of the relatively large uncertainty in the absolute value of $R_{\rm K}$, however, the

---

[2] This point was already stressed in von Klitzing's discovery article [659], entitled "*New method for high-accuracy determination of the fine-structure constant based on quantized Hall resistance.*"

quantum Hall effect does not give, at present, the most accurate value for the fine structure constant.

2.5 The Tsui–Stormer–Gossard discovery

The next revolution [627, 650, 651] occurred with the discovery of the "fractional quantum Hall effect" (FQHE), or the "Tsui–Stormer–Gossard (TSG) effect," which refers to the observation of plateaus on which the Hall resistance is quantized at

$$R_H = \frac{h}{fe^2},\qquad(2.20)$$

where f is a rational fraction. The observation of a plateau at $R_H = h/fe^2$ is referred to as the observation of the fraction f. The $R_H = h/fe^2$ plateau is seen in the magnetic field region where approximately a fraction f of the lowest Landau level is full (i.e., $\nu \approx f$). Similarly to the IQHE, the longitudinal resistance exhibits an Arrhenius behavior, vanishing in the limit $T \to 0$. The signature of FQHE in R_{xx}, through a minimum at $\nu = f$, is often more robust than the plateau in R_H, and is tentatively regarded as the observation of a fraction; the definitive observation of a fraction requires, however, a reasonably well defined plateau at $R_H = h/fe^2$.

The first fraction ($f = 1/3$) was observed in 1982 (Tsui, Stormer and Gossard [650]) in GaAs heterostructures; the original discovery is shown in Fig. 1.2 and the first published result in Fig. 2.6. An improvement of experimental conditions (higher mobilities, higher magnetic fields, lower temperatures) has led to the observation of a large number of fractions since then. Figures 2.7 and 1.3 show resistance traces from increasingly better quality samples, to be contrasted with the data in Fig. 2.6. The full picture is simply stunning in its richness. At the time of the writing of this book, the number of observed fractions, counting only fractions below unity, is more than 50. Many new fractions are reported in Refs. [61, 89, 132, 204, 440, 441, 496, 620, 680].

Surprisingly little effort has been expended toward establishing the accuracy of the fractional Hall quantization, perhaps because of the more rigorous experimental conditions required for this purpose.[3] Nonetheless, the fractional Hall quantization is also widely accepted to be exact and universal, as corroborated by its insensitivity to many sample parameters.

Every fraction has a distinct quantum mechanical state associated with it. Listing all the observed fractions will not be especially meaningful at this stage. As we see in Chapter 7, a theoretical understanding of the origin of the FQHE allows an arrangement of all fractions and integers together in a unified, periodic-table-like structure (Table 7.1), which reveals a deep and beautiful connection between the fractional and the integral effects, as well as

---

[3] M. Cage, private communication.

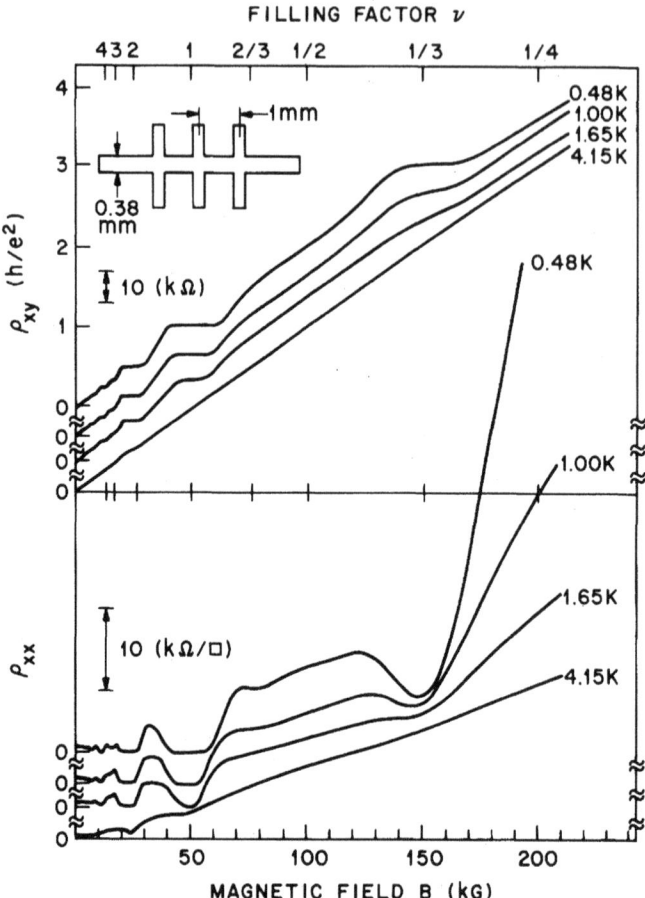

Fig. 2.6. The magnetic field dependence of Hall and longitudinal resistances, ρ_{xy} and ρ_{xx}, for a two-dimensional electron system at the GaAs–AlGaAs interface. Source: D. C. Tsui, H. L. Stormer, and A. C. Gossard, *Phys. Rev. Lett.* **48**, 1559 (1982). (Reprinted with permission.)

between various fractions. The classification of fractions depends on the type ("flavor") of the emergent "particles" and the number of "Landau-level-like shells" they occupy.

2.6 Role of technology

The importance of materials science in driving the field of the quantum Hall effect cannot be overstated. An important development was the seminal idea of "modulation doping," introduced by Stormer and collaborators in 1978 [119, 618, 619], which spatially separates dopant atoms from conduction electrons. The reduced scattering of electrons results in very high mobilities, which has laid the foundation for much of the progress in the field.

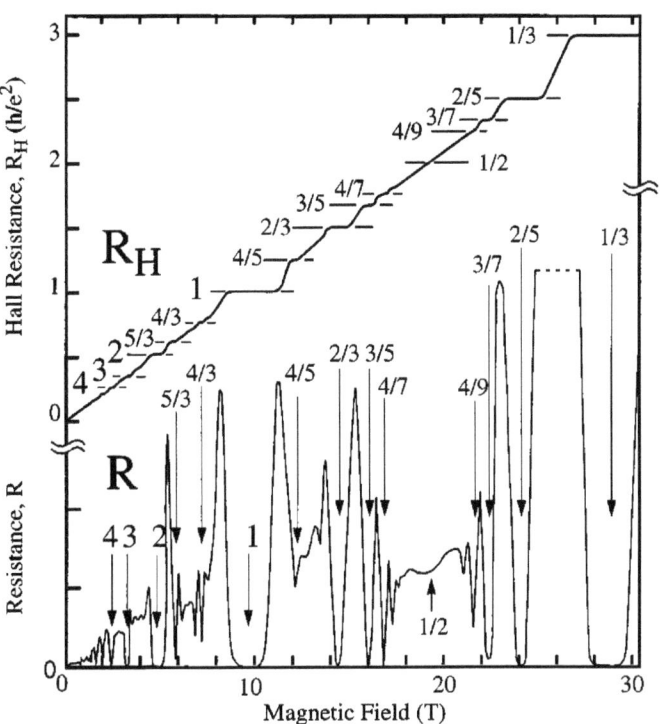

Fig. 2.7. Overview of Hall and longitudinal resistances, R_H and R, respectively. (No plateau is associated with $f = 1/2$.) Source: H. L. Stormer, *Rev. Mod. Phys.* **71** 875–889 (1999). (Reprinted with permission.)

(Such a separation is not feasible in three-dimensional doped semiconductors.) Various improvements in MBE technology (Fig. 2.8 [514]) have resulted in the rise of sample mobility from 100 thousand in 1980 to above 20 million in 2000 (in the conventional units of cm²/Vs) in GaAs based 2D systems. QHE has also been observed in several other materials including: strained Si quantum wells in Si–SiGe heterostructures [268, 362, 452]; single graphite layers, also known as graphene [480, 745]; and perhaps even 2D multilayer organic metals [20]. The discussion in the rest of this book assumes parameters appropriate for GaAs 2D systems, either electron or hole doped, which, to date, have revealed the most extensive phenomenology.

Exercises

2.1 The self-consistent potential experienced by the electrons in the heterostructure geometry can often be approximated by a triangular potential well. The Schrödinger equation for the z component of the wave function for a single electron

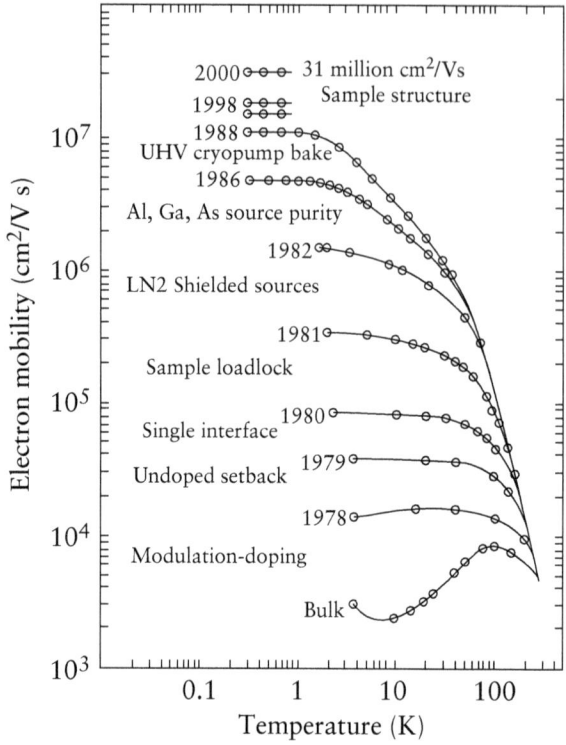

Fig. 2.8. Timeline of electron mobility in 2D GaAs systems. Various technical innovations are also indicated on the figure. Source: L. N. Pfeiffer and K. W. West, *Physica E: Low-dimensional Systems and Nanostructures* **20**, 57–64 (2003). (Reprinted with permission.)

is then given by

$$\left[-\frac{\hbar^2}{2m_b}\frac{d^2}{dz^2} + V(z)\right]\xi_i(z) = E_i\xi_i(z), \tag{E2.1}$$

where the potential energy is taken to be

$$V(z) = \begin{cases} \infty & z < 0 \\ eFz & z > 0. \end{cases}$$

The Schrödinger equation can be solved exactly for this potential [10, 611]. The eigenfunctions are given by Airy functions [2]:

$$\zeta_i(z) = Ai\left[\frac{2m_b eF}{\hbar^2}\left(z - \frac{E_i}{eF}\right)\right] \tag{E2.2}$$

with

$$E_i = \left(\frac{\hbar^2}{2m_b}\right)^{1/3}\left[\frac{3\pi eF}{2}\alpha_i\right]^{2/3}, \tag{E2.3}$$

where $\alpha_0 = 0.7587$, $\alpha_1 = 1.7540$, $\alpha_2 = 2.7575$, and $\lim_{i \to \infty} \alpha_i = i + 3/4$. The average values of z and z^2 are $2E_i/3eF$ and $8E_i^2/15e^2F^2$. This exercise concerns the Fang–Howard variational wave function for the ground subband [150]:

$$\xi(z) = N_b \, z \, \exp\left[-\frac{z}{2b}\right], \tag{E2.4}$$

where N_b is the normalization constant.

(i) Determine b by minimizing the energy expectation value.

(ii) Compare the ground state energy and the average values of z and z^2 with the exact result.

(iii) The parameter F in the potential energy must be determined self-consistently, as the combined electric field from the donor ions as well as electrons themselves. Neglecting the latter for simplicity, F is the electric field due to a positively charged sheet with density $e\rho$ in a delta doped layer at $z = -d$. Derive the formula for the density dependence of the width ($\sim$ average z) of the wave function.

2.2 Obtain the effective interaction $V^{\mathrm{eff}}(r)$ for the Fang–Howard trial wave function considered in the previous problem. Evaluation of the Fourier transform is more straightforward. In which limit (large or small density) does the effective interaction approach the Coulomb interaction? Is this consistent with what you would have anticipated?

2.3 Interacting *bosons* are described by the Hamiltonian

$$H = \sum_j \left(\frac{p^2}{2m} + \frac{1}{2}m\omega^2 z^2\right) + \sum_{j<k} V^{3D}(r_{jk}), \tag{E2.5}$$

$$V^{3D}(r) = \frac{4\pi\hbar^2 a_s}{m} \delta^{3D}(r), \tag{E2.6}$$

where a_s is the s-wave scattering length. Show that when the harmonic confinement in the z direction is strong enough that only the lowest transverse band is occupied, the effective two-dimensional interaction between bosons is given by

$$V^{2D}(r) = \sqrt{8\pi}\,\frac{\hbar^2 a_s}{m\ell_z}\delta^{2D}(r), \tag{E2.7}$$

where $\ell_z = \sqrt{\hbar/(m\omega)}$.

3

Landau levels

The appearance of the Planck constant h in the formula for R_H is an indication of the inherently quantum mechanical nature of the effect. In this chapter, we study a single electron confined to two dimensions and exposed to a magnetic field.[1] This problem was solved exactly soon after the invention of quantum mechanics (Darwin [106]; Fock [169]; Landau [364]), because it is merely a one-dimensional simple harmonic oscillator problem in disguise. The most remarkable aspect of the solution is that the electron kinetic energy is quantized. The discrete kinetic energy levels are called "Landau levels."

The Landau level is the workhorse of the quantum Hall problem. The integral quantum Hall effect is seen below as a direct consequence of the Landau level formation. The explanation for the fractional quantum Hall effect, which is caused by interactions, requires further insights, but Landau levels again provide the key analogy: the effect arises because the lowest Landau level splits into *Landau-like* energy levels (called Λ levels).

This chapter deals with Landau levels in two geometries: planar and spherical. Two gauges are used in the planar geometry, the Landau and the symmetric gauges. The spherical geometry considers an electron moving on the surface of a sphere, subjected to a radial magnetic field. Wrapping the plane on to the surface of a sphere is simply a choice of boundary conditions, which should not affect the bulk properties of the state. Periodic boundary conditions, which are equivalent to a toroidal geometry, are also considered briefly in this chapter, but not used in the rest of the book.

One might ask, "Why consider so many geometries and gauges?" While the physics, in principle, is independent of the choice of gauge or the boundary conditions, certain formulations of the problem are more natural, even crucial, for a description of the relevant physics. Arguably, the physics of the FQHE would not have revealed itself in a gauge other than the symmetric gauge of the planar geometry, and its quantitative confirmation would not have been possible without the spherical geometry.

The periodic potential due to the host lattice is of no relevance to the quantum Hall problem, because the size of the electron wave packet in a magnetic field, on the order of 10–30 nm for typical experimental parameters, is much larger than the lattice period. The

---

[1] Sections 3.9, 3.10.7, 3.11.3, 3.13, 3.14, 3.15 may be omitted. They are not used elsewhere in the book.

periodic potential due to the lattice is, therefore, neglected in studies of the quantum Hall effect.

Finally, we stress that although simply derivable, Landau levels are a nonperturbative consequence of the magnetic field. None of the physics presented in this book would have been possible if, for some reason, we decided to treat the magnetic field perturbatively.

3.1 Gauge invariance

The Hamiltonian for a nonrelativistic electron[2] moving in two dimensions in a perpendicular magnetic field is given by

$$H = \frac{1}{2m_b} \left(p + \frac{eA}{c} \right)^2. \tag{3.1}$$

Here, e is defined to be a positive quantity, the electron's charge being $-e$. For a uniform magnetic field,

$$\nabla \times A = B\hat{z}. \tag{3.2}$$

The vector potential A is a linear function of the spatial coordinates. It follows that H is a generalized two-dimensional harmonic oscillator Hamiltonian which is quadratic in both the spatial coordinates and in the canonical momentum $p = -i\hbar\nabla$, and, therefore, can be diagonalized exactly.

The Schrödinger equation

$$H\Psi = E\Psi \tag{3.3}$$

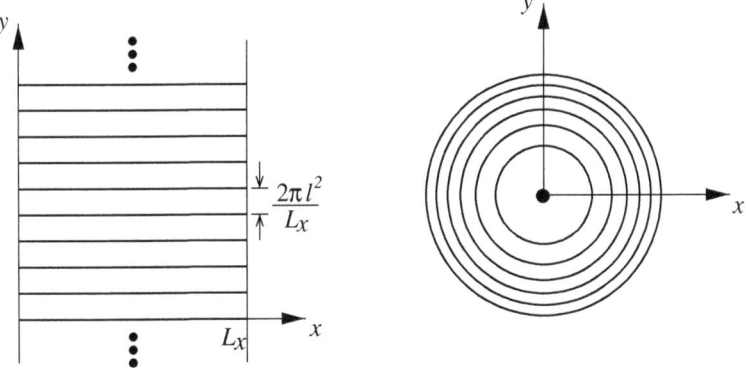

Fig. 3.1. Left panel: Single particle orbitals in the Landau gauge on a strip of width L_x. Periodic boundary conditions are assumed in the x direction (giving the topology of a cylinder). The lines trace the maximum probability for orbitals with different wave vectors k_x; the y location is related to k_x. Right panel: Single particle orbitals in the symmetric gauge for an infinite system. The lines depict the maximum probability for orbitals with different angular momenta.

---

[2] Relativistic effects are neglected throughout.

is invariant under the gauge transformation:

$$A(r) \rightarrow A(r) + \nabla \xi(r), \tag{3.4}$$

$$\Psi(r) \rightarrow \exp\left[-\frac{ie}{\hbar c}\xi(r)\right]\Psi(r). \tag{3.5}$$

This is called gauge invariance.

Calculations are often performed with a specific choice of gauge for the vector potential. We discuss two popular gauges, namely the Landau and the symmetric gauges. The eigenfunctions will look quite different in the two gauges (Fig. 3.1), but are related, for an infinite system, by a unitary transformation. Many essential features can be derived in a gauge independent manner.

3.2 Landau gauge

For the Landau gauge,

$$A = B(-y, 0, 0), \tag{3.6}$$

the Hamiltonian contains no x, and therefore commutes with p_x. That implies that $p_x = \hbar k_x$ is a good quantum number. The convenient unit of length is the magnetic length, defined as

$$\ell = \sqrt{\frac{\hbar c}{eB}}. \tag{3.7}$$

In terms of dimensionless quantities

$$y' = \frac{y}{\ell} - \ell k_x \tag{3.8}$$

and

$$p'_y = \frac{\ell p_y}{\hbar}, \tag{3.9}$$

we get

$$H = \hbar \omega_c \left[\frac{1}{2}y'^2 + \frac{1}{2}(p'_y)^2\right], \tag{3.10}$$

which is the familiar Hamiltonian of a one-dimensional harmonic oscillator. Here, $\hbar \omega_c$ is the cyclotron energy, with $\omega_c = eB/m_b c$. The energy eigenvalues are quantized at

$$E_n = \left(n + \frac{1}{2}\right)\hbar \omega_c, \tag{3.11}$$

with $n = 0, 1, \ldots$, called Landau levels.[3] The continuous energy levels of the zero magnetic field thus combine to produce degenerate Landau levels (Fig. 3.2). The associated eigenvectors are

$$\eta_{n,k_x}(r) = [\pi 2^{2n}(n!)^2]^{-1/4} e^{ik_x x} \exp\left[-\frac{1}{2}\left(\frac{y}{\ell} - \ell k_x\right)^2\right] H_n\left(\frac{y}{\ell} - \ell k_x\right), \tag{3.12}$$

where H_n are Hermite polynomials.

---

[3] We refer to the $n = 0$ Landau level as the *lowest* Landau level, the $n = 1$ Landau level as the *second* Landau level, and so on.

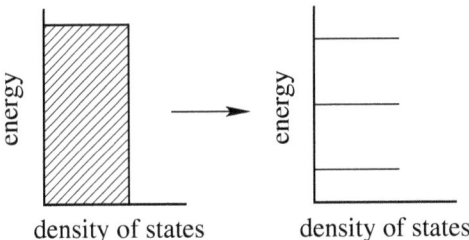

density of states   density of states

Fig. 3.2. The Fermi sea splits into equally spaced, degenerate Landau levels upon the application of a magnetic field.

Two points are worth noting:

- The energy does not depend on k_x. The eigenstates with different k_x in a given Landau level are degenerate.
- The y position depends on k_x. An eigenfunction is Gaussian-localized in a narrow strip of width $\sim \ell$ centered at $y = k_x \ell^2$.

3.3 Symmetric gauge

The symmetric gauge refers to the choice

$$\boldsymbol{A} = \frac{\boldsymbol{B} \times \boldsymbol{r}}{2} = \frac{B}{2}(-y, x, 0). \tag{3.13}$$

In terms of dimensionless lengths and energies, the Hamiltonian can be expressed as

$$H = \frac{1}{2}\left[\left(-i\frac{\partial}{\partial x} - \frac{y}{2}\right)^2 + \left(-i\frac{\partial}{\partial y} + \frac{x}{2}\right)^2\right]. \tag{3.14}$$

For notational convenience, we use the same notation for dimensionless x, y, and H; the correct units can be restored by introducing factors of ℓ and $\hbar \omega_c$. We next transform to new variables z and $\bar{z}$:[4]

$$z = x - iy = r\,\mathrm{e}^{-i\theta}, \qquad \bar{z} = x + iy = r\,\mathrm{e}^{i\theta}. \tag{3.15}$$

The reason for the unconventional choice (as opposed to $z = x + iy$), as we see below, is that the lowest Landau level (LLL) wave functions involve only z's, and not $\bar{z}$'s, apart from a Gaussian factor. The above definition will save us from writing everywhere $\bar{z}$'s later on. The derivatives are related as:

$$\frac{\partial}{\partial x} = \frac{\partial}{\partial z} + \frac{\partial}{\partial \bar{z}} \tag{3.16}$$

and

$$\frac{\partial}{\partial y} = -i\left(\frac{\partial}{\partial z} - \frac{\partial}{\partial \bar{z}}\right). \tag{3.17}$$

---

[4] The symbol θ is used for the azimuthal angle in the disk geometry, whereas θ and ϕ are used, respectively, for the polar and the azimuthal angles in the spherical geometry.

With z and $\bar{z}$ as independent variables, the Hamiltonian becomes:

$$H = \frac{1}{2}\left[-4\frac{\partial^2}{\partial z \partial \bar{z}} + \frac{1}{4}z\bar{z} - z\frac{\partial}{\partial z} + \bar{z}\frac{\partial}{\partial \bar{z}} \right].$$ (3.18)

We define the following sets of ladder operators:

$$b = \frac{1}{\sqrt{2}}\left(\frac{\bar{z}}{2} + 2\frac{\partial}{\partial z} \right),$$ (3.19)

$$b^\dagger = \frac{1}{\sqrt{2}}\left(\frac{z}{2} - 2\frac{\partial}{\partial \bar{z}} \right),$$ (3.20)

$$a^\dagger = \frac{1}{\sqrt{2}}\left(\frac{\bar{z}}{2} - 2\frac{\partial}{\partial z} \right),$$ (3.21)

$$a = \frac{1}{\sqrt{2}}\left(\frac{z}{2} + 2\frac{\partial}{\partial \bar{z}} \right),$$ (3.22)

which have the property that

$$[a, a^\dagger] = 1, \qquad [b, b^\dagger] = 1,$$ (3.23)

and all other commutators are zero. In terms of these operators the Hamiltonian can be written as

$$H = a^\dagger a + \frac{1}{2}.$$ (3.24)

The LL index n is the eigenvalue of $a^\dagger a$. The z component of the angular momentum operator is defined as

$$L = -i\hbar\frac{\partial}{\partial \theta}$$

$$= -\hbar\left(z\frac{\partial}{\partial z} - \bar{z}\frac{\partial}{\partial \bar{z}} \right)$$ (3.25)

$$= -\hbar(b^\dagger b - a^\dagger a).$$

(In the disk geometry only the the z component of the angular momentum is relevant, so we have suppressed the subscript of L_z.) Exploiting the property $[H, L] = 0$, the eigenfunctions are chosen to diagonalize H and L simultaneously. The eigenvalue of L is denoted by $-m\hbar$; with this definition the quantum number m takes values

$$m = -n, -n+1, \ldots, 0, 1, \ldots$$ (3.26)

in the nth Landau level. The application of $b^\dagger$ increases m by one unit while preserving n, whereas $a^\dagger$ simultaneously increases n and decreases m by one unit.

The analogy to the harmonic oscillator problem immediately gives the solution

$$H|n, m\rangle = E_n|n, m\rangle, \tag{3.27}$$

$$E_n = \left(n + \frac{1}{2}\right), \tag{3.28}$$

$$|n, m\rangle = \frac{(b^\dagger)^{m+n}}{\sqrt{(m+n)!}} \frac{(a^\dagger)^n}{\sqrt{n!}} |0, 0\rangle, \tag{3.29}$$

where $m = -n, -n+1, \ldots$ is the angular momentum quantum number defined above. The single particle orbital at the bottom of the two ladders defined by the two sets of raising and lowering operators is

$$\langle r|0, 0\rangle \equiv \eta_{0,0}(r) = \frac{1}{\sqrt{2\pi}} e^{-\frac{1}{4}z\bar{z}}, \tag{3.30}$$

which satisfies

$$a|0, 0\rangle = b|0, 0\rangle = 0. \tag{3.31}$$

The single particle states are especially simple in the lowest Landau level ($n = 0$):

$$\eta_{0,m} = \langle r|0, m\rangle = \frac{(b^\dagger)^m}{\sqrt{m!}} \eta_{0,0} = \frac{z^m e^{-\frac{1}{4}z\bar{z}}}{\sqrt{2\pi 2^m m!}}, \tag{3.32}$$

where we have used Eqs. (3.20) and (3.30). Aside from the ubiquitous Gaussian factor, a general state in the lowest Landau level is simply given by a polynomial of z. It does not involve any $\bar{z}$. In other words, apart from the Gaussian factor, the lowest Landau level wave functions are analytic functions of z.[5]

The eigen function for a general n and m is given by

$$\eta_{n,m}(r) = \langle r|n, m\rangle \tag{3.36}$$

$$= \frac{1}{\sqrt{2\pi 2^{m+2n} n!(m+n)!}} \left(\frac{\bar{z}}{2} - 2\frac{\partial}{\partial z}\right)^n \left(\frac{z}{2} - 2\frac{\partial}{\partial \bar{z}}\right)^{m+n} e^{-\frac{z\bar{z}}{4}}.$$

This can be expressed in terms of standard functions by writing $e^{-z\bar{z}/4} = e^{z\bar{z}/4} e^{-z\bar{z}/2}$, and noting that

$$e^{-\frac{1}{4}z\bar{z}} \left(\frac{\bar{z}}{2} - 2\frac{\partial}{\partial z}\right)^n \left(\frac{z}{2} - 2\frac{\partial}{\partial \bar{z}}\right)^{m+n} e^{\frac{1}{4}z\bar{z}} = \left(-2\frac{\partial}{\partial z}\right)^n \left(-2\frac{\partial}{\partial \bar{z}}\right)^{m+n}, \tag{3.37}$$

---

[5] A polynomial which is only a function of z is an analytic function because, for a function $f(x, y) = u(x, y) + iv(x, y)$, the condition

$$\frac{\partial f}{\partial \bar{z}} = 0 \tag{3.33}$$

can be seen, with the help of

$$\frac{\partial}{\partial \bar{z}} = \frac{1}{2}\left(\frac{\partial}{\partial x} - i\frac{\partial}{\partial y}\right), \tag{3.34}$$

to be equivalent to the famous Cauchy–Riemann analyticity conditions (appropriate to our nonstandard definition of $z = x - iy$)

$$\frac{\partial u}{\partial x} = -\frac{\partial v}{\partial y}, \quad \frac{\partial u}{\partial y} = \frac{\partial v}{\partial x}. \tag{3.35}$$

which gives

$$\eta_{n,m}(\boldsymbol{r}) = \frac{1}{\sqrt{2\pi\, 2^{m+2n}n!(m+n)!}}\, e^{\frac{z\bar{z}}{4}} \left(-2\frac{\partial}{\partial z}\right)^{n} \left(-2\frac{\partial}{\partial \bar{z}}\right)^{m+n} e^{-\frac{z\bar{z}}{2}}. \tag{3.38}$$

The derivative $-2\partial/\partial\bar{z}$ acting on the Gaussian produces a factor of z. Inserting appropriate powers of $\bar{z}$ allows us to write

$$\eta_{n,m}(\boldsymbol{r}) = \frac{1}{\sqrt{2\pi\, 2^{m+2n}n!(m+n)!}}(-1)^{n}2^{m+n}$$

$$\times \bar{z}^{-m}\, e^{\frac{z\bar{z}}{4}} \left(\frac{\partial}{\partial\frac{\bar{z}z}{2}}\right)^{n} \left(\frac{\bar{z}z}{2}\right)^{m+n} e^{-\frac{z\bar{z}}{2}}. \tag{3.39}$$

Defining $t = z\bar{z}/2 = r^2/2$, and using the Rodrigues definition of the associated Laguerre polynomial,

$$L_n^\alpha(t) = \frac{1}{n!}\, e^t t^{-\alpha}\, \frac{d^n}{dt^n}\left(e^{-t}t^{n+\alpha}\right), \tag{3.40}$$

yields

$$\eta_{n,m}(\boldsymbol{r}) = \frac{(-1)^n}{\sqrt{2\pi}}\sqrt{\frac{n!}{2^m(m+n)!}}\, e^{-\frac{r^2}{4}} z^m L_n^m\left(\frac{r^2}{2}\right). \tag{3.41}$$

3.4 Degeneracy

Each Landau level has degenerate orbitals labeled by the quantum numbers k_x and m in the Landau and symmetric gauges. The degeneracy per unit area is the same in each Landau level, but depends on the magnetic field.

In the Landau gauge, the orbital labeled by k_x is localized at $y = k_x\ell^2$. To facilitate the counting of states, we take a sample of length L_x along the x-direction, and impose periodic boundary conditions in the x direction (Fig. 3.1, left panel):

$$e^{ik_x(x+L_x)} = e^{ik_x x}. \tag{3.42}$$

The allowed values of k_x, then, are

$$k_x = 2\pi\frac{n_x}{L_x}. \tag{3.43}$$

We now count the number of states in a given area, say the region of area $L_x L_y$ defined by $y = 0$ and $y = L_y$. (The sample itself extends infinitely in the y-direction. We do not wish to complicate the issue with real edges here.) The state at $y = 0$ is labeled by $n_x = 0$ and the one at $y = L_y$ by wave vector $k_x = L_y/\ell^2$, or by the integer

$$N_x = \frac{L_x L_y}{2\pi\ell^2}. \tag{3.44}$$

Neglecting $O(1)$ effects, N_x is the number of states in the area $L_x L_y$, yielding for the degeneracy per unit area:

$$G = \frac{1}{2\pi \ell^2} = \frac{B}{\phi_0}. \tag{3.45}$$

The last term tells us that there is precisely one state per flux quantum ($\phi_0 = hc/e$) in each Landau level.

The LL degeneracy can also be obtained readily in the symmetric gauge. Consider a disk of radius R centered at the origin, and ask how many states lie inside it in a given Landau level. Taking, for simplicity, the lowest Landau level, the eigenstate $|0, m\rangle$ has its weight located at the circle of radius $r = \sqrt{2m} \cdot \ell$. Thus the largest value of m for which the state falls inside the disk is given by $m = R^2/2\ell^2$, which is also the total number of eigenstates in the lowest Landau level that fall inside the disk (ignoring order one corrections). Thus, the degeneracy per unit area is $(2\pi l^2)^{-1} = eB/hc$. Of course, the degeneracy is the same everywhere; we could, for example, have counted all states from R_1 to R_2.

In the symmetric gauge the eigenstates are localized along a circle (Fig. 3.1, left panel). This circle should not be confused with the cyclotron orbit of classical physics. The energy does not depend on the radius (in the lowest Landau level). The electron is making very tiny cyclotron orbits, of radius $\sim \ell$, and the whole object is going around very slowly in a larger circle. The energy depends only on the tiny orbits. Of course, the shape of the single particle eigenstates is gauge dependent. Other bases can be constructed with other shapes.

In the presence of disorder the degeneracy of the states in a Landau level is lifted, the density of states is broadened, and a Landau band is produced. The belief is that there are localized states in the tails and extended states at the center of each Landau level. This is discussed further in Chapter 4.

3.5 Filling factor

The filling factor is defined by

$$\nu = \frac{\rho}{G} = 2\pi l^2 \rho = \frac{\rho}{B/\phi_0}, \tag{3.46}$$

where ρ is the 2D density of electrons. The filling factor is equal to the number of electrons per flux quantum penetrating the sample. This quantity plays a crucial role in the physics of the QHE.

The quantity ν is called the filling factor because it equals the number of occupied Landau levels for *noninteracting* electrons at a given magnetic field. In general, interactions and disorder cause some amount of LL mixing; higher LL states are occupied with a nonzero probability even for $\nu < 1$.

The filling factor is inversely proportional to the magnetic field. As the magnetic field is increased, each Landau level can accommodate more and more electrons, and, as a result, fewer and fewer Landau levels are occupied.

3.6 Wave functions for filled Landau levels

For a general filling factor, for noninteracting electrons in the absence of disorder, the ground state is highly degenerate, because electrons in the partially filled topmost Landau level can be arranged in a large number of ways. For an integral filling factor, $\nu = n$, the ground state is unique, containing n fully occupied Landau levels. Its wave function is denoted by Φ_n. In the disk geometry, a Landau level will be viewed as "full" if all states inside a radius R are occupied.

The wave function of one filled Landau level, Φ_1, has a simple form in the symmetric gauge, given by (apart from a normalization factor):

$$\Phi_1 = \begin{vmatrix} 1 & 1 & 1 & \cdot & \cdot \\ z_1 & z_2 & z_3 & \cdot & \cdot \\ z_1^2 & z_2^2 & z_3^2 & \cdot & \cdot \\ \cdot & \cdot & \cdot & \cdot & \cdot \\ \cdot & \cdot & \cdot & \cdot & \cdot \end{vmatrix} \exp\left[-\frac{1}{4}\sum_i |z_i|^2\right] \tag{3.47}$$

$$= \prod_{j<k}(z_j - z_k) \exp\left[-\frac{1}{4}\sum_i |z_i|^2\right]. \tag{3.48}$$

We have used the fact that the determinant, which is the well-known Vandermonde determinant,[6] is simply equal to $\prod_{j<k}(z_j - z_k)$ (apart from an unimportant overall sign). Let us derive the result in two ways. The first one is to consider the determinant as a function of a specific coordinate, say z_1. The determinant is a polynomial of z_1 of degree $N - 1$, N being the number of particles. From the "fundamental theorem of algebra," we know that a polynomial of degree $N - 1$ has precisely $N - 1$ zeroes in the complex plane. At the same time, we also know that the polynomial must vanish whenever z_1 coincides with the position of another particle z_j, $j > 1$. Thus, the zeroes must sit on the remaining $N - 1$ particles. This implies a factor $\prod_k'(z_1 - z_k)$, where the prime denotes the condition $k \neq 1$. The antisymmetry of the determinant tells us that the factor must actually appear in the form $\prod_{j<k}(z_j - z_k)$. This argument also proves that any LLL wave function must have the form $\prod_{j<k}(z_j - z_k)F[\{z_i\}]$, where $F[\{z_i\}]$ is a symmetric polynomial. In the second proof, we multiply each row by z_1 and subtract from the next row. The factor $\prod_j'(z_1 - z_j)$ can be pulled out and what remains is precisely the Vandermonde determinant for the remaining $N - 1$ particles. An iteration of the process yields the desired result.

The wave function Φ_1 describes a translationally invariant state, apart from the choice of a center of mass, as can be made transparent by writing it as (Laughlin [370])

$$\Phi_1 = \prod_{j<k}\left[(z_j - z_k)\exp\left(-\frac{1}{4N}|z_j - z_k|^2\right)\right]\exp\left(-\frac{N}{4}\left|\frac{1}{N}\sum_{l=1}^N z_l\right|^2\right). \tag{3.49}$$

---

[6] Named after Alexandre-Theophile Vandermonde.

Except for the last factor, which depends only on the center of mass coordinates, the wave function is now explicitly invariant under $z_j \rightarrow z_j + \eta$.

The wave function for the state with n filled Landau levels is again a determinant formed from the appropriate single particle states. There is some ambiguity, in the disk geometry (but not in the spherical), regarding the number of particles in each Landau level, but a natural choice is to put N/n particles in each Landau level in the innermost single particle orbitals. (Another choice could have been to fill orbitals up to a given angular momentum in each Landau level.) The form of the determinant can be simplified to some extent by using row operations, which we illustrate with the example of Φ_2. The single particle states in the second Landau level are given by

$$\langle r|1, m\rangle = -[2\pi 2^{m+2}(m+1)!]^{-1/2} z^m (2m+2-z\bar{z}) \exp\left[-\frac{z\bar{z}}{4}\right]. \tag{3.50}$$

In the determinant for Φ_2, the term proportional to z^m in the second-LL single particle wave function can be eliminated with the help of the LLL row containing the same term, and the wave function simplifies to

$$\Phi_2 = \begin{vmatrix} 1 & 1 & 1 & . & . \\ z_1 & z_2 & z_3 & . & . \\ z_1^2 & z_2^2 & z_3^2 & . & . \\ . & . & . & . & . \\ . & . & . & . & . \\ z_1^{N/2-1} & z_2^{N/2-1} & z_3^{N/2-1} & . & . \\ \bar{z}_1 & \bar{z}_2 & \bar{z}_3 & . & . \\ \bar{z}_1 z_1 & \bar{z}_2 z_2 & \bar{z}_3 z_3 & . & . \\ \bar{z}_1 z_1^2 & \bar{z}_2 z_2^2 & \bar{z}_3 z_3^2 & . & . \\ . & . & . & . & . \\ . & . & . & . & . \\ \bar{z}_1 z_1^{N/2-1} & \bar{z}_2 z_2^{N/2-1} & \bar{z}_3 z_3^{N/2-1} & . & . \end{vmatrix} \exp\left[-\frac{1}{4}\sum_i |z_i|^2\right]. \tag{3.51}$$

Similar wave functions can be written for $n \geq 3$ filled Landau levels. Further simplification of Φ_n has not been possible for $n \geq 2$.

Filled Landau levels have the special property of being incompressible. Can we see that from the wave functions? The state Φ_1 has a remarkably simple nodal structure: When viewed as a function of one particle, say z_1, with all other particles held fixed, the wave function has no wasted zeroes; the only zeroes are on the other particles, as mandated by the Pauli principle. It is natural to wonder if a more general class of nodal structures can be associated with incompressibility. No special properties are known about the zeroes of the wave function Φ_n for $n \neq 1$, due, in part, to the feature that the polynomial now is a function of two variables, z and $\bar{z}$. No direct connection between the nodal property of the wave function and incompressibility has been established so far.

3.7 Lowest Landau level projection of operators

When applied to a LLL state, an operator $V(\bar{z}, z)$, in general, produces states with amplitude in higher Landau levels. It is convenient to work with its LLL projection, $V_p(\bar{z}, z)$, which satisfies

$$\langle\phi|V_p(\bar{z}, z)|\phi'\rangle = \langle\phi|V(\bar{z}, z)|\phi'\rangle, \tag{3.52}$$

and for which $V_p(\bar{z}, z)|\phi\rangle$ resides within the lowest Landau level, where $|\phi\rangle$ and $|\phi'\rangle$ are arbitrary LLL wave functions. Girvin and Jach [190] have shown that:

Theorem 3.1 The projected operator is given by

$$V_p(\bar{z}, z) = e^{-\frac{1}{4}z\bar{z}} : V\left(\bar{z} \rightarrow 2\frac{\partial}{\partial z}, z\right) : e^{\frac{1}{4}z\bar{z}} \tag{3.53}$$

where the normal ordering symbol : : *on the right hand side refers to bringing all factors of $\bar{z}$ to the left of the z's in each term. It is customary to drop the Gaussian factors on the right hand side of Eq. (3.53), with the convention that the derivatives do not act on the Gaussian part of the wave function.*

Proof   We begin by expanding the operator $V(\bar{z}, z)$ as

$$V(\bar{z}, z) = \sum_{j,k} C_{jk}\bar{z}^j z^k. \tag{3.54}$$

The projected operator according to the above prescription is given by

$$V_p(\bar{z}, z) = e^{-\frac{1}{4}z\bar{z}} \sum_{j,k} C_{jk}\left(2\frac{\partial}{\partial z}\right)^j z^k e^{\frac{1}{4}z\bar{z}}. \tag{3.55}$$

When V_p acts on any LLL wave function of the form $z^n e^{-\frac{1}{4}z\bar{z}}$, it produces a LLL wave function, because the derivatives do not produce any factors containing $\bar{z}$. It remains to show that $V(\bar{z}, z)$ and $V_p(\bar{z}, z)$ have identical matrix elements in the lowest Landau level. Let us consider the matrix element of one of the terms in the expansion of $V(\bar{z}, z)$ between two arbitrary angular momentum eigenstates:

$$\langle m|V(\bar{z}, z)|n\rangle = \sum_{j,k} C_{jk} \int d^2r\, \bar{z}^m e^{-\frac{1}{4}z\bar{z}} (\bar{z}^j z^k) z^n e^{-\frac{1}{4}z\bar{z}}$$

$$= \sum_{j,k} C_{jk} \int d^2r\, \bar{z}^m z^k z^n \left(-2\frac{\partial}{\partial z}\right)^j e^{-\frac{1}{2}z\bar{z}}$$

$$= \sum_{j,k} C_{jk} \int d^2r \, e^{-\frac{1}{2}z\bar{z}} \bar{z}^m \left(2\frac{\partial}{\partial z}\right)^j z^k z^n$$

$$= \sum_{j,k} C_{jk} \int d^2r \, \bar{z}^m \, e^{-\frac{z\bar{z}}{4}}$$

$$\times \left[e^{-\frac{z\bar{z}}{4}} \left(2\frac{\partial}{\partial z}\right)^j z^k \, e^{\frac{z\bar{z}}{4}} \right] z^n \, e^{-\frac{z\bar{z}}{4}}$$

$$= \langle m|V_p(\bar{z},z)|n\rangle.$$

$\blacksquare$

Projection of the coordinates x and y into the lowest Landau level gives

$$x_p = e^{-\frac{1}{4}z\bar{z}} \left(\frac{\partial}{\partial z} + \frac{1}{2}z\right) e^{\frac{1}{4}z\bar{z}}, \tag{3.56}$$

$$y_p = e^{-\frac{1}{4}z\bar{z}} \left(-i\frac{\partial}{\partial z} + \frac{i}{2}z\right) e^{\frac{1}{4}z\bar{z}}. \tag{3.57}$$

They satisfy the commutator

$$[x_p, y_p] = i\ell^2, \tag{3.58}$$

where we have restored the units on the right hand side. Thus, after projection into the lowest Landau level, the x and y coordinates of an electron act like canonically conjugate variables.[7] From the Heisenberg uncertainty principle, it follows that

$$\Delta x_p \Delta y_p \geq \frac{1}{2}\ell^2, \tag{3.59}$$

where the Δx_p and Δy_p are the standard deviations in the two coordinates for a given wave function. It is therefore not possible to write a wave function within the lowest Landau level for which both the x and y coordinates are sharply defined.

3.8 Gauge independent treatment

Many of the above results can be derived in a gauge invariant fashion. In terms of the kinetic momentum,

$$\boldsymbol{\pi} = \boldsymbol{p} + \frac{e}{c}\boldsymbol{A}, \tag{3.60}$$

the Hamiltonian is given by

$$H = \frac{1}{2m_b}(\pi_x^2 + \pi_y^2). \tag{3.61}$$

---

[7] The space becomes "noncommutative" in the lowest Landau level.

From the commutator,

$$[\pi_x, \pi_y] = \frac{-i\hbar e}{c} \hat{z} \cdot (\nabla \times A) = \frac{-i\hbar^2}{\ell^2}, \tag{3.62}$$

we see that π_x and π_y are canonically conjugate variables, and the Hamiltonian is formally identical to that of a harmonic oscillator. Ladder operators can be defined in the standard fashion:

$$a^\dagger \equiv \frac{\ell/\hbar}{\sqrt{2}} (\pi_x + i\pi_y), \tag{3.63}$$

$$a \equiv \frac{\ell/\hbar}{\sqrt{2}} (\pi_x - i\pi_y), \tag{3.64}$$

so that

$$[a, a^\dagger] = 1, \tag{3.65}$$

and

$$H = \frac{\hbar\omega_c}{2} (aa^\dagger + a^\dagger a). \tag{3.66}$$

It follows from Eq. (3.66) that the eigenenergies for the Schrödinger equation of a free particle are quantized at values

$$E_n = \hbar\omega_c \left(n + \frac{1}{2} \right). \tag{3.67}$$

Following Johnson [306], let us introduce another set of operators:

$$r_0 \equiv r + \frac{\ell^2}{\hbar} \hat{z} \times \pi \tag{3.68}$$

$$= \left(x - \frac{\ell^2}{\hbar}\pi_y, y + \frac{\ell^2}{\hbar}\pi_x \right)$$

$$= \left(x - \frac{1}{m_b\omega_c}\pi_y, y + \frac{1}{m_b\omega_c}\pi_x \right)$$

$$\equiv (x_0, y_0).$$

r_0 is a constant of motion, as can be verified by a direct calculation of the commutators $[x_0, H]$ and $[y_0, H]$. Alternatively, it is instructive to derive the following equations of motion:

$$\dot{\pi}_x = \frac{i}{\hbar}[H, \pi_x] = -w_c\pi_y, \tag{3.69}$$

$$\dot{\pi}_y = \frac{i}{\hbar}[H, \pi_y] = w_c\pi_x, \tag{3.70}$$

$$m_b\dot{x} = \frac{i}{\hbar}m_b[H, x] = \pi_x, \tag{3.71}$$

$$m_b\dot{y} = \frac{i}{\hbar}m_b[H, y] = \pi_y. \tag{3.72}$$

It follows that

$$\dot{x}_0 = \dot{x} - \frac{1}{m_b \omega_c} \dot{\pi}_y = 0 \tag{3.73}$$

and

$$\dot{y}_0 = \dot{y} + \frac{1}{m_b \omega_c} \dot{\pi}_x = 0. \tag{3.74}$$

These equations of motion are identical to those for a classical cyclotron orbit. Therefore, (x_0, y_0) is naturally identified with the center of the orbit, also called the guiding center. In agreement with that identification is the feature that the expectation values of x_0 and y_0 with respect to an eigenstate of energy, ϕ_E, are identical to those of x and y. That follows because the expectation value of any time derivative with respect to ϕ_E vanishes:

$$\langle \phi_E | \dot{O} | \phi_E \rangle = \frac{i}{\hbar} \langle \phi_E | [H, O] | \phi_E \rangle = 0. \tag{3.75}$$

In particular, $\langle \phi_E | \pi_x | \phi_E \rangle = 0$ and $\langle \phi_E | \pi_y | \phi_E \rangle = 0$, because π_x and π_y are equal to time derivatives of certain operators, as seen above. Therefore,

$$\langle \phi_E | r_0 | \phi_E \rangle = \langle \phi_E | r | \phi_E \rangle. \tag{3.76}$$

The radius defined through

$$r_c^2 = (x - x_0)^2 + (y - y_0)^2 = \frac{\pi_x^2 + \pi_y^2}{(m_b \omega_c)^2} = \frac{2H}{m_b \omega_c^2} \tag{3.77}$$

is analogous to the radius of the classical orbit, given by $r_c^2 = 2E/m_b\omega_c^2$, with $E = m_b v^2/2$. The coordinates of the center of the orbit are canonically conjugate variables:

$$[x_0, y_0] = i\ell^2 \tag{3.78}$$

which implies the uncertainty relation

$$\Delta x_0 \Delta y_0 \geq \frac{\ell^2}{2}. \tag{3.79}$$

Thus, the radius of an orbit is known exactly for an energy eigenstate, according to Eq. (3.77), but its center cannot be located precisely.

The guiding center coordinates x_0 and y_0 commute with π_x and π_y, but not with one another. The commutator in Eq. (3.78) allows us to define another set of ladder operators:

$$b = \frac{1}{\sqrt{2}\ell} (x_0 + iy_0), \tag{3.80}$$

$$b^\dagger = \frac{1}{\sqrt{2}\ell} (x_0 - iy_0), \tag{3.81}$$

satisfying

$$[b, b^\dagger] = 1. \tag{3.82}$$

The operators b and $b^\dagger$ commute with a, $a^\dagger$ and H. Thus, given an eigenstate, new eigenstates at the same energy can be obtained by the application of b or $b^\dagger$. The full set of eigenstates can be generated by using raising operators starting from the bottom of the ladder:

$$|n, m\rangle = \frac{(a^\dagger)^n (b^\dagger)^m}{\sqrt{n! m!}} |0, 0\rangle, \tag{3.83}$$

$$E_n = \hbar \omega_c \left(n + \frac{1}{2}\right). \tag{3.84}$$

The state $|0, 0\rangle$ is annihilated by both a and b. The different energies, characterized by an integer $n = 0, 1, \ldots$, give Landau levels. States with the same $a^\dagger a$ but different $b^\dagger b$ are degenerate and belong to the same Landau level.

3.9 Magnetic translation operator

The operator

$$T_R = e^{\frac{i}{\hbar} R \cdot p} \tag{3.85}$$

is the generator of translations

$$T_R f(r) = f(r + R). \tag{3.86}$$

This can be seen by noting that an infinitesimal translation $r \to r + R/N$, where $N \to \infty$, is generated by

$$f(r + R/N) = f(r) + \frac{R}{N} \cdot \nabla_r f(r)$$

$$= \left(1 + \frac{i}{\hbar} \frac{R}{N} \cdot p\right) f(r). \tag{3.87}$$

Therefore,

$$f(r + R) = \left(1 + \frac{i}{\hbar} \frac{R}{N} \cdot p\right)^N f(r)$$

$$= e^{\frac{i}{\hbar} R \cdot p} f(r). \tag{3.88}$$

Alternatively, the equation

$$e^{\frac{i}{\hbar} \alpha R \cdot p} f(r) = f(r + \alpha R) \tag{3.89}$$

can be proved by noting that both sides satisfy the same differential equation with respect to α, and have the same initial condition for $\alpha = 0$. The set of all translations forms a group

$$T_R T_{R'} = T_{R + R'}. \tag{3.90}$$

For a translationally invariant Hamiltonian, which satisfies $H(r+R) = H(r)$, T_R commutes with the Hamiltonian

$$T_R H(r)\Psi(r) = H(r+R)\Psi(r+R)$$
$$= H(r)T_R\Psi(r), \tag{3.91}$$

implying

$$T_R H(r) = H(r)T_R. \tag{3.92}$$

One can therefore look for eigenstates that simultaneously diagonalize T_R and H.

In the presence of a magnetic field, the Hamiltonian

$$H = \frac{1}{2m_b}\left(p + \frac{eA}{c}\right)^2 \tag{3.93}$$

is not translationally invariant, because

$$A(r+R) \neq A(r), \tag{3.94}$$

and, therefore, does not commute with T_R. A translation operator commuting with the Hamiltonian can be constructed (Zak [735]). It is called the magnetic translation operator, denoted by $\mathcal{T}_R$.

Let us begin by noting that the uniformity of the magnetic field implies, with $r' \equiv r+R$,

$$\nabla_{r'} \times A(r') = \nabla_r \times A(r') = B = \nabla_r \times A(r). \tag{3.95}$$

Therefore, we must have

$$A(r+R) = A(r) + \nabla\xi(r). \tag{3.96}$$

(The R dependence of $\xi(r)$ is not displayed for simplicity.) Translation of A is thus equivalent to a gauge transformation. This suggests that $\mathcal{T}_R$ must be a combination of translation and gauge transformation. From

$$T_R H\left(p + \frac{e}{c}A(r)\right) = H\left(p + \frac{e}{c}A(r+R)\right)T_R$$
$$= H\left(p + \frac{e}{c}A + \frac{e}{c}\nabla\xi(r)\right)T_R \tag{3.97}$$

and

$$H\left(p + \frac{e}{c}A + \frac{e}{c}\nabla\xi(r)\right) = e^{-\frac{ie}{\hbar c}\xi(r)}H\left(p + \frac{e}{c}A\right)e^{\frac{ie}{\hbar c}\xi(r)}, \tag{3.98}$$

it follows that

$$\mathcal{T}_R \equiv e^{\frac{ie}{\hbar c}\xi(r)}T_R \tag{3.99}$$

commutes with the Hamiltonian:

$$\mathcal{T}_R H = H\mathcal{T}_R. \tag{3.100}$$

The magnetic translation operators are useful for imposing periodic boundary conditions, and for the quantum mechanics of electrons in a periodic potential. Let us construct the explicit form of $\mathcal{T}_R$ for the symmetric and the Landau gauges.

Symmetric gauge   With $A = (B \times r)/2$, we have

$$\nabla \xi(r) = \frac{1}{2}(B \times R). \tag{3.101}$$

This gives

$$\xi(r) = \frac{1}{2}(B \times R) \cdot r, \tag{3.102}$$

$$\mathcal{T}_R = \exp\left(\frac{ie}{2\hbar c}B \times R \cdot r\right) T_R = \exp\left(\frac{i}{2\ell^2}\hat{z} \cdot (R \times r)\right) T_R. \tag{3.103}$$

Landau gauge   With $A = -By\hat{x}$ and $R = (X, Y)$, we have $\nabla \xi = -BY\hat{x}$, which is solved by $\xi(r) = -BYx$. The magnetic translation operator is therefore given by

$$\mathcal{T}_R = \exp\left(-\frac{ie}{\hbar c}BYx\right) T_R = \exp\left(-\frac{i}{\ell^2}Yx\right) T_R. \tag{3.104}$$

In general, the magnetic translation operators do not commute and do not form a group. For example, in the Landau gauge:

$$\mathcal{T}_R \mathcal{T}_{R'} = e^{-\frac{i}{\ell^2}Yx} e^{\frac{i}{\hbar}R \cdot p} e^{-\frac{i}{\ell^2}Y'x} e^{\frac{i}{\hbar}R' \cdot p}$$

$$= e^{-\frac{i}{\ell^2}XY'} e^{-\frac{i}{\ell^2}Yx} e^{-\frac{i}{\ell^2}Y'x} e^{\frac{i}{\hbar}R \cdot p} e^{\frac{i}{\hbar}R' \cdot p}$$

$$= e^{-\frac{i}{\ell^2}XY'} \mathcal{T}_{R+R'}, \tag{3.105}$$

where we have used results from Appendix B.

3.10 Spherical geometry

In the spherical geometry (Fig. 3.3), the two-dimensional sheet containing electrons is wrapped around the surface of a sphere, and a perpendicular (radial) magnetic field is generated by placing a Dirac magnetic monopole at the center of the sphere. Two reasons for the popularity of this compact geometry are: First, it does not have edges, which makes it suitable for an investigation of the bulk properties. Second, Landau levels have a finite degeneracy (for a finite magnetic field), which is useful in identifying incompressible states for finite systems. In particular, filled Landau levels are unambiguously defined. This ought to be contrasted with the planar geometry for which, without an external confining potential, a Landau level extends from $+\infty$ to $-\infty$, containing an infinite number of states, and the notion of a filled Landau level is ill-defined for a finite number of particles. The spherical geometry has been instrumental in establishing the validity of the theory of the FQHE, and provides the cleanest proofs for many properties.

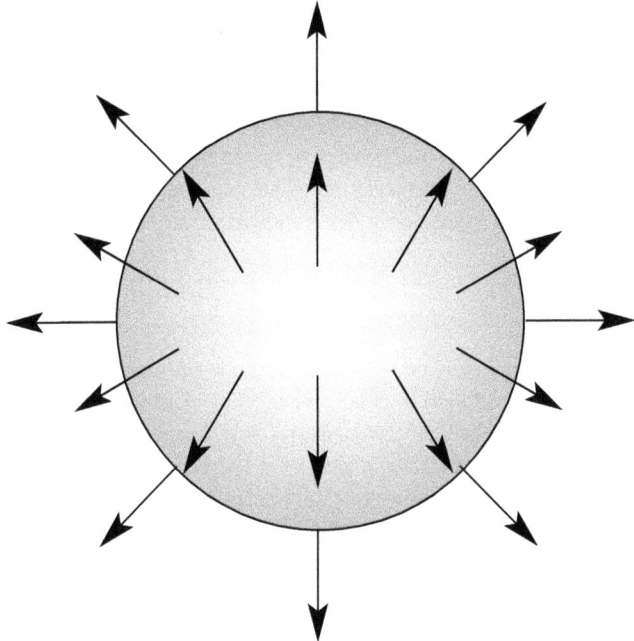

Fig. 3.3. Spherical geometry. Electrons move on the surface of the sphere under the influence of a radial magnetic field. The total flux through the surface of a sphere is $2Q\phi_0$, with $2Q$ being an integer.

The quantum mechanics of a single particle in the presence of a Dirac monopole[8] was studied in detail by Wu and Yang [701, 702] in the 1970s. The spherical geometry was introduced into the FQHE physics by Haldane in 1983 [221]. The discussion in this section draws from these references, as well as from the work of Fano, Ortolani, and Colombo [151].[9]

3.10.1 Basic spherical facts

We make extensive use of the spherical geometry in this book. Because the solutions of the Schrödinger equation in the spherical geometry are not as well known as those in the planar geometry (the latter being familiar from the harmonic oscillator problem), we summarize, before going to a more complete treatment, some facts that are crucial (and often sufficient) for an understanding of many results in the rest of the book.

- A magnetic field is produced (theoretically) by placing a magnetic monopole at the center of the sphere. The magnetic flux through the surface of the sphere, measured in units of the flux quantum, is quantized to be an integer for consistency, denoted by $2Q$. The quantity Q, called the

---

[8] The Dirac monopole remains a theoretical construct so far. Experimental efforts to observe it have not yet borne fruit.
[9] The eigenvalue equation for an electron on a compact torus (parallelogram with periodic boundary conditions) with a magnetic field perpendicular to the surface has been worked out by Haldane and Rezayi [223]. This geometry has also been used in the FQHE literature, but is not considered in this book.

monopole strength, can be a positive or a negative integer or a half integer. The filling factor is defined by

$$v = \lim_{N \to \infty} \frac{N}{2|Q|}.$$ (3.106)

The filling factor of a finite system may be slightly shifted from its thermodynamic limit.

- As evident from the geometry, the *orbital* angular momentum and its z component are good quantum numbers, denoted by l and m, respectively. Their allowed values are

$$l = |Q|, |Q| + 1, \ldots,$$ (3.107)

$$m = -l, -l + 1, \ldots, l.$$ (3.108)

Each combination occurs precisely once. In contrast to the situation without magnetic fields, the minimum value of l is $|Q|$, and l can be either an integer or a half-integer.

- Different angular momentum shells are the Landau levels in the spherical geometry. The degeneracy of each Landau level is equal to the total number of m values, i.e., $2l + 1$, increasing by two units for each successive Landau level. In contrast to the planar geometry, the degeneracy is finite in the spherical geometry, due to its compact nature. For the lowest Landau level, the degeneracy is $2|Q| + 1$; for the next it is $2|Q| + 3$; and so on. The lowest filled Landau level is obtained when $N = 2|Q| + 1$. In general, the state with n filled Landau levels is obtained when Q and N are related as

$$Q = \pm \frac{N - n^2}{2n}.$$ (3.109)

- The single particle eigenstates are called monopole harmonics, denoted by $Y_{Qlm}(\mathbf{\Omega})$, which are a generalization of the familiar spherical harmonics. (The former reduces to the latter for $Q = 0$.) $\mathbf{\Omega}$ represents the angular coordinates θ and ϕ on the sphere. In the lowest Landau level, the monopole harmonics are given by $Y_{QQm} \sim u^{Q+m} v^{Q-m}$, with $u = \cos(\theta/2) e^{i\phi/2}$ and $v = \sin(\theta/2) e^{-i\phi/2}$. The wave function for the lowest full Landau level is $\Phi_1 = \prod_{j<k} (u_j v_k - v_j u_k)$. The state with n filled Landau levels will be denoted by Φ_n.

- The product of two single particle wave functions at Q and Q' produces a wave function at $Q + Q'$. In other words, we have

$$Y_{Qlm}(\mathbf{\Omega}) Y_{Q'l'm'}(\mathbf{\Omega}) = \sum_{l''} C_{l''} Y_{Q+Q', l'', m+m'}(\mathbf{\Omega}).$$ (3.110)

- Complex conjugation is equivalent to changing the sign of Q (or the magnetic field).

3.10.2 Dirac monopole and Dirac string

We consider a sphere of radius R with a flux $2Q\phi_0$ extending radially outward through the surface. The flux corresponds to a magnetic field

$$\mathbf{B} = \frac{2Q\phi_0}{4\pi R^2} \hat{r},$$ (3.111)

which is produced by the vector potential

$$\mathbf{A} = -\frac{\hbar c Q}{eR} \cot \theta \, \hat{\phi}.$$ (3.112)

A closer inspection shows that this vector potential produces more than the uniform radial magnetic field, as hinted by the singularities at the north and south poles (for $\theta = 0, \pi$). To understand their meaning, we calculate the line integral $\int A \cdot dr$ around infinitesimal closed loops encircling the two poles. Using

$$A \cdot dr = -\frac{\hbar c Q}{e} \cos\theta \, d\phi, \tag{3.113}$$

the integrals are given by

$$\oint A \cdot dr = \mp Q\phi_0, \tag{3.114}$$

where the upper (lower) sign is for the loop around the north (south) pole. Thus, the vector potential actually describes two "Dirac strings" of strength $Q\phi_0$ entering into the poles, and then emanating out uniformly in the radial direction. The Dirac monopole apparently has strings attached to it. The singularities cannot be intrinsic, however, because the space around the monopole is spherically symmetric and free of singularities. It should, therefore, be possible to describe a Dirac monopole with no strings attached. One of the singularities can be eliminated by considering the vector potentials (Wu and Yang [701, 702])

$$A_\pm = -\frac{\hbar c Q}{eR} \left(\cot\theta \pm \frac{1}{\sin\theta} \right) \hat{\phi}, \tag{3.115}$$

which have singular behavior only at one of the two poles. The entire magnetic field $2Q\phi_0$ now enters through one of the poles. Any choice of a vector potential for a Dirac monopole must possess at least one singularity [700], as is also the case for a coordinate system on a sphere. The physical lack of singular behavior can be made manifest by working with "sections" rather than ordinary functions [701, 702], a concept encountered in the context of the mathematics of "fiber bundles." The idea is to divide the sphere into two (or more) overlapping regions, and define a singularity-free vector potential in each region and a singularity-free transformation in the overlapping region. Such considerations will not be necessary for our purposes, because the singularities do not affect the Coulomb matrix elements. We work below with the vector potential of Eq. (3.112).

Acceptable quantum mechanical solutions are obtained only for integral values of $2Q$. Some understanding into this restriction can be gained by considering a gauge in which the "Dirac string" carrying a magnetic flux $\Phi = -2Q\phi_0$ enters through the north pole. Since the Dirac string is not real, it must not have any observable effect. In particular, it must not make any contribution to the Aharonov–Bohm phase factor associated with a tiny closed loop around the singularity. That requires $\exp[2\pi i\Phi/\phi_0] = \exp[-i4\pi Q] = 1$, which, in turn, implies $2Q =$ integer.

3.10.3 Angular momentum algebra

The kinetic energy operator is given by

$$H = \frac{\hbar^2}{2mR^2}|\mathbf{\Lambda}|^2, \tag{3.116}$$

where

$$\mathbf{\Lambda} = \mathbf{R} \times \left(-i\nabla + \frac{e}{\hbar c}A(\mathbf{\Omega})\right) \tag{3.117}$$

is proportional to the canonical momentum tangential to the surface of the sphere, with the unit vector $\mathbf{\Omega} = \mathbf{R}/R$ denoting the angular coordinates of the electron position on the sphere. Since electrons are confined to move on the surface of the sphere, we fix $r = R$. (It may be verified that $[r^2, H] = 0$, so we can diagonalize r^2 and study the problem for a fixed $r^2 = R^2$. In other words, the value of r is not changed by the Hamiltonian.)

In spherical coordinates, we have

$$\begin{aligned}
\mathbf{\Lambda} &= \mathbf{R} \times \left[-i\nabla + \frac{e}{\hbar c}A(\mathbf{\Omega})\right] \\
&= \mathbf{\Omega} \times \left[-i\left(\hat{\boldsymbol{\theta}}\frac{\partial}{\partial\theta} + \hat{\boldsymbol{\phi}}\frac{1}{\sin\theta}\frac{\partial}{\partial\phi}\right) - Q\cot\theta\,\hat{\boldsymbol{\phi}}\right] \\
&= -i\left(\hat{\boldsymbol{\phi}}\frac{\partial}{\partial\theta} - \hat{\boldsymbol{\theta}}\frac{1}{\sin\theta}\frac{\partial}{\partial\phi}\right) + Q\cot\theta\,\hat{\boldsymbol{\theta}},
\end{aligned} \tag{3.118}$$

where we have used the standard expression (for fixed $r = R$)

$$R\nabla = \hat{\boldsymbol{\theta}}\frac{\partial}{\partial\theta} + \hat{\boldsymbol{\phi}}\frac{1}{\sin\theta}\frac{\partial}{\partial\phi}, \tag{3.119}$$

and also $\mathbf{\Omega} \times \hat{\boldsymbol{\theta}} = \hat{\boldsymbol{\phi}}$ and $\mathbf{\Omega} \times \hat{\boldsymbol{\phi}} = -\hat{\boldsymbol{\theta}}$. One now conveniently switches to the coordinate independent Cartesian unit vectors with the help of the expressions

$$\hat{\boldsymbol{\phi}} = -\sin\phi\,\hat{\boldsymbol{x}} + \cos\phi\,\hat{\boldsymbol{y}}, \tag{3.120}$$

$$\hat{\boldsymbol{\theta}} = \cos\theta\cos\phi\,\hat{\boldsymbol{x}} + \cos\theta\sin\phi\,\hat{\boldsymbol{y}} - \sin\theta\,\hat{\boldsymbol{z}}, \tag{3.121}$$

$$\hat{\boldsymbol{r}} = \mathbf{\Omega} = \sin\theta\cos\phi\,\hat{\boldsymbol{x}} + \sin\theta\sin\phi\,\hat{\boldsymbol{y}} + \cos\theta\,\hat{\boldsymbol{z}}. \tag{3.122}$$

Some algebra yields for $2mR^2H/\hbar^2$

$$|\mathbf{\Lambda}|^2 = -\frac{1}{\sin\theta}\frac{\partial}{\partial\theta}\sin\theta\frac{\partial}{\partial\theta} + \left(Q\cot\theta + \frac{i}{\sin\theta}\frac{\partial}{\partial\phi}\right)^2. \tag{3.123}$$

To fully exploit the spherical symmetry of the problem, it is useful to introduce an angular momentum operator. It may be verified that

$$[\Lambda_i, \Lambda_j] = i\epsilon_{ijk}(\Lambda_k - Q\Omega_k) \tag{3.124}$$

and

$$[\Lambda_i, \Omega_j] = i\epsilon_{ijk}\Omega_k. \tag{3.125}$$

Here, the subscripts i, j, k stand for x, y, or z of the Cartesian coordinate system. These relations make it straightforward to see that the operator

$$L = \Lambda + Q\Omega \tag{3.126}$$

satisfies the angular momentum algebra

$$[L_i, L_j] = i\epsilon_{ijk}L_k. \tag{3.127}$$

The raising and lowering operators are defined as

$$L_\pm = L_x \pm iL_y. \tag{3.128}$$

They satisfy the commutation relations

$$[L_z, L_\pm] = \pm L_z, \tag{3.129}$$

which implies that they raise or lower the L_z eigenvalue by one unit. Explicit expressions can be obtained for the angular momentum operator L. We quote here L_z and $L_\pm = L_x \pm iL_y$:

$$L_z = -i\frac{\partial}{\partial\phi}, \tag{3.130}$$

$$L_\pm = e^{\pm i\phi}\left[\pm\frac{\partial}{\partial\theta} + i\cot\theta\frac{\partial}{\partial\phi} + \frac{Q}{\sin\theta}\right]. \tag{3.131}$$

3.10.4 Constraints on quantum numbers

Noting that $\Lambda \cdot \Omega = \Omega \cdot \Lambda = 0$, and using Eq. (3.126), we have

$$|\Lambda|^2 = |L - Q\Omega|^2 = L^2 - Q^2. \tag{3.132}$$

The Hamiltonian, therefore, commutes with the angular momentum operators L_i. Because $[L^2, L_z] = 0$, we choose eigenfunctions that simultaneously diagonalize H, L^2 and L_z. These eigenfunctions are called "monopole harmonics," denoted by $Y_{Q,l,m}$, the explicit form of which is derived below. They satisfy

$$L^2 Y_{Q,l,m} = l(l+1)Y_{Q,l,m}, \tag{3.133}$$

$$L_z Y_{Q,l,m} = mY_{Q,l,m}. \tag{3.134}$$

The last equation implies that we can write

$$Y_{Q,l,m}(\theta, \phi) = e^{im\phi}P_{Q,l,m}(\theta). \tag{3.135}$$

We now derive the allowed values for the quantum numbers l, m and Q. From standard quantum mechanics, we know that the angular momentum commutation relations imply that l must have values $l = 0, \frac{1}{2}, 1, \frac{3}{2}, \ldots$, and $m = -l, -l+1, \ldots, l$.[10] It may be tempting to impose single-valuedness on $Y_{Q,l,m}(\theta, \phi)$ and conclude from Eq. (3.135) that m is an integer (which is what eliminates nonintegral values of l and m in quantum mechanics without monopoles), but that would be incorrect. The gauge we have used has singularities on *both* poles, and therefore the wave function is not necessarily single valued. We see in Exercise 3.11 that, for the gauges $A_{\pm}$, which have only one singularity on the sphere, the ϕ dependence of the wave function is given by $\exp[i(m \pm Q)\phi]$. Requiring the wave function to be single valued thus gives the weaker condition

$$m - Q = \text{integer}. \tag{3.136}$$

Half-integral values for m are thus not disallowed. The monopole strength Q is either an integer or a half integer. Furthermore, Eq. (3.132) implies that

$$l(l+1) \geq |Q|^2, \tag{3.137}$$

which restricts l to the values given in Eq. (3.107).

The eigenvalues of $|\Lambda|^2$ are $l(l+1) - Q^2$, giving the energy eigenvalues

$$E_{Q,l,m} = \hbar^2 \frac{l(l+1) - Q^2}{2mR^2} = \frac{l(l+1) - Q^2}{2|Q|} \hbar\omega_c, \tag{3.138}$$

where we have used

$$2|Q|\phi_0 = 4\pi R^2 |B| \tag{3.139}$$

for the flux through the surface of the sphere. The angular momentum shells (labeled by l) are the Landau levels in the spherical geometry. The lowest Landau level has $l = |Q|$, and the nth Landau level has $l = |Q| + n$. For any given n, we recover the familiar result, $E = (n + 1/2)\hbar\omega_c$, in the limit $|Q| \to \infty$.

3.10.5 Explicit solution

Next we proceed to obtain the explicit expression for the single particle eigenstate, $Y_{Q,l,m}$, which satisfies the Schrödinger equation[11]

$$|\Lambda|^2 Y_{Q,l,m} = [l(l+1) - Q^2] Y_{Q,l,m}. \tag{3.141}$$

[10] The argument, briefly, goes as follows. Consider an eigenstate of L^2 and L_z with eigenvalues $l(l+1)$ and m. We can raise or lower the m eigenvalue by successive operations of $L_{\pm}$. On the other hand, the equation $L^2 = L_x^2 + L_y^2 + L_z^2$ implies the condition $l(l+1) \geq m^2$. Therefore, there must exist a state with the highest m which is annihilated by L_+ and another state with the lowest m which is annihilated by L_-. We denote the corresponding m values by m_+ and m_-, respectively. Using the equation $L^2 = L_{\pm}L_{\mp} + L_z^2 \mp L_z$, the eigenvalues of L^2 for the highest and the lowest states are $m_{\pm}(m_{\pm} \pm 1)$, which fixes $m_{\pm} = \pm l$. Furthermore, repeated applications of L_+ on the m_- state must eventually produce m_+; otherwise we would violate the condition $l(l+1) \geq m^2$. That implies that l must be an integer or a half integer.

[11] Many useful results were derived by Wu and Yang [701, 702]. The solutions below are related to their solutions by the equation

$$Y_{Q,l,m} = \left[Y_{Q,l,-m}^{\text{WY}} e^{-iQ\phi} \right]^*, \tag{3.140}$$

With Eq. (3.135) the Schrödinger equation becomes

$$\left[-\frac{1}{\sin\theta}\frac{\partial}{\partial\theta}\sin\theta\frac{\partial}{\partial\theta}+\frac{1}{\sin^2\theta}(Q\cos\theta-m)^2\right]P_{Q,l,m}=[l(l+1)-Q^2]P_{Q,l,m}. \quad (3.142)$$

Substituting $\cos\theta=x$ gives

$$\left[2x\frac{\partial}{\partial x}-(1-x^2)\frac{\partial^2}{\partial x^2}+\frac{(m-Qx)^2}{1-x^2}\right]P_{Q,l,m}=[l(l+1)-Q^2]P_{Q,l,m}. \quad (3.143)$$

The $(1-x^2)$ in the denominator can be eliminated by substituting

$$P_{Q,l,m}=(1+x)^{\frac{a+b}{2}}(1-x)^{\frac{b-a}{2}}R_{Q,l,m} \quad (3.144)$$

and making an appropriate choice for a and b. The Schrödinger equation for $R_{Q,l,m}$ is

$$\left[(1-x^2)\frac{\partial^2}{\partial x^2}+[2a-2(b+1)x]\frac{\partial}{\partial x}-\frac{C}{(1-x^2)}\right]R_{Q,l,m}$$
$$=\left[Q^2-l(l+1)\right]R_{Q,l,m}, \quad (3.145)$$

where

$$C=(-b^2-b+Q^2)x^2+2(ab-mQ)x+(m^2+b-a^2). \quad (3.146)$$

We set $a=Q$ and $b=m$, giving $C=(m+m^2-Q^2)(1-x^2)$. The Schrödinger equation now has the same form as the hypergeometric equation:

$$(1-x^2)y''+(\beta-\alpha-(\alpha+\beta+2)x)y'+n(n+\alpha+\beta+1)y=0, \quad (3.147)$$

the solution of which is the Jacobi polynomial $P_n^{\alpha,\beta}$, given by

$$P_n^{\alpha,\beta}(x)=\frac{(-1)^n}{2^n n!}(1-x)^{-\alpha}(1+x)^{-\beta}\frac{d^n}{dx^n}[(1-x)^{\alpha+n}(1+x)^{\beta+n}]$$
$$=\frac{1}{2^n}\sum_{s=0}^{n}\binom{n+\alpha}{s}\binom{n+\beta}{n-s}(x-1)^{n-s}(x+1)^s. \quad (3.148)$$

Comparing the two differential equations gives $\alpha=m-Q$, $\beta=m+Q$, and $n=l-m$. Putting everything together and normalizing, the single particle eigenstates are given by

$$Y_{Q,l,m}(\mathbf{\Omega})=N_{Qlm}2^{-m}(1-x)^{\frac{-Q+m}{2}}(1+x)^{\frac{Q+m}{2}}P_{l-m}^{-Q+m,Q+m}(x)e^{im\phi}, \quad (3.149)$$

$$x=\cos\theta, \quad (3.150)$$

---

where $Y_{Q,l,m}^{WY}$ is the Wu–Yang solution in their "a" gauge. The difference arises because of the different gauge choices, and also because of a slightly different convention for the Hamiltonian.

$$N_{Qlm} = \left(\frac{(2l+1)}{4\pi} \frac{(l-m)!(l+m)!}{(l-Q)!(l+Q)!} \right)^{1/2}. \tag{3.151}$$

A useful form for the single particle eigenfunctions is obtained by first writing

$$Y_{Qlm} = N_{Qlm} 2^{-m} (1-x)^{\frac{Q-m}{2}} (1+x)^{\frac{Q+m}{2}} e^{im\phi}$$

$$\times \sum_{s=0}^{l-m} (-1)^{l-m-s} \binom{l-Q}{s} \binom{l+Q}{l-m-s} (1-x)^{l-Q-s} (1+x)^s. \tag{3.152}$$

In terms of the "spinor" variables (Haldane [221]),

$$u = \cos(\theta/2) \, e^{i\phi/2}, \tag{3.153}$$

$$v = \sin(\theta/2) \, e^{-i\phi/2}, \tag{3.154}$$

we have

$$1 - x = 2\sin^2(\theta/2), \qquad 1 + x = 2\cos^2(\theta/2), \qquad x = uu^* - vv^*, \tag{3.155}$$

and the single particle eigenstates become

$$Y_{Qlm} = N_{Qlm} (-1)^{l-m} v^{Q-m} u^{Q+m}$$

$$\times \sum_{s=0}^{l-m} (-1)^s \binom{l-Q}{s} \binom{l+Q}{l-m-s} (v^*v)^{l-Q-s} (u^*u)^s. \tag{3.156}$$

(Note that the series on the right hand side can be expressed in different forms, using the identity $u^*u + v^*v = 1$, but the form above will be most useful later on.) Here and elsewhere, we follow the convention that $\binom{j}{k}$ is to be set to zero if either $k < 0$ or $k > j$.

The above equations are valid for positive as well as negative values of Q. For $Q \geq 0$, the upper limit in the sum in Eq. (3.156) can be set equal to $l - Q$.

3.10.6 Lowest Landau level

The wave functions in the lowest Landau level are obtained by setting $l = Q$ in the preceding displayed equation. Only the $s = 0$ term contributes in the sum, yielding

$$Y_{QQm} = \left[\frac{2Q+1}{4\pi} \binom{2Q}{Q-m} \right]^{1/2} (-1)^{Q-m} v^{Q-m} u^{Q+m}. \tag{3.157}$$

Note that the sum in the preceding equation contributes a factor $\binom{2Q}{Q-m}$.

The wave functions for filled LL states are unique, and can be written as Slater determinants. The form of the wave function for the lowest filled Landau level is again especially nice and will play an important role later. Apart from the normalization, the one filled LL state is given by

$$\Phi_1 = \prod_{j<k} (u_j v_k - v_j u_k). \tag{3.158}$$

To see this, note that the LLL eigenstates are proportional to $v^{2Q}(u/v)^{Q+m}$, with $Q + m = 0, 1, \ldots, 2Q$. The wave function of one filled Landau level at $2Q + 1 = N$ is then

$$\Phi_1 = \left(\prod_{j=1}^{N} v_j^{2Q} \right) \begin{vmatrix} 1 & 1 & 1 & . & . \\ z_1 & z_2 & z_3 & . & . \\ z_1^2 & z_2^2 & z_3^2 & . & . \\ . & . & . & . & . \\ . & . & . & . & . \end{vmatrix} \tag{3.159}$$

with $z_j = u_j/v_j$. The Slater determinant is the familiar Vandermonde determinant, equal to $\prod_{j<k}(z_j - z_k)$; the factor in front can be rewritten as

$$\prod_{j=1}^{N} v_j^{2Q} = \prod_{j=1}^{N} v_j^{N-1} = \prod_{j<k} v_j v_k. \tag{3.160}$$

That gives $\Phi_1 = \prod_{j<k}[v_j v_k (u_j/v_j - u_k/v_k)] = \prod_{j<k}(u_j v_k - v_j u_k)$.

In the LLL subspace, a simple representation exists for the angular momentum operator in terms of the spinor variables:

$$L_+ = -u \frac{\partial}{\partial v}, \tag{3.161}$$

$$L_- = -v \frac{\partial}{\partial u}, \tag{3.162}$$

$$L_z = \frac{1}{2} \left(u \frac{\partial}{\partial u} - v \frac{\partial}{\partial v} \right), \tag{3.163}$$

the correctness of which can be verified by noting

$$L_\pm |m\rangle = \sqrt{Q(Q+1) - m(m \pm 1)} |m \pm 1\rangle \tag{3.164}$$

$$L_z |m\rangle = m|m\rangle, \tag{3.165}$$

where $|m\rangle$ denotes the LLL wave function Y_{QQm}. The expression for L^2 is given by

$$L^2 = L_x^2 + L_y^2 + L_z^2 = L_+ L_- + L_z^2 - L_z. \tag{3.166}$$

3.10.7 Representation of spin

In analogy to the spherical harmonics that generate a representation for integer spins, the monopole harmonics with a magnetic monopole of strength $Q = S$ generate a representation for an arbitrary spin S, in particular for $S = 1/2$, because Q can take both integral and half-integral values. We have the following correspondence:

$$Y_{SSm} \Longleftrightarrow |S, m\rangle, \tag{3.167}$$

$$L_{\pm} \Longleftrightarrow S_{\pm}, \tag{3.168}$$

$$L_z \Longleftrightarrow S_z. \tag{3.169}$$

Here, Y_{SSm} are the monopole harmonics for an electron in the lowest Landau level ($l = S$).

A spin pointing in the direction $\mathbf{\Omega}$ can be represented as an electron on the sphere with angular coordinates $\mathbf{\Omega}$. The rotation of spin then maps into the motion of the electron on the sphere. To see this, we note that the operator for rotation by an angle α around the axis $\mathbf{\Omega}$ in the spin space is given by $\exp[-i\alpha\mathbf{\Omega} \cdot \mathbf{L}]$, and that for the rotation in real space is given by $\exp[-i\alpha\mathbf{\Omega} \cdot \mathbf{\Lambda}]$. Equation (3.126) implies

$$e^{-i\alpha\mathbf{\Omega}\cdot\mathbf{L}} = e^{-i\alpha\mathbf{\Omega}\cdot\mathbf{\Lambda}-iQ\alpha}, \tag{3.170}$$

indicating that the rotation operators in the spin and real space are identical, modulo a phase factor. A Berry phase is known to be associated with the rotation of spin; its origin can be understood in the above mapping in terms of the more familiar Aharonov–Bohm phase of the electron.

3.11 Coulomb matrix elements

Let us consider many interacting electrons in a given Landau level. Exact diagonalization requires enumerating all basis functions, evaluating the matrix elements of the Hamiltonian (i.e., the Coulomb interaction), and then diagonalizing the Hamiltonian matrix. This is most easily done in the second quantized formulation (Appendix F). Labeling the single particle orbitals by m, the L_z quantum number, the basis functions (suppressing the LL index), are given by

$$|m_1, m_2, \ldots, m_N\rangle = a^\dagger_{m_1} \ldots a^\dagger_{m_N} |0\rangle, \tag{3.171}$$

where m_j is the angular momentum of the jth particle, $|0\rangle$ is the vacuum state with no particles, and we have taken $m_1 \leq m_2 \leq \cdots \leq m_N$. The interaction Hamiltonian is given by

$$H = \frac{1}{2} \sum_{m_1,m_2,m_3,m_4} a^\dagger_{m_1} a^\dagger_{m_2} a_{m_4} a_{m_3} \langle m_1, m_2 | V | m_3, m_4 \rangle. \tag{3.172}$$

The expectation value of H with respect to any two basis functions is a sum of terms, with each term being a product of (a) a two-body matrix element

$$\langle m_1, m_2 | V(|\mathbf{r}_1 - \mathbf{r}_2|) | m_3, m_4 \rangle, \tag{3.173}$$

where

$$V(|\mathbf{r}_1 - \mathbf{r}_2|) = \frac{e^2}{\epsilon|\mathbf{r}_1 - \mathbf{r}_2|}, \tag{3.174}$$

and (b) the expectation value of a product of annihilation and creation operators with respect to the vacuum state. The latter can be evaluated with the help of what is known as the Wick theorem. The remainder of this section deals with the two-body matrix element. As usual, we take the unit of distance to be the magnetic length, ℓ, and the units of interaction energy to be $e^2/\epsilon\ell$.

3.11.1 Disk geometry

In disk geometry, the two-particle Coulomb matrix element in the lowest Landau level is defined as

$$\langle r, s | V | t, u \rangle = \int \int d^2 \mathbf{r}_1 \, d^2 \mathbf{r}_2 \, \eta^*_{0,r}(\mathbf{r}_1) \eta^*_{0,s}(\mathbf{r}_2) \frac{1}{|\mathbf{r}_1 - \mathbf{r}_2|} \eta_{0,t}(\mathbf{r}_1) \eta_{0,u}(\mathbf{r}_2)$$

$$= [2^{r+s+t+u} \, (2\pi)^4 \, r!s!t!u!]^{-1/2}$$

$$\times \int d^2 z_1 \, d^2 z_2 \frac{z_1^{*r} z_2^{*s} z_2^u z_1^t}{|z_1 - z_2|} \, e^{-\frac{1}{2}(|z_1|^2 + |z_2|^2)}, \tag{3.175}$$

where r, s, t, and u are angular momentum quantum numbers. For the two-body problem, as always, we transform to the relative and center of mass coordinates:

$$Z = \frac{z_1 + z_2}{2}, \qquad z = z_1 - z_2, \tag{3.176}$$

$$z_1 = Z + \frac{z}{2}, \qquad z_2 = Z - \frac{z}{2}, \tag{3.177}$$

and similar expressions for the complex conjugates. The Jacobian for the transformation is unity. After switching to the center of mass and relative coordinates and making a binomial expansion of the powers of $(Z \pm z/2)$, we are left with Gaussian integrals that can be performed using the formulas given in Appendix A.

The Coulomb interaction conserves the total angular momentum, that is, $r + s = t + u$. To see that, note that after transforming into the center of mass and relative coordinates, a typical term in the integrand has the form

$$\bar{Z}^{\mu_1} \bar{z}^{\mu_2} Z^{\mu_3} z^{\mu_4},$$

with $\mu_1 + \mu_2 = r + s$ and $\mu_3 + \mu_4 = t + u$. The angular integrals vanish unless $\mu_1 = \mu_3$ and $\mu_2 = \mu_4$, which implies $r + s = t + u$. We see below that the total angular momentum operator commutes with the Hamiltonian.

Explicit expressions for the Coulomb matrix elements can be found in the literature (Dev and Jain [116]; Stone, Wyld and Schultz [617]; Tsiper [649]). Particularly useful for numerical studies is the form obtained by Tsiper [649], which we quote here:

$$\langle s+r, t|V|s, t+r \rangle = \sqrt{\frac{(s+r)!(t+r)!}{s!t!}} \frac{\Gamma(r+s+t+\frac{3}{2})}{\pi \, 2^{r+s+t+2}} [A_{st}^r B_{ts}^r + B_{st}^r A_{ts}^r], \quad (3.178)$$

where

$$A_{st}^r = \sum_{i=0}^{s} \binom{s}{i} \frac{\Gamma(\frac{1}{2}+i)\Gamma(\frac{1}{2}+r+i)}{(r+i)!\Gamma(\frac{3}{2}+r+t+i)}, \quad (3.179)$$

$$B_{st}^r = \sum_{i=0}^{s} \binom{s}{i} \frac{\Gamma(\frac{1}{2}+i)\Gamma(\frac{1}{2}+r+i)}{(r+i)!\Gamma(\frac{3}{2}+r+t+i)} \left(\frac{1}{2}+r+2i\right). \quad (3.180)$$

It expresses the matrix element as a sum of positive terms, which makes it numerically more stable, especially for large angular momenta.

3.11.2 Haldane pseudopotentials

Haldane [221] introduced a useful parametrization of the Coulomb interaction, which we describe following MacDonald [412]. Consider the kinetic energy of two particles

$$H_K = \frac{1}{2m_b} \left(\mathbf{p}_1 + \frac{e}{c}\mathbf{A}(\mathbf{r}_1)\right)^2 + \frac{1}{2m_b} \left(\mathbf{p}_2 + \frac{e}{c}\mathbf{A}(\mathbf{r}_2)\right)^2. \quad (3.181)$$

Each kinetic energy term can be converted into the form Eq. (3.18) by going from the variables x_1, y_1, x_2, y_2 to $z_1, \bar{z}_1, z_2, \bar{z}_2$. Further transforming to the relative and center of mass coordinates, and noting

$$\frac{\partial}{\partial z_1} = \frac{1}{2}\frac{\partial}{\partial Z} + \frac{\partial}{\partial z}, \quad (3.182)$$

$$\frac{\partial}{\partial z_2} = \frac{1}{2}\frac{\partial}{\partial Z} - \frac{\partial}{\partial z}, \quad (3.183)$$

we get, after a little algebra,

$$H_K = \frac{1}{2}\left[-4\ell_R^2 \frac{\partial^2}{\partial Z \partial \bar{Z}} + \frac{1}{4\ell_R^2}Z\bar{Z} - Z\frac{\partial}{\partial Z} + \bar{Z}\frac{\partial}{\partial \bar{Z}}\right]$$
$$+ \frac{1}{2}\left[-4\ell_r^2 \frac{\partial^2}{\partial z \partial \bar{z}} + \frac{1}{4\ell_r^2}z\bar{z} - z\frac{\partial}{\partial z} + \bar{z}\frac{\partial}{\partial \bar{z}}\right], \quad (3.184)$$

with

$$\ell_R = \frac{1}{\sqrt{2}}\ell, \quad (3.185)$$

$$\ell_r = \sqrt{2}\ell, \quad (3.186)$$

and with $\hbar\omega_c = 1$. Thus the center of mass and relative degrees of freedom decouple, and each has the same form for the kinetic energy as a single particle in a magnetic field. As in Eq. (3.22), we can define ladder operators for the center of mass:

$$b_R = \frac{1}{\sqrt{2}} \left(\frac{\bar{Z}}{2\ell_R} + 2\ell_R \frac{\partial}{\partial Z} \right), \tag{3.187}$$

$$b_R^\dagger = \frac{1}{\sqrt{2}} \left(\frac{Z}{2\ell_R} - 2\ell_R \frac{\partial}{\partial \bar{Z}} \right), \tag{3.188}$$

$$a_R^\dagger = \frac{1}{\sqrt{2}} \left(\frac{\bar{Z}}{2\ell_R} - 2\ell_R \frac{\partial}{\partial Z} \right), \tag{3.189}$$

$$a_R = \frac{1}{\sqrt{2}} \left(\frac{Z}{2\ell_R} + 2\ell_R \frac{\partial}{\partial \bar{Z}} \right), \tag{3.190}$$

and an analogous set of operators for the relative coordinate, with the R, Z and ℓ_R replaced by r, z and ℓ_r, respectively.

A complete set of two-particle states in the lowest Landau level is given by

$$|M, m\rangle = \frac{(b_R^\dagger)^M (b_r^\dagger)^m}{\sqrt{M! m!}} |0, 0\rangle. \tag{3.191}$$

In coordinate representation

$$\eta_{M,m}(r, R) = \eta_M^R(R)\eta_m^r(r)$$

$$= \frac{(Z/\ell_R)^M}{\sqrt{2\pi \ell_R^2 2^M M!}} \frac{(z/\ell_r)^m}{\sqrt{2\pi \ell_r^2 2^m m!}} \exp\left[-\frac{z\bar{z}}{4\ell_r^2} - \frac{Z\bar{Z}}{4\ell_R^2} \right]. \tag{3.192}$$

(The two bases $|m_1, m_2\rangle$ and $|M, m\rangle$ are distinguished below by their arguments.) The Gaussian factor is equal to the usual $\exp\left[-\frac{z_1\bar{z}_1}{4\ell^2} - \frac{z_2\bar{z}_2}{4\ell^2} \right]$.

Assuming that the interaction is isotropic, i.e., depends only on the distance between the particles, the interaction Hamiltonian takes a simple form in this basis

$$\langle M', m' | V(r) | M, m \rangle = \langle M' | M \rangle \langle m' | V(r) | m \rangle$$

$$= \delta_{M'M} \delta_{m'm} \langle m | V(r) | m \rangle, \tag{3.193}$$

because $V(r)$ acts only on the relative coordinates, and the angular integral in $\langle m | V(r) | m' \rangle$ produces zero unless $m = m'$. Thus, the interaction conserves both the total ($M + m$) and the relative (m) angular momenta. The interaction Hamiltonian is then given by

$$V = V(|r_1 - r_2|) \tag{3.194}$$

$$= \sum_{m,M} \sum_{m',M'} |M', m'\rangle \langle M', m' | V(r) | M, m \rangle \langle M, m|$$

$$= \sum_{m,M} |M,m\rangle \langle m|V(r)|m\rangle \langle M,m|$$

$$= \sum_{m,M} V_m P_{M,m}.$$

Here,

$$P_{M,m} = |M,m\rangle \langle M,m| \tag{3.195}$$

is the projection operator for a state of relative angular momentum m and total angular momentum $M + m$. It satisfies the property

$$P_{M,m}|M',m'\rangle = \delta_{m,m'}\delta_{M,M'}|M,m\rangle, \tag{3.196}$$

that is, it is unity when applied on $|M,m\rangle$ and zero otherwise. The parameters V_m, called Haldane pseudopotentials, are the energies of two particles in a state of relative angular momentum m. These can be evaluated readily for any given interaction. For the Coulomb interaction, we have

$$V_m = \langle m|V|m\rangle$$

$$= \int d^2r \frac{1}{r} |\eta_m^r(r)|^2$$

$$= \frac{1}{2\ell} \frac{\Gamma(m+\frac{1}{2})}{\Gamma(m+1)}, \tag{3.197}$$

where we have made use of the Gaussian integral in Appendix A. The general matrix element $V_{rs,tu} = \langle r,s|V|t,u\rangle$ is expressed in terms of the Haldane pseudopotentials as

$$\langle r,s|V|t,u\rangle = \sum_{m=0}^{\infty} V_m \sum_{M=0}^{\infty} \langle r,s|M,m\rangle \langle M,m|t,u\rangle. \tag{3.198}$$

The pseudopotentials V_m completely specify the interaction in the LLL problem. All real-space interactions that produce the same value for V_m are identical for the LLL problem. (Of course, they have different matrix elements in higher Landau levels.) It is sometimes convenient to define a model interaction in terms of the pseudopotentials rather than a real-space form $V(r)$. For example, a "hard-core" interaction can be chosen as

$$V_0 = 1, \qquad V_1 = V_2 = V_3 = \cdots = 0. \tag{3.199}$$

3.11.3 Higher Landau levels

We now consider electrons confined to the nth Landau level (with $n = 0$ being the lowest Landau level). The lower Landau levels are taken as full and inert, and it is assumed that the interaction does not cause any LL mixing, as appropriate when the cyclotron energy is very large compared to the interaction strength. We need matrix elements

$$\langle n, m_1'; n, m_2' | V(|\boldsymbol{r}_1 - \boldsymbol{r}_2|) | n, m_1; n, m_2 \rangle. \tag{3.200}$$

The situation in which all electrons do not have the same LL index is also straightforward but more cumbersome, and will not be considered here.

A note on notation: We use a slightly different convention in this subsection, with

$$|n, m\rangle \equiv \frac{(a^\dagger)^n}{\sqrt{n!}} \frac{(b^\dagger)^m}{\sqrt{m!}} |0\rangle. \tag{3.201}$$

The symbol m now takes values $0, 1, \ldots$ for any Landau level and does not represent the eigenvalue of L; the angular momentum in this notation is $m - n = -n, -n+1, \ldots$ In general, it is the angular momentum that is conserved, but for a problem restricted to a given Landau level, $m_1 + m_2$ is also conserved. Also, we suppress the n quantum number whenever it is $n = 0$.

The problem of electrons interacting with the Coulomb interaction (or for that matter any given interaction) in a higher Landau level is mathematically equivalent to the problem of electrons in the lowest Landau level interacting with an effective interaction $V_{\text{eff}}(r)$, defined by

$$\langle n, m_1'; n, m_2' | V(|\boldsymbol{r}_1 - \boldsymbol{r}_2|) | n, m_1; n, m_2 \rangle = \langle m_1', m_2' | V_{\text{eff}}(|\boldsymbol{r}_1 - \boldsymbol{r}_2|) | m_1, m_2 \rangle. \tag{3.202}$$

Another way of stating this is that solving the problem in the lowest Landau level for the effective interaction gives, by replacing the Fock states $|m_1, \ldots, m_N\rangle$ by $|n, m_1; \ldots; n, m_N\rangle$, the solution for the Coulomb problem in the nth Landau level. This correspondence is extremely useful in the studies of higher Landau levels, because various trial wave functions, which form the basis for our understanding of the FQHE, are most readily written in the lowest Landau level. If we could express the LLL trial wave functions as a linear superposition of Fock states, then we could translate them into higher LL wave functions and calculate with them. That program, however, becomes impossible beyond a few particles, because the dimension of the Fock space quickly grows larger than what any computer can store. In contrast, LLL calculations with an effective interaction can be performed for fairly large systems.

The evaluation of $V_{\text{eff}}(r)$ is best accomplished in the Fourier space. The Fourier transform of the interaction is given by

$$V(|\boldsymbol{r}_1 - \boldsymbol{r}_2|) = \int \frac{d^2k}{(2\pi)^2} V(k) e^{i\boldsymbol{k} \cdot (\boldsymbol{r}_1 - \boldsymbol{r}_2)}, \tag{3.203}$$

where

$$V(k) = \frac{2\pi}{k}. \tag{3.204}$$

The identity proven below implies that the effective interaction in the nth Landau level is given by

$$V_{\text{eff}}(k) = \left[L_n\left(\frac{k^2}{2}\right) \right]^2 V(k), \tag{3.205}$$

where L_n are the Laguerre polynomials.

Theorem 3.2 *The matrix elements of $e^{ik\cdot(r_1-r_2)}$ in the nth Landau levels are related to those in the lowest Landau level by the relation*

$$\langle n, m_1'; n, m_2'|e^{ik\cdot(r_1-r_2)}|n, m_1; n, m_2\rangle$$

$$= \left[L_n\left(\frac{k^2}{2}\right) \right]^2 \langle m_1'; m_2'|e^{ik\cdot(r_1-r_2)}|m_1; m_2\rangle. \tag{3.206}$$

Proof   We first consider

$$\langle n, m'|e^{ik\cdot r}|n, m\rangle = \langle n|e^{\frac{i}{\sqrt{2}}(\bar{k}a+ka^\dagger)}|n\rangle \langle m'|e^{\frac{i}{\sqrt{2}}(\bar{k}b^\dagger+kb)}|m\rangle$$

$$= e^{-\frac{k\bar{k}}{2}} \langle n|e^{\frac{i}{\sqrt{2}}ka^\dagger} e^{\frac{i}{\sqrt{2}}\bar{k}a}|n\rangle \langle m'|e^{\frac{i}{\sqrt{2}}\bar{k}b^\dagger} e^{\frac{i}{\sqrt{2}}kb}|m\rangle. \tag{3.207}$$

Here we have substituted

$$k \cdot r = \frac{k\bar{z} + \bar{k}z}{2}$$

$$= \frac{k(a^\dagger + b) + \bar{k}(a + b^\dagger)}{\sqrt{2}}, \tag{3.208}$$

where $k = k_x - ik_y$, $\bar{k} = k_x + ik_y$, $z = \sqrt{2}(a + b^\dagger)$, $\bar{z} = \sqrt{2}(a^\dagger + b)$, and Eq. (B.4) has been used. (Note that $k^2 = |k|^2$.) One of the factors in Eq. (3.207) is evaluated as

$$\langle n|e^{\frac{i}{\sqrt{2}}ka^\dagger} e^{\frac{i}{\sqrt{2}}\bar{k}a}|n\rangle = \sum_{s=0}^{n} \frac{1}{(s!)^2} \langle n| \left(\frac{i}{\sqrt{2}}ka^\dagger\right)^s \left(\frac{i}{\sqrt{2}}\bar{k}a\right)^s |n\rangle$$

$$= \sum_{s=0}^{n} \left(-\frac{k\bar{k}}{2}\right)^s \frac{n!}{(s!)^2(n-s)!}$$

$$= L_n\left(\frac{k\bar{k}}{2}\right), \tag{3.209}$$

where we have made use of

$$a^s|n\rangle = \left(\frac{n!}{(n-s)!}\right)^{1/2} |n-s\rangle, \quad (n \geq s) \tag{3.210}$$

and

$$L_m^n(x) = \sum_{s=0}^{m} \frac{(-x)^s}{s!} \binom{m+n}{m-s}$$ (3.211)

with

$$L_m^0(x) \equiv L_m(x).$$ (3.212)

We thus get

$$\langle n, m' | e^{i\mathbf{k}\cdot\mathbf{r}} | n, m \rangle = e^{-\frac{k\bar{k}}{2}} L_n \left(\frac{k^2}{2}\right) \langle m' | e^{\frac{i}{\sqrt{2}}\bar{k}b^\dagger} e^{\frac{i}{\sqrt{2}}kb} | m \rangle$$

$$= L_n \left(\frac{k^2}{2}\right) \langle m' | e^{i\mathbf{k}\cdot\mathbf{r}} | m \rangle,$$ (3.213)

with the second equality resulting because $L_0(x) = 1$. Equation (3.206) follows because the expression on the left hand side of that equation contains two such factors, producing $[L_n(k^2/2)]^2$. $\qquad\square$

3.11.4 Spherical geometry

The following properties of the monopole harmonics, quoted here without derivation, are useful (Wu and Yang [701, 702]):

$$Y^*_{Q,l,m} = (-1)^{Q-m} Y_{-Q,l,-m},$$ (3.214)

$$\int d\Omega \, Y_{Q_1 l_1 m_1} Y_{Q_2 l_2 m_2} Y_{Q_3 l_3 m_3} = S(\{Q_j, l_j, m_j\}),$$ (3.215)

$$Y_{Q_1 l_1 m_1} Y_{Q_2 l_2 m_2} = (-1)^{Q_3 - m_3} \sum_{l_3} S(\{Q_j, l_j, m_j\}) Y_{-Q_3, l_3, -m_3}.$$ (3.216)

Here, $d\Omega$ is the integral over the angular variables,

$$Q_3 \equiv -Q_1 - Q_2, \qquad m_3 \equiv -m_1 - m_2,$$ (3.217)

$$S(\{Q_j, l_j, m_j\}) \equiv (-1)^{l_1+l_2+l_3} \left(\frac{(2l_1+1)(2l_2+1)(2l_3+1)}{4\pi}\right)^{1/2}$$

$$\times \begin{pmatrix} l_1 & l_2 & l_3 \\ -m_1 & -m_2 & -m_3 \end{pmatrix} \begin{pmatrix} l_1 & l_2 & l_3 \\ Q_1 & Q_2 & Q_3 \end{pmatrix},$$ (3.218)

and the round brackets are 3j symbols, related to Clebsch–Gordon coefficients through the equation

$$\begin{pmatrix} l_1 & l_2 & l_3 \\ m_1 & m_2 & m_3 \end{pmatrix} = \frac{(-1)^{l_1-l_2-m_3}}{\sqrt{2l_3+1}} \langle l_1, m_1; l_2, m_2 | l_3, -m_3 \rangle.$$ (3.219)

For exact diagonalizations on finite systems, we need to know the Coulomb matrix elements. The two-body matrix element is given by

$$\langle l_1\, m_1, l_2\, m_2 | V(|\boldsymbol{r}_1 - \boldsymbol{r}_2|)| l_1'\, m_1', l_2'\, m_2' \rangle =$$

$$\int d\boldsymbol{\Omega}_1\, d\boldsymbol{\Omega}_2\, Y^*_{Ql_1 m_1}(\boldsymbol{r}_1) Y^*_{Ql_2 m_2}(\boldsymbol{r}_2) \frac{e^2}{|\boldsymbol{r}_1 - \boldsymbol{r}_2|} Y^*_{Ql_2' m_2'}(\boldsymbol{r}_2) Y^*_{Ql_1' m_1'}(\boldsymbol{r}_1). \qquad (3.220)$$

A common practice is to use the chord distance between electrons given by $|\boldsymbol{r}_1 - \boldsymbol{r}_2| = \sqrt{2}|u_1 v_2 - u_2 v_1|R$. Using the "addition theorem," the interaction can be expressed as

$$\frac{1}{|\boldsymbol{r}_1 - \boldsymbol{r}_2|} = \frac{4\pi}{R} \sum_{l=0}^{\infty} \frac{1}{2l+1} \sum_{m=-l}^{l} Y^*_{0,l,m}(\theta_1, \phi_1) Y_{0,l,m}(\theta_2, \phi_2). \qquad (3.221)$$

With this substitution, the two integrals decouple and can be evaluated using Eq. (3.215). Alternatively, the Coulomb interaction could have been defined in terms of the arc distance between electrons; the two definitions (using chord or arc distance) are identical in the thermodynamic limit, and produce approximately the same eigenfunctions for finite systems.

The Haldane pseudopotentials can be defined in the spherical geometry as

$$\langle m_1', m_2' | V(r) | m_1, m_2 \rangle = \sum_{L=0}^{2Q} \sum_{M=-L}^{L} \langle Qm_1', Qm_2' | LM \rangle V_L \langle LM | Qm_1, Qm_2 \rangle, \qquad (3.222)$$

where we have used

$$\langle L'M' | V(r) | LM \rangle = \delta_{MM'} \delta_{LL'} V_L. \qquad (3.223)$$

The pseudopotential V_L is thus the energy of two electrons in the relative angular L multiplet. We quote here the result for the lowest Landau level (Fano, Ortolani, and Colombo [151]):

$$V_L = \frac{2}{R} \frac{\binom{4Q-2L}{2Q-L}\binom{4Q+2L+2}{2Q+L+1}}{\binom{4Q+2}{2Q+1}^2}. \qquad (3.224)$$

The translational symmetry of the planar geometry, which shows up in the two-body problem through a degeneracy of states that differ only by the center of mass part of the wave function, is equivalent to the rotational symmetry in the spherical geometry, which, in turn, is exhibited through the $(2L+1)$ degeneracy of each L multiplet.

The total orbital angular momentum and its z component commute with the interaction Hamiltonian, and are, therefore, good quantum numbers. For a many-particle system, one can diagonalize within a sector with definite values of L and L_z. In practice, constructing basis states with a given L_z is easiest; choosing the smallest value ($L_z = 0$ or $1/2$) produces the entire spectrum.

3.12 Disk geometry/parabolic quantum dot

This section concerns interacting electrons in a high magnetic field forming a droplet in the two-dimensional plane. This geometry is not very useful for understanding the origin of the quantum Hall effect, because, for a finite and not very large number of electrons, the charge density is everywhere nonuniform and the notion of filling factor is not well defined. (The FQHE can of course be understood in the planar geometry, but would require hard work: the study of a very large number of electrons in the presence of a hard-wall confinement should produce the FQHE.) Nonetheless, electron droplets are interesting in their own right, as they can be produced and studied experimentally.

How do we keep electrons from flying away to minimize their Coulomb energy? They can be confined by an electrostatic potential, which is usually taken to be parabolic. This is an example of a "quantum dot." Alternatively, we can prepare a state with a definite angular momentum, which is a good quantum number because the Coulomb interaction commutes with L_z. Angular momentum conservation guarantees a spatial confinement of electrons. We see below that the solution of the latter problem also gives, with minimal modification, the solution of the parabolic quantum dot problem.

Even though no FQHE occurs in a parabolic quantum dot, composite fermions are formed therein and provide insight into its strongly correlated states. How large must the system be before composite fermions can be realized? As seen below, the physics of the droplet is well described by the composite fermion theory even for very few (e.g., three) particles.

3.12.1 Fock–Darwin levels

Let us begin with a single electron Hamiltonian:

$$H = \frac{1}{2m_b}\left(p + \frac{e}{c}A\right)^2 + \frac{1}{2}m_b\omega_0^2(x^2 + y^2),$$

(3.225)

where ω_0 is a measure of the strength of the confinement. This model is popular because it yields exact solutions for a single particle (Darwin [106]; Fock [169]), and also because it is often a reasonable approximation to the actual confinement close to the minimum. We exploit the rotational symmetry of the problem to map it into a previously solved problem.

The natural gauge for this problem is the symmetric gauge $A = \frac{B}{2}(-y, x, 0)$. In this gauge, the Hamiltonian reduces to

$$H = \frac{1}{2m_b}\left[p^2 + \frac{1}{4}m_b^2\Omega^2 r^2 + \frac{eB}{c}L_z\right],$$

(3.226)

where

$$\Omega^2 = \omega_c^2 + 4\omega_0^2,$$

(3.227)

and

$$L_z = xp_y - yp_x$$
$$= \hat{z} \cdot r \times p$$
$$= -i\hbar \hat{z} \cdot r \times \nabla$$
$$= -i\hbar \hat{z} \cdot r \times \left(\hat{r} \frac{\partial}{\partial r} + \frac{\hat{\theta}}{r} \frac{\partial}{\partial \theta} \right)$$
$$= -i\hbar \frac{\partial}{\partial \theta}. \tag{3.228}$$

L_z can be seen to commute with the Hamiltonian, as expected from the circular symmetry of the problem.

The natural length ℓ_c for this problem is defined by

$$\ell_c = \left(\frac{\hbar}{m_b \Omega} \right)^{1/2}, \tag{3.229}$$

which reduces to the usual magnetic length in the limit of no confinement. Expressing all lengths in units of ℓ_c we get

$$H = \frac{1}{2} \hbar \Omega \left[-\nabla^2 + \frac{r^2}{4} \right] + \frac{\omega_c}{2} L_z$$
$$= \frac{1}{2} \hbar \Omega \left[-\nabla^2 + \frac{r^2}{4} + \frac{L_z}{\hbar} \right] - \frac{\Omega - \omega_c}{2} L_z. \tag{3.230}$$

From Chapter 3, we know the solutions of the problem in the absence of the confining potential, which is defined by the Hamiltonian

$$H(\omega_0 = 0) = \frac{1}{2} \hbar \omega_c \left[-\nabla^2 + \frac{r^2}{4} + \frac{L_z}{\hbar} \right]. \tag{3.231}$$

The first term on the right hand side in Eq. (3.230) is equivalent to this problem, except that the energy scale is $\hbar \Omega$ and the length scale is ℓ_c. The eigenfunctions are given by

$$\eta_{n,m}(r) = \frac{(-1)^n}{\sqrt{2\pi}} \sqrt{\frac{2^n n!}{2^m m!}} \, e^{-\frac{r^2}{4}} z^{m-n} L_n^{m-n} \left(\frac{r^2}{2} \right), \tag{3.232}$$

which are also the solutions of the full Hamiltonian of Eq. (3.230), as they are already eigenstates of L_z. With $z = x - iy = r e^{-i\theta}$ we have

$$L_z z^{m-n} = -i\hbar \frac{\partial}{\partial \theta} (r e^{-i\theta})^{m-n}$$
$$= -(m - n) z^{m-n}, \tag{3.233}$$

and the eigen energies, therefore, are

$$E_{n,m} = \frac{1}{2}\hbar\Omega\left(n + \frac{1}{2}\right) + \frac{1}{2}\hbar(\Omega - \omega_c)(m - n). \tag{3.234}$$

The second term vanishes in the absence of confinement ($\omega_c = \Omega$) and the result of Section 3.3 is reproduced. With parabolic confinement, the states with a given n are not degenerate, and the energy ordering of states depends on both n and m.

High magnetic field limit   In the spirit of this book, we restrict the discussion to the high magnetic field limit

$$\frac{\omega_0}{\omega_c} \to 0. \tag{3.235}$$

In this case, the second term of Eq. (3.234) is small, and the low energy states (for not too large m) are given by $n = 0$, which we call the lowest Landau level. Then,

$$E_{0,m} = \frac{1}{4}\hbar\Omega + \frac{\hbar}{2}(\Omega - \omega_c)m. \tag{3.236}$$

The first term is constant, and the second term, proportional to the L_z quantum number, is interpreted as the confinement energy.

3.12.2 Many electrons: exact diagonalization

Next we consider several interacting electrons in the quantum dot, described by the Hamiltonian

$$H = \sum_j \frac{1}{2m_b}\left(\mathbf{p}_j + \frac{e}{c}\mathbf{A}(\mathbf{r}_j)\right)^2 + \sum_j \frac{1}{2}m_b\omega_0^2(x_j^2 + y_j^2) + \sum_{j<k} \frac{e^2}{\epsilon|\mathbf{r}_j - \mathbf{r}_k|}. \tag{3.237}$$

The Coulomb interaction conserves the total angular momentum

$$L = \sum m_i, \tag{3.238}$$

which can, therefore, be used to label the many-body eigenstates. (L here is the eigenvalue of the total L_z, and ought not to be confused with the orbital angular momentum in the spherical geometry, for which we used the same symbol.) The energy, measured from $\frac{N}{2}\hbar\Omega$, is given by

$$E(L) = E_c(L) + V(L). \tag{3.239}$$

The first term is the confinement energy, which has the simple form

$$E_c(L) = \frac{\hbar}{2}(\Omega - \omega_c)L. \tag{3.240}$$

The second term is the Coulomb interaction energy, which can be obtained from the corresponding energy for electrons without the confinement, with a trivial change of units

$$\frac{e^2}{\ell} \to \frac{e^2}{\ell_c}. \tag{3.241}$$

Consideration of interacting electrons in the absence of confinement is thus sufficient. With a change of the interaction energy scale and an addition of the known confinement term, the results apply to parabolic quantum dots in sufficiently high magnetic fields.

3.12.3 Center of mass excitation

We prove a general result in the absence of confinement, which allows construction of an infinite number of eigenstates from a given eigenstate by application of ladder operators for the center of mass, which we now construct. Switch variables as follows:

$$\{z_1, \ldots, z_N\} \to \{w_0, \ldots, w_{N-1}\}, \tag{3.242}$$

where

$$w_0 \equiv \frac{1}{N} \sum_i z_i, \qquad w_j \equiv z_j - \frac{1}{N} \sum_i z_i, \tag{3.243}$$

w_0 is the center of mass coordinate and w_j are positions measured from the center of mass. A little algebra shows that

$$B \equiv \frac{1}{\sqrt{N}} \sum_j b_j = \frac{1}{\sqrt{2}} \left(\sqrt{N} \frac{\bar{w}_0}{2} + \frac{2}{\sqrt{N}} \frac{\partial}{\partial w_0} \right), \tag{3.244}$$

$$B^\dagger \equiv \frac{1}{\sqrt{N}} \sum_j b_j^\dagger = \frac{1}{\sqrt{2}} \left(\sqrt{N} \frac{w_0}{2} - \frac{2}{\sqrt{N}} \frac{\partial}{\partial \bar{w}_0} \right), \tag{3.245}$$

$$A \equiv \frac{1}{\sqrt{N}} \sum_j a_j = \frac{1}{\sqrt{2}} \left(\sqrt{N} \frac{w_0}{2} + \frac{2}{\sqrt{N}} \frac{\partial}{\partial \bar{w}_0} \right), \tag{3.246}$$

$$A^\dagger \equiv \frac{1}{\sqrt{N}} \sum_j a_j^\dagger = \frac{1}{\sqrt{2}} \left(\sqrt{N} \frac{\bar{w}_0}{2} - \frac{2}{\sqrt{N}} \frac{\partial}{\partial w_0} \right), \tag{3.247}$$

where $a, a^\dagger, b,$ and $b^\dagger$ were defined in Eqs. (3.19)–(3.22). It follows that

$$[A, A^\dagger] = 1, \qquad [B, B^\dagger] = 1, \tag{3.248}$$

and the commutators between $A, A^\dagger$ and $B, B^\dagger$ vanish. Furthermore, we have

$$[H, A] = -\hbar\omega_c A, \qquad [H, A^\dagger] = \hbar\omega_c A^\dagger, \qquad [H, B] = 0, \qquad [H, B^\dagger] = 0, \tag{3.249}$$

where $H = V + K$, V is the interaction energy,

$$K = \sum_j \left(a_j^\dagger a_j + \frac{1}{2} \right) \hbar \omega_c,$$ (3.250)

and we used that the Coulomb interaction V, being independent of center of mass coordinates, commutes with the ladder operators. Given an eigenstate

$$H \Psi = E \Psi,$$ (3.251)

we have

$$HA^\dagger \Psi = ([H, A^\dagger] + A^\dagger H) \Psi = (E + \hbar \omega_c) A^\dagger \Psi$$ (3.252)

and

$$HB^\dagger \Psi = ([H, B^\dagger] + B^\dagger H) \Psi = EB^\dagger \Psi.$$ (3.253)

By use of the ladder operators, an infinite set of eigenstates can be constructed at E and $E + n\hbar \omega_c$. The above result is true for arbitrary interactions, but assumes a translational invariance system.

To see how the ladder operators affect the wave function, let us note that the center of mass dependent part of the wave function comes from the Gaussian factors:

$$\exp \left(-\frac{1}{4} \sum_j |z_j|^2 \right) = \exp \left(-\frac{1}{4N} \sum_{j<k} |z_j - z_k|^2 \right) \exp \left(-\frac{N}{4} w_0 \bar{w}_0 \right).$$ (3.254)

Application of $[A^\dagger]^n$ and $[B^\dagger]^m$ on $\exp(-\frac{N}{4} w_0 \bar{w}_0)$ produces the LL wave functions of the center of mass. The application of $B^\dagger$ increases the angular momentum by one unit, which implies that for a given eigenstate at some angular momentum, degenerate eigenstates appear at all higher angular momenta. This feature is explicitly confirmed in the exact spectra.

3.13 Torus geometry

This section is devoted, for completeness, to a brief discussion of an electron on a torus, that is, a rectangular sample of sides L_x and L_y with periodic boundary conditions in both directions. (More generally, one can work with a sample in the shape of a parallelogram [223].) One's first impulse might be to look for wave functions satisfying the boundary conditions $\Psi(x+L_x, y) = T_{L_x} \Psi(x, y) = \Psi(x, y)$ and $\Psi(x, y+L_y) = T_{L_y} \Psi(x, y) = \Psi(x, y)$. These boundary conditions, however, cannot be imposed consistently for an eigenstate of H, because, as a result of $[T_R, H] \neq 0$, $T_R \Psi$ is, in general, not an eigenstate of H. The proper way of imposing periodic boundary conditions is through the magnetic translation operator:

$$T_{L_x \hat{x}} \Psi(x, y) = \Psi(x, y),$$ (3.255)

$$T_{L_y \hat{y}} \Psi(x, y) = \Psi(x, y).$$ (3.256)

Another way to see this is to note that the effect of $\mathcal{T}$ is to translate the wave function, followed by a gauge transformation which sets the vector potential at $\boldsymbol{r} + \boldsymbol{R}$ equal to its original value at $\boldsymbol{r}$. (Life would have been much simpler if only we had available a vector potential periodic with the desired periodicity.) In the Landau gauge, using the explicit form of $\mathcal{T}_{\boldsymbol{R}}$ derived earlier, we get

$$\Psi(x + L_x, y) = \Psi(x, y), \tag{3.257}$$

$$e^{-\frac{i}{\ell^2} L_y x} \Psi(x, y + L_y) = \Psi(x, y). \tag{3.258}$$

Further, a physical wave function must satisfy

$$\mathcal{T}_{L_x \hat{x}} \mathcal{T}_{L_y \hat{y}} \Psi(x, y) = \mathcal{T}_{L_y \hat{y}} \mathcal{T}_{L_x \hat{x}} \Psi(x, y). \tag{3.259}$$

However, from Eq. (3.105)

$$\mathcal{T}_{L_x \hat{x}} \mathcal{T}_{L_y \hat{y}} = e^{-\frac{i}{\ell^2} L_x L_y} \mathcal{T}_{L_y \hat{y}} \mathcal{T}_{L_x \hat{x}}. \tag{3.260}$$

A consistent imposition of periodic boundary conditions therefore requires that

$$\frac{L_x L_y}{\ell^2} = 2\pi N_\phi, \tag{3.261}$$

where N_ϕ is an integer. This condition is equivalent to the requirement that the flux through the rectangle be an integral number (N_ϕ) of flux quanta:

$$B L_x L_y = N_\phi \phi_0. \tag{3.262}$$

Thus, just as with the spherical geometry, the torus geometry also allows only for discrete values of magnetic field.

The wave functions on the torus are linear superpositions of the planar LL wave functions discussed earlier. For the lowest Landau level, the single particle wave functions in the Landau gauge are given by (without worrying about normalization)

$$\eta_{k_x} = e^{-\frac{1}{2}(y - k_x)^2} e^{i k_x x}, \tag{3.263}$$

where we have set $\ell = 1$ and used $H_0(x) = 1$. One can see that the LLL wave functions for the torus are given by

$$\Psi_{k_x} = \sum_{n=-\infty}^{\infty} e^{-\frac{1}{2}(y - n L_y - k_x)^2} e^{i(k_x + n L_y)x}. \tag{3.264}$$

The periodic condition in the x direction,

$$\mathcal{T}_{L_x \hat{x}} \Psi(x, y) = \Psi(x + L_x, y) = \Psi(x, y) \tag{3.265}$$

is guaranteed by the choice

$$k_x = 2\pi \frac{n_x}{L_x}, \tag{3.266}$$

where n_x is an integer. One can confirm that

$$T_{L_y\hat{y}} \Psi(x, y) = e^{-iL_y x} \Psi(x, y + L_y) = \Psi(x, y). \tag{3.267}$$

The degeneracy of the lowest Landau level is equal to the number of k_x values that give distinct wave functions. The substitution

$$n_x \to n_x + N_\phi, \tag{3.268}$$

with $N_\phi = L_x L_y / 2\pi$, which is equivalent to

$$k_x \to 2\pi \frac{n_x + N_\phi}{L_x} = k_x + L_y, \tag{3.269}$$

leaves the wave function invariant. Therefore, the degeneracy of the lowest Landau level is equal to the number of magnetic flux quanta passing through the surface of the torus.

3.14 Periodic potential: the Hofstadter butterfly

To study the effect of a periodic potential on an electron in the presence of a magnetic field, we consider a potential $U(r)$ that is periodic under translations by

$$R = na + mb, \tag{3.270}$$

(n and m are integers) called Bravais lattice vectors. (We used the symbol R earlier for an arbitrary translation, but, in this section, it is reserved for Bravais lattice vectors.) For simplicity, we consider a rectangular lattice for which the lattice vectors a and b are mutually perpendicular.

First consider the standard problem of a particle in a periodic potential without magnetic field. Then, all translation operators T_R commute with the Hamiltonian and with each other. Therefore, we look for eigenfunctions that simultaneously diagonalize H and $\{T_R\}$. Noting that

$$T_R T_{R'} \psi(r) = T_{R+R'} \psi(r), \tag{3.271}$$

we must have

$$T_R \psi_k(r) = e^{ik \cdot R} \psi_k(r), \tag{3.272}$$

or

$$\psi_k(r + R) = e^{ik \cdot R} \psi_k(r), \tag{3.273}$$

where k is the crystal momentum. This is known as Bloch's theorem, also stated as

$$\psi_k(r) = e^{ik \cdot r} u_k(r), \tag{3.274}$$

where $u_k(r)$ reflects the periodicity of the external potential

$$u_k(r + R) = u_k(r). \tag{3.275}$$

In the presence of a magnetic field, T_R no longer commutes with H. That is taken care of by working with the magnetic translation operators $\mathcal{T}_R$ instead. However, $\mathcal{T}_R$ do not, in general, commute with one another, as seen in Eq. (3.105). Further progress can be made by assuming that the flux through a unit cell, expressed in units of the flux quantum, is a rational number

$$\frac{\phi}{\phi_0} = \frac{Bab}{\phi_0} = \frac{p}{q}, \tag{3.276}$$

where p and q are relatively prime. (An irrational value of ϕ/ϕ_0 can presumably be approximated by a rational number.) In that case, a subset of translations commute with one another. These correspond to magnetic translations

$$R = nqa + mb, \tag{3.277}$$

which defines a new Bravais lattice, with lattice vectors qa and b defining a "magnetic unit cell." The flux through a magnetic unit cell is equal to an integral number (p) of flux quanta. (One could have equally well chosen the magnetic unit cell as (a, qb).) These operators form a group

$$\mathcal{T}_{R_1}\mathcal{T}_{R_2} = e^{-\frac{i}{\ell^2}n_1 m_2 qab}\mathcal{T}_{R_1+R_2}$$
$$= e^{-in_1 m_2 q(\phi/\phi_0)}\mathcal{T}_{R_1+R_2}$$
$$= \mathcal{T}_{R_1+R_2}. \tag{3.278}$$

The reader can verify the gauge independence of this property by evaluating the commutator for the symmetric gauge. Furthermore, from

$$\mathcal{T}_{R_1}\mathcal{T}_{R_2} = \mathcal{T}_{R_1+R_2} = \mathcal{T}_{R_2+R_1} = \mathcal{T}_{R_2}\mathcal{T}_{R_1} \tag{3.279}$$

it follows that

$$[\mathcal{T}_{R_1}, \mathcal{T}_{R_2}] = 0. \tag{3.280}$$

Having identified a set of commuting operators, we can look for eigenfunctions that simultaneously diagonalize them, which gives, as before,

$$\mathcal{T}_R \psi_{\alpha,k}(r) = e^{ik \cdot R}\psi_{\alpha,k}(r), \tag{3.281}$$

where α is the magnetic subband index, and k is the generalized crystal momentum, confined to the magnetic Brillouin zone $0 \le k_x \le 2\pi/qa$, $0 \le k_y \le 2\pi/b$. One conveniently defines

$$\psi_{\alpha,k}(r) = e^{ik \cdot r}u_{\alpha,k}(r), \tag{3.282}$$

but $u_{\alpha,k}(r)$ is not periodic under $r \to r + \mathcal{R}$. Substituting in Eq. (3.281), it satisfies the boundary condition

$$\mathcal{T}_{\mathcal{R}} e^{ik \cdot r} u_{\alpha,k}(r) = e^{ik \cdot (r + \mathcal{R})} u_{\alpha,k}(r). \tag{3.283}$$

Using the explicit form for the magnetic translation operator derived in Section 3.9, we get

$$u_{\alpha,k}(r + \mathcal{R}) = e^{\frac{i}{\ell^2} mbx} u_{\alpha,k}(r) \tag{3.284}$$

for the Landau gauge, and

$$u_{\alpha,k}(r + \mathcal{R}) = e^{-\frac{i}{2\ell^2}(qnay - mbx)} u_{\alpha,k}(r) \tag{3.285}$$

for the symmetric gauge. The Schrödinger equation assumes the form

$$\left[\frac{1}{2m_b} \left(-i\hbar \nabla + \hbar k + \frac{e}{c} A(r) \right)^2 + U(r) \right] u_{\alpha,k}(r) = E_{\alpha,k} u_{\alpha,k}(r). \tag{3.286}$$

To proceed further, we must work with a specific periodic potential. Fortunately, many essential properties of the band structure can be deduced without the actual solution. Two ways of looking at the problem depend on whether the cyclotron energy or the periodic potential dominates. (i) For small magnetic fields, we begin with a periodic lattice in a zero magnetic field and ask how the magnetic field splits the Bloch bands. In zero magnetic field, each Bloch band has M eigenstates, where M is the number of cells. Because the magnetic unit cell contains q cells, each Bloch band splits into q magnetic subbands, with M/q eigenstates in each subband. (ii) When the external potential is weak compared to the cyclotron energy, we ask how switching on the periodic potential affects the Landau levels. Because each Landau level has Mp/q single particle states (which is the number of magnetic flux quanta penetrating the sample), it splits into p magnetic subbands.

Let us take some special cases. For $p = 1$, each Bloch band of zero magnetic field splits into q magnetic bands, each of which corresponds to a single Landau level. This is convenient for calculations that study the integral quantum Hall effect, for example the role of disorder. On the other hand, for $q = 1$, each single Landau level breaks into p magnetic subbands. The plot of allowed energies as a function of the flux per plaquette is known as the "Hofstadter butterfly" [259].

3.15 Tight binding model

For many situations, especially in studying the effect of disorder, one works with a discretized Schrödinger equation on a lattice [259], which is a special case of the periodic potential discussed above. In the absence of a magnetic field, the Schrödinger equation for an electron on a square lattice is given by

$$H\Psi(r) = -t[\Psi(x+a,y) + \Psi(x-a,y)$$
$$+ \Psi(x,y+a) + \Psi(x,y-a) - 4\Psi(x,y)]$$
$$= E\Psi(r), \qquad (3.287)$$

where t is the tunneling matrix element, a is the lattice constant, and $(x,y) = (ma, na)$. Its solutions are given by

$$\Psi_k(r) = e^{ik\cdot r} \qquad (3.288)$$
$$E_k = -2t[\cos(k_x a) + \cos(k_y a) - 2]. \qquad (3.289)$$

Equation (3.287) reduces to the familiar eigenvalue equation in the continuum limit, as can be seen by Taylor expanding to order a^2, which gives

$$\frac{p^2}{2m_b}\Psi(r) = E'\Psi(r), \qquad (3.290)$$

with

$$E' = \frac{\hbar^2 E}{2m_b a^2 t}. \qquad (3.291)$$

For a given k, the energy eigenvalue is given by

$$E_k' = \frac{\hbar^2 E_k}{2m_b a^2 t} = \frac{\hbar^2 k^2}{2m_b} \qquad (3.292)$$

in the limit $ka \to 0$, where we have made use of

$$E_k = 4t\left[\sin^2\left(\frac{k_x a}{2}\right) + \sin^2\left(\frac{k_y a}{2}\right)\right] \approx tk^2 a^2. \qquad (3.293)$$

The Schrödinger equation is generalized to include the magnetic field, assuming the Landau gauge $A = -By\hat{x}$, as follows:

$$H\Psi(r) = -t[e^{-ie\frac{aBy}{\hbar c}}\Psi(x+a,y) + e^{ie\frac{aBy}{\hbar c}}\Psi(x-a,y)$$
$$+ \Psi(x,y+a) + \Psi(x,y-a) - \Psi(x,y)]$$
$$= E\Psi(r). \qquad (3.294)$$

The tunneling matrix element now contains a phase factor that includes the effect of the magnetic field through the Aharonov–Bohm phase associated with the path that connects two neighboring sites between which tunneling is taking place. The absence of extra phase factors in the last two terms in the first step results because, in our gauge choice, no phase is associated with paths along the y direction. The consistency of Eq. (3.294) can be confirmed by taking the continuum limit as before (by expanding to second order in a; the terms linear in a vanish because the expressions are an even function of a), which gives the familiar eigenvalue equation:

$$\frac{1}{2m_b}\left(p + \frac{e}{c}A\right)^2 \Psi(r) = E'\Psi(r). \tag{3.295}$$

Equation (3.294) has been studied in detail by Hofstadter [259] and produces a band structure consistent with the qualitative discussion at the end of the previous section.

Exercises

3.1 Verify that the Schrödinger equation remains invariant under the gauge transformation defined in Eqs. (3.4) and (3.5).

3.2 Derive Eq. (3.50). Show that the eigenstates in the nth Landau level do not contain $\bar{z}^p$ with $p > n$ in the polynomial part of the wave function.

3.3 Express $r^2 = z\bar{z}$ in terms of the ladder operators, and show that

$$\langle n, m|r^2|n, m\rangle = 2\ell^2(m + 2n + 1), \tag{E3.1}$$

where the symmetric gauge has been assumed. Using this result, prove, neglecting $O(1)$ corrections, that the degeneracy per unit area is the same for all Landau levels.

3.4 The coherent states of a harmonic oscillator are eigenstates of the lowering operator. In this exercise, it is shown that

$$\langle r|\rho\rangle = \phi_\rho(r) = \frac{1}{\sqrt{2\pi}}\exp\left[\frac{1}{2}\bar{\rho}z - \frac{1}{4}|z|^2 - \frac{1}{4}|\rho|^2\right] \tag{E3.2}$$

are the coherent states in the lowest Landau level, and many of their properties are derived. In particular, it is seen that they form a nonorthogonal, overcomplete basis. We have defined $\rho = (\rho_x, \rho_y)$ and $\rho = \rho_x - i\rho_y$. Show the following:

(i) $\phi_\rho(r)$ is confined to the lowest Landau level.
(ii) It is normalized.
(iii) It is an eigenstate of the angular momentum lowering operator:

$$b\phi_\rho(r) = \frac{\bar{\rho}}{\sqrt{2}}\phi_\rho(r). \tag{E3.3}$$

(iv) It can be expressed as:

$$\phi_\rho(r) = \sqrt{2\pi}\sum_{m=0}^{\infty}\bar{\eta}_{0,m}(\rho)\eta_{0,m}(r). \tag{E3.4}$$

(v) The coherent state in the nth Landau level is given by

$$\phi_\rho^{(n)}(r) \sim (\bar{z} - \bar{\rho})^n \exp\left[\frac{1}{2}\bar{\rho}z - \frac{1}{4}|z|^2 - \frac{1}{4}|\rho|^2\right]. \tag{E3.5}$$

(Hint: Use ladder operators.)

(vi) The coherent state wave function can be expressed as

$$\phi_\rho(r) = \frac{1}{\sqrt{2\pi}} \exp\left[-\frac{1}{4}|r - \rho|^2 + \frac{i}{2}(r \times \rho) \cdot \hat{z}\right].$$ (E3.6)

This form shows that $\phi_\rho(r)$ is Gaussian-localized at ρ.

(vii) The coherent states are not orthogonal:

$$\langle \rho_1 | \rho_2 \rangle = \exp\left[-\frac{1}{4}|\rho_1 - \rho_2|^2 + \frac{i}{2}(\rho_1 \times \rho_2) \cdot \hat{z}\right].$$ (E3.7)

(viii) The operator

$$P_0 \equiv \frac{1}{2\pi} \int d^2\rho |\rho\rangle\langle\rho|$$ (E3.8)

is an identity operator in the lowest Landau level, i.e., it satisfies

$$\langle \rho_1 | P_0 | \rho_2 \rangle = \langle \rho_1 | \rho_2 \rangle.$$ (E3.9)

This operator can be used to define a coherent state Feynman path integral. (Source: Kivelson *et al.* [342].)

3.5 The completeness property implies that

$$\sum_{n,m} \eta_{n,m}(r')\bar{\eta}_{n,m}(r) = \delta^{(2)}(r' - r).$$ (E3.10)

Show that the LLL projection of the Dirac delta function is given by

$$\delta_p^{(2)}(r' - r) \equiv \sum_m \eta_{0m}(r')\bar{\eta}_{0m}(r)$$ (E3.11)

$$= \frac{1}{2\pi} \exp\left[\frac{1}{2}\bar{z}z' - \frac{1}{4}|z|^2 - \frac{1}{4}|z'|^2\right].$$ (E3.12)

Note its similarity to the LLL coherent state $\phi_{r'}(r)$. Verify that it satisfies

$$\int d^2r\, \delta_p^2(r' - r)\phi(r) = \phi(r'),$$ (E3.13)

where $\phi(r')$ is a LLL function. (Source: Girvin and Jach [190].)

3.6 An alternative derivation of the LLL projection of a wave function is as follows: Consider a wave function of the type $\bar{z}^j \phi$, where ϕ is an arbitrary LLL wave function. Express $\bar{z}$ in terms of ladder operators defined in Section 3.3 and show that the LLL projection produces $(\sqrt{2}b)^j \phi$. Demonstrate that this is identical to the result in Section 3.7. (This method was suggested by P. E. Lammert.)

3.7 Show that Eq. (E3.2) is a maximally localized wave packet in the lowest Landau level, i.e., Eq. (3.59) is satisfied as an equality.

3.8 Derive the eigenfunctions of Eq. (3.41) by expressing the appropriate Schrödinger equation in polar coordinates, making the substitution $\Psi = e^{-r^2/4} r^m e^{-im\theta} f(r)$, and noting that $f(r)$ satisfies a known differential equation.

3.9 For the symmetric gauge, show that the ladder operators in Section 3.8 reduce to the ones discussed in Section 3.3. Construct ladder operators for the Landau gauge.

3.10 Because the guiding center coordinates commute with the kinetic energy, they do not mix different Landau levels. This implies, in particular, that they produce a LLL wave function when applied to a LLL wave function. Show that for an arbitrary LLL wave function $\phi(r)$,

$$x_0 \, \phi(r) = x_p \, \phi(r) \tag{E3.14}$$

$$y_0 \, \phi(r) = y_p \, \phi(r), \tag{E3.15}$$

where (x_p, y_p) and (x_0, y_0) are defined in Eqs. (3.56), (3.57) and (3.69).

3.11 (i) Verify that Eq. (3.112) produces Eq. (3.111).
(ii) Show that the vector potentials in Eq. (3.115) are related to that in Eq. (3.112) by a gauge transformation. Determine how the wave functions transform.

3.12 Verify Eqs. (3.124) and (3.125).

3.13 Show that the operators defined in Eqs. (3.130) and (3.131) satisfy the usual angular momentum commutation relations $[L_z, L_\pm] = \pm L_\pm$, $[L_+, L_-] = 2L_z$.

3.14 The expressions for $L_\pm$ in Section 3.10.6 are valid only in the lowest Landau level, but L_z is valid generally. Show, by a change of variables from θ and ϕ to u and v:

$$L_z = -i\frac{\partial}{\partial \phi} = \frac{1}{2}\left(u\frac{\partial}{\partial u} - v\frac{\partial}{\partial v}\right). \tag{E3.16}$$

3.15 Consider Y_{QQm} in Eq. (3.157).
(i) Verify, by direct substitution into the Schrödinger equation, that the Y_{QQm} are the LLL solutions of the spherical geometry.
(ii) Apply $L_\pm$ in Eq. (3.131) to Y_{QQm} to verify that they change the m quantum number by unity, and annihilate the wave function for $m = \pm Q$.
(iii) Plot $|Y_{QQm}|^2$ as a function of θ for $Q = 10$ and all possible values of m. For what values of m is the probability most localized at the poles?

3.16 The completeness of Y_{Qlm} implies

$$\sum_{l,m} Y^*_{Qlm}(\mathbf{\Omega}) Y_{Qlm}(\mathbf{\Omega}') = \delta(\mathbf{\Omega} - \mathbf{\Omega}'). \tag{E3.17}$$

Show that the LLL projected Dirac delta function is given by

$$\delta_p(\mathbf{\Omega} - \mathbf{\Omega}') = \sum_m Y^*_{QQm}(\mathbf{\Omega}) Y_{QQm}(\mathbf{\Omega}')$$

$$= \frac{2Q+1}{4\pi}(v^*v' + u^*u')^{2Q}. \tag{E3.18}$$

3.17 A useful, dimensionless parameter for interacting electrons is $r_s = r_0/r_B$, which is the distance between electrons, $r_0 = 1/\sqrt{\pi\rho}$, expressed in units of the Bohr radius $a_B = \hbar^2\epsilon/m_b e^2$. Show that

$$r_s = \sqrt{\frac{2}{\nu}\frac{e^2/\epsilon\ell}{\hbar\omega_c}}. \tag{E3.19}$$

The LLL approximation is equivalent to taking $r_s = 0$. Increasing r_s at a given filling factor implies increased mixing between Landau levels due to the Coulomb interaction.

3.18 In the disk geometry, the matrix elements for the Coulomb interaction involving arbitrary Landau levels can be expressed in terms of the LLL matrix elements $\langle r, s|V|t, u\rangle$. Do this explicitly for

$$\langle 1r, 0s|V|0t, 0u\rangle = \int\int d^2r_1\, d^2r_2 \frac{\eta^*_{1,r}(\mathbf{r}_1)\eta^*_{0,s}(\mathbf{r}_2)\eta_{0,t}(\mathbf{r}_1)\eta_{0,u}(\mathbf{r}_2)}{|\mathbf{r}_1 - \mathbf{r}_2|}.$$

3.19 Show that the relative and the total angular momentum operators are given by $-\hbar$ times the following quantities:

$$b_r^\dagger b_r - a_r^\dagger a_r = z\frac{\partial}{\partial z} - \bar{z}\frac{\partial}{\partial\bar{z}}, \tag{E3.20}$$

$$b_r^\dagger b_r - a_r^\dagger a_r + b_R^\dagger b_R - a_R^\dagger a_R = z\frac{\partial}{\partial z} - \bar{z}\frac{\partial}{\partial\bar{z}} + Z\frac{\partial}{\partial Z} - \bar{Z}\frac{\partial}{\partial\bar{Z}}. \tag{E3.21}$$

Z and z are the center of mass and relative coordinates, and the various raising and lowering operators have been defined in Section 3.11.2.

3.20 You will determine in this exercise the relative angular momentum for the state $a_1^\dagger a_2^\dagger|M, m\rangle$, where $|M, m\rangle$ is the LLL state of two particles with the center of mass and relative angular momenta M and m, respectively. First derive, using the explicit expressions of the operators, the relations

$$a_1^\dagger = \frac{1}{\sqrt{2}}(a_R^\dagger + a_r^\dagger), \tag{E3.22}$$

$$a_2^\dagger = \frac{1}{\sqrt{2}}(a_R^\dagger - a_r^\dagger). \tag{E3.23}$$

Then evaluate the commutator of $a_1^\dagger a_2^\dagger$ with the relative angular momentum operator obtained in the previous exercise:

$$[b_r^\dagger b_r - a_r^\dagger a_r, a_1^\dagger a_2^\dagger] = a_r^\dagger a_r. \tag{E3.24}$$

Act with this commutator on $|M, m\rangle$ to show that $a_1^\dagger a_2^\dagger|M, m\rangle$ has relative angular momentum m.

3.21 Calculate the Haldane pseudopotentials for the following real-space interactions:
   (i) $V(r) = 4\pi V_0 \delta^{(2)}(r)$,
   (ii) $V_{TK}(r) = 2\pi V_1 \nabla^2 \delta^{(2)}(r)$,
   (iii) $V(r) = \alpha\, e^{-r^2}$.
   For the second interaction (Trugman and Kivelson [646]), write

$$\nabla^2 = \frac{1}{r}\frac{\partial}{\partial r} r \frac{\partial}{\partial r},$$

   and perform integration by parts.

3.22 Show that, for $n' \geq n$,

$$\langle n'| e^{-\frac{i}{\sqrt{2}} k a^\dagger} e^{-\frac{i}{\sqrt{2}} \bar{k} a} |n\rangle = \left(-\frac{i}{\sqrt{2}} k\right)^{n'-n} \sqrt{\frac{n!}{n'!}}\, L_n^{n'-n}\left(\frac{k\bar{k}}{2}\right). \tag{E3.25}$$

   For $n > n'$, the answer is obtained by the replacement $n \Longleftrightarrow n'$. (Hint: Use results from Section 3.11.3.)

3.23 This exercise presents another expression for the pseudopotentials (Haldane [221]). Consider the relative and center of mass coordinates, with

$$|m\rangle = \frac{(b_r^\dagger)^m}{\sqrt{m!}} |0\rangle. \tag{E3.26}$$

   Using results from Section 3.11.3, derive the following:

$$V_m = \langle m|V|m\rangle$$

$$= \int \frac{d^2q}{(2\pi)^2} V(q) \langle m|e^{iq\cdot r}|m\rangle$$

$$= \int \frac{d^2q}{(2\pi)^2} V(q) e^{-q^2} L_m(q^2). \tag{E3.27}$$

   Show that this reduces to Eq. (3.197) for the Coulomb interaction. In higher Landau levels, we have

$$V_m^{(n)} = \int \frac{d^2q}{(2\pi)^2} V(q) e^{-q^2} \left[L_n\left(\frac{q^2}{2}\right)\right]^2 L_m(q^2). \tag{E3.28}$$

3.24 Confirm that in the thermodynamic limit $Q \to \infty$, the pseudopotential V_L of Eq. (3.224) reduces to V_m of the planar geometry with the correspondence $L = 2Q - m$. That is,

$$V_m^{\text{disk}} = V_{2Q-L}^{\text{sphere}}. \tag{E3.29}$$

   Explain physically why the pair with the largest angular momentum on the sphere corresponds to the smallest m in the plane. (Source: Sitko *et al.* [596].)

3.25 Stone [615] has introduced symmetric functions

$$S_k = \sum_{j=1}^{N} z_j^k. \tag{E3.30}$$

Multiplication by S_k raises the angular momentum of a state in the disk geometry by k units, creating excitations at the edge of the system when $k \ll N$. Show:
(i) Apart from an overall normalization factor, multiplying a LLL wave function by S_k is equivalent, in the second quantized language, to application of the operator

$$\hat{S}_k^\dagger = \sum_{n=0}^{\infty} \sqrt{\frac{(n+k)!}{n!}} \, c_{n+k}^\dagger c_n. \tag{E3.31}$$

Here, c_m and $c_m^\dagger$ are the destruction and creation operators in the lowest Landau level.
(ii) $\hat{S}_0^\dagger$ is proportional to the center of mass angular momentum raising operator $B^\dagger$ (Eq. 3.245).
(iii) Oaknin *et al.* [483] introduce a new set of operators

$$\hat{S}_k'^\dagger = \sum_{n=0}^{\infty} \sqrt{\frac{n!}{(n+k)!}} \, c_{n+k}^\dagger c_n, \tag{E3.32}$$

which also increase the total angular momentum by k units. Show that

$$\hat{S}_0 \hat{S}_k'^\dagger |\Phi_1\rangle = 0, \tag{E3.33}$$

where Φ_1 is the $\nu = 1$ ground state. (Hint: Calculate the commutator $[\hat{S}_0, \hat{S}_k'^\dagger]$.) Thus, $\hat{S}_k'^\dagger |\Phi_1\rangle$ is also an eigenstate of the center of mass angular momentum operator, a property not satisfied by $\hat{S}_k^\dagger |\Phi_1\rangle$. It is shown in Ref. [483] that $\hat{S}_k'^\dagger |\Phi_1\rangle$ provides a better representation of the actual ground state in the appropriate total angular momentum sector than $\hat{S}_k^\dagger |\Phi_1\rangle$.

4

Theory of the IQHE

The theory of the IQHE, namely integral quantization of the Hall resistance, must answer the following questions: What is the origin of the quantization of the Hall resistance? What determines its value? Why is the quantization precise?

Although not predicted, the IQHE is well understood. At its origin lies the simple fact that the ground state at filling factor $v = n$ consists of n Landau levels fully occupied, with a gap to excitations, as shown in Fig. 4.1. The IQHE is thus a dramatic manifestation of the long-known Landau level quantization of the electron kinetic energy. Quantized plateaus in the Hall resistance result when the LL formation is combined with disorder-induced Anderson localization of most single particle orbitals. The phenomenon is explained within a model in which electron–electron interactions are neglected, and then it is argued that they do not alter the essential features. In this chapter, we neglect interactions for the most part, and also take the temperature to be zero whenever not specified.

4.1 The puzzle

Let us take the classical formula for the Hall resistance

$$R_H = \frac{B}{\rho ec} .$$ (4.1)

At integral filling factors $v \equiv \rho \phi_0 / B = n$, this equation reduces to

$$R_H = \frac{h}{ne^2} ,$$ (4.2)

which is precisely the value of quantized Hall resistance. Is this the explanation of the Hall resistance? No, because it does not explain the plateaus, which would require showing that R_H has exactly this value in a range of filling factors near $v = n$. For any general filling factor the classical value of the Hall resistance can be recast as

$$R_H = \frac{h}{ve^2} ,$$ (4.3)

which has no plateaus.

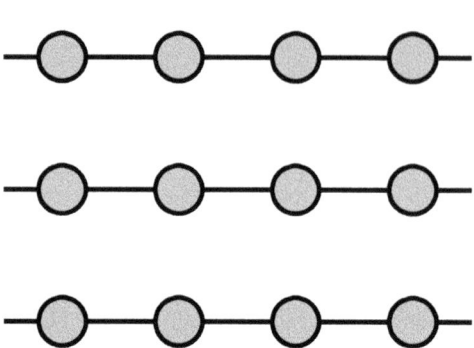

Fig. 4.1. The state with an integral number of filled Landau levels can be thought of as a giant noble gas (filled-shell) atom.

To deepen the mystery, we recall Section 2.1, wherein it was shown that boosting the observer to a moving frame of reference to eliminate the electric field allows a calculation of the current in the laboratory frame of reference, yielding the classical formula for the Hall resistance. The calculation is valid quantum mechanically, because, in the moving frame, the current is zero independent of whether we use classical or quantum mechanics to calculate it.

How can we reconcile this apparently general (and correct) result with experiment? There happens to be a way out. An assumption made above, that of translational invariance, is not satisfied in the actual experimental system because of the unavoidable presence of disorder. Disorder, surprisingly, is crucial for the establishment of plateaus.[1]

4.2 The effect of disorder

In the absence of disorder, the energy spectrum of a single electron (which is all we need in the assumed absence of interactions) only has states at $E = (n + 1/2)\hbar\omega_c$. Disorder obviously modifies the spectrum. Much work has been done on how the single electron eigenstates are affected by the presence of disorder. The resulting picture is shown schematically in Fig. 4.2. Noteworthy are the following features:

- The degenerate LL states broaden into bands in the presence of disorder. In general, the energy gaps disappear.
- The single particle states near the unperturbed energies $E = (n + 1/2)\hbar\omega_c$ form bands of *extended* states. The extended bands are separated by localized states. Because localized states do not contribute to transport, we say that there is a "mobility gap."

---

[1] In this context we recall that for a type-II superconductor, zero resistance is obtained only in the presence of disorder, needed to pin the Abrikosov vortex lattice.

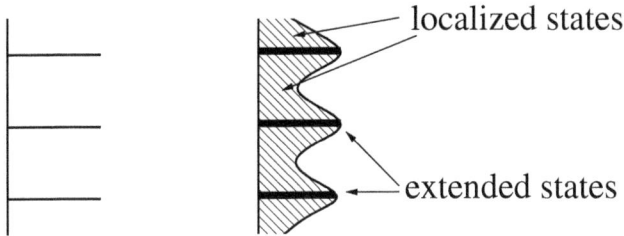

Fig. 4.2. The degenerate Landau level states (left) broaden into bands in the presence of disorder (right), with extended states near the center (dark), separated by localized states (shaded). The x-axis is the density of states and the y-axis is the energy.

To develop a feel for the effect of disorder on single particle eigenstates, let us consider a disorder potential that is smooth on the scale of the magnetic length, which will allow us to use our semiclassical intuition. These ideas were developed by Iordansky [267], Kazarinov and Luryi [329], Prange and Joynt [527], and Trugman [645]. We imagine a potential landscape with hills and valleys. Landau levels are locally well defined for such a potential. In a given Landau level, the energy is then given by

$$E = \left(n + \frac{1}{2} \right) \hbar \omega_c + V(\boldsymbol{R}) , \qquad (4.4)$$

where $\boldsymbol{R}$ is the position of the guiding center of the electron. This equation makes it possible to see why the electron wave function, in general, gets localized by the disorder. For a given energy, an electron moves along an equipotential contour. Another way to see this is that the $\boldsymbol{E} \times \boldsymbol{B}$ drift, with $\boldsymbol{E} = -\nabla V(r)$, is along lines of constant V. Thus, the disorder potential traps the electron into a closed orbit. This, interestingly, is true independent of whether the potential is attractive (a potential well) or repulsive (a potential hill).

Not all such classical trajectories are allowed, though. The allowed single particle orbitals are given by the solution of the Schrödinger equation for a given disorder, but we can gain further insight from simpler, Bohr–Sommerfeld-like semiclassical considerations. Only those equipotential orbitals are allowed quantum mechanically for which the electron's phase matches after a complete loop. The Aharonov–Bohm phase associated with a closed path is given by $2\pi\phi/\phi_0$, which, according to our semiclassical condition, must be $2\pi m$, where m is an integer. Thus, in a semiclassical picture, each closed trajectory with $V(\boldsymbol{R}) =$ constant that encloses an integral number of flux quanta corresponds to a single particle eigenstate. Figure 4.3 illustrates the picture for a single potential hill. The correct degeneracy, i.e., one state per flux quantum, follows because each successive state encloses precisely one additional flux quantum. The actual landscape has many potential hills and valleys. In a full quantum mechanical treatment, the eigenstates are linear superpositions of contours on *different* hills or valleys, but at the present level of approximation, which neglects tunneling, the states are localized on a single hill or valley. This simple picture will suffice for the present purpose.

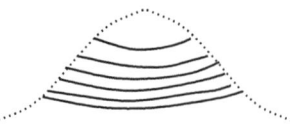

Fig. 4.3. Semiclassical view of the localized single particle states on a potential hill. Each closed contour represents the intersection of an equipotential plane and the potential hill, and encloses an integral number of flux quanta.

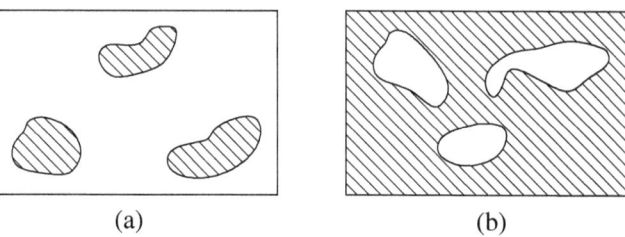

(a)   (b)

Fig. 4.4. (a) The electron occupied regions (shaded lakes) do not percolate. (b) The electron occupied regions percolate. The unshaded regions have no electrons.

Let us begin with $V = 0$, when all states are extended. Now we decorate it with a few hills and valleys here and there. Clearly, the states trapped on the hills and valleys have energy away from $V = 0$. The disorder thus creates localized states that fill the region where there was a gap, thus broadening Landau levels into Landau bands. At $V = 0$ we can still draw equipotential lines across the sample, indicating the presence of extended states. Extended states survive only at one energy, at the center of the Landau level. The vanishing width of the extended band in Fig. 4.2 is a property of several models of disorder.

A more realistic model works with a random potential (still taken to be slowly varying). Filling all states up to some energy (the Fermi energy) is akin to pouring water into a landscape with hills and valleys up to a given height (Fig. 4.4). Initially, when the filling factor is small, a few lakes form in the otherwise solid ground, and no current may flow at the Fermi energy. As more and more water is added, at some point the character of the landscape changes. Now we have islands in an ocean. This is called the percolation transition. Water percolates on one side whereas land does on the other. The localization length at a given energy is the size of the largest connected piece of lake or land at that energy. An extended state, i.e., an infinite island or lake, occurs only at one energy (or a band of energies for a finite sized sample), which is the energy at the percolation transition. The width of the extended band in Fig. 4.2 is again zero.

The existence of extended bands separated by localized states is crucial for the IQHE. The only way we know of obtaining such a spectrum is to introduce weak disorder in a state that has bands separated by gaps in the absence of disorder. Disorder creates localized states in the gap, turning it into a mobility gap.

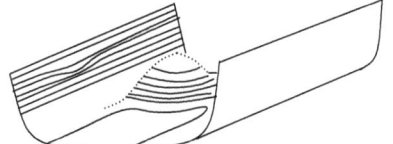

Fig. 4.5. The figure on the left shows the potential landscape and a few typical equipotential contours. A hill in the interior captures an electron in a closed orbit, but a hill on the confining potential does not. The figure on the right shows the bird's eye view of some equipotential contours.

4.3 Edge states

An aspect crucial for the explanation of the IQHE is missing in Fig. 4.4: the edge states. Electrons feel a steep potential at the edge of the sample, which serves to keep them inside. With the confining potential (Fig. 4.5), for energies above a minimum, an equipotential plane always intersects the confinement potential. An electron on an edge equipotential contour is not trapped into a closed orbit but goes across the entire length of the sample. This remains true even with the presence of small hills on the steep edge; they distort the edge equipotential contour but fail to bind an electron. The classical analog of an edge state is a skipping orbit. Consider $V = 0$ for $x > 0$ and $V = \infty$ for $x < 0$. An electron deep in the interior ($x > R$, R is the cyclotron radius) completes its cyclotron orbit and keeps circling around the same point; an electron near a wall bounces off the wall and moves alongside it.

The following points are noteworthy:

- Due to the confinement potential, the states at the edges typically go from one end of the sample all the way to the other.
- Electrons move in different directions on two opposite edges. This follows simply because the confining electric fields are antiparallel, giving antiparallel drifts.
- Since edge states do not enclose any area, they form a continuum. Edge states exist even at energies inside the bulk gap.

The phrase "extended state" often refers to an eigenstate that is spread over the entire *two*-dimensional space. An edge state, on the other hand, is extended along one direction only. The meaning should be clear from the context.

4.4 Origin of quantized Hall plateaus

The presence of localized bands explains the plateaus. To see this, let us consider changing the filling factor by adding electrons at a fixed magnetic field rather than varying the magnetic field at fixed electron density. This way, the single particle eigenstates do not change. Now, when the Fermi level lies in the localized states, the additional electrons go into localized states and do not affect the transport. As a result, the Hall resistance remains constant as a function of the filling factor so long as the Fermi level moves through the localized states. In other words, the localized states provide a reservoir for electrons that do not contribute to transport.

While this is essentially the correct physics of the plateaus, the story is far from complete, because we have not explained why the Hall resistance has the value that it does on the plateaus. Consider, for example, the plateau $R_H = h/ne^2$ at $v \approx n$. Naively, we might expect the Hall resistance to be given by $R_H = B/\rho_{ex}ec$, where ρ_{ex} is the density of electrons in the extended states. Since ρ_{ex} is small, R_H could be very large, with no reason for it to have a simple, universal value. However, the Hall resistance is given precisely by $R_H = h/ne^2$, as though n Landau levels were carrying full current, without disorder. This describes the enigma of the IQHE: if most of the electrons are sitting in localized states, then why does transport behave as though n Landau levels were fully carrying current? On the one hand, we need localized states to explain the plateaus, but on the other, we do not really see any effect of localization in the value of the Hall current. It is as though nature wishes to have its cake and eat it too.

Following an important insight into this problem by Prange [526], this puzzle was resolved by Laughlin [368], who showed that the Hall resistance is quantized at $R_H = h/ne^2$, independent of the filling factor, so long as the Fermi level lies in the mobility gap between the nth and $(n+1)$th Landau levels. We now present several related ways of understanding the origin of Hall plateaus at precisely quantized values. The trick will be to argue, *by using the principle of current continuity*, that a sample with Fermi level in the mobility gap between the nth and $(n+1)$th Landau levels behaves exactly as a disorder-free sample at $v = n$.

4.4.1 Intuitive explanation

We begin with a disorder-free system at $v = n$, with Hall resistance $R_H = h/ne^2$. Switching on a sufficiently weak disorder does not cause any mixing with states across the gap. In that case, the disorder has no effect on the wave function, because only one wave function is available. The Hall resistance thus remains unchanged by the addition of weak disorder. The filling is now moved away from $v = n$ by adding some electrons or holes. In a perfect system, the additional particles would also be free to carry current, but in the disordered sample, they are trapped by the localized states created by the disorder, and consequently do not contribute to transport. Therefore, the Hall resistance retains its value $R_H = h/ne^2$. That, crudely, is how the presence of a gap at filling $v = n$ in a pure system becomes manifest, in the presence of disorder, through a quantized Hall plateau at $R_H = h/ne^2$.

4.4.2 Current operator

In terms of the particle density

$$\rho(r) = \sum_j \delta^{(2)}(r - r_j) , \tag{4.5}$$

the current density is expressed as

$$J(r) = -e\rho(r)v(r) . \tag{4.6}$$

The current in the x direction is given by

$$
\begin{aligned}
I_x &= \int \boldsymbol{J}(\boldsymbol{r}) \cdot \hat{\boldsymbol{x}} \, dy \\
&= \frac{1}{L} \int \int J_x(x, y) \, dy \, dx \\
&= -\frac{e}{L} \int d^2 r \, \rho(\boldsymbol{r}) v_x(\boldsymbol{r}) \\
&= -\frac{e}{L} \sum_j v_x(\boldsymbol{r}_j).
\end{aligned}
\tag{4.7}
$$

Here, the integral in the first step is along a constant x line across the width of the sample. In the second step we have made use of the continuity property that the current does not depend on the choice of x. L is the length of the sample in the x-direction.

The current operator is

$$
\hat{I}_x = -\frac{e}{L} \sum_j \hat{v}_{jx},
\tag{4.8}
$$

where $\hat{v}_{jx}$ is the x component of the velocity operator for the jth particle:

$$
\hat{v}_j = \frac{1}{m_b} \left(\boldsymbol{p}_j + \frac{e}{c} \boldsymbol{A}(\boldsymbol{r}_j) \right).
\tag{4.9}
$$

In the second quantized notation, we have

$$
\hat{I}_x = -\frac{e}{L} \sum_p v_{px} \hat{n}_p.
\tag{4.10}
$$

Here, v_{px} is the expectation value of $\hat{v}_x$ for the eigenstate labeled by p. In the absence of disorder, the label p can be taken to be the momentum, and the current operator becomes

$$
\hat{I}_x = -e \int \frac{dp_x}{2\pi \hbar} v_{px} \hat{n}_p.
\tag{4.11}
$$

4.4.3 Landauer-type explanation

The Landauer approach [366] relates the resistance of a sample to the probability of transmission through it. It provides a powerful way of understanding the physics of the Hall plateaus (Büttiker [43]; Jain and Kivelson [269, 270]; Streda, Kucera, and MacDonald [628]). The discussion below also serves as an introduction to the Landauer approach to transport.

To begin with, let us consider a sample with translational invariance in the x-direction. Along the y-direction, the electrons feel an electrostatic potential produced by a combination of the confinement potential at the edges and the applied Hall voltage. Detailed shape of the potential is not important at the moment. The only assumption is that the potential depends only on y, i.e., $\phi = \phi(y)$.

The geometry of the problem suggests using the Landau gauge, $A = B(-y, 0, 0)$. The Hamiltonian is

$$H = \frac{1}{m_b} \left(p + \frac{e}{c} A \right)^2 - e\phi(y). \tag{4.12}$$

Since it does not contain x, H commutes with p_x, which therefore is a good quantum number and can be used to label the eigenstates. The single particle wave function depends on $\phi(y)$, but will not be needed. It will suffice to know that the eigenstates are localized at $y = y(p_x)$ (i.e., the value of p_x determines the position of the state in the y direction), with y increasing monotonically with p_x. For potentials that vary smoothly over the scale of the magnetic length, we have $y(p_x) = p_x \ell^2 / \hbar$.

Let us now consider a "full" Landau level, i.e., a state in which all single particle energy levels of the Landau level in the range $p_- < p_x < p_+$ are occupied.[2] The current in the x direction is obtained by adding the individual currents (using Eq. 4.8):

$$I_x = -e \int \frac{dp_x}{2\pi \hbar} \langle v_x \rangle$$

$$= -e \int \frac{dp_x}{2\pi \hbar} \frac{1}{m_b} \left\langle p_x + \frac{e}{c} A_x \right\rangle$$

$$= -e \int \frac{dp_x}{2\pi \hbar} \left\langle \frac{\partial H}{\partial p_x} \right\rangle$$

$$= -e \int \frac{dp_x}{2\pi \hbar} \frac{\partial E}{\partial p_x}$$

$$= -\frac{e}{h} \int_-^+ dE$$

$$= -\frac{e}{h} (\mu_+ - \mu_-)$$

$$= \frac{e^2}{h} V_H. \tag{4.13}$$

Because the energy E includes both the kinetic and potential energies, the quantity

$$\mu_+ - \mu_- = -eV_H \tag{4.14}$$

gives directly the electrochemical potential difference, precisely what the voltmeter measures. The details of how the potential drop is distributed within the sample are not relevant for the value of net current. We also did not have to specify the LL index. When n Landau levels are occupied, we have

$$I = \frac{ne^2}{h} V_H, \tag{4.15}$$

---

[2] The Landau level is full inside the strip $y(p_-) < y < y(p_+)$, which defines our Hall bar.

giving

$$R_{\mathrm{H}} = \frac{V_{\mathrm{H}}}{I} = \frac{h}{ne^2}. \tag{4.16}$$

So far, we have only considered filling factor $\nu = n$ and obtained the classical result $R_{\mathrm{H}} = \frac{h}{\nu e^2}$. We still need to show what causes plateaus.

We now introduce disorder. The single particle eigenstates become extremely complicated. The sharply defined Landau levels are replaced by disorder-broadened Landau bands, with localized states occupying what was an energy gap in the absence of disorder. For reasons that will soon become clear, we attach to our disordered sample long, disorder-free, ideal leads on both the left and the right. The leads, in turn, are connected to reservoirs, with chemical potentials μ_+ (left reservoir) and μ_- (right reservoir), as shown in Fig. 4.6. The Hall bar has three regions now: the nonideal region, referred to as the "sample," and two ideal leads. If you object to the ideal leads, which are indeed not present in experiments, please be patient; they will be dispensed with shortly. The reservoir on the left fills all states moving rightward up to μ_+ on the upper left edge, and the reservoir on the right fills all states moving leftward up to μ_- on the lower right edge. The two chemical potentials are taken to be infinitesimally close (as appropriate for the linear response regime). We take $\mu_+ > \mu_-$ with no loss of generality.

Voltmeters are connected to the system through additional leads at the upper and lower edges of the Hall bar. To keep from cluttering the figure with too many leads, we do not show the voltage leads explicitly. As explained in Fig. 4.7, if a voltmeter is attached between two edges at chemical potentials μ_1 and μ_2, the measured voltage difference V is simply related to the chemical potential difference by the expression $\mu_1 - \mu_2 = -eV$.

Let us now make a crucial *assumption*:

- *An electron impinging on the sample from the left is transmitted with probability one. In other words, backscattering is completely suppressed.*

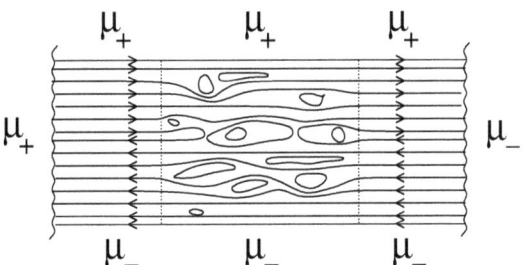

Fig. 4.6. The disordered sample (middle region) is connected through ideal leads (outer regions) to reservoirs at chemical potential μ_+ (left reservoir) and μ_- (right reservoir). Only the occupied states are shown, with the states at the edge being at the Fermi energy. (For simplicity of illustration, only states of a single Landau level are shown.)

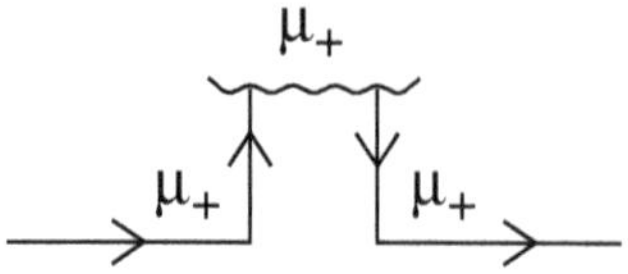

Fig. 4.7. This figure shows an edge at chemical potential μ_+ to which a voltmeter (shown as a reservoir) is attached. Because a voltage lead draws no current, the chemical potentials of the incoming and outgoing edges must be the same. The latter is also equal to the chemical potential of the reservoir. The voltmeter attached to two edges measures the chemical potential difference between two reservoirs.

This assumption is satisfied when the the single particle eigenstates at the Fermi energy (more precisely, in the energy range $\mu_- < E < \mu_+$) are (i) extended all the way from left to right at the edge, (ii) localized in the interior, and (iii) the Hall bar is sufficiently wide. The left reservoir fills all states up to μ_+ on the upper left edge, and the right reservoir fills all states up to μ_- on the lower right edge. Which electrons can scatter? Because all states below μ_- are completely occupied, electrons with energies below μ_- have nowhere to scatter (for an elastic scattering) and are inert. All action takes place in the thin energy slice $\mu_+ \geq E \geq \mu_-$ at the Fermi energy. Here, the left reservoir supplies electrons at the upper edge, which are the only electrons that we need to consider. A classical electron moving on an edge equipotential contour is perfectly transmitted. Quantum mechanically, the incoming electron couples with some of the closed contours in the interior, but such mixing is restricted to orbits within approximately one localization length of the edge. When the sample width is large compared to the localization length, the coupling between the upper and lower edges is exponentially small, and the electron has no choice but to be fully transmitted.

In the absence of backscattering, the chemical potential is μ_+ on the *entire* upper edge and μ_- on the *entire* lower edge. In the leads, *all* single particle states are occupied from μ_- at the lower edge to μ_+ at the upper edge. Therefore, the filling factor in the leads, away from the edge, is always an integer, independent of the filling factor in the disordered sample region. For a simple model with symmetric disorder (with equal preference to hills and valleys), a filling factor $\nu = n$ occurs in the leads for the filling factor range $n + 1/2 > \nu > n - 1/2$ in the nonideal sample.

The quantization of the Hall resistance follows from two simple observations.

- The current in the leads is given by $I = \frac{ne^2}{h} V_H$, according to our earlier calculation for an ideal sample. Because of current continuity, this is the current everywhere, including in the sample.
- The Hall voltage in the ideal leads is also the Hall voltage in the sample.

Therefore, voltage probes attached to the nonideal sample measure $R_H = h/ne^2$ and $R_{xx} = 0$, even though the filling factor is not an integer and many single particle states are localized. The ideal leads have served their purpose and can now be removed. This completes the explanation of the quantized Hall plateaus.

The IQHE results from a complete suppression of backscattering, which itself is a result of a spatial separation between the right and the left moving electrons at the Fermi energy. This is a remarkably simple, but nontrivial scattering problem, with unit transmission coefficient.

In a two-probe measurement, in which the reservoirs on the left and right in Fig. 4.6 are used to measure voltage, the resistance is $R = (\mu_+ - \mu_-)/(-eI) = h/ne^2$. (It is identical to the Hall resistance in the geometry of Fig. 4.6.) It has a nonzero value even in the absence of scattering. (Alternatively, the conductance $G \equiv R^{-1} = n(e^2/h)$ does not diverge even in the ideal limit.) This is referred to as the "contact resistance," and is a standard feature of two-terminal measurements. For a four-terminal measurement, in which additional leads are attached to the upper or lower edge for the measurement of the voltage, the longitudinal resistance vanishes as shown above. (A controversy on whether the conductance of a one-dimensional system diverges or remains finite in the limit of no disorder was resolved when Büttiker [42] realized that the voltage leads ought to be separate from the current leads.)

A hypothetical situation is when the background potential increases monotonically from μ_- to μ_+. Then, all electrons in the ideal leads move in the $+x$ direction, and backscattering is suppressed for the trivial reason that no scattering states move backward. (This was shown by Prange in an explicit calculation by introducing a delta function scatterer that binds an electron [526].) This model is useful for the physics at an individual edge.

4.4.4 Corrections due to backscattering

An explanation of plateaus is unsatisfactory unless it clarifies what destroys the quantization. That is naturally accomplished in the above framework by allowing for the possibility of backscattering (Buttiker [43]; Jain and Kivelson [269,270]; Streda, Kucera, and MacDonald [628]). We assume in this subsection, for simplicity, that only one Landau level is occupied in the leads. The problem is then formally equivalent to a one-dimensional transport problem [269].

Let us represent the entire sample as a single scatterer (Fig. 4.8) with reflection coefficient R and transmission coefficient $T = 1 - R$; i.e., the electron coming in the energy range $\mu_+ \geq E \geq \mu_-$ from the upper left edge goes to the upper right edge with probability T and

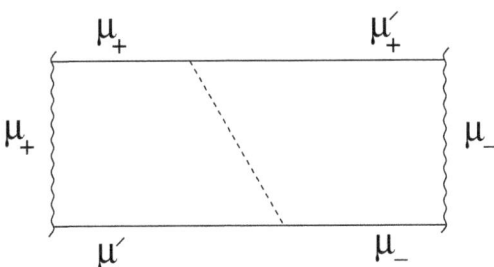

Fig. 4.8. Inter-edge scattering affects transport coefficients.

to the lower left edge with probability R. T and R are taken to be energy independent, which is a valid approximation because we have taken μ_+ and μ_- to be infinitesimally close. The chemical potentials at the upper left edge and the lower right edge remain unaffected, but those on the upper right and lower left edges are modified due to scattering. The new chemical potentials, called μ'_+ and μ'_-, are related to the current according to Eq. (4.13). (Scattering into an edge may create a locally nonequilibrium situation, but we assume that electrons rapidly equilibrate due to *intra*-edge scatterings to attain a well-defined chemical potential. The same comment applies to the contact between the sample and a reservoir.)

Let us begin by noting that now the current is

$$I = -\frac{e}{h}T(\mu_+ - \mu_-), \tag{4.17}$$

since, of all the incident electrons in the energy range $\mu_+ > E > \mu_-$ on the upper left, only a fraction T makes it through. The chemical potential on the upper right is therefore given by the condition

$$I = -\frac{e}{h}(\mu'_+ - \mu_-), \tag{4.18}$$

which gives

$$\mu'_+ = \mu_- + T(\mu_+ - \mu_-). \tag{4.19}$$

Similarly, by noting that the reflected current is

$$I_R = -\frac{e}{h}R(\mu_+ - \mu_-) = -\frac{e}{h}(\mu'_- - \mu_-), \tag{4.20}$$

we get

$$\mu'_- = \mu_- + R(\mu_+ - \mu_-). \tag{4.21}$$

The longitudinal and the Hall resistances, as measured by voltage contacts on the ideal leads, are modified to

$$R_{xx} = -\frac{\mu_+ - \mu'_+}{eI} = -\frac{\mu'_- - \mu_-}{eI} = \frac{h}{e^2}\frac{R}{T}, \tag{4.22}$$

and

$$R_H = -\frac{\mu_+ - \mu'_-}{eI} = -\frac{\mu'_+ - \mu_-}{eI} = \frac{h}{e^2}. \tag{4.23}$$

That the Hall resistance is not affected by scattering is an artifact of the geometry in Fig. 4.8, which does not consider scattering in the region containing the two probes used to measure the Hall resistance. Such a scattering also leads to a correction to the Hall resistance (reader, please check), but experiments do show that the Hall resistance is more robust to disorder than the longitudinal resistance; well-defined plateaus are often seen even when the longitudinal resistance has a sizable value.

The above formulas can be generalized to include many Landau levels (Büttiker [43]). The equipotential contours in different Landau levels at a given *total* energy (say, the Fermi

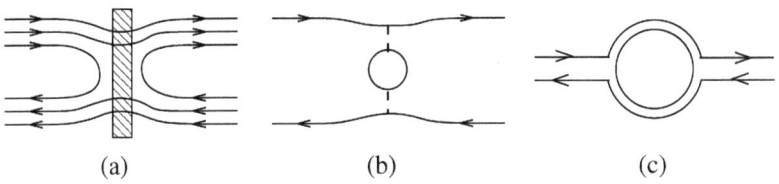

Fig. 4.9. (a) Scattering from a potential barrier across the width of the sample. Two Landau levels are fully transmitted while one is fully reflected. (b) Resonant tunneling through a bound state on a potential hill. (c) Hall effect in ring geometry. Only the equipotential contours at the Fermi energy are depicted.

energy) are spatially separated, and therefore may behave dissimilarly. We mention here a few simple types of scattering.

(i) **Quantized longitudinal resistance** An impediment across the current path can be constructed which fully transmits j of the n incident LL edge states, while fully reflecting the rest (Fig. 4.9(a)). This can be arranged by imposing a potential barrier along the current path, the height of which can be varied by application of a gate potential. Landau levels can be switched off one by one (Haug *et al.* [241]; Washburn *et al.* [668]). The longitudinal resistance can be obtained straightforwardly in this situation [44, 273]. When n Landau levels are occupied, the current is n-fold, giving

$$R_{\mathrm{L}} = \frac{h}{ne^2} \frac{R}{1 - R}.$$
(4.24)

The reflection coefficient, however, is simply $R = (n - j)/n$, which gives:

$$R_{\mathrm{L}} = \frac{h}{ne^2} \frac{n - j}{j}.$$
(4.25)

The longitudinal resistance is quantized in this situation. This relation has been verified experimentally by Haug *et al.* [241] and by Washburn *et al.* [668]. Figure 4.10 reproduces data from Ref. [241].

Equation (4.25) can be generalized to a situation where an arbitrary incompressible QHE state at fraction f_1 is interrupted by a barrier with another incompressible state at fraction f_2. If the inner state carries current to its maximum capacity, then the barrier produces a transmission coefficient $T = f_2/f_1$. With $R = 1 - f_2/f_1$ and n replaced by f_1, Eq. (4.24) becomes

$$R_{\mathrm{L}} = \frac{h}{e^2} \left(\frac{1}{f_2} - \frac{1}{f_1} \right) = \Delta R_{xy}.$$
(4.26)

(Equation (4.25) is a special case with $f_1 = n$ and $f_2 = j$.) A quantized plateau in R_{L} can thus give us information about the filling factor in the barrier region.

(ii) **Resonant tunneling** Let us consider tunneling through a potential hill, as shown in Fig. 4.9(b) (Jain and Kivelson [270]). Each bound state on the hill produces a resonant tunneling peak in the resistance as a function of the Fermi energy (which can be varied by

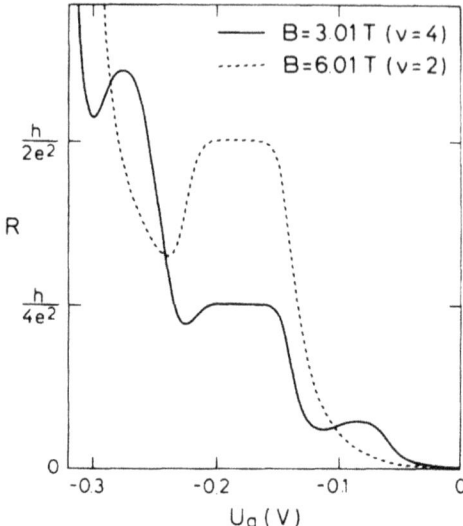

Fig. 4.10. The resistance across a gate as a function of the barrier height, controlled by the gate voltage U_g. (See Fig. 4.9(a) for reference.) As the barrier height is increased, the edge channels switch off (i.e., are fully reflected) one by one. The well-defined plateau on the dashed curve represents the situation when one of the two edge channels is transmitted with the other fully reflected. The solid plateau occurs when two of the four edge channels are transmitted and the remaining two reflected. In the latter case, the less well-developed plateaus at $h/12e^2$ and $3h/4e^2$ correspond to the reflection of one or three of the four incoming edge channels; these plateaus are more difficult to resolve because the spin splitting is small compared to the LL spacing. Source: R. J. Haug, A. H. MacDonald, P. Streda, and K. von Klitzing, *Phys. Rev. Lett.* **61**, 2797 (1988). (Reprinted with permission.)

changing the magnetic field). When resonant tunneling is dominated by a single potential hill in the sample, one expects resistance peaks that are approximately periodic in the magnetic field, with the period corresponding to a flux change through the hill by one flux quantum. When the tunneling can occur through internal quasi-bound states on many hills, a more complicated, nonperiodic sequence of resonant tunneling peaks is expected in the transition region from one plateau to another. That is consistent with experiments on mesoscopic samples. (See, for example, experimental studies by Peled *et al.* [508, 509] and the first principles theoretical analysis by Zhou and Berciu [748, 749].)

(iii) **Hall effect in a ring** A ring-type structure shown in Fig. 4.9(c) shows Aharonov–Bohm oscillations at low magnetic fields, with the period related to the average area of the ring. At very high magnetic fields, when transport occurs only at the outer edges that do not enclose any area, Aharonov–Bohm oscillations should be suppressed [271]. This is demonstrated experimentally by Timp *et al.* [637]. At intermediate fields, when there is some coupling between the outer and the inner edges of the ring, the area controlling the period of the (weak) Aharonov–Bohm oscillations is the area enclosed by the inner circle [637].

4.4.5 Laughlin's explanation
Hall bar geometry

We now present a closely related explanation of the quantized Hall plateaus by Laughlin [368] (also, see Halperin [228]), which initially clarified the physics of the problem. This method introduces a "test" vector potential along the x direction, whose job will be to help us calculate the current, but which will be set to zero in the end. Write

$$H(\phi_t) = \sum_j \frac{1}{2m_b} \left(p_j + \frac{e}{c} A + \frac{e}{c} A_t \right)^2 + V, \tag{4.27}$$

where

$$A_t \equiv \frac{\phi_t}{L} \hat{x} \equiv \frac{\alpha \phi_0}{L} \hat{x}. \tag{4.28}$$

The quantity V includes everything other than the kinetic energy, namely the coupling to impurity potentials and the external fields, and also the interelectron interaction. Then Eq. (4.8) can be expressed as

$$\hat{I}(\phi_t) = -c \frac{dH(\phi_t)}{d\phi_t}. \tag{4.29}$$

The expectation value of the current operator is

$$I(\phi_t) = \langle \hat{I}(\phi_t) \rangle = -c \left\langle \frac{dH(\phi_t)}{d\phi_t} \right\rangle = -c \frac{d\langle H(\phi_t) \rangle}{d\phi_t} = -c \frac{dU(\phi_t)}{d\phi_t}. \tag{4.30}$$

The current can be calculated by knowing how the ground state energy U changes as a function of ϕ_t. We note that Eq. (4.30) is valid for interacting electrons.

The test vector potential has a nice interpretation when the sample is wrapped into a cylinder in the x direction, as in Fig. (4.11). Then A_t produces a flux

$$\oint A_t \cdot dl = \int_0^L A_t \, dx = \phi_t \tag{4.31}$$

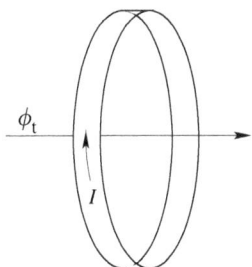

Fig. 4.11. A Hall bar with periodic boundary conditions has the topology of a cylinder. The test vector potential corresponds to a flux along the axis of the cylinder.

passing through the hole of the cylinder, pointing along its axis. (The curl of A_t defined in Eq. (4.28) is zero. However, bending the Hall strip into a cylinder gives a spatial dependence to the direction of $\hat{x}$, which generates a nonzero flux.)

We have apparently made the problem more complicated, because we now need to know how the ground state energy evolves as a function of ϕ_t. It turns out that the change in energy can be calculated for a change in ϕ_t that is precisely equal to one quantum of flux (Laughlin [368]). We therefore approximate

$$I = -c\frac{dU}{d\phi_t} \approx -c\frac{\Delta U}{\phi_0}. \tag{4.32}$$

ΔU is calculated by keeping track of how each single particle eigenstate evolves with ϕ_t. The philosophy will again be to connect ideal regions to the disordered sample, in this case ideal annuli at the inner and outer edges of the sample, and to show that disorder does not change the result.

Let us first consider the sample without disorder. The single particle problem in the presence of the test flux,

$$H\Psi = E\Psi, \tag{4.33}$$

$$H = \frac{1}{2m_b}\left(p + \frac{e}{c}A + \frac{e}{c}A_t\right)^2, \tag{4.34}$$

can be solved by eliminating A_t from the Hamiltonian through the gauge transformation

$$\Psi = \exp\left[-i\frac{e}{\hbar c}\int^x A_t \cdot dl\right]\Psi' = \exp\left[-i2\pi\alpha\frac{x}{L}\right]\Psi'. \tag{4.35}$$

The Schrödinger equation then becomes

$$H'\Psi' = E\Psi', \tag{4.36}$$

with

$$H' = \frac{1}{2m_b}\left(p + \frac{e}{c}A\right)^2. \tag{4.37}$$

In the Landau gauge, $A = B(-y, 0, 0)$, the x dependence of the solutions is given as

$$\Psi'(x, y) = \Phi(y)e^{-ik_x x}, \tag{4.38}$$

which implies

$$\Psi(x, y) = \Phi(y)e^{-i(k_x + \frac{2\pi\alpha}{L})x}. \tag{4.39}$$

We impose periodic boundary conditions, remembering that it is Ψ and not Ψ' that is periodic in the x direction. Invariance of Ψ under $x \to x + L$ gives the condition

$$k_x = \frac{2\pi}{L}(j - \alpha), \tag{4.40}$$

where j is an integer.

At integral values of α, the single particle eigenstates are exactly the same as for $\alpha = 0$. As α is adiabatically (very slowly) changed from 0 to 1, each j changes by unity, indicating that each k_x state moves to the next one. (The change in the test flux must be adiabatic, otherwise it will couple to states across the gap in higher Landau levels.) Because k_x actually tells us where the state is localized along the y direction ($y \approx k_x \ell^2$), the states physically move along the y-axis with variation in α.

Let us consider a system with n Landau levels fully occupied. (The chemical potential lies in the gap above the nth Landau level.) As ϕ_t changes from 0 to ϕ_0, each single particle state moves to the next one. Because all states are occupied, they carry their electrons with them, so, in the end, n electrons have been transported from one edge of the sample to the other edge. For a potential difference V_H between the two edges, we have $\Delta U = neV_H$, giving

$$I = c\frac{neV_H}{\phi_0},\tag{4.41}$$

which yields

$$R_H = \frac{V_H}{I} = \frac{1}{n}\frac{h}{e^2} \quad (\nu = n).\tag{4.42}$$

This is simply the classical result $R_H = h/\nu e^2$ derived by a complicated method.

Disorder modifies the single particle eigenstates. Two kinds of states are relevant: those that go around the periodic cylinder, and those that do not, i.e., are localized. (The first kind of state is not necessarily extended in *two* dimensions.) From our general discussion earlier, the localized states fill the energy region between the unperturbed Landau levels. Because the localized states do not enclose ϕ_t, they are unaffected by a variation in it. On the other hand, the states that are extended along the length (or around the cylinder) evolve to accommodate the Aharonov–Bohm phase due to ϕ_t.

Let us now consider a nonintegral filling factor in a disordered sample. We assume that: (i) all *interior* states *at* the Fermi energy are localized; (ii) the states at the edge are extended (due to the confining potential). We assume that there are n branches of edge states, associated with n Landau levels below the Fermi energy. No constraint is placed on the filling factor, but it is likely to be in the range $n + 1/2 > \nu > n - 1/2$.

We further attach two disorder-free regions, of widths large on the scale of the magnetic length, to the two edges. These are smooth continuations of the extended edge states. In these ideal regions, all states are extended across the sample and the filling factor is exactly $\nu = n$. We thus again have three regions: two ideal ones, and the nonideal "sample."

As the test flux is varied, the extended states in the ideal regions move in the way explained above, carrying their electrons with them. When the flux is changed by ϕ_0, each extended state in the ideal region moves into the next one. Even though the states in the disordered sample evolve in a complicated manner, the net effect is that precisely n electrons are transported from one edge to the other, because they have nowhere else to go. (This would not be the case if there were unoccupied extended states in the interior, which may be the case when there are extended states *at* the Fermi energy.) We thus continue to have

$$R_H = \frac{V_H}{I} = \frac{1}{n}\frac{h}{e^2} \quad (\nu \approx n),$$ (4.43)

where n is the number of Landau levels whose edges cross the Fermi level. The result remains valid as the widths of the ideal regions are shrunk to zero, establishing a quantized Hall resistance plateau in the vicinity of $\nu \approx n$.

Corbino geometry

Halperin [228] has reformulated Laughlin's construction in the Corbino geometry, shown in Fig. 4.12. Here we attach ideal, disorder-free regions at the inner and the outer boundaries of the Corbino sample, representing the extended edge states. A Hall voltage V_H between the inner and the outer edges induces an azimuthal current I. The current can be calculated by considering the response of the system to an additional test flux, ϕ_t, through the origin, created by the vector potential A_t:

$$A_t = \frac{\phi_t}{2\pi r}\hat{\theta}.$$ (4.44)

In the presence of the test flux, the Hamiltonian has the form

$$H = \frac{1}{2m_b}\sum_{j=1}^{N}\left(p_j + \frac{e}{c}A(r_j) + \frac{e}{c}A_t(r_j)\right)^2 + V.$$ (4.45)

Fig. 4.12. The Corbino geometry. The flux ϕ_t through the center is in addition to the flux due to the uniform magnetic field. The lines depict equipotential contours. Ideal, disorder-free regions have been attached at the inner and the outer edges.

The current, as before, is given by Eq. (4.30). To derive that equation for the Corbino geometry, note that the azimuthal current is given by

$$
\begin{aligned}
I &= \int_{R_-}^{R_+} dr \, J_\theta(r, \theta) \\
&= \frac{1}{2\pi} \int_{R_-}^{R_+} dr \int_0^{2\pi} d\theta \, J_\theta(r, \theta) \\
&= -\sum_j \frac{ev_\theta(r_j)}{2\pi r_j},
\end{aligned} \tag{4.46}
$$

where the integral in the first step is evaluated for a fixed θ. The second step exploits the fact that the current does not depend on θ, and the last step follows from the expression for the current density in polar coordinates:

$$
\begin{aligned}
J_\theta(r) &= -e\rho(r)v_\theta(r) \\
&= -e \sum_j \frac{1}{r} \delta(r - r_j)\delta(\theta - \theta_j)v_\theta(r). \tag{4.47}
\end{aligned}
$$

In operator form, we recover Eq. (4.29) for the current operator:

$$
\begin{aligned}
\hat{I} &= -e \sum_j \frac{\hat{v}_\theta(r_j)}{2\pi r_j} \\
&= -e \sum_j \frac{1}{2\pi r_j} \frac{1}{m_b} \left(p_j + \frac{e}{c} A(r_j) + \frac{e}{c} A_t(r_j) \right)_\theta \\
&= -c \frac{\partial H}{\partial \phi_t}. \tag{4.48}
\end{aligned}
$$

Its expectation value yields Eq. (4.30).

As before, we write $I \approx -c\Delta U/\phi_0$, where ΔU is the change in energy as the test flux is changed from 0 to ϕ_0. During this process, each single particle orbital in the outer, ideal regions moves to the next orbital, as shown in Appendix D, carrying its electron with it. The localized orbitals in the interior region are not affected by the test flux. When the Fermi energy lies in the localized region, the extended states in the interior are a nonzero energy gap away from the Fermi energy; all extended states below the Fermi energy are occupied and those above are unoccupied. As a result, as the test flux is changed from 0 to ϕ_0, one electron in each Landau level is transferred from the inner edge to the outer edge (or vice versa), giving $\Delta U = neV_H$, which produces $R_H = V_H/I = h/ne^2$.

The correction due to the approximation $dU/d\phi \approx \Delta U/\phi_0$ can be shown to be of order unity, thus negligible in the thermodynamic limit. For finite systems, the current $I(\phi_t)$ has $O(1)$ oscillations around the mean value obtained above. The origin of such oscillations can be understood by considering what happens when the chemical potential at the two edges is fixed by connecting them to two reservoirs through a tunnel barrier. For a finite sample the single particle eigenstates at the edge are discrete. As the states move up or down under the adiabatic change of ϕ_t, electrons are transferred between the reservoirs and the sample at some discrete points, producing oscillatory behavior.

4.4.6 Landauer vs. Laughlin

The Landauer-type and Laughlin's derivations of the Hall quantization are closely related. Both require the Fermi level to lie in the localized states in the interior. Both also require extended edge states. Laughlin's derivation implicitly assumes that the inner and outer edges are not coupled by disorder, which is equivalent to assuming a suppression of backscattering in the other formulation.

The Landauer-type formulation is more general. (i) It more closely resembles experiments. (ii) Laughlin's derivation assumes that edge states are phase coherent over the entire length of the sample, which is not the case at nonzero temperatures. The Landauer-type formulation clarifies that intra-edge scattering may destroy phase coherence but does not affect the Hall quantization. (iii) The Landauer-type formulation sheds light on the breakdown of the Hall quantization. (iv) In the presence of disorder, there is, in general, a nonzero density of states at the Fermi energy. Therefore, strictly speaking, the notion of "*adiabatic* flux change" is not well defined (as it would be in the the presence of a real gap).

4.4.7 Are edge states necessary for Hall quantization?

The above explanation for the IQHE depends crucially on extended edge states. That the extended edge states carry current in the plateau region can be seen in many ways. It follows theoretically because the edge states are the only extended states at the Fermi energy. Many experimental manifestations of edge states are considered in Section 4.4.4 in the context of Fig. 4.9. Perhaps the most compelling experimental proof is simply to note that, in the incompressible regions, no current flows unless the incoming and outgoing leads are connected by an edge. An interior contact does not conduct in the plateau region. A related effect is seen in the Corbino geometry, where current transport from the inner edge to the outer edge is exponentially suppressed in the plateau region.

When we say that the current flows at the edge, we refer only to the *injected* current. Even when no net current flows through the sample, a complex pattern of local currents is generated by the disorder and confining potentials. Disorder generates closed current loops in the interior. The confining potential generates an edge current, which, in the absence of a net current through the sample, is canceled by an equal amount of edge current on the opposite edge.

4.5 IQHE in a periodic potential

The periodic potential due to the background semiconductor material is irrelevant for the physics of the quantum Hall effect, but it is possible to engineer superlattices with periods comparable to the interelectron separation. As shown in Section 3.14, for flux per plaquette equal to $\phi = (p/q)\phi_0$, the principal effect of the periodic potential is to split a Landau level into p magnetic subbands. Given that a single Landau level contributes a Hall conductance of e^2/h (Section 3.14), one might at first expect that each of the p magnetic subbands would contribute only e^2/ph to the conductance, and a Hall resistance of $h/(s/p)e^2$ would result if s subbands were occupied. If so, this could contain the germ of a possible "mean-field" explanation for the *fractional* quantization of the Hall resistance, with a periodic potential presumably generated self-consistently through the formation of a charge density wave or a Wigner crystal ground state. (A filling factor $\nu = s/p$ would be produced for s/q electrons per plaquette.) A detailed study by Thouless *et al.* [634] showed, however, that the Hall resistance for an integral number of subbands is h/ie^2, where i is an *integer* that depends on p, q, and the number of occupied subbands in a complicated manner and can be either positive or negative. The Hall resistance thus fluctuates wildly as a function of the magnetic field. The Hall resistance of a full Landau level (p magnetic subbands) remains h/e^2 even in the presence of a periodic potential. For further treatment of the quantum Hall effect in periodic potentials, we refer the reader to the excellent review by Thouless [635]. Significant experimental advances have been made on this topic in recent years. Albrecht *et al.* [5], Geisler *et al.* [181], and Melinte *et al.* [438] have studied magnetotransport in samples with periodic "superlattice" modulation, and have seen structures in the magnetoresistance traces which they analyze in terms of the Hofstadter spectrum. The role of disorder in the last experiment has been studied theoretically by Zhou, Berciu, and Bhatt [750].

4.6 Two-dimensional Anderson localization in a magnetic field

Not long before the discovery of the IQHE, Abrahams *et al.* [1] formulated a scaling theory of localization, which leads to the conclusion that random disorder, however weak, localizes all single particle states in two dimensions. This implies that metallic behavior is not possible in two dimensions (for noninteracting quasiparticles). In the early days of the quantized Hall effect, what surprised some theorists was the existence of extended states, as implied by the exponentially small longitudinal resistance. The magnetic field was suggested to be responsible for creating such truly extended states. While interesting in its own right, the question of how the localization length diverges is not central to the physics of the quantum Hall effect, which only requires bands in energy in which all states are localized or extended. We provide here a brief overview of our current understanding. Further discussion and relevant references can be found in the review articles by Das Sarma [107], Huckestein [261], and Tsui [651].

The most reliable studies in this context are direct numerical solutions of the Schrödinger equation for a single electron in random disorder, which have demonstrated convincingly that the localization length diverges as the energy approaches the center as $\xi \sim |E - E_c|^{-\alpha}$, with the localization length exponent[3] $\alpha \approx 2.3$ for the short-range white noise disorder

---

[3] The standard notation for the correlation length exponent is ν, but we use α instead, reserving ν for the filling factor.

(Huckestein [260]; Mieck [450]). A comparison with the localization length exponent $\alpha = 4/3$ in a classical percolation model (Trugman [645]) indicates the importance of quantum tunneling; inclusion of tunneling at the saddle points suggests a quantum percolation network model which changes the classical exponent to $\alpha \approx 2.5$ (Chalker and Coddington [58]). Bhatt and Huo [30] identify the extended states in a disordered potential through their first Chern number and obtain a localization length exponent of $\alpha \approx 2.4$. From scaling arguments, the exponent is naturally expected to be independent of the Landau level index, but nonuniversal corrections to scaling can be strongly dependent on the Landau level index (Huckestein [261]). All these calculations imply, to the extent the noninteracting electron model is valid, that extended states exist only at a single energy, and that the transition from one integral plateau to the next would be infinitely sharp at $T = 0$.

Wei *et al.* [669] studied LLL plateau transitions in InGaAs–InP samples, which have relatively low mobilities ($34\,000\ \text{cm}^2/\text{V s}$), and found that the width of the transition region, as measured by the width of the R_{xx} peak, vanishes over a range of temperature as $\Delta B \sim T^\kappa$ with $\kappa = 0.42 \pm 0.04$. This exponent has two interpretations. (i) The size of a phase-coherent region, which behaves as $L_\phi \sim T^{-p/2}$, can be treated as the effective size of the sample. A finite longitudinal resistance results when this is equal to the localization length $\xi \sim (\Delta E)^{-\alpha} \sim (\Delta B)^{-\alpha}$, which gives the width of the transition region $\Delta B \sim T^\kappa$, with $\kappa = p/2\alpha$. (ii) Ideas from the theory of quantum critical phenomena [562] suggest that $\kappa = 1/\alpha z$, where z is the dynamical exponent. Briefly, the logic is as follows (see, for example, Ref. [608]): A quantum phase transition occurs at $T = 0$ as a function of some parameter, which can be taken as the magnetic field, the Fermi energy, or the filling factor for our problem. Defining a parameter $\delta \sim |B - B_c|/B_c \sim |E - E_c|/E_c \sim |\nu - \nu_c|/\nu_c$, the localization length behaves as $\xi \sim \delta^{-\alpha}$. At nonzero temperatures, the system effectively behaves like a three-dimensional slab, of infinite extent in the two spatial directions, but of length $L_\tau \equiv \hbar/k_B T$ in the third, imaginary time (τ) direction. The correlation length in the imaginary time direction, called ξ_τ, also diverges as $T \to 0$, and the dynamical exponent z is defined as $\xi_\tau \sim \xi^z$. Now, because the dimension of the system is finite in the τ direction, we must appeal to the theory of finite size scaling, which tells us that in the scaling region the magnetoresistance is a function of the ratio L_τ/ξ_τ [200], which can be expressed as $L_\tau/\xi_\tau \sim 1/T\xi^z \sim \delta^{\alpha z}/T$. In other words, the magnetoresistance does not depend on T and δ independently but only through a special combination. The width of the R_{xx} peak, therefore, is proportional to $T^{1/\alpha z}$. In this interpretation, the experimental κ is to be identified with $1/\alpha z$.

Determination of the localization length exponent α is not possible from the temperature dependence of R_{xx} peak width alone. Koch *et al.* [347] determined α directly by studying Hall bar geometries with different sizes, and found its value to be $\alpha = 2.3 \pm 0.1$, consistent with the numerical estimates. Alternatively, one can investigate scaling as a function of other variables. As a function of frequency, the scaling functions depend on the ratio $\delta^{\alpha z}/\omega$, which depends on the same combination αz. Scaling as a function of the electric field (or current) involves the ratio $\alpha(z + 1)$, which allows one to deduce α and ν separately. The experiments of Wei *et al.* [670] are consistent with $\alpha \approx 2.3$ and $z \approx 1$.

While satisfactory at first sight, the above results appear somewhat puzzling upon closer inspection. The closeness of the experimental α with the theoretical value for *non*interacting electrons is surprising, especially in light of $z = 1$, which has been argued (Fisher, Grinstein, and Girvin [164]) to be a generic feature of *Coulomb* systems. This can be seen in the work of Polyakov and Shklovskii [523], who approach the problem from the point of view of variable range hopping, believed to be the dominant conduction mechanism near the resistivity minima. In the presence of the Efros–Shklovskii Coulomb gap [133], the temperature dependence of resistance in two dimensions has the form $R_{xx} = R_0 \exp[-(T_0/T)^{1/2}]$, where $T_0(\nu) \sim e^2/\epsilon\xi(\nu)$. At a given temperature, the concept of variable range hopping breaks down at a characteristic value of ν where $T_0 \sim T$. With $\xi(\nu) \sim (\Delta\nu)^{-\alpha}$, that condition is equivalent to $\Delta\nu \sim T^{1/\alpha}$, giving $\kappa = 1/\alpha$. These explanations mix results from noninteracting and interacting models. A properly combined treatment of interactions and disorder would be more satisfying, but is currently unavailable.

The universality of the critical exponent is not universally accepted. Measurements in silicon MOSFET and GaAs–AlGaAs systems have given values of κ ranging from 0.2 to 0.8 (Koch *et al.* [348]; Wakabayashi *et al.* [666]), systematically increasing with decreasing mobility. The scaling hypothesis itself has been questioned by several groups. Balaban, Meirav, and Bar-Joseph [18] ascertain the width of the $2 \to 1$ transition region as a function of frequency by measuring transmission of edge magnetoplasmons from one side of the sample to the other, which is related to the conductivity tensor. They conclude that the width remains nonzero in the zero-frequency zero-temperature limit. Shahar *et al.* [579] argue that even though ρ_{xx} appears to exhibit scaling behavior in certain ranges of parameters, it is also well described by the expression

$$\rho_{xx} = \exp\left(-\frac{\Delta\nu}{aT + b}\right), \tag{4.49}$$

where $\Delta\nu = |\nu - \nu_c|$, and a and b are nonzero empirical constants. This form is inconsistent with the scaling theory, and implies a continuous crossover rather than a second-order quantum phase transition at $T = 0$ as a function of ν (because ρ_{xx} is a continuous function of ν in Eq. (4.49)). Ilani *et al.* [266] find (more below) that the extended states are not confined to a single energy but form a band, which is also inconsistent with the scaling hypothesis. High mobility samples appear to behave differently from those with low mobility.

Li *et al.* [392] study the plateau transition in $Al_xGa_{1-x}As–Al_{0.33}Ga_{0.67}As$ heterostructures as a function of the Al concentration x. This allows a determination of how the transition depends on the nature of the scattering potential, in this case alloy scattering, which is of short range character. From the temperature dependence of the slope of the Hall resistance, they find good scaling behavior, with the critical exponent $\kappa = 0.42 \pm 0.01$, for a range of Al concentrations (x ranging from 0.65% to 1.6%), which they identify with the alloy scattering dominated region. (At small x the ionized impurities dominate the scattering, while for very large x the mechanism of scattering changes because Al atoms form clusters.) These studies demonstrate that the exponent characterizing the plateau transition is affected by

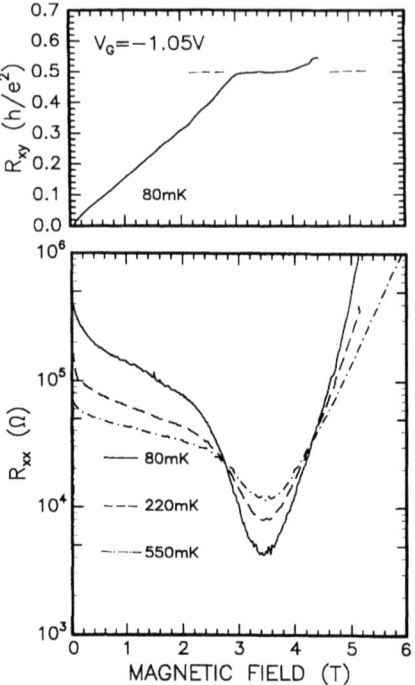

Fig. 4.13. The Hall and longitudinal resistances for a low mobility sample as a function of the magnetic field (the upper and lower panels, respectively). Near the minimum in R_{xx} the resistance decreases with temperature, indicative of a quantized Hall state; outside this region, the resistance increases with temperature, as expected for an insulator. The top panel shows a well-developed $h/2e^2$ state. The mobility of the sample is 2.5×10^4 cm^2/V s. Source: H. W. Jiang, C. E. Johnson, K. L. Wang, and S. T. Hannahs, *Phys. Rev. Lett.* **71**, 1439 (1993). (Reprinted with permission.)

the nature of the disorder, but a "universal" scaling behavior is obtained when short-range scattering is dominant.

The above discussion is based on the notion that the magnetic field creates extended states in what was otherwise an insulator. The reality for many experimental systems is quite the opposite. The best quality samples behave not as insulators but as excellent metals in the absence of a magnetic field. They exhibit Shubnikov–de Haas oscillations, the hallmark of a Fermi-liquid metal, and have resistivities that approach a constant value down to the lowest temperatures available. (An Anderson insulator would have a logarithmically divergent resistivity at very low temperatures.) The best 2D samples that exhibit the IQHE and the FQHE have some of the highest known mobilities. For such samples, it is perfectly valid to take all states to be extended in zero magnetic field. What is responsible for the quantization of the Hall resistance in such systems is that the magnetic field *localizes* most of the states.

Of course, one can intentionally prepare a heavily disordered sample which is insulating at zero magnetic field, and ask if it would exhibit IQHE upon the application of a transverse magnetic field. Such behavior has been seen by Jiang *et al.* [303], who observe a transition from an Anderson insulator at zero magnetic field into the $h/2e^2$ state at a high field, as shown in Fig. 4.13.

4.7 Density gradient and R_{xx}

Pan *et al.* [497] have made the unexpected observation in their ultra-low temperature experiments ($\sim$6 mK), that the peaks in R_{xx} *generically* have a flat top, as seen in the upper left inset of Fig. 4.14 for the R_{xx} peak between the $f = 3$ and $f = 8/3$ plateaus. Furthermore, the R_{xx} peak values are close to those predicted from the difference between the two adjacent Hall plateaus using Eq. (4.26), shown by solid dots in Fig. 4.14. For example, the R_{xx} plateau in the upper left inset occurs at $R_{xx} = 1097\ \Omega$, whereas Eq. (4.26) yields (with $f_1 = 8/3$ and $f_2 = 3$) the value 1076 Ω. This suggests that, at sufficiently low temperatures, all of the R_{xx} peaks may develop a quantized plateau at the top. Unlike in Fig. 4.10, no intentional macroscopic barrier is present in this sample; in fact, it has the currently record mobility of

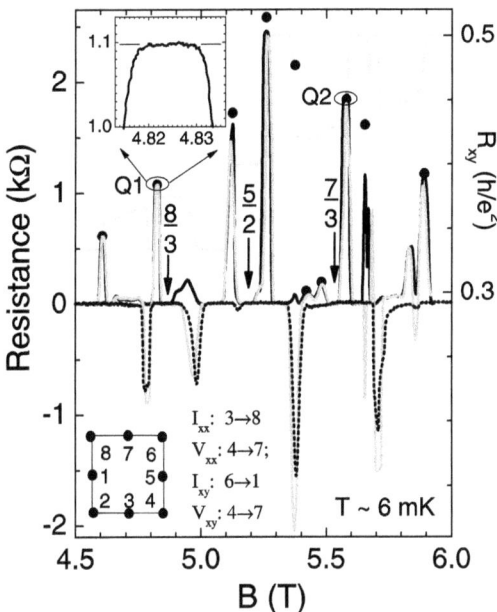

Fig. 4.14. The black trace depicts $R_{xx}(B+)$ and the dotted line, $-R_{xx}(B-)$. The gray line is $\Delta R_{xy} = R_{xy}(\rho) - R_{xy}(\rho - 0.005\rho)$. The inset in the upper left magnifies the peak Q1. Quantized R_{xx} is also seen for the peak labeled Q2. Dots mark the positions of the difference, ΔR_{xy}, between the two quantized plateaus on either side. The inset in the bottom left shows the sample geometry and the contacts used for the measurement of R_{xx} and R_{xy}. Source: W. Pan, J. S. Xia, H. L. Stormer, D. C. Tsui, C. L. Vicente, E. D. Adams, N. S. Sullivan, L. N. Pfeiffer, K. W. Baldwin, and K. W. West, *Phys. Rev. Lett.* **95**, 066808 (2005). (Reprinted with permission.)

31×10^6 cm^2/V s. Pan *et al.* explain these observations quantitatively by assuming a tiny ($\sim$1%/cm) gradient in the density, and taking $R_{xx} = \Delta R_{xy} = R_{xy}(\rho) - R_{xy}(\rho + \Delta\rho)$. They determine $R_{xy}(\rho + \Delta\rho)$ from the experimental R_{xy} by appropriately rescaling the filling factor; the difference, $R_{xy}(\rho) - R_{xy}(\rho + \Delta\rho)$, is obtained numerically and plotted as the gray curve in Fig. 4.14. It shows excellent quantitative agreement with the actual R_{xx}.

The model of density gradient also provides an explanation [497] for the previously puzzling phenomenological resistivity rule (Chang and Tsui [62]; Rötger *et al.* [556]; Stormer *et al.* [621]):

$$R_{xx} \propto B \times \frac{dR_{xy}}{dB} . \tag{4.50}$$

At the same time, the experimental result is "disturbing," through its implication that R_{xx} is not a measure of the longitudinal resistivity, ρ_{xx}, as assumed in previous analyses, but is determined entirely from R_{xy}. Further work will show how this understanding affects various quantities determined from R_{xx}, such as activation gaps and scaling behavior.

4.8 The role of interaction

The picture presented above is incomplete. Interactions bring about profound changes in the nature of the state in several ways.

FQHE For certain parameters interelectron interactions destroy the integrally quantized plateaus to produce new, fractionally quantized plateaus, the main topic of this book.

Nature of localization The single particle picture for localization appears to be valid only for strong disorder, and several experiments observe deviation from it. Ilani *et al.* [266] have used a scanning single-electron transistor to measure the local electrostatic potential of an IQHE system as a function of the magnetic field and the electron density. This allows them to distinguish between localized and extended states, as the charging of the former alters the charge in increments of a single electron charge, causing oscillations in the local electrostatic potential. The detailed analysis shows that the nature of the localized states is incompatible with predictions of the single particle theory, except in the limit of strong disorder, and is strongly affected by the Coulomb interaction. The experiment also suggests that the bands of extended states have a nonzero extent.

Koulakov–Fogler–Shklovskii (KFS) phases Interactions also create new phases that do not destroy the integral plateaus but are superimposed over them. Close to integral fillings, the extra electrons or holes are expected to form a Wigner crystal, at least for weak disorder. Koulakov, Fogler and Shklovskii [170, 350] predict theoretically, within a Hartree–Fock formulation, that the extra electrons or holes in a high Landau level form a Wigner crystal, a "bubble" crystal (a Wigner crystal of bubbles of electrons), or stripes of alternating fillings, depending on their filling factor. Experiments support this picture. When a high index Landau level is half full (i.e., the filling factor is $\nu = n + 1/2$ with n being a sufficiently

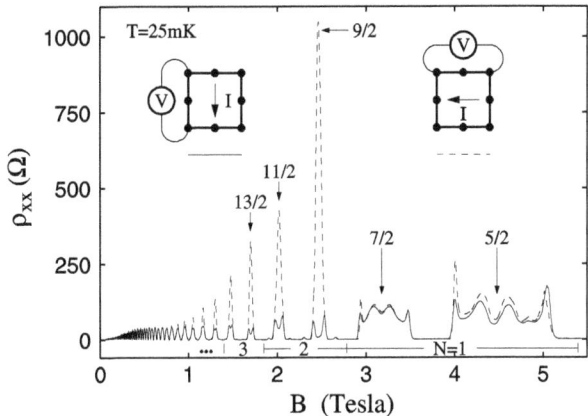

Fig. 4.15. The longitudinal resistance ρ_{xx} at fillings of the form $\nu = n + 1/2$ depends strongly on the direction of the current for $n \geq 4$ (i.e., in the third and higher Landau levels). Source: M. P. Lilly, K. B. Cooper, J. P. Eisenstein, L. N. Pfeiffer, and K. W. West, *Phys. Rev. Lett.* **82**, 394 (1999). (Reprinted with permission.)

big integer), a strong transport anisotropy is observed, as shown in Fig. 4.15, consistent with the formation of stripes (Du *et al.* [128]; Lilly *et al.* [398]). The direction of the anisotropy aligns with one of the crystal axes, perhaps due to the weak symmetry breaking potential of the host lattice, and flips by $90°$ as a function of the in-plane magnetic field (Lilly *et al.* [399]; Pan *et al.* [493]) or the density (Zhu *et al.* [751]). As seen in Fig. 4.16, many re-entrant transitions are seen in higher Landau levels (Eisenstein *et al.* [140]; Gervais *et al.* [184]; Xia *et al.* [706]) which indicate competition between many approximately degenerate ground states, and find a natural interpretation in terms of a bubble-crystal phase. A direct observation of these phases (for example, through a scanning tunneling microscopy experiment) has not been possible so far.

No matter what the state of the *additional* electrons or holes, integrally quantized Hall plateaus result so long as they are localized and do not contribute to transport. The KFS crystal or stripe phases can be expected to be pinned by disorder, and, therefore, do not affect the integral quantization. Nonetheless, they add to the richness of the phenomenon, and show that the physics of localization can be much more intricate than the single particle picture suggests.

Plateau transitions The transition between integral plateaus is also affected by interaction. For a model that neglects interelectron interactions, such a transition would occur at $\nu = n + 1/2$. For weak disorder, experiments clarify that such a model is inadequate for addressing the nature of the transition. For high quality samples, there is no direct transition between two integrally quantized plateaus in the lowest two Landau levels; interactions produce a Fermi-sea-like state in the lowest Landau level (at $\nu = 1/2$ and $3/2$, Chapter 10), and a FQH state in the second Landau level ($\nu = 5/2$, Section 7.4.2. In higher Landau levels, an anisotropic state, perhaps the stripe phase, occurs in the transition region.

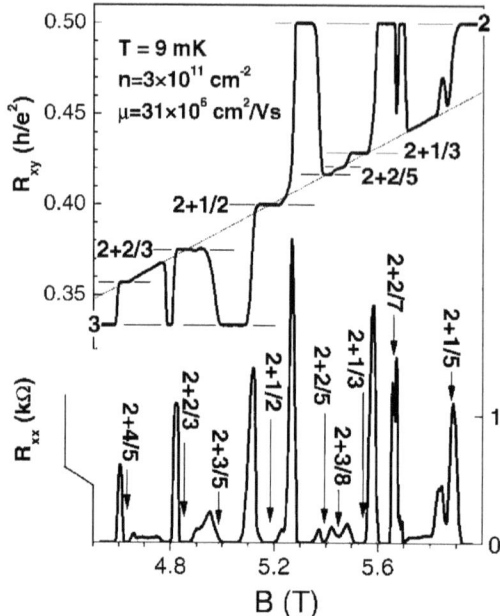

Fig. 4.16. The Hall and the longitudinal resistances R_{xy} and R_{xx} in the filling factor range $2 \leq \nu \leq 3$ (second Landau level). The $h/2e^2$ and $h/3e^2$ integral plateaus are interrupted several times when the system goes into a fractional state, indicated by well-defined fractional plateaus. The re-entrant integral plateaus at $h/2e^2$ or $h/3e^2$ are presumably due to the pinning of a correlated crystal state of electrons or holes, respectively, in the second Landau level. Source: J. -S. Xia, W. Pan, C. L. Vicente, E. D. Adams, N. S. Sullivan, H. L. Stormer, D. C. Tsui, L. N. Pfeiffer, K. W. Baldwin, and K. W. West, *Phys. Rev. Lett.* **93**, 176809 (2004). (Reprinted with permission.)

We have already mentioned the possible role of Coulomb interaction in the context of the scaling exponent for Anderson localization. An interesting development has been the surprising observation of a metal–insulator transition in two-dimensional systems with very low densities. We refer the interested reader to a review by Kravchenko and Sarachik [353]. The basic experimental observation is an unusually strong temperature dependence of the resistivity at a very low temperature of order 1 K for carrier densities close to a certain very low critical density. A substantial resistance drop with decreasing temperature indicates that the system behaves as a metal at $T = 0$. A proper understanding of the true nature of this state is not yet available, although a widely held belief is that interelectron interactions play a crucial role in its physics. A true metallic state in two dimensions at zero magnetic field has implications on the theory of localization in the presence of a magnetic field.

5

Foundations of the composite fermion theory

This chapter introduces the basic principles of the composite fermion theory. It should really be called the "composite fermion model" or the "composite fermion hypothesis" in this chapter. The extensive scrutiny and testing that elevate it to the status of a "theory" are topics of subsequent chapters.

5.1 The great FQHE mystery

Not often does nature present us with a mystery as well defined as the phenomenon of the fractional quantum Hall effect. The questions that theory is challenged to answer could not be more sharply posed.

- What is the physics of this quantum fluid? The appearance of precise quantum numbers and dissipationless transport in a dirty solid state system containing many electrons is a signature of cooperative behavior. What are the correlations in the FQHE state and why do they manifest themselves in such a rich, yet stunningly simple fashion? Why do gaps open at certain fractional fillings of the lowest Landau level? Experiments point to something unique and special about ground states at certain special filling factors. What order brings about this uniqueness?
- A tremendous amount of factual information is contained in the fractions that are observed and the order in which they appear. That imposes rigorous constraints on theory. With the proliferation of fractions, certain striking patterns have emerged. Many fractions are conspicuous by their absence. For example, the simplest fraction $f = 1/2$ has not been observed. In fact, no sub-unity fractions with even denominators have been observed. Another remarkable aspect is that fractions are not isolated but belong to certain sequences. For example, in Fig. 2.7, the fractions $1/3, 2/5, 3/7, 4/9, 5/11, \ldots$ follow the sequence $f = n/(2n + 1)$. These observations lead to the following questions: Why are some fractions seen but not others? Why do they appear in sequences? What determines the order of their stability? Why are even denominator fractions absent (with one exception)?
- What is the microscopic description of this state? What theory will provide quantitative predictions for various experimentally measurable quantities?

Any theory that attempts to explain the FQHE must answer these basic questions. Many subsidiary questions can be added to the above list.

- What is the nature of the state at even denominator fractions? An explanation of the odd-denominator "rule" would be unsatisfactory and surely incomplete in the absence of such understanding.
- A FQHE has been observed at $\nu = \frac{5}{2}$, the only even denominator fraction observed so far (in a single layer system). What is its physics? Why is there FQHE at the half-filled second Landau level, but none at half-filled lowest Landau level?
- In the limit of very high magnetic fields, the spin degree of freedom is completely frozen. The spin does play a role, however, in the experimentally studied parameter regime. What kinds of structures occur when the spin degree of freedom can fluctuate?
- Why is there no FQHE at small filling factors? When does the FQHE disappear and what causes its disappearance?
- The abundance of FQHE in the lowest Landau level is in stark contrast to a scarcity of FQHE in higher Landau levels. Why?
- What is the nature of excitations of the FQHE states? This is a rather complicated question because there are a myriad FQHE states, often with several possible spin polarizations, with their excitations depending on parameters such as the wave vector, the Zeeman energy, etc.

The list of questions appears daunting at first sight, but, encouragingly, they are not all independent and are not going to be answered one at a time. The physical origin of the fractions must be the same, the appreciation of which should tell us why they belong to certain sequences of fractions that exclude even denominator fractions. An understanding of what causes a gap must surely offer an insight into the nature of excitations, and knowing the physics of the fully polarized FQHE will likely give a clue into the role of spin. The questions listed above are so inextricably intertwined that the correct physical principle must answer many of them at once, and show the way for the rest.

A new principle invariably has implications beyond the questions that initially inspire it. New questions arise that could not have been asked before, and new phenomena become apparent that could not have been envisioned previously. So, finally:

- What other phenomena does this state exhibit?

5.2 The Hamiltonian

The goal is to identify the solutions to the Schrödinger equation *and* the physics they signify. The relevant Schrödinger equation is given by

$$H\Psi = E\Psi, \tag{5.1}$$

where

$$H = \sum_j \frac{1}{2m_b} \left[\frac{\hbar}{i} \nabla_j + \frac{e}{c} A(r_j) \right]^2 + \frac{e^2}{\epsilon} \sum_{j<k} \frac{1}{|r_j - r_k|} + \sum_j U(r_j)$$

$$+ g\mu B \cdot S. \tag{5.2}$$

The first term on the right hand side is the kinetic energy in the presence of a constant external magnetic field $B = \nabla \times A$, the second term is the Coulomb interaction energy, the third term is a one-body potential incorporating the effects of the uniform positive background and disorder, and the last term is the Zeeman energy.

To get a feel for the relative importance of the various terms, we consider GaAs–AlGaAs heterostructures, on which most quantum Hall experiments have been performed. We use the following values: band mass of electrons $m_b = 0.067 m_e$ (where m_e is the electron mass in vacuum); dielectric constant $\epsilon = 12.6$; Landé g factor $g = -0.44$. Relevant parameters are the cyclotron energy

$$\hbar \omega_c = \hbar \frac{eB}{m_b c} \approx 20 B[\text{T}] \text{ K}, \tag{5.3}$$

the typical Coulomb energy

$$V_C \equiv \frac{e^2}{\epsilon \ell} \approx 50 \sqrt{B[\text{T}]} \text{ K}, \tag{5.4}$$

the Zeeman splitting

$$E_Z = 2g \mu_B \mathbf{B} \cdot \mathbf{S} = \frac{g}{2} \frac{m_b}{m_e} \hbar \omega_c \approx 0.3 B[\text{T}] \text{ K}, \tag{5.5}$$

and the magnetic length

$$\ell = \left(\frac{\hbar c}{eB} \right)^{1/2} \approx \frac{25}{\sqrt{B[\text{T}]}} \text{ nm}. \tag{5.6}$$

Here, the last terms in Eqs. (5.3), (5.4), and (5.5) give the energy (in kelvin) for parameters appropriate for GaAs, with $B[\text{T}]$ in tesla. The last term in Eq. (5.6) quotes the magnetic length in nm. The Zeeman splitting is defined as the energy required to flip a spin. Another possible unit for the Coulomb interaction is $e^2/\epsilon r_0$, where r_0 is defined by $\pi r_0^2 \rho = 1$; this scale is related to $e^2/\epsilon \ell$ through $r_0/\ell = \sqrt{2/\nu}$.

Following the standard practice, we often use ℓ as the unit of length and $e^2/\epsilon \ell$ as the unit of interaction energy. This amounts to setting

$$\ell = 1, \qquad \frac{e^2}{\epsilon \ell} = 1. \tag{5.7}$$

Our first and foremost concern is with conceptual foundations of the FQHE. With that goal in mind, we simplify the problem to the maximum extent possible, without, of course, throwing away the essential physics of the FQHE. We (i) switch off disorder, (ii) take a vanishing transverse width for the electron wave function, (iii) neglect LL mixing, and (iv) assume that electrons are fully polarized, i.e., are effectively spinless. The simplifications (iii) and (iv) are equivalent to taking the limit $B \rightarrow \infty$, because that implies

$$\frac{e^2/\epsilon \ell}{\hbar \omega_c} \rightarrow 0, \tag{5.8}$$

$$\frac{e^2/\epsilon \ell}{E_Z} \rightarrow 0. \tag{5.9}$$

In this limit, the Coulomb interaction is too weak to cause either LL mixing or spin reversal. The limit $B \to \infty$ is taken *at a fixed filling factor*, which also requires $\rho = \nu B/\phi_0 \to \infty$. (If ρ were held fixed instead, the limit $B \to \infty$ would imply $\nu \to 0$, which is a trivial limit with no FQHE; the distance between electrons becomes large compared to their size ($\sim$ the magnetic length), and the system becomes classical.)

The justification for these simplifying approximations is as follows. (i) As in the IQHE, disorder is crucial for establishing plateaus, but is not needed for an understanding of the underlying physics, which has to do with the appearance of gaps. We see in Chapter 7 that the presence of a gap at $\nu = f$ in a disorder-free system leads to a plateau at $R_H = h/fe^2$ when a weak disorder is introduced. (ii) Qualitative features of the phenomenon are not affected by the transverse width, provided it is not too large. (iii) The qualitative phenomenology of the FQHE (e.g., the value of the quantized Hall resistance, or which fractions are observed) is independent of the amount of LL mixing, at least when it is sufficiently small, as can be seen from experiments on samples with different parameters (density, band mass, ϵ, etc.). (iv) Equation (5.9) is not satisfied in most experiments, and not all observed FQHE states are fully spin polarized. However, fully spin-polarized FQHE states do occur. We first focus on these states, coming to the role of spin in Chapter 11.

The simplifying assumptions will have to be relaxed before a detailed *quantitative* comparison with experiment becomes feasible. Neglecting LL mixing and nonzero thickness corrections, however, is often not too unrealistic even when it comes to comparing numbers. The idealized model explains major trends seen in experiments and provides zeroth order estimates for experimental numbers for fully spin-polarized FQHE states. We see in Chapter 11 that the electron spin produces much interesting physics at relatively low magnetic fields. Specifically, many FQHE states with different spin polarizations can occur at a given filling factor. These states can be explained by a generalization of the physics of fully polarized FQHE states.

When electrons are confined in the lowest Landau level, with their spin degree of freedom frozen, the kinetic and Zeeman energies are irrelevant constants which we throw away and forget. They are the same for all eigenstates, and do not contribute to any energy differences. We thus end up with the idealized model of "effectively spinless" electrons in the lowest Landau level, restricted to an ideal two-dimensional plane, with the Hamiltonian given by

$$H = \mathcal{P}_{\mathrm{LLL}} \frac{e^2}{\epsilon} \sum_{j<k} \frac{1}{|r_j - r_k|} \mathcal{P}_{\mathrm{LLL}}, \tag{5.10}$$

which must be solved with the LLL restriction, as explicitly indicated by the LLL projection operator $\mathcal{P}_{\mathrm{LLL}}$. This is the simplest and the cleanest model containing the essential physics of the FQHE, stripped of all inessential features. Study of this model is sufficient for establishing the physical principles of the FQHE.

Note: Without the LLL restriction, the right hand side of Eq. (5.10) is classical, as it contains no noncommuting operators. The LLL projection turns it into a nontrivial quantum mechanical problem. The Coulomb interaction can be formally projected into the lowest

Landau level using the method discussed in Section 3.7. For that purpose, working with the Fourier transform is convenient [192]:

$$\sum_{j<k} V(|\boldsymbol{r}_j - \boldsymbol{r}_k|) = \int \frac{d^2\boldsymbol{q}}{(2\pi)^2} V(q) \sum_{j<k} \exp\left[i\boldsymbol{q} \cdot (\boldsymbol{r}_j - \boldsymbol{r}_k)\right]$$

$$= \int \frac{d^2\boldsymbol{q}}{(2\pi)^2} V(q) \sum_{j<k} \exp\left[\frac{i}{2}q(\bar{z}_j - \bar{z}_k)\right] \exp\left[\frac{i}{2}\bar{q}(z_j - z_k)\right],$$

where $q = q_x - iq_y$. We have already normal-ordered the expression on the right, with $\bar{z}$'s preceding z's. The LLL projection is obtained by making the replacement $\bar{z}_j \rightarrow 2\partial/\partial z_j$, which gives

$$\mathcal{P}_{\text{LLL}} \sum_{j<k} V(|\boldsymbol{r}_j - \boldsymbol{r}_k|) \mathcal{P}_{\text{LLL}} = \int \frac{d^2\boldsymbol{q}}{(2\pi)^2} V(q)$$

$$\times \sum_{j<k} \exp\left[iq\left(\frac{\partial}{\partial z_j} - \frac{\partial}{\partial z_k}\right)\right] \exp\left[\frac{i}{2}\bar{q}(z_j - z_k)\right]$$

with the proviso that derivatives do not act on the Gaussian part of the wave function. While it brings out the quantum mechanical nature of the problem, this form has not proved to be amenable to further manipulations. In practice, one implements the LLL restriction on wave functions. Once the wave functions are restricted to the LLL sector, projection of the Hamiltonian becomes unnecessary, because, by definition, the projected and the unprojected interactions have identical matrix elements in the LLL subspace.

5.3 Why the problem is hard

The simplicity of the statement of the problem is deceptive. Equation (5.10) already reveals the fundamental difficulty. In many-body problems, the starting point for a theoretical investigation of a phenomenon is often obtained by switching off the interaction altogether. But the interaction is not small compared to anything else in our problem, because there is nothing else. The interaction is the only energy in our problem, and therefore cannot be neglected. Equation (5.10) contains no small parameter. Taking the magnetic length ℓ as the unit of length, and $e^2/\epsilon\ell$ as the unit of energy, we can rewrite Eq. (5.10) as

$$H = \mathcal{P}_{\text{LLL}} \sum_{j<k} \frac{1}{|\boldsymbol{r}_j - \boldsymbol{r}_k|} \mathcal{P}_{\text{LLL}}, \tag{5.11}$$

where all quantities are dimensionless. This form shows that the problem has no real parameters whatsoever.[1] The usual quantity characterizing the strength of correlations,

[1] The eigenfunctions of this problem do not contain any sample specific parameters.

namely the ratio of interaction energy to kinetic energy, is infinite, because the latter is absent. The FQHE state is one of the most strongly correlated systems in the world.

Let us ask what happens if, just for a moment, we set the interaction to zero. That produces a large number of ground states, because the number of available single particle orbitals greatly exceeds the number of electrons. A simple calculation gives a glimpse into the enormous complexity of the problem. A typical 1mm $\times$ 1mm sample contains 10^9 electrons. Assuming 2.5×10^9 single particle orbitals in the lowest Landau level (which corresponds to a filling factor $\nu = 0.4$), the number of distinct ground states is $10^{7 \times 10^8}$, which is an unimaginably large number. Even a toy system containing only 100 electrons in 250 single particle orbitals has 10^{72} distinct ground state configurations, which is roughly equal to the number of atoms in the Universe.

In the absence of interaction, no single state is picked out at any fractional filling factor. When the interaction is switched back on, the degeneracy is lifted and one linear combination of the allowed states becomes the nondegenerate ground state. The interaction causes a spectacular reorganization of the low-energy Hilbert space, thereby producing remarkable phenomena. The difficulty is that, on purely theoretical grounds, with no small parameter to guide our thinking, we have no clue where to start looking for the ground state.[2] Standard perturbative methods are of no use and the problem appears hopelessly intractable.

In other words: The FQHE has no "normal state." The FQHE cannot be understood as an instability of a familiar state, caused by turning on a weak interaction of the appropriate kind. In the absence of such a reference state, computing deviations from a known state is not possible, and one must calculate *full* answers. That would be akin to solving the Schrödinger equation to explain superconductivity without assuming a normal state with weakly interacting quasiparticles.

It is thus not possible to *first* solve the Schrödinger equation and *then* try to understand what physics the solution signifies. Our approach will be first to identify the physics, with help from experiment, which then will lead us to the solution of the Schrödinger equation.

5.4 Condensed matter theory: solid or squalid?

Solid state physics was famously ridiculed, allegedly by Murray Gell-Mann, as the "squalid state physics." It is worth understanding why that might appear to be the case, and why that is not true (any more than for any other branch of physics).

The reader is surely aware of the crucial difference between mathematics and physics: A mathematical theorem, once proven, is forever true. A physical theory, in contrast, is always subject to revision in light of new information. Examples abound where a once widely accepted theory is discarded or superseded by a better theory. Not that the old theory does not work any more. It continues to explain whatever it explained before. But another theory, based on a fundamentally new approach, explains a wider body of phenomena while

---

[2] *A priori*, a crystalline state might appear an attractive starting point, but experiments rule it out in the region where FQHE is seen.

obtaining the previous theory as a limiting case. Progress in physics can be viewed as such "paradigm shifts" at various levels (Kuhn [355]).

While no physical theory can have the tidiness or the finality of a mathematical theorem, the situation can be especially confusing in condensed matter physics. An advantage here is that we know exactly what problem needs to be solved: the nonrelativsitic Schrödinger equation for the many-particle system of interacting electrons and ions. (As far as condensed matter physics is concerned, this equation can be taken as the "exact" starting point; relativistic effects are often of no consequence.) The difficulty is that finding its exact solution is impossible. We do not know how to solve the general problem containing even three particles, let alone a macroscopic number. No condensed matter theory, therefore, can be one hundred percent correct. We attempt either to find an approximate solution of the exact problem defined by the aforementioned Schrödinger equation, or to solve exactly an approximate model which, hopefully, contains the essential physics of the phenomenon in question. Most of the time, we have to be content with approximate solutions of approximate models.

How, then, do we come to accept a condensed matter theory? The answer appears easy: It must explain nature. An approximate model or an approximate solution may appear compellingly elegant to its protagonists, but experiments are the ultimate judge for distinguishing a physical theory from a mere mathematical model. What makes this criterion difficult to employ in practice is that, being approximate, no theory of a many-body system will ever explain *all* experimental facts. Every theory will have aspects that deviate from experiments, and the question, therefore, boils down to what one considers the "essential" features. That is the reason why the validity of a given theory is often a topic of an intense and bitter debate; the situation is like the proverbial blind men and the elephant, with scientists focusing on different aspects drawing dissimilar conclusions.

Condensed matter theories are – and should be – viewed with great skepticism. They can go wrong in many places. The effective model being considered may not contain the physics we are seeking to describe. Some of the approximations made in deriving consequences of the model may be invalid. And even if the theory describes some phenomena, it will fail in other aspects. When certain consequences of a theory are inconsistent with experiment, it is not clear if the error lies in our solution of the effective model, in the model itself, or both, and whether the inconsistency is substantial. This explains why theoretical condensed matter physics might appear to be a singularly messy enterprise. No new principle seems to be at stake (we know the starting point precisely), the exact solution is hopelessly complicated, and arriving at any kind of consensus is difficult.

Lest the reader should become overly pessimistic, we hasten to add that the above stated views are flawed in many respects. While condensed matter physics surely has its own share of dubious theoretical attempts, the most successful condensed matter theories are highly nontrivial, stunningly elegant, and so well confirmed by experiment that their essential correctness is beyond doubt.

Perhaps counterintuitively, the exact solution of the exact problem is not what we are after. Suppose we are able to calculate the exact eigenfunctions and eigenvalues of the

Schrödinger equation for interacting electrons in the lowest Landau level by a brute force diagonalization on a huge computer. Of course, given the dimension of the matrix to be diagonalized, it would be impossible for a computer to produce the answer for a realistic system. Let us ignore that problem for a moment and imagine that we have a "quantum computer" that can calculate the exact eigenstates. The computer would disgorge a long list of numbers, which are projections of the ground and excited state vectors along various directions of the very large dimensional Hilbert space. Given all eigenfunctions, we can insert them into the Kubo formula to calculate magnetoresistances, which would surely reproduce the FQHE in its full glory. Is the problem solved? Not really. The computer does not tell us *what* the numbers mean, and *why* nature is behaving the way it is. The computer does not supply an insight into the physics. That is hardly a satisfying state of affairs.

So, what are we after? As enunciated in a landmark 1972 article by P. W. Anderson entitled "More is different" [8] (also see a delightful recent book by Laughlin [373] on this topic), even though we may know the defining theory, many-body systems collectively behave in remarkably surprising ways, an understanding of which requires entirely new sets of concepts. The goal of condensed matter theory is to identify simple "emergent" principles that enable us to unify, explain, predict, and calculate. These principles should facilitate an exact account of the universal properties, and a reasonably accurate quantitative description of nonuniversal observables. While numbers are important, physical insights are more so, especially in condensed matter physics, where our goal is not to test the starting point (we believe in the Schrödinger equation) but to understand beautiful structures that emerge when many particles interact. To that end, an approximate solution that reveals the physics of the problem plays a more important role than the exact solution.

A marvelous example is the Bardeen–Cooper–Schrieffer theory of ordinary super-conductors, based on the principle of pairing of electrons. It explained the zero resistance, the Meissner effect, the isotope effect, flux quantization, and many other mysterious phenomena. It led to the prediction of the Josephson effect. And, its quantitative consequences have been verified in great detail for myriad phenomena for weakly coupled superconductors.

How do we determine the emergent principle? Answer: The only method, really, is to make an intelligent *guess* guided by experiment. We then build that principle into a theoretical formulation and deduce its consequences to the best of our ability, which then are put to the test against experiment. A theoretical treatment may often give the misleading impression of "deriving" the principle, but the reader must beware that the principle can only be motivated, not derived. Certain essential assumptions are made at the very first step of every condensed matter theory about the qualitative nature of the state; the rest of the theoretical effort is toward gaining a better quantitative description. For example, in the Landau Fermi liquid theory one may calculate many complicated Feynman diagrams, but the nature of the state is assumed as soon as one decides to perturb around the noninteracting solution. The theory does not *prove* that the actual state is a Landau Fermi liquid, but only gives an approximate description of it were that to be the case. It is sometimes possible to deduce, theoretically, that a hypothesized phase is unstable, but it is never possible to prove that it is not. Our intention here is not to minimize the importance of quantitative calculations, which play an all important role in the validation of the basic principle, but to urge the reader to make an effort to distinguish between what is being assumed and what is being derived, and not to confuse mathematical formalism with physics.

5.5 Laughlin's theory

A crucial clue into the physics of the FQHE problem came soon after the observation of $f = 1/3$, when Laughlin [369] made a brilliant ansatz for the ground state wave function at $\nu = 1/3$, which we present in this section.

Single particle states in the lowest Landau level are given by

$$\eta_l(z) = (2\pi 2^l l!)^{-1/2} z^l \, e^{-\frac{1}{4}|z|^2}, \tag{5.12}$$

where $z = x - iy$ and l is the angular momentum (not to be confused with the magnetic length, which has been set to unity). A LLL wave function for many electrons must necessarily have the form

$$\Psi = F_A[\{z_j\}] \exp\left[-\frac{1}{4}\sum_i |z_i|^2\right], \tag{5.13}$$

where $F_A[\{z_j\}]$ is a polynomial of z's antisymmetric under exchange of two coordinates. (The spin part of the wave function, which is fully symmetric under exchange of particles, is not shown explicitly.) The task of theory is to determine the polynomial $F_A[\{z_j\}]$ for low-energy many-body states of interacting electrons as a function of the filling factor. Laughlin proceeded as follows:

(i) Motivated by the success of Jastrow-type variational wave functions [31, 289] for superfluid helium, which have pairwise correlations, let us assume the form

$$F_A[\{z_j\}] = \prod_{j<k} f(z_j - z_k). \tag{5.14}$$

(ii) In order for the full wave function to be an eigenstate of the total angular momentum, which commutes with the Coulomb interaction, the product $\prod_{j<k} f(z_j - z_k)$ must be a polynomial of $z_1, z_2, \ldots, z_N$ of degree L, i.e., the replacement $z_i \to z_i e^{-i\theta}$ is equivalent to multiplication by $e^{-iL\theta}$. This is possible only if $f(z_j - z_k)$ itself has a definite angular momentum.

(iii) At the same time, antisymmetry under exchange requires that $f(-z) = -f(z)$. The only form that satisfies these two criteria (and also analyticity in z as required by the LLL constraint) is

$$f(z) = z^m, \tag{5.15}$$

where m is an odd integer. This produces what is known as the Laughlin wave function (see also Ref. [45]):

$$\Psi_{1/m} = \prod_{j<k} (z_j - z_k)^m \exp\left[-\frac{1}{4}\sum_i |z_i|^2\right]. \tag{5.16}$$

As seen later in this chapter, and again in Chapter 12, this wave function describes a uniform density state at filling factor $\nu = 1/m$. In particular, for $m = 3$, it describes a state at $\nu = 1/3$, the fraction that had been observed at the time of Laughlin's work. Detailed calculations (e.g., Haldane and Rezayi [222]) confirm that the Laughlin wave function is a valid and accurate representation of the actual ground states at $f=1/3$ and $f=1/5$. The fractions $f=1/m$ are referred to as the Laughlin fractions.

Had only the fractions $f = 1/m$ and their particle–hole symmetric counterparts $f = 1 - 1/m$ been observed, this would have been the end of the story; it would only remain to evaluate the properties of the Laughlin wave function. The subsequent appearance of a large number of other fractions, to which Laughlin's theory does not apply,[3] tells us that this theory is incomplete. For example, referring to Fig. 1.3, the Laughlin wave function is pertinent to $1/3$ and $2/3$ but not to the $\sim$30 other fractions seen on this figure in the range $\nu < 1$.[4]

The remainder of this chapter is concerned with a new principle: the formation of topological particles called composite fermions (to be defined below). They bring out the general physics of the FQHE, giving a physical understanding of why gaps open at fractional fillings. The CF theory explains all fractions in a unified fashion, and also reveals an underlying structure that encompasses phenomena other than the FQHE. While the CF theory also leads to wave functions, the essential physics is independent of wave functions (Section 5.8). Composite fermions represent a paradigm that is more general than a particular theoretical model or a wave function. (Short, non-technical reviews on composite fermions can be found in Refs. [283, 284].)

What about the Laughlin wave function? The CF theory recovers it as a special case (Section 5.8.6). The CF "derivation" of the Laughlin wave function proceeds along a different logical route from the one described above, enriching this wave function with a new physical interpretation and making manifest that it belongs to a larger conceptual structure. In contrast, composite fermions are neither a part of, nor can be derived from, the Laughlin wave function.

Laughlin also demonstrated the striking property, using a general flux insertion argument, that incompressibility at a fractional filling factor implies the existence of fractionally charged excitations.[5] That is one of the topics addressed in Chapter 9.

---

[3] The exponent m in the Laughlin wave function must be an odd integer. An even m gives a wave function for bosons, whereas a fractional m fails to produce a single-valued wave function.

[4] Each FQHE state represents a distinct state that cannot be continuously deformed into another, as evident from the fact that the Hall resistance, being quantized at *discrete* values, cannot be modified *continuously* from one plateau to another.

[5] The possibility of fractionally charged quasiparticles was first mentioned in the experimental discovery paper (Tsui, Stormer, and Gossard [650]). In the context of the quantized Hall resistance at $3h/e^2$, it states: "If we attribute it to the presence of a gap at E_F when $1/3$ of the lowest Landau level is occupied, [Laughlin's] argument will lead to quasiparticles with fractional

5.6 The analogy

Experience has taught us that, in physics, life invariably becomes more complicated before it becomes simple. As more and more facts are gathered, the situation grows more confusing at first, but eventually new patterns become apparent that give a clue into the deeper structure, leading to a simplification. For example, the Lyman–Balmer–Paschen spectral lines of the hydrogen atom motivated the simple Bohr model. As another example, the discovery of a large number of hadrons in the 1950s led to an identification of structures such as the SU(3) flavor octet, which took us one level deeper into the realm of quarks, enabling a more fundamental description of hadrons as bound states of quarks. Experimental facts, through patterns they reveal, provide a hint into an underlying connection between the mysterious phenomenon at hand and some other familiar problem. The trick is to identify patterns and the correct analogy.

With a rapid growth in the number of fractions, the situation became, at first, similarly bewildering in FQHE, but eventually two crucial patterns became apparent:

- Fractions do not appear in isolation but are members of certain sequences.
- Even denominator fractions are absent.

Thus, rather than trying to understand one fraction at a time, we should search for a principle that explains whole sequences at once. The correct explanation should produce the observed sequences of fractions, clarifying, at the same time, why they have odd denominators.

What is the key analogy here? Let us again look at Fig. 2.7 and, to avoid distraction by too much detail, do the mental exercise of erasing all the numbers. What then jumps out is that telling the fractional plateaus from the integral ones is impossible. The search for an analogy thus brings us to the IQHE. Can the well-understood integral quantum Hall effect teach us something about the mysterious fractional quantum Hall effect? As we see below, the answer is in the affirmative, and the relation between the two is profound and far-reaching, forming the very basis of our understanding of the FQHE. In hindsight, that is not surprising at all. The history of physics is replete with examples where analogies crop up in unexpected places. Here, the two phenomena, the integral and the fractional Hall effects, are so similar that the possibility of their being unrelated would be inconceivable.

In view of the remarkable parallel between the FQHE and the IQHE, it is tempting to postulate that the FQHE is really an IQHE in disguise. That raises the question: Integral quantum Hall effect of *what*? What objects in the FQHE are analogous to the electrons of the IQHE? These objects must be some kind of fermion, because, after all, the fermionic nature of the electrons is what makes integral fillings special. The pursuit of the analogy between the FQHE and the IQHE thus points inescapably to a new kind of fermionic particles in the FQHE.

---

electronic charge of $1/3$." As recounted in Ref. [627], D. C. Tsui exclaimed "Quarks!" upon first noticing (Fig. 1.2) that the $1/3$ plateau occurs at roughly three times the magnetic field as the $\nu = 1$ plateau.

5.7 Particles of condensed matter

The appearance of new particles in condensed matter physics may seem surprising at first, but it should not. A short digression is worthwhile to discuss what we mean by a "particle."[6] Of course, we know all the particles that go into the Hamiltonian of a condensed matter system, namely electrons and ions (only electrons for our problem), but of concern here are the particles that come out. A most profound fact of nature – indeed the very reason why physics can make progress at many different levels – is that strongly interacting particles reorganize themselves to become more weakly coupled particles of a new kind. These new particles are, in a deep sense, the "true" particles of the system in question, because it is reasonable to reserve the title "particle" for nearly independent objects. The old particles were the particles of the problem, and the new particles are the particles of the solution.

Identifying the relevant particles is equivalent to mapping the strongly correlated system of one kind of particles into a weakly coupled system of a different kind of particles. Mathematically, we have expressed the initial Hamiltonian as

$$H = H_0^{\text{new}} + V^{\text{new}}, \tag{5.17}$$

where H_0^{new} describes a system of certain independent particles (not necessarily the particles that define H). V^{new}, the interaction between these particles, can be treated perturbatively; it makes quantitative corrections, but does not change the qualitative behavior, i.e., does not cause a phase transition. Because theorists know what to do with weakly interacting particles – perhaps the only thing they really know – the problem is, in essence, solved as soon as such objects are identified. The qualitative phenomenology can be understood by switching off V^{new} altogether and working with a system of free particles. A quantitative understanding can be obtained by working sufficiently hard on perturbation theory. The primary goal is thus to identify the relevant particles.

At times, what objects are weakly interacting is intuitively obvious, and the choice of true particles is made without much fanfare. We know when to start with quarks and when with ^{4}He atoms. At other times, the identification of true particles may be subtle and nontrivial. Their rigorous derivation from first principles is generally not possible, especially when their formation entails nonperturbative physics. The clue often lies in the experimental phenomenology.

It is magical when the mysterious properties of a system suddenly become obvious, even inevitable, when viewed through the frame of reference of the new particles. Phenomena that were difficult or impossible to understand in terms of the original particles become simply comprehensible as properties of almost free particles.

In many familiar examples outside condensed matter physics, the new particles are simple bound states of old ones. The emergent particles of condensed matter, however, can be fantastically complex objects. Sometimes, only with a great deal of work can their existence be recognized and established. We cite here a few well-known examples.

---

[6] The views discussed in this section are widely appreciated in the condensed matter community. See, for example, Refs. [282,372].

- **Phonons** Viewing a crystal lattice as a system of almost free atoms makes no sense. The Einstein model considers a system of independent atoms, but with each atom trapped in a harmonic oscillator potential created by the other atoms. This model is also inadequate, for example, for the low-temperature specific heat, because all excitations have the same energy. The coupling between neighboring atoms cannot be neglected. If only terms that are of second order in displacement are kept, the problem can be solved by Fourier transformation, which brings us to the basis of phonons. Of course, phonons are also interacting, due to higher-order terms in the coupling, but the interaction between them is much weaker than that between atoms, and is negligible for most purposes.
- **Magnons** These are the particles of a magnet. Instead of individual spin flips, which are coupled strongly to nearby spin flips, we discuss the physics in terms of spin waves, or magnons, which are weakly interacting.
- **Landau quasiparticles** Electrons in a normal metal interact via the Coulomb interaction, which is strong. Nonetheless, a neglect of the interaction is a valid approximation for many quantities, e.g., the specific heat. The origin of such behavior remained a puzzle for many years, until Landau clarified that the weakly interacting objects are not electrons but "Landau quasiparticles," which are, roughly, screened electrons. Landau quasiparticles share their quantum numbers – charge, spin, and statistics – with electrons, and, in fact, are perturbatively connected to electrons. Here, in principle, it *is* possible to treat the electrons as the "particles," and to extract the correct physics in a perturbative treatment of the interaction. Nonetheless, Landau quasiparticles are better particles because they are more weakly interacting.
- **Cooper pairs** In superconductors, the phonon-mediated attractive interaction between electrons causes a nonperturbative rearrangement of the system. The particles of a superconductor are Cooper pairs. An understanding of the basic physics is obtained in terms of noninteracting, i.e., uncorrelated, Cooper pairs. In this case, the true particles (Cooper pairs) bear no resemblance to the constituent particles, namely electrons. The charge of a Cooper pair is $2e$, its spin is zero, and it is (crudely) a boson. This kind of qualitative distinction between the constituent and the true particles is perhaps the most direct sign that nonperturbative physics is in play.

One characteristic that makes the particles of condensed matter nontrivial is that they live only within the condensed matter background. We cannot pull a phonon out of a crystal, a magnon out of a magnet, or a Cooper pair out of a superconductor. (In the last case, we end up with two electrons, not a Cooper pair.) In that sense, the particles of condensed matter are more akin to quarks: they cannot live in isolation, but are perfectly legitimate particles by virtue of being weakly interacting (in some limit).

Another difference is that the particles of condensed matter physics are never *completely* noninteracting. That is not surprising. What is surprising is that out of the complexity emerge particles which are weakly interacting to the extent they often are.

What exactly is weakly interacting depends on the experimental phenomenon in question, and the energy with which the system is probed. For example, while it is appropriate to view the O_2 molecule as a particle when one has the properties of oxygen gas in mind, one must worry about electrons and nuclei separately in chemical reactions, and even about nucleons inside the the nucleus in high-energy collisions. A hierarchy of phenomena exists, with new particles associated with each level.

5.8 Composite fermion theory

We now return to the FQHE problem. What are the relevant fermions here? They are certainly not electrons, because a system of noninteracting electrons would only exhibit the IQHE. Motivated by the Cooper pairs of superconductivity, we might envision new fermions that are bound states of an odd number of electrons. However, the occurrence of such bound states appears unlikely because the interaction between electrons is strongly repulsive. A little further thought shows this scenario to be grossly inconsistent with the known phenomenology, thus ruling it out as a viable possibility.

The new fermions, which have the unfortunate[7] name "composite fermions" [272], turn out to be unlike any other particles known in nature. They are complex entities from the vantage point of electrons, yet they behave as more or less independent particles in many situations, with well-defined charge, spin, statistics, and other properties that we attribute to particles. This section elucidates the fundamental concepts of the composite fermion theory, and also alerts the reader to some common misinterpretations and misconceptions.

5.8.1 The definition of composite fermion

Topological objects known as "vortices" occur in many contexts in condensed matter physics. An order j vortex is defined by the property that a complete loop around it produces a change of $2j\pi$ in the phase of the wave function or the order parameter. Perhaps the best known example is the Abrikosov vortex [3], which occurs in type-II superconductors in the form of a vortex line. The order parameter of a superconductor is a complex function

$$\Delta(r) = |\Delta(r)|e^{i\chi(r)}. \tag{5.18}$$

For type-I superconductors the phase $\chi(r)$ is single valued. For type-II superconductors, on the other hand, a vortex line solution becomes relevant for which the phase of the order parameter changes by 2π around the vortex.

In the context of the quantum Hall liquid the term "vortex" is used in a slightly different sense. Because the magnetic field breaks time reversal invariance, the wave function is not real. In very high magnetic fields, when electrons are confined in the lowest Landau level, the wave function is a polynomial of $z_j = x_j - iy_j$, apart from a real Gaussian factor. We know from the property of complex functions that $z = re^{-i\theta}$ has a vortex at the origin, because a complete loop around the origin changes θ by -2π. Similarly, $z - z_0$ has a vortex at z_0. This is what is meant by the term vortex in the FQHE. We encounter below factors like $(z_1 - z_2)^{2p}$, in which particle one sees $2p$ vortices on particle two, and vice versa.

The vortices in superconductors and the FQHE are not identical. (i) The former refer to a structure in the order parameter, whereas the FQHE vortices occur in the microscopic wave function. (ii) The phase gradient around a vortex in a superconductor induces a macroscopic supercurrent circulating around it, which, in turn, produces an experimentally detectable

---

[7] The name "composite fermion" is used in a generic sense in atomic or high energy physics, e.g., for a proton or a ^{3}He atom. It now has acquired a definite meaning in condensed matter physics.

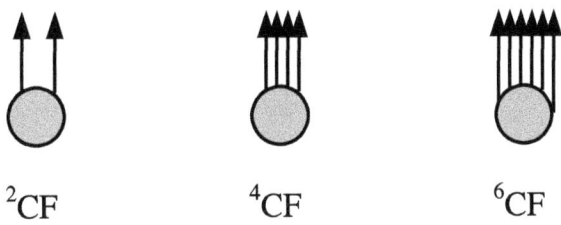

^{2}CF ^{4}CF ^{6}CF

Fig. 5.1. Schematic view of three flavors of composite fermion carrying (a) two, (b) four and (c) six vortices. These are denoted by ^{2}CF, ^{4}CF and ^{6}CF, respectively. Each arrow represents a quantized vortex. Vortices are modeled as flux quanta in a mean-field description.

magnetic "flux tube" carrying a quantized magnetic flux of magnitude $hc/2e$ ($2e$ is the charge of a Cooper pair). In contrast, no real flux is associated with the FQHE vortex; the magnetic field does not bunch near it, nor does any current circulate around it.

Now we are ready to define the composite fermion:

Definition   *A composite fermion is the bound state of an electron and an even number of quantized vortices.*

A pictorial representation of several flavors of composite fermions is shown in Fig. 5.1, where each arrow represents a vortex. Sometimes composite fermions are pictured as bound states of electrons and an even number of magnetic flux quanta (a flux quantum is defined as $\phi_0 = hc/e$), which is how they were first introduced [272].[8] This model for the composite fermion derives its justification from the property that a point flux quantum and a vortex are "topologically" similar, in the sense that they both produce the same winding phases. A vortex, by definition, produces a phase of 2π for a closed path around it, which is also the Aharonov–Bohm phase produced by a flux quantum for a closed electron loop encircling it. The electron–flux bound state, however, is only a crude model for the "true" composite fermion. It is intuitively useful because it provides a nice picture, but it must not be taken literally. No real fluxes are bound to electrons, and the physical magnetic field is uniform.[9]

5.8.2 From IQHE to FQHE: the mean-field approximation

This subsection walks us through the golden path connecting the integral quantum Hall effect to the fractional quantum Hall effect, following Ref. [272]. For illustration, we begin by considering the special filling factors $\nu^* = n$. The connection is established through the following steps:

Step I   Let us consider noninteracting electrons at $\nu^* = n$. The system is incompressible, i.e., the ground state is nondegenerate, separated from the other eigenstates by a gap (equal

---

[8] This is reminiscent of, and was inspired by, the ingenious idea of Leinaas–Myrheim–Wilczek braiding statistics [386, 677] (Section 9.8), wherein particles obeying fractional braiding statistics are modeled as fermions or bosons carrying a fictitious flux of magnitude $\alpha\phi_0$. As α is increased to an even integer, the statistics comes back to being fermionic or bosonic.

[9] The name "Chern–Simons fermion" is sometimes used for the electron–flux composite. We refer to both electron–flux and electron–vortex composites as composite fermions to avoid unnecessary proliferation of names.

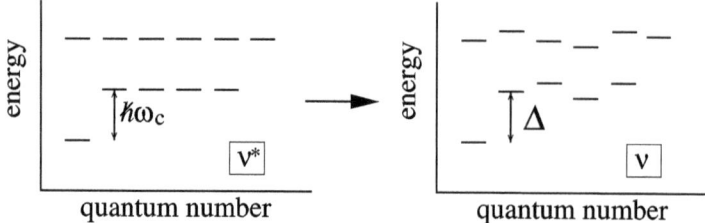

Fig. 5.2. The general structure of the energy spectrum of the many-body system at an integral filling, $v^* = n$. The ground state is nondegenerate, containing n Landau levels fully occupied (shown in Fig. 5.3). The excited states form bands separated by the cyclotron energy. The x-axis label is a convenient quantum number (wave vector in the periodic geometry, or the orbital angular momentum in the spherical geometry). The CF *mean-field* theory predicts that the low-energy spectrum at fractional fillings $v = n/(2pn \pm 1)$ has identical structure, except that states at v are only quasi-degenerate and the cyclotron gap evolves into a gap Δ (the determination of which requires a microscopic theory). Explicit calculations beyond the mean-field theory show that the one-to-one correspondence displayed in this figure is valid for the lowest band, but the mean-field theory predicts spurious states in higher bands at v.

to the cyclotron energy). The *many-particle* energy spectrum is shown in Fig. 5.2. The ground state has n full Landau levels, shown schematically in the left column of Fig. 5.3. The lowest energy excited state is a particle–hole pair, or an exciton, shown in the left column of Fig. 5.4(d) for $v^* = 3$. These diagrams have precise wave functions associated with them. We denote the magnetic field by B^*, which can be either positive or negative. It is related to the filling factor by $v^* = \rho\phi_0/|B^*| = n$.

The long-range rigidity in the system at an integral filling factor (manifested by the presence of the gap) is caused solely by the Fermi statistics. Thinking in the standard Feynman path-integral language [163, 575] is useful. The partition function gets contributions from all closed paths in the configuration space for which the initial and the final positions of electrons are identical, although the paths may involve fermion exchanges. Some examples are: a path in which one electron moves in a loop while others are held fixed; or a cooperative ring exchange path [342] in which the closed loop involves many electrons, with each electron moving to the position earlier occupied by the next one. The phase associated with each closed path has two contributions: the Aharonov–Bohm phase, which depends on the flux enclosed by the loop; and the statistical phase, which is $+1$ or -1, depending on whether the final state is related to the initial one by an even or an odd number of pairwise electron exchanges. The incompressibility at integral fillings is presumably caused by some special correlations built into the phase factors of various paths. These correlations are not easily identifiable in the path integral language, but we know they are present.

Step II Now we attach to each electron a massless, infinitely thin magnetic solenoid carrying $2p$ flux quanta pointing in the $+z$ direction. This converts electrons into composite fermions. The flux added in this manner is unobservable, because it does not alter the

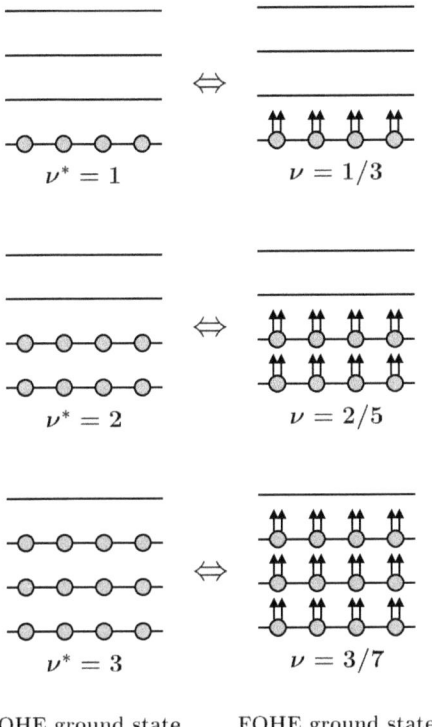

IQHE ground state   FQHE ground state

Fig. 5.3. Schematic view of the electron ground state at $\nu^* = n$ (left column). The columns on the right show the CF view of the ground states at $\nu = n/(2n + 1)$, as n filled Λ levels of composite fermions. (The Λ levels of composite fermions are analogous to Landau levels of electrons at ν^*, but lie within the lowest Landau level of electrons at ν.) Horizontal lines in the left column depict Landau levels of electrons, and those on the right depict Λ levels of composite fermions.

phase factors associated with any closed Feynman paths. The excess or deficit of an integral number of flux quanta through any closed path is physically unobservable, and the fermionic nature of particles guarantees that the phase factors of paths involving particle exchanges also remain intact. In other words, the new problem defined in terms of composite fermions is identical to the original problem of noninteracting electrons at B^*. We have thus transformed an incompressible state of electrons into an incompressible state of composite fermions.

Step III This exact reformulation prepares the problem for a mean-field approximation that was not available in the original language. Let us adiabatically (i.e., slowly compared to $\hbar/\Delta$, where Δ is the gap) spread the flux attached to each electron until it becomes a part of the uniform magnetic field.[10] (Because the initial state has a uniform electron density,

---

[10] This innovative mean-field theory was first introduced by Zhang, Hansson, and Kivelson [743] to map the FQHE at $\nu = 1/m$ into bosons at zero magnetic field, and also employed by Laughlin and collaborators [159, 371] to map a system of anyons (particles obeying fractional statistics; more in Section 9.8) into fermions or bosons in a magnetic field.

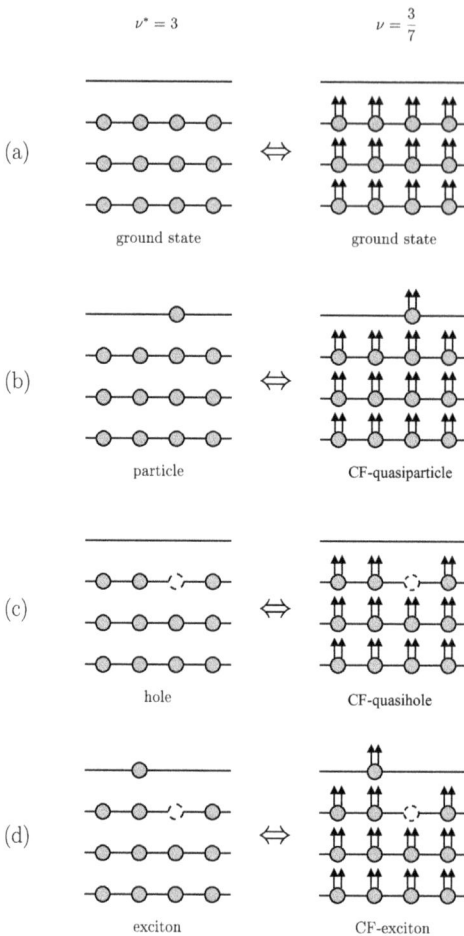

Fig. 5.4. Schematic view of the correspondence between the ground state and excitations at $\nu = 3/7$, viewed as the $\nu^* = 3$ of composite fermions, with the analogous states of electrons at $\nu^* = 3$. Shown on the right are: (a) the CF ground state; (b) a CF-quasiparticle; (c) a CF-quasihole; and (d) a CF-exciton. Horizontal lines in the left column depict Landau levels of electrons, and those on the right depict Λ levels of composite fermions.

the additional flux, tied to the density, produces a uniform magnetic field.) At the end, we obtain particles moving in an enhanced magnetic field B, given by

$$B = B^* + 2p\rho\phi_0. \tag{5.19}$$

The relation $|B^*| = \rho\phi_0/n$ implies that B is always positive (pointing in the $+z$ direction). The corresponding filling factor is given by

$$\nu = \frac{n}{2pn \pm 1}, \tag{5.20}$$

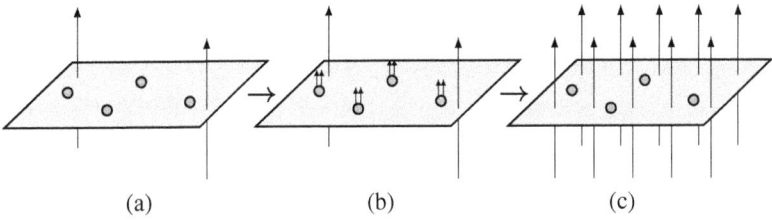

Fig. 5.5. The golden path from the IQHE to the FQHE. We begin with an IQHE state (a); attach to each electron two magnetic flux quanta to convert it into a composite fermion (b); and spread out the attached flux to obtain electrons in a higher magnetic field, which is a FQHE state (c).

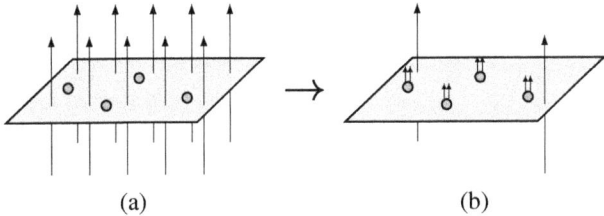

Fig. 5.6. Each electron captures two flux quanta to turn into a composite fermion. Composite fermions sense the residual magnetic field, which is much smaller than the applied magnetic field.

which follows from the relations $\nu = \rho \phi_0 / B$ and $\nu^* = n = \rho \phi_0 / |B^*|$. The $+ (-)$ sign in the denominator corresponds to B^* pointing in the $+z$ $(-z)$ direction.[11]

Let us now make the crucial assumption that the gap does not close during the flux diffusion process, i.e., there is no phase transition. To be sure, *quantitative* changes will occur. The gap and the wave functions will undergo a complex evolution. Nonetheless, if our assumption is correct, then Fig. 5.2 also represents, *qualitatively*, the spectrum at B.

The absence of a phase transition is an assumption that remains to be verified, and will surely not be valid for all n and p. If it is valid for some parameters, however, then the above construction gives a possible way of seeing how a gap can result at the fractions of Eq. (5.20). Three remarkable features already provide a strong hint that we are on the right track. First, these fractions are precisely the observed fractions. Second, they have odd denominators. Third, we naturally obtain sequences of fractions.

The three steps are depicted schematically in Fig. 5.5. The net effect, in a manner of speaking, is that each electron has absorbed $2p$ flux quanta from the external magnetic field to transform into a composite fermion. Composite fermions experience the residual magnetic field B^*. This is shown in Fig. 5.6. See Fig. 5.7 for a humorous portrayal of composite fermions.

Step IV Quantitative theory   The CF physics described above is sufficient for an explanation of much of the phenomenology of the FQHE. Can it also help us write microscopic wave functions for the FQHE state? As depicted in Fig. 5.8, solving directly for

---

[11] The fractions in Eq. (5.20) have been referred to as the Jain sequences or the Jain fractions in the FQHE literature.

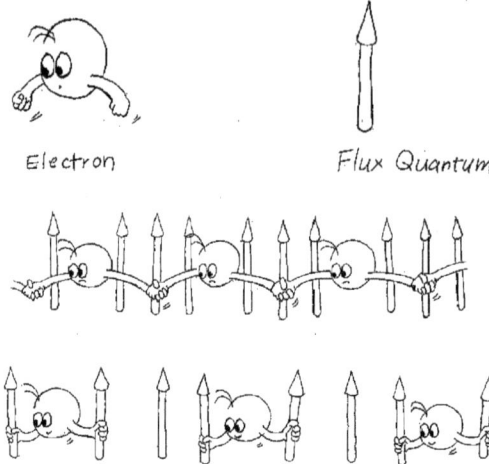

Fig. 5.7. A humorous view of composite fermions. Source: Kwon Park.

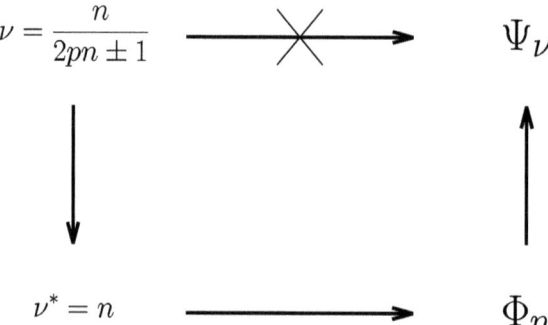

Fig. 5.8. The composite fermion route to the FQHE wave function Ψ_ν.

the wave function Ψ_ν of the FQHE state (the horizontal link) is not possible. However, the CF theory provides a new possible route. We have now mapped the problem of interacting electrons at $\nu = n/(2pn + 1)$ into that of weakly interacting fermions at $\nu^* = n$, the ground state of which is known (Φ_n). We now complete, following the reasoning of Ref. [272], the last leg of the route (from Φ_n to Ψ_ν in Fig. 5.8). What follows is not a rigorous derivation but an attempt to construct variational wave functions based on a physical picture. The wave functions obtained at the end must be confirmed by independent means. Some readers may find the alternative treatment in the next section more satisfying, where we first arrive at the wave functions by postulating composite fermions, and then derive from them B^* and ν^*.

Let us consider the mean-field description indicated in Fig. 5.5(b), in which particles sense a uniform external magnetic field B^* and also have $2p$ flux quanta tied to them. The vector potential $\boldsymbol{a}$, which binds flux quanta to electrons, is given by

$$\boldsymbol{a}(\boldsymbol{r}_i) = \frac{2p}{2\pi}\phi_0 \sum_j{}' \nabla_i \theta_{ij}, \tag{5.21}$$

where the prime denotes the condition $j \neq i$, and

$$\theta_{jk} = i \ln \frac{z_j - z_k}{|z_j - z_k|} \tag{5.22}$$

is the relative angle between particles j and k. Note that $\theta_{jk} = \theta_{kj} \pm (2m+1)\pi$, m integer, but below we only need the gradient of this angle, which satisfies

$$\nabla_j \theta_{jk} = \nabla_j \theta_{kj}. \tag{5.23}$$

As shown in Appendix C, $\boldsymbol{a}(\boldsymbol{r}_i)$ generates a magnetic field

$$\boldsymbol{b}_i = \nabla_i \times \boldsymbol{a}(\boldsymbol{r}_i) = 2p\phi_0 \sum_l{}' \delta^{(2)}(\boldsymbol{r}_i - \boldsymbol{r}_l), \tag{5.24}$$

i.e., each electron sees a flux tube of strength $2p\phi_0$ on every other electron. This suggests the following mean-field Hamiltonian for composite fermions in B^*:

$$H_{\mathrm{MF}} = \frac{1}{2m_{\mathrm{b}}} \sum_i \left(\boldsymbol{p}_i + \frac{e}{c}\boldsymbol{A}^*(\boldsymbol{r}_i) + \frac{e}{c}\boldsymbol{a}(\boldsymbol{r}_i) \right)^2, \tag{5.25}$$

where $\boldsymbol{A}^*$ produces a uniform magnetic field B^*. We assume that composite fermions are free; the current discussion is too crude for an explicit treatment of the interaction between particles.

Let us now consider the Schrödinger equation

$$H_{\mathrm{MF}}\Psi^{\mathrm{MF}} = E\Psi^{\mathrm{MF}}. \tag{5.26}$$

The vector potential $\boldsymbol{a}$ can be eliminated by making the gauge transformation

$$\Psi^{\mathrm{MF}} = \Phi\, e^{-i2p\sum_{j<k}\theta_{jk}} = \Phi \prod_{j<k} \left(\frac{z_j - z_k}{|z_j - z_k|} \right)^{2p}, \tag{5.27}$$

with Φ determined by

$$\frac{1}{2m_{\mathrm{b}}} \sum_i \left(\boldsymbol{p}_i + \frac{e}{c}\boldsymbol{A}^*(\boldsymbol{r}_i) \right)^2 \Phi = E\Phi. \tag{5.28}$$

This is the Schrödinger equation for noninteracting electrons at $\nu^* = n$, for which we know the solution. In particular, the ground state wave function, denoted by $\Phi_{\pm n}(B^*)$, is a single

Slater determinant with the lowest n Landau levels fully occupied, where the sign $\pm$ refers to the direction $\pm z$ for B^*. Because switching the direction of the magnetic field is equivalent to complex conjugation, we have, in general,

$$\Phi_{-\nu^*}(B^*) = \left[\Phi_{\nu^*}(B^*) \right]^* . \tag{5.29}$$

Thus, we obtain for the ground state at $\nu = n/(2pn \pm 1)$:

$$\Psi^{\mathrm{MF}}_{\frac{n}{2pn\pm1}}(B) = \Phi_{\pm n}(B^*) \prod_{j<k} \left(\frac{z_j - z_k}{|z_j - z_k|} \right)^{2p} . \tag{5.30}$$

This "mean-field" wave function is unsatisfactory for the following reasons [272]:

(i) It does not build favorable correlations between electrons. In fact, Ψ^{MF}_{ν} has the same probability amplitude as the uncorrelated IQHE wave function $\Phi_{\pm n}$. As seen in Table 5.1, the interaction energy for the mean-field wave function is significantly higher than that of the LLL Hartree–Fock Wigner crystal, which itself is not an especially good state in the filling factor region where the FQHE is seen.

(ii) Ψ^{MF}_{ν} also involves significant mixing with higher Landau levels. The wave function consists of large powers of $\bar{z}_j$'s through both the factors $\Phi_{\pm n}(B^*)$ and $|z_j - z_k|^{2p} = (z_j - z_k)^p (\bar{z}_j - \bar{z}_k)^p$. The kinetic energy per particle, shown in Table 5.1, is a measure of the amount of LL mixing.

(iii) At $\nu = 1/m$, m odd, Ψ^{MF}_{ν} does not reduce to the Laughlin wave function.

Reference [272] notes that these problems can be remedied, to a great extent, by throwing away the factor $|z_j - z_k|^{2p}$ in the denominator, which yields

$$\Psi_{\frac{n}{2pn\pm1}}(B) = \Phi_{\pm n}(B) \prod_{j<k} (z_j - z_k)^{2p} . \tag{5.31}$$

(As explained in Section 5.8.13, the wave function $\Phi_{\pm n}$ must now be evaluated at B in Eq. (5.31) to ensure the correct filling factor for Ψ.) The Jastrow factor now explicitly builds repulsive correlations between electrons, which results in a substantially lower interaction energy, as seen in Table 5.1. At the same time, admixture with higher Landau levels is substantially reduced (Table 5.1), although not completely eliminated. For many quantitative purposes, a wave function strictly confined to the lowest Landau level is very useful. Such a wave function is obtained by an explicit projection of Eq. (5.31) into the lowest Landau level:

$$\Psi_{\frac{n}{2pn\pm1}} = \mathcal{P}_{\mathrm{LLL}} \Phi_{\pm n} \prod_{j<k} (z_j - z_k)^{2p} , \tag{5.32}$$

where $\mathcal{P}_{\mathrm{LLL}}$ denotes projection into the lowest Landau level (more details in Section 5.14). These wave functions[12] do not contain any adjustable parameters (for a given filling factor),

---

[12] Wave functions of the form given in Eqs. (5.31) and (5.32) have been referred to in the literature as the Jain wave functions.

Table 5.1. *Interaction and kinetic energies for various trial wave functions*

ν	1/3	2/5	3/7	4/9
V^{MF}	−0.3619(9)	−0.3848(16)	−0.3947(15)	−0.4007(16)
$E_{\mathrm{K}}^{\mathrm{MF}}$	0.335(4)	0.408(9)	0.430(7)	0.459(17)
V^{up}	−0.4098	−0.4489(10)	−0.4644(20)	−0.4734(15)
$E_{\mathrm{K}}^{\mathrm{up}}$	0	0.0403(6)	0.0575(15)	0.0701(36)
V^{p}	−0.4098	−0.4328	−0.4423	−0.4474
$E_{\mathrm{K}}^{\mathrm{p}}$	0	0	0	0
V^{WC}	−0.3885	−0.4130	−0.4225	−0.4275
$E_{\mathrm{K}}^{\mathrm{WC}}$	0	0	0	0

Notes: V and E_K per particle for the mean-field wave function, the Laughlin wave function (Eq. 5.16) at $\nu = 1/3$, and unprojected and projected wave functions of Eqs. (5.31) and (5.32) at $\nu = 2/5$, 3/7 and 4/9.

The energies of the mean-field, unprojected, and projected wave functions are labeled by superscripts "MF," "up" and "p," respectively.

As a reference, the interaction energy of the LLL Hartree–Fock Wigner crystal state, V^{WC}, is also given.

The interaction energy V is quoted in units of $e^2/\epsilon\ell$ and includes the contribution from the positively charged background. The kinetic energy E_K is expressed in units of the cyclotron energy.

The zero point energy $\hbar\omega_c/2$ has been subtracted from the kinetic energy, so E_K is zero for a strictly LLL wave function. E_K is a measure of LL mixing in the wave function.

We note that the kinetic energy per particle for Φ_n is $E_K = (n - 1)/2$. All energies represent thermodynamic limits; the uncertainty in the last digit(s) from Monte Carlo sampling and extrapolation to $N^{-1} \to 0$ is shown in parentheses.

Source: Liquid energies are from Kamilla and Jain [313,314]; some of the numbers are improvements on the earlier work by Trivedi and Jain [644], Levesque *et al.* [389] and Morf and Halperin [455]. The interaction energy of the Hartree–Fock Wigner crystal state is taken from Levesque, Weiss and MacDonald [389] (using the interpolation formula given in Ref. [363]).

because $\Phi_{\pm n}$ are parameter free.[13] They can be generalized to excited states (Eqs. 5.60, 5.61, 5.62, 12.35, 12.37), to partially spin polarized FQHE (Eq. 11.59), to arbitrary fractions (Eq. 5.36), to bilayer systems (Eq. 13.7), and to quantum dots (Eq. 6.4). Additionally, wave functions have also been constructed for paired states of composite fermions (Eq. 7.11) and composite fermion crystals (Eq. 15.8).

We see in Section 5.14 that the unprojected and projected wave functions of Eqs. (5.31) and (5.32) differ in quantitative detail, but describe the same qualitative physics. Many essential features of the unprojected wave function survive projection into the LLL. The

---

[13] These wave functions are sometimes called "variational" wave functions, which might appear to be a misnomer in view of the lack of adjustable parameters. The variational freedom here is the form of the wave function. Once the form is fixed by the CF physics, the wave function is fully determined.

projected wave function is used for quantitative calculations, but the unprojected wave function is useful for a derivation of the universal properties of the CF state.

Composite-fermionization Constructing a wave function Ψ from a given wave function Φ according to Eq. (5.32) is called composite-fermionization. It involves multiplication by an appropriate Jastrow factor, followed by projection into the lowest Landau level.

Vortex vs. flux tube We earlier defined composite fermions in two somewhat different ways: as bound states of electrons and flux quanta, and as bound states of electrons and quantized vortices. In going from Eq. (5.30) to Eq. (5.31), we have gone from the former to the latter. The factor

$$\prod_{j<k} \left(\frac{z_j - z_k}{|z_j - z_k|} \right)^{2p} \tag{5.33}$$

binds, by construction, a *point* flux of strength $2p\phi_0$ to each electron. On the other hand, the Jastrow factor

$$\prod_{j<k} (z_j - z_k)^{2p} \tag{5.34}$$

binds $2p$ *vortices* to each electron. (More precisely, each electron sees $2p$ vortices on every other electron.) Throwing away the denominator converts the flux tubes into vortices.[14] In other words, we are postulating that flux quanta turn into vortices during the adiabatic process in going from Fig. 5.5(b) to 5.5(c), and electron–flux composites evolve into electron–vortex composites.

5.8.3 Generalization to arbitrary fillings

The "law of corresponding states" [276] motivated by the above mean-field approximation can be generalized to an arbitrary filling v^* as follows [115, 272, 276]:

(i) We begin with noninteracting electrons at a magnetic field B^* at a filling factor $n + 1 > v^* > n$. The energy spectrum now looks like that shown in Fig. 5.9. It contains many degenerate many-body ground states corresponding to distinct configurations of electrons in the partly occupied $(n + 1)$th Landau level. These are denoted $\Phi^\alpha_{\pm v^*}$, where α labels different states. Excited states are separated by at least one unit of cyclotron energy.

(ii) We take *all* degenerate ground states $\Phi^\alpha_{\pm v^*}$, and attach to each electron a massless, infinitely thin magnetic solenoid carrying $2p$ flux quanta, pointing in the $+z$ direction.

(iii) The attached flux is slowly spread until it turns into a uniform magnetic field. States will now mix with one another, if allowed by symmetry, and the degeneracy will be lifted in some complicated

---

[14] Strictly speaking, the factor in Eq. (5.33) also produces $2p$ vortices, in that it produces a phase of $2p \times 2\pi$ for a closed loop around it. It is customary in the FQHE literature to reserve the term "vortex" for factors of the type given in Eq. (5.34), and refer to Eq. (5.33) as attaching (point) "flux quanta."

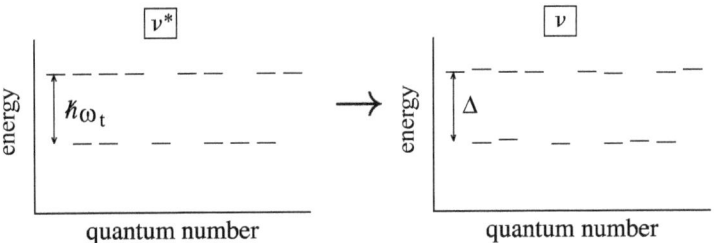

Fig. 5.9. Schematic structure of the energy spectrum of the many-body system at a nonintegral filling, $v^* \neq n$; it contains many degenerate ground states (separated from excited states by the cyclotron energy). The CF theory predicts that the low-energy spectrum at fractional fillings $v = v^*/(2pv^* \pm 1)$ has a band of *quasi*-degenerate ground states that has a one-to-one correspondence with the degenerate ground states at v^*. The last statement in the caption of Fig. 5.2 applies here as well.

fashion. But it is possible that the gap separating these states from the rest will not close. At the end we obtain particles moving in a larger magnetic field $B = B^* + 2p\rho\phi_0$, corresponding to

$$v = \frac{v^*}{2pv^* \pm 1}. \tag{5.35}$$

As before, if no phase transition occurs (i.e., if the gap does not close), then the "band structure" at v^* carries over to v, implying, as depicted in Fig. 5.9, a one-to-one correspondence between the low-energy states at v^* and v.

(iv) The wave function for each final state is constructed by composite-fermionization (cf. Eq. 5.32):

$$\Psi^\alpha_{\frac{v^*}{2pv^* \pm 1}} = \mathcal{P}_{\mathrm{LLL}} \Phi^\alpha_{\pm v^*} \prod_{j<k} (z_j - z_k)^{2p}. \tag{5.36}$$

Now we end up with many wave functions. These constitute a correlated basis for the low-energy band of quasi-degenerate "ground states" at v. A determination of the fine structure within the ground band requires a diagonalization of the full Hamiltonian in this basis, referred to as "CF diagonalization."

5.8.4 CF diagonalization

The wave functions and energies for low-energy states at any given v are obtained by the method outlined in Fig. 5.10. We choose the value $2p$ that gives the largest v^*. The distinct degenerate ground states at v^* (for noninteracting electrons) form a low-energy band spanned by a basis $\{\Phi^\alpha_{\pm v^*}\}$ (with magnetic field pointing in $\pm z$ direction). A basis for low-energy states at v, $\{\Psi^\alpha_v\}$, is constructed by composite-fermionization of $\{\Phi^\alpha_{\pm v^*}\}$, as in Eq. (5.36). These states are in general not orthogonal, but an orthogonal basis can be generated through the standard Gram–Schmidt method. The spectrum at v is obtained by diagonalizing the *full* Hamiltonian in the space defined by correlated basis functions $\{\Psi^\alpha_v\}$. This entire set of steps is termed "composite fermion diagonalization."

The LLL projection of basis functions and the evaluation of the Hamiltonian matrix are technically challenging, but mathematically well defined, and CF diagonalization has been

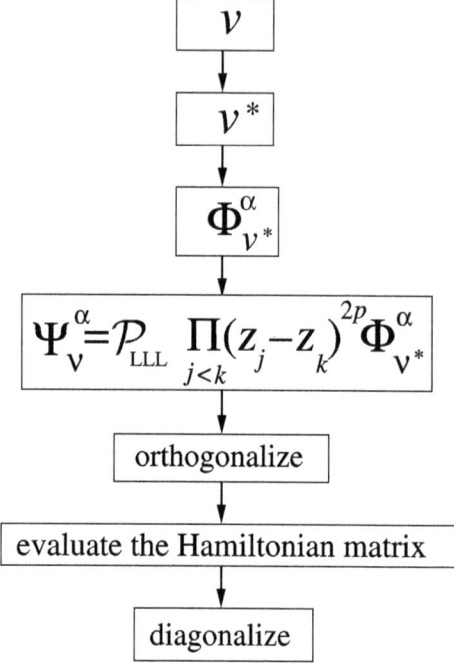

Fig. 5.10. CF diagonalization.

explicitly carried through for many nontrivial cases. Various multi-dimensional integrals required for this purpose do not reduce to known functions catalogued in standard books. But that is merely a technical point and should not be taken to indicate a gap in our understanding. The integrals are well defined, and can be evaluated numerically, usually by the Metropolis Monte Carlo method, to produce energies to desired accuracy (typically four or five significant figures) without making any approximations. Systems with as many as 100 composite fermions have been studied. See Appendix L for further technical details.

5.8.5 Λ *levels and energy level diagrams*

For each initial state at ν^* in the left column of Figs. 5.3 and 5.4, the corresponding final state at ν is depicted by the diagram in the right column. Each diagram on the right has a precise wave function associated with it, obtained by composite-fermionization of the wave function of the corresponding state in the left column.

The CF theory suggests the following interpretation: We picture that composite fermions form Landau-*like* levels in the reduced magnetic field B^*. These are called "Λ levels."[15] While Λ levels are analogous to Landau levels of electrons at B^*, the two are not the

---

[15] They have also been called "composite fermion Landau levels," "pseudo-Landau levels," or "quasi-Landau levels." These names have been a source of confusion, however, because the term "Landau level" has the universally accepted meaning as

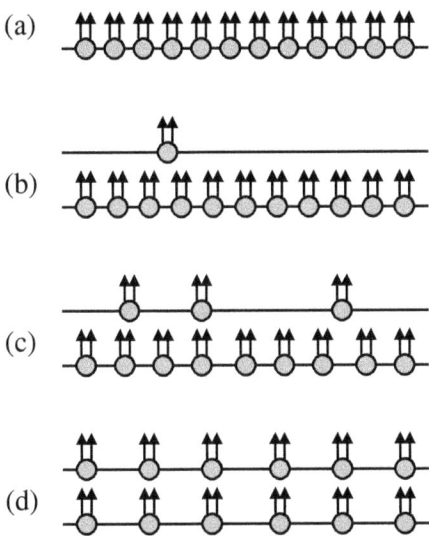

Fig. 5.11. Schematic view of the evolution of the state from $\nu = 1/3$ to $\nu = 2/5$. (The filling factor is changed by varying the magnetic field, altering the degeneracy of each Λ level.) (a) The 1/3 state, i.e., one filled Λ level of composite fermions. (b) A CF-quasiparticle, i.e., a composite fermion in an otherwise empty Λ level. (c) An intermediate state. (d) The 2/5 ground state, which is equivalent to two filled Λ levels. While the ground states at 1/3 and 2/5 are nondegenerate, many quasi-degenerate ground states exist at intermediate fillings, which correspond to different configurations of composite fermions in the second Λ level in (b) and (c); for example, the degeneracy of the CF-quasiparticle is equal to the degeneracy of the second Λ level.

same; the Λ levels reside within the lowest Landau level. In other words, when interacting electrons at ν transform into composite fermions, the lowest Landau level splits into Λ levels. In particular, the fractions $\nu = n/(2pn \pm 1)$ of Eq. (5.20) map into integral fillings of composite fermions; the ground state here fills an integral number of Λ levels (Fig. 5.3) and excited states are CF excitons (Fig. 5.4). Figure 5.11 depicts how the system evolves as the filling factor changes from $\nu = 1/3$ to $\nu = 2/5$.

The right hand side of the projected wave function in Eq. (5.32) or (5.36) is interpreted as describing composite fermions at filling ν^*. At the same time, it is also a wave function for interacting electrons at ν. There is no inconsistency. Such a "dual" interpretation is not dissimilar to that for the BCS wave function for superconductors: The BCS wave function describes electrons, but, since electrons always appear paired, it is also interpreted as a pair wave function. Similarly, because an even number of vortices are bound to each electron in

---

the quantized kinetic energy level of an *electron* in a magnetic field. Because the kinetic energy levels of composite fermions are a crucial and nontrivial concept in the theory of the FQHE, they deserve their own name and we call them "Λ levels." The term "Landau level" refers exclusively to the familiar electronic Landau level in this book.

the wave function of Eq. (5.32) or (5.36), it can be viewed as a wave function for composite fermions.

5.8.6 A new derivation of the Laughlin wave function

We now show how the Laughlin wave function can be "derived" from the CF theory. For the ground state at $\nu = \frac{1}{2p+1}$, the projected wave function of Eq. (5.32) reduces to

$$\Psi_{\frac{1}{2p+1}} = \mathcal{P}_{\text{LLL}} \prod_{j<k} (z_j - z_k)^{2p} \Phi_1. \tag{5.37}$$

With

$$\Phi_1 = \prod_{j<k} (z_j - z_k) \exp\left[-\frac{1}{4} \sum_i |z_i|^2 \right], \tag{5.38}$$

it reduces to

$$\Psi_{\frac{1}{2p+1}} = \mathcal{P}_{\text{LLL}} \prod_{j<k} (z_j - z_k)^{2p+1} \exp\left[-\frac{1}{4} \sum_i |z_i|^2 \right], \tag{5.39}$$

which is identical to the Laughlin wave function of Eq. (5.16) with $m = 2p + 1$. (The projection operator $\mathcal{P}_{\text{LLL}}$ does not do anything because the wave function is already in the lowest Landau level.) This was part of the motivation for throwing away the denominator in the mean-field wave function of Eq. (5.30), leading to the unprojected wave function of Eq. (5.31). Written as in Eq. (5.37), the wave function acquires the physical interpretation of one filled Λ level of composite fermions. The form in Eq. (5.37) reveals a close connection between the FQHE and the IQHE, gives new insight into the excitations at $\nu = 1/(2p+1)$ by analogy to the known excitations of the one filled Landau level state (Sections 12.5 and 12.6), and shows that this wave function is part of a more general structure.[16] The ground state wave function at $\nu = 1/(2p+1)$ happens to have a particularly simple form because of the fortunate coincidence that Φ_1 has a simple Jastrow form. Although written in a compact form, the wave functions of Eq. (5.32) are, in general, extremely complicated after projection into the lowest Landau level, and could not have been arrived at without guidance from the composite fermion principle.

5.8.7 Remarks

(i) No rigorous theoretical method is known at present for implementing the adiabatic flux delocalization process to test directly whether or not the gap closes during the process. Several facts suggest that this is likely to be a complicated issue. Whether the final state is incompressible or not depends on the detailed form of the interaction. At any given filling factor, say $\nu = 2/5$, the system can be incompressible for some interactions (say,

---

[16] The simple equation $m = 2p + 1$ contains much physics.

the interaction corresponding to electrons in the lowest Landau level) but compressible for others (e.g., for the interaction that simulates electrons in the 17th Landau level). Evolution of the Landau level gap (the cyclotron energy) at integral fillings into the Λ level gap (sometimes called the "CF-cyclotron energy") is also bound to be a messy problem, because the two have different parametric dependences. The former is proportional to B and inversely proportional to the band mass of the electron. The latter is determined entirely by the Coulomb interaction, the only energy scale in the LLL problem; it is proportional to $e^2/\ell \sim \sqrt{B}$ and contains no electron mass.

Our philosophy will be not to worry about the adiabatic *process* of going from ν^* to ν, but directly to make statements at ν that can be tested by independent means. Various energies and energy gaps can be calculated directly at ν with the help of projected wave functions of Eqs. (5.32) and (5.62) for ground and excited states. Because these are restricted, by construction, to the lowest Landau level, their energies are guaranteed to have the correct units of e^2/ℓ.

(ii) The CF vortex-attachment construction can produce new *possible* incompressible QHE states from any given incompressible QHE state. Its most important use lies in generating new FQHE states from the known *integral* QHE states. Starting from the *fractional* QHE states, we can produce yet more incompressible states.

(iii) We have described above the original version of the composite fermion mean-field approximation [272]. A more powerful formulation, pioneered by Lopez and Fradkin [401], is described in Section 5.16.

5.8.8 The composite fermion principle

Although the mean-field approach of the previous subsection serves as inspiration for writing wave functions, and also illustrates much of the physics, some of its aspects are not literally correct. It is instructive to see how the entire theory follows from the assumption of composite fermions. The logical order in this subsection is approximately the reverse of what we had in the previous subsection: the wave functions appear first, from which B^* and ν^* are deduced.

The fundamental postulate *Strongly interacting electrons turn into weakly interacting composite fermions, where a composite fermion is a bound state of an electron and an even number of quantized vortices.*

The microscopic meaning of the formation of composite fermions is that the wave function of interacting electrons at ν has the form

$$\Psi_\nu = \Phi \prod_{j<k} (z_j - z_k)^{2p}, \tag{5.40}$$

where Φ is an antisymmetric wave function for electrons. The Jastrow factor, $\prod_{j<k} (z_j - z_k)^{2p}$, binds $2p$ vortices to each electron to convert it into a composite fermion.

Equation (5.40) has no content as it stands, because *any* wave function can be written in this form at the expense of making Φ sufficiently complicated. We further demand that Φ

be the wave function for *weakly interacting* electrons. That is the precise meaning of the statement that composite fermions are weakly interacting. In other words, the interaction enters only through the Jastrow factor, that is, only through the formation of composite fermions.

One more observation completes the basic conceptual story. Section 5.11 derives by several methods,[17] starting from the form in Eq. (5.40), that Φ is a wave function for electrons at ν^* given by Eq. (5.35). With Φ replaced by $\Phi_{\pm\nu^*}$, Eq. (5.40) reduces to the earlier unprojected wave function. In Section 5.11 we also see a better way of understanding why the magnetic field sensed by composite fermions is B^* rather than B: the Berry phase due to the bound vortices partly cancels the Aharonov–Bohm phase due to the external magnetic field, with the result that composite fermions experience a reduced magnetic field

$$B^* = B - 2p\rho\phi_0. \tag{5.41}$$

This also implies the filling factor ν^* given by Eq. (5.35). We thus arrive at the following corollary of the fundamental postulate:

Corollary *Composite fermions experience a reduced magnetic field B^* and fill ν^* Landau-like levels.*

5.8.9 CF theory in a nutshell

The next three paragraphs summarize the defining statements of the CF theory. The first two list the *exact* consequences of the CF theory; these are sufficient for an understanding of the universal phenomenology of the FQHE and many other phenomena. The third paragraph gives a mathematical formulation of composite fermions, which enables quantitative calculations.

The physics The CF theory postulates the formation of composite fermions. This is a nonperturbative consequence of the repulsive interactions, because the number of vortices, quantized to be an even integer, cannot be changed continuously. (Wave functions with nonintegral values of p are not valid wave functions for electrons.) The CF theory further asserts that composite fermions are weakly interacting, and that their interaction can be neglected for many qualitative purposes. The most important outcome of the formation of composite fermions is that the vortices bound to them produce Berry phases that partly cancel the Aharonov–Bohm phase due to the external magnetic field. Consequently, composite fermions experience a reduced magnetic field $B^* = B - 2p\rho\phi_0$. They form Landau-like levels, called Λ levels, in the reduced magnetic field, and their filling factor ν^* is given by $\nu = \nu^*/(2p\nu^* \pm 1)$.

Qualitative consequences for the energy spectrum The mapping of strongly interacting electrons at B (ν) into weakly interacting composite fermions at B^* (ν^*) implies a one-to-one correspondence between the low-energy spectra of the two. The CF theory

---

[17] The reason for postponing the proofs to Section 5.11 is that some of the derivations use wave functions in the spherical geometry, introduced in Section 5.9.

thus predicts formation of bands in the energy spectrum of interacting electrons at ν, and further asserts that the lowest band resembles the lowest band of noninteracting fermions at ν^*. That implies incompressibility at the the fractions $\nu = \frac{n}{2pn\pm1}$.

Quantitative theory   The low-energy spectrum is obtained quantitatively by the method of CF diagonalization, outlined in Section 5.8.4 and Fig. 5.10.

The CF theory thus tells us how nature takes advantage of the enormous degeneracy to produce new nonperturbative structures. The fundamental effect of the repulsive interaction between electrons is to produce composite fermions; the interaction between composite fermions themselves is much weaker and can be neglected altogether to a good first approximation.[18] The formation of bands and incompressibility at certain special fractions are nonperturbative consequences of the formation of composite fermions. All LLL states are degenerate in the absence of interaction, and a gap opens as soon as the interaction is switched on, no matter how small its strength. (One may manipulate the interaction strength by varying the dielectric function of the host material, but that merely affects the overall energy scale.)

The correspondence between interacting electrons at ν and weakly interacting composite fermions at ν^* is immensely powerful and beautifully constraining. For the important special cases of the fractions $\nu = n/(2pn \pm 1)$, it not only predicts incompressibility but picks out unique, parameter-free wave functions for the ground states out of the astronomically large number of possibilities. The lack of adjustable parameters may appear to be a serious deficiency at first sight, raising doubts regarding the applicability of these wave functions to the actual states. It, however, turns out to be one of the strengths of the CF theory. Aside from the obvious aesthetic appeal of a parameter-free theory, it enables an unbiased testing of the CF theory (Chapter 6). Substantial simplification is achieved away from the special fractions also, because the dimension of the CF basis is exponentially small compared to the dimension of the full LLL Fock space at ν. To take an example, the number of degenerate ground states for 100 particles at $\nu^* = 5/2$ is smaller by a factor of 2×10^{58} compared to that at $\nu = 5/12$ (with spin neglected and the interaction switched off).

5.8.10 Why composite fermions form

Why electrons *might* want to capture vortices to turn into composite fermions is easy to see. A typical wave function satisfying the Pauli principle vanishes as r when two particles approach one another, r being the distance separating them, but the unprojected wave functions $\Phi\Phi_1^{2p}$ vanish as r^{2p+1}, with the Jastrow factor contributing $2p$ to the exponent and Φ contributing the rest. The Jastrow factor is very effective in keeping particles apart from one another, producing favorable correlations. These correlations should survive,

---

[18] (a) This is analogous to the BCS theory of superconductivity. There, Cooper pairs are produced due to an attractive interaction between electrons. Cooper pairs themselves are taken as noninteracting, however, as seen from the fact that the BCS wave function has *all* Cooper pairs in the same state, as appropriate for uncorrelated bosons.
(b) Bound states are always more weakly interacting than the constituent particles, because much of the interaction has been "spent" in making the bound states. Some familiar examples are: atoms, neutrons, Cooper pairs.

to some extent, projection into the lowest Landau level, given that the unprojected wave functions are already predominantly in the lowest Landau level.

5.8.11 $1 = 1 + 1$

A consequence of the formation of composite fermions is a dynamical generation of a new energy scale. Two energy scales emerge out of one. The Coulomb interaction energy of electrons splits into a "large" effective cyclotron energy and a "small" residual inter-CF interaction. Both have the same units $(e^2/\epsilon\ell)$, however, and their relative strengths cannot be changed for a given filling factor.[19]

5.8.12 Caveats

While the systems of composite fermions at ν^* and noninteracting electrons at ν^* share many qualitative features, they are not identical in every respect, and one must be careful not to overinterpret their correspondence. We note two instances where a naive application of the CF theory leads to a wrong conclusion.

The energy of a particle–hole pair of *electrons* at $\nu^* = n$ is equal to the cyclotron energy $\hbar eB^*/m_b c = \hbar eB/(2pn + 1)m_b c$. (We have used $B^* = B/(2pn + 1)$ for $\nu = n/(2pn + 1)$.) That is *not* the energy of a particle–hole pair of composite fermions at $\nu^* = n$. The latter must be proportional to $e^2/\ell \sim \sqrt{B}$, and also independent of the electron band mass. The analogy between electrons at ν^* and composite fermions at ν^* extends to the *structure* of the low-energy spectrum, but not to the energies. The energies are to be determined by CF diagonalization.

The second example is that of the Hall resistance. The mapping of interacting electrons at ν into free composite fermions at ν^* might suggest that the Hall resistance is $R_H = h/\nu^* e^2$. That is obviously incorrect; for a disorder-free system at ν, one can derive $R_H = h/\nu e^2$, independent of interactions, by the general trick of boosting to a moving frame of reference of velocity $v = cE/B$. The resolution of this apparent paradox is given in Section 5.16.3.

5.8.13 Another representation of wave functions

In Eq. (5.31), (5.32), or (5.36) all of the Gaussian factors are incorporated into the definition of $\Phi_{\nu^*}^\alpha$ by evaluating it at the *external* magnetic field *but at the effective filling factor*. Multiplication by $\prod_{j<k}(z_j - z_k)^{2p}$ expands the size of the disk containing electrons, reducing the density and producing the filling factor ν.

An alternative form for the wave functions of Eq. (5.36) is given by

$$\Psi_\nu^\alpha = \mathcal{P}_{\text{LLL}}\Phi_{\pm\nu^*}^\alpha \Phi_1^{2p}. \tag{5.42}$$

---

[19] That should be contrasted with the IQHE, where the Coulomb and the cyclotron energies can be varied independently relative to one another by controlling various parameters (e.g., the electron mass, the dielectric constant, or the density).

Here Φ_1 is the wave function of one filled Landau level:

$$\Phi_1 = \prod_{j<k}(z_j - z_k)\exp\left[-\frac{1}{4\ell_1^2}\sum_i |z_i|^2\right], \tag{5.43}$$

where ℓ_1 is the magnetic length at $\nu = 1$. The wave function $\Phi_{\pm\nu*}^\alpha$ in Eq. (5.42) is constructed at the effective magnetic field B^*, i.e., has the Gaussian factor $\exp[-\frac{1}{4\ell^{*2}}\sum_i |z_i|^2]$. The Gaussian factors in Φ_1^{2m} and $\Phi_{\nu*}^\alpha$ combine to produce a Gaussian factor corresponding to the external magnetic field B, since

$$\frac{2p}{\ell_1^2} + \frac{1}{\ell^{*2}} = \frac{1}{\ell^2}. \tag{5.44}$$

(This is nothing but the formula $2p\rho\phi_0 + B^* = B$.) In Eq. (5.42), multiplication by Φ_1^{2p} leaves the size of the disk unchanged (Exercise 5.2), but increases the magnetic field, thereby decreasing the filling factor.

Of course, this is only a difference of representation; the wave function in Eq. (5.42) is identical to that in Eq. (5.36). We use either form below; it should be clear from the context whether $\Phi_{\pm\nu*}^\alpha$ is to be evaluated at the effective or the real magnetic field. The form in Eq. (5.42) will be more natural when we generalize the wave function to the spherical geometry.

5.8.14 Simplifying notation

In Eqs. (5.35) and (5.36) we have defined ν^* to be positive. Remembering that complex conjugation is equivalent to the reversal of magnetic field direction, we have

$$\Phi_{-\nu*}^\alpha = [\Phi_{+\nu*}^\alpha]^*. \tag{5.45}$$

We sometimes use a convention in which ν^* can take either positive or negative values, with the understanding that negative values refer to magnetic field in the $-z$ direction. That removes the $\pm$ signs in Eqs. (5.35) and (5.36). We sometimes also suppress the superscript α, and the subscripts ν and ν^*, remembering that Ψ refers to interacting electrons at ν and Φ to noninteracting electrons at ν^*. In the simplest form, the equations reduce to

$$\Psi = \mathcal{P}_{\text{LLL}}\Phi\prod_{j<k}(z_j - z_k)^{2p} = \mathcal{P}_{\text{LLL}}\Phi\Phi_1^{2p} \tag{5.46}$$

$$\nu = \frac{\nu^*}{2p\nu^* + 1}. \tag{5.47}$$

5.9 Wave functions in the spherical geometry

Generalizing to the spherical geometry, the wave function for interacting electrons at Q, Ψ_Q, is obtained from that of noninteracting electrons at Q^*, Φ_{Q^*}, as

$$\Psi_Q = \mathcal{P}_{LLL}\Phi_{Q^*}\Phi_1^{2p}. \tag{5.48}$$

The Jastrow factor is given by

$$\Phi_1^{2p} = \prod_{j<k}(u_j v_k - v_j u_k)^{2p}, \tag{5.49}$$

where Φ_1 is the wave function of the lowest filled Landau level (Eq. 3.158).

5.9.1 Relationship between Q and Q^*

The monopole strength Q is related to Q^* by the property that the monopole strength of the product is the sum of monopole strengths. The monopole strength for Φ_1 is $Q_1 = (N-1)/2$, because the LLL degeneracy here is $2Q_1 + 1 = N$. Adding monopole strengths for factors on the right hand side of Eq. (5.48) gives

$$Q = Q^* + p(N-1). \tag{5.50}$$

5.9.2 Quantum numbers

In the spherical geometry, eigenstates of the many-particle system are also eigenstates of the orbital angular momentum, with the angular momentum quantum number denoted by L. One of the most important properties of the wave function $\Psi_Q = \Phi_{Q^*}\Phi_1^{2p}$, with or without LLL projection, is that it has the same angular momentum quantum numbers as Φ_{Q^*}. (The same is true for spin quantum numbers, as we see in Chapter 11.) In other words, composite-fermionization of a wave function does not alter its angular momentum quantum numbers.

Proof Let us assume that Φ_{Q^*} is an eigenstate of the total angular momentum operators L^2 and L_z with eigenvalues $L(L+1)$ and M. Write

$$L^2 = L_z^2 + \frac{1}{2}(L_+^2 + L_-^2) \tag{5.51}$$

where $L_+ = L_x + iL_y$ and $L_- = L_x - iL_y$. The wave function Φ_1 has $L = 0$, so it satisfies

$$L_z\Phi_1 = L_+\Phi_1 = L_-\Phi_1 = 0. \tag{5.52}$$

Noting that all L_z, L_+, and L_- involve at most first-order derivatives, we can commute them through the factor Φ_1. Thus,

$$L_z\Phi_1^2\Phi_{Q^*} = \Phi_1^2 L_z\Phi_{Q^*} = M\Phi_1^2\Phi_{Q^*} \tag{5.53}$$

and

$$L^2 \Phi_1^2 \Phi_{Q^*} = \Phi_1^2 L^2 \Phi_{Q^*} = L(L+1) \Phi_1^2 \Phi_{Q^*}. \tag{5.54}$$

To prove that the angular momentum is not altered upon projection into the lowest Landau level, we write the projection operator, following Rezayi and MacDonald [548], as

$$\mathcal{P}_{\text{LLL}} = \prod_{i=1}^{N} \mathcal{P}_{\text{LLL}}^i \tag{5.55}$$

$$\mathcal{P}_{\text{LLL}}^i = \prod_{l=Q+1}^{\infty} \frac{l(l+1) - L_i^2}{l(l+1) - Q(Q+1)} \tag{5.56}$$

where L_i^2 is the angular momentum operator for the ith electron. That $\mathcal{P}_{\text{LLL}}$ is the projection operator is seen by noting that $\mathcal{P}_{\text{LLL}}^i$ produces a zero when applied to any single particle state in a higher Landau level, and one when applied to any state in the lowest Landau level. Given that the total angular momentum operators L^2 and L_z commute with the L_i^2 of an individual electron, they also commute with the projection operator $\mathcal{P}_{\text{LLL}}$. □

5.9.3 Wave functions for filled Λ levels

The filling factor $\nu = n/(2pn+1)$ of electrons corresponds to $\nu^* = n$ of composite fermions. The wave function for the state containing n filled Λ levels (Fig. 5.3) is given by

$$\Psi_{\frac{n}{2pn+1}} = \mathcal{P}_{\text{LLL}} \Phi_n \Phi_1^{2p}, \tag{5.57}$$

where Φ_n is the Slater determinant wave function of n filled Landau levels of electrons. In the spherical geometry, n Landau levels are fully occupied when $2Q^* = N/n - n$. The corresponding state $\Psi_{n/(2pn+1)}$ occurs at the real monopole strength

$$2Q = \frac{(2pn+1)}{n} N - (2p+n). \tag{5.58}$$

(In general, this equation does not have a solution for all N. Only those integral N values are meaningful for which Q is either an integer or half-integer.) In the thermodynamic limit, $N \to \infty$, the expected filling factor is obtained:

$$\nu \equiv \frac{N}{2Q} = \frac{n}{2pn+1}. \tag{5.59}$$

5.9.4 Wave functions for CF-quasiparticle and CF-quasihole

Excitations of the FQHE state at $\nu = n/(2pn \pm 1)$ are understood by analogy to excitations of the IQHE state at $\nu^* = n$. For the latter, adding an electron to the lowest unoccupied Landau level produces a "particle," with charge equal to the charge of an electron. Similarly, the removal of an electron from the topmost occupied Landau level creates a "hole." Let us denote the wave functions of these states by Φ_n^{qp} and Φ_n^{qh}. The left column of Fig. 5.4 depicts (a) the ground state, (b) a particle, and (c) a hole for $\nu^* = 3$.

The images of IQHE particle and hole are called CF-quasiparticle and CF-quasihole, shown schematically in the right column of Fig. 5.4 for $\nu^* = 3$ of composite fermions ($\nu = 3/7$). A CF-quasiparticle is a solitary composite fermion in an otherwise empty Λ level, and a CF-quasihole is the state in which a single composite fermion is missing from an otherwise full Λ level. Their wave functions are given by

$$\Psi_{\frac{n}{2pn+1}}^{\text{CF-qp}} = \mathcal{P}_{\text{LLL}} \Phi_n^{qp} \Phi_1^{2p} \tag{5.60}$$

and

$$\Psi_{\frac{n}{2pn+1}}^{\text{CF-qh}} = \mathcal{P}_{\text{LLL}} \Phi_n^{qh} \Phi_1^{2p}. \tag{5.61}$$

These do not contain any adjustable parameters.

5.9.5 Wave function for CF-exciton

A CF-exciton is a particle–hole pair of composite fermions, also shown schematically in Fig. 5.4. Microscopic wave function for a CF-exciton is constructed by analogy to the wave function Φ_n^{ex} of the exciton of the IQHE state:

$$\Psi_{\frac{n}{2pn+1}}^{\text{CF-ex}} = \mathcal{P}_{\text{LLL}} \Phi_n^{\text{ex}} \Phi_1^{2p}. \tag{5.62}$$

In the spherical geometry, Φ_n^{ex} is taken to be an eigenstate of the total orbital angular momentum. Its wave function can be written straightforwardly, because it is effectively a two-particle problem. Let us denote the orbital angular momentum quantum numbers of an electron by $(\bar{l}, m)$, with $\bar{l} = q + n - 1$ for the topmost occupied LL shell, q being the monopole strength and n being the number of filled shells. Let us denote by $\Phi^{(m,m')}$ the Slater determinant for the state in which an electron is removed from the orbital $(\bar{l}, m)$, creating a hole with quantum numbers $(\bar{l}, -m)$, and placed in the orbital $(\bar{l} + 1, m')$. The z component of the total angular momentum can be chosen to be zero with no loss of generality, so we set $m = m'$. The exciton state with a definite total angular momentum L is then given by

$$\Phi_{q,L}^{\text{ex}} = \sum_m \langle \bar{l}, -m; \bar{l} + 1, m | L, 0 \rangle \Phi^{(m,m)}, \tag{5.63}$$

with $m = -\bar{l}, \ldots, \bar{l}$. The wave function for the CF-exciton with a well defined L is then constructed as

$$\Psi_{Q,L}^{\text{CF-ex}} = \mathcal{P}_{\text{LLL}} \Phi_{q,L}^{\text{ex}} \Phi_1^{2p}, \tag{5.64}$$

with $Q = q + p(N - 1)$. $\Psi_{Q,L}^{\text{CF-ex}}$ contains no adjustable parameters. The relative amplitudes of various Slater determinants ("coherence factors") remain unchanged in going from the electron-exciton at q to the CF-exciton at Q according to the preceding displayed equation.

5.10 Uniform density for incompressible states

The filled Λ level states have uniform electron density. That is most easily proven in the spherical geometry. Translational invariance of the planar geometry is equivalent to rotational invariance in the spherical geometry. Consequently, the states with angular momentum quantum number $L = 0$ have uniform density, independent of microscopic details of the wave function. The state Φ with n filled Landau levels has $L = 0$. It follows that the n filled Λ level state also has $L = 0$, and, therefore, a uniform density.

No similarly simple proof is available for the planar geometry. One reason is that for any finite N, the state, strictly speaking, does *not* have uniform density, because it represents electrons inside a disk of finite radius, with the density vanishing outside. Here, we need to show that the density is uniform inside the disk. For the Laughlin wave function, a mapping into a one-component plasma allows demonstration of uniform density (Section 12.4). The density has also been calculated for the wave functions in Eq. (5.32) numerically, explicitly demonstrating that they describe a state with a constant density in the interior. Of course, that is hardly surprising in light of the result from the spherical geometry.

5.11 Derivation of v^* and B^*

We now derive, as promised, the relation between v and v^* (Eq. 5.35) and that between B and B^* (Eq. 5.41) directly from the wave function of Eq. (5.36). The two relations are equivalent; one can be derived from the other with the help of

$$v^* = \frac{\rho\phi_0}{|B^*|} \tag{5.65}$$

and

$$v = \frac{\rho\phi_0}{B}. \tag{5.66}$$

Here, B is always assumed to be positive; B^* can be either parallel (for the $+$ sign) or antiparallel (for the $-$ sign) to B.

5.11.1 Derivation I

In the spherical geometry, the CF and electron filling factors are given by

$$v^* = \lim_{N\to\infty} \frac{N}{2|Q^*|}, \tag{5.67}$$

$$v = \lim_{N\to\infty} \frac{N}{2|Q|}. \tag{5.68}$$

Substituting in $Q = Q^* + p(N-1)$ produces Eq. (5.35). The relation between B^* and B can be derived by noting that

$$B = \frac{2Q\phi_0}{A}, \tag{5.69}$$

where $2Q\phi_0$ is the flux through the sphere and A is the surface area. Similarly,

$$B^* = \frac{2Q^*\phi_0}{A}. \tag{5.70}$$

Substituting in Eq. (5.50) and taking the limit $N \to \infty$ gives Eq. (5.41), with $\rho = N/A$.

5.11.2 Derivation II

In planar geometry the simplest way of determining the filling factor of the wave function on the right hand side of $\Psi_\nu = \Phi_{\nu^*} \prod_{j<k}(z_j - z_k)^{2p}$ is to ask what is the total number of single particle orbitals with nonzero occupation. Neglecting $O(1)$ corrections, it is given by the largest power of one of the coordinates, say z_1, in the polynomial part of the wave function. In Φ_{ν^*} this power is N/ν^*, as required to give the filling factor ν^*. In the Jastrow factor, the largest power of z_1 is $2p(N-1)$. The largest power of z_1 in Ψ, therefore, is $N\nu^{*-1} + 2p(N-1)$, which yields the filling factor

$$\nu = \frac{N}{N\nu^{*-1} + 2p(N-1)} = \frac{\nu^*}{2p\nu^* + 1}. \tag{5.71}$$

5.11.3 Derivation III

In planar geometry, the filling factor of *uniform* density states is related to the total angular momentum. Remembering that each factor of z_j in the polynomial in front of the Gaussian factor contributes $+1$ to the total angular momentum, and each factor of $\bar{z}_j$ contributes -1, the total angular momentum is given by the difference between the total power of z's and the total power of $\bar{z}$'s. We neglect $O(1)$ corrections below, which do not affect the thermodynamic value of the filling factor.

In a uniform state at ν, each orbital is occupied with a probability ν, with a total of N/ν orbitals occupied within a disk defined by the outermost orbital with angular momentum $\nu^{-1}N - 1$. The total angular momentum is given by

$$L = \sum_{m=0}^{\nu^{-1}N-1} f_m m$$

$$= \frac{N^2}{2\nu}, \tag{5.72}$$

where $f_m = \nu$ is the average occupation of the angular momentum m orbital, and subdominant terms have been dropped in the last step. This gives the relation $\nu = N^2/2L$.

The total angular momentum of $\Psi_\nu = \mathcal{P}_{LLL} \prod_{j<k}(z_j - z_k)^{2p}\Phi_{\pm\nu^*}$ can be calculated before projection, because projection into the lowest Landau level conserves the angular

momentum. Φ_{v^*} contributes $N^2/2v^*$, as required to produce the correct filling, and the Jastrow factor contributes $pN(N-1)$. Thus, the total angular momentum of Ψ_v is

$$L = pN(N-1) \pm \frac{N^2}{2v^*}, \tag{5.73}$$

where we have used that complex conjugation changes the sign of the angular momentum. (Φ_{v^*} also has some powers of $\bar{z}$ but the largest power is finite, limited by the number of occupied Landau levels, and, therefore, does not affect the filling factor in the thermodynamic limit.) The filling factor of Ψ_v is given by

$$v = \frac{N^2}{2L} = \frac{v^*}{2pv^* \pm 1}. \tag{5.74}$$

5.11.4 Derivation IV

We now derive the effective magnetic field experienced by a composite fermion from a Berry phase calculation. For this purpose, we first show that when a composite fermion, i.e., an electron along with its vortices, is taken in a closed loop enclosing an area A (in the counterclockwise direction), it acquires a Berry phase

$$\Phi^* = -2\pi \frac{BA}{\phi_0} + 2\pi 2pN_{\text{enc}}, \tag{5.75}$$

where N_{enc} is the number of composite fermions inside the loop. The first term is the familiar Aharonov–Bohm (AB) phase due to a charge going around in a loop. The second, as proven in the following paragraph, is the Berry phase due to the $2p$ vortices going around N_{enc} particles, with each particle producing a phase of 2π.

The Berry phase of a vortex is derived following Arovas, Schrieffer, and Wilczek [13]. A vortex at

$$\eta = R e^{-i\theta} \tag{5.76}$$

is defined by the wave function

$$\Psi_\eta = N_R \prod_j (z_j - \eta)\Psi, \tag{5.77}$$

where Ψ is the wave function for the incompressible ground state in question. Electrons avoid the point η, creating a hole there, which has a positive charge relative to the incompressible state. We have explicitly included the normalization factor N_R, which depends on the amplitude of η, but can be chosen to be independent of the angle θ. Let us now take η in a circular loop of radius R by slowly changing θ from 0 to 2π, while holding

R constant. The Berry phase (Appendix E) associated with this path is given by

$$\gamma = \oint dt \left\langle \Psi_\eta | i \frac{d}{dt} \Psi_\eta \right\rangle$$

$$= \oint d\theta \left\langle \Psi_\eta | i \frac{d}{d\theta} \Psi_\eta \right\rangle$$

$$= \oint d\theta (-i) \frac{d\eta}{d\theta} \left\langle \Psi_\eta | \sum_j \frac{1}{z_j - \eta} | \Psi_\eta \right\rangle$$

$$= \oint (-i) d\eta \int d^2r \frac{1}{z - \eta} \left\langle \Psi_\eta | \hat{\rho}(r) | \Psi_\eta \right\rangle$$

$$= 2\pi \int_{r<R} d^2r \, \rho_\eta(r)$$

$$= 2\pi N_{enc}. \tag{5.78}$$

In the above, we have used: $z = x - iy$, $r = (x, y)$, $\hat{\rho}(r) = \sum_j \delta^{(2)}(r_j - r)$, $\rho_\eta(r) = \langle \Psi_\eta | \hat{\rho}(r) | \Psi_\eta \rangle$, and N_{enc} is the number of particles inside the closed loop. Because of the definition $z = x - iy$, the residue for a contour integral differs from the usual by a sign; this can be seen, for example, by evaluating $\oint d\eta / 2\pi i \eta = -1$ for a conterclockwise contour. The Berry phase of a vortex thus simply counts the number of particles inside the loop, with each particle contributing 2π.

For uniform density states, we replace N_{enc} in Eq. (5.75) by its average value ρA, where ρ is the electron or the CF density (this is a mean-field approximation), and equate the entire phase to the AB phase due to an effective magnetic field, B^*:

$$\Phi^* = -2\pi \left(\frac{BA}{\phi_0} - 2p\rho A \right) \equiv -2\pi \frac{B^* A}{\phi_0}. \tag{5.79}$$

Composite fermions thus experience an effective magnetic field given by

$$B^* = B - 2p\rho\phi_0. \tag{5.80}$$

In essence, *the Berry phases originating from the vortices bound to electrons partly cancel the AB phase of the external magnetic field*, and as a consequence, composite fermions behave as though they were in a much smaller effective magnetic field. The field B^* can be antiparallel to B.

The relative sign in Eq. (5.75)   The two terms on the right hand side of Eq. (5.75) come with opposite signs, leading to a partial *cancellation* of the magnetic field. In other words, the phases due to vortices are equivalent to an effective magnetic field pointing in the $-z$ direction. This can also be seen by eliminating the phases of the Jastrow factor in favor of a vector potential. Let us consider the Schrödinger equation

$$\left[\frac{1}{2m_b} \sum_i \left(p_i + \frac{e}{c} A(r_i) \right)^2 + V \right] \prod_{j<k} (z_j - z_k)^{2p} \Phi_{\nu^*} = E \prod_{j<k} (z_j - z_k)^{2p} \Phi_{\nu^*}, \tag{5.81}$$

where V is the interaction. Our focus in the following is on the kinetic energy term. It is convenient for our purpose to display the phases due to the Jastrow factor explicitly:

$$\prod_{j<k}(z_j - z_k)^{2p} = e^{-i2p\sum_{j<k}\theta_{jk}} \prod_{j<k} |z_j - z_k|^{2p},$$ (5.82)

where θ_{jk} was defined in Eq. (5.22). We have been careful to keep track of the definition $z = re^{-i\theta}$, as appropriate for an external magnetic field in the $+z$ direction. Using

$$\boldsymbol{p}_i e^{-i2p\sum_{j<k}\theta_{jk}} = e^{-i2p\sum_{j<k}\theta_{jk}} \left(\boldsymbol{p}_i - 2p\hbar\sum_{j}{}' \nabla_i\theta_{ij} \right),$$ (5.83)

where the prime denotes the condition $j \neq i$, we write

$$\left[\frac{1}{2m_b} \sum_i \left(\boldsymbol{p}_i + \frac{e}{c}\boldsymbol{A}(\boldsymbol{r}_i) - \frac{e}{c}\boldsymbol{a}(\boldsymbol{r}_i) \right)^2 + V \right] \prod_{j<k} |z_j - z_k|^{2p} \Phi_{\nu*}$$

$$= E \prod_{j<k} |z_j - z_k|^{2p} \Phi_{\nu*},$$ (5.84)

with $\boldsymbol{a}$ defined in Eq. (5.21). The additional vector potential $(-\boldsymbol{a})$ simulates the effect of the phases of the Jastrow factor. Using the formulas in Appendix C, the corresponding magnetic field is

$$\boldsymbol{b}_i = \nabla_i \times (-\boldsymbol{a}_i) = -2p\phi_0 \sum_l{}' \delta^{(2)}(\boldsymbol{r}_i - \boldsymbol{r}_l).$$ (5.85)

Thus, the phase of the Jastrow factor is equivalent to each electron seeing a flux tube of strength $-2p\phi_0$ on every other electron; the minus sign indicates that the flux tube points in the $-z$ direction, opposite to the direction of the external field $\boldsymbol{B}$.

The singular vector potential does not take care of *all* of the phases in the wave function, as $\Phi_{\nu*}$ has additional vortices and antivortices. What about their effect? That is incorporated in the CF theory by saying that composite fermions fill ν^* Λ levels. Also, the Jastrow factor does more than produce flux tubes. Flux tubes are only a result of the phase part of the Jastrow factor; the amplitude $\prod_{j<k} |z_j - z_k|^{2p}$ builds repulsive correlations between electrons.

5.12 Reality of the effective magnetic field

The picture in Fig. 5.6 ought not to be taken literally. No real magnetic flux binds to electrons. The external magnetic field does not bundle into individual flux quanta as it passes through the 2D surface (as it does in type-II superconductors). The currents required to create a localized flux would incur a prohibitively high kinetic energy cost. An external magnetometer will measure a uniform magnetic field of magnitude B. The effective magnetic field is a purely quantum mechanical effect, produced because the Berry phases due to vortices partly cancel the Aharonov–Bohm phases. The effective magnetic field is fully internal to composite fermions.

If the effective magnetic field could not be measured, a legitimate question would be: "Is it real or a mere mathematical construct?" Fortunately, it *can* be measured, except that composite fermions themselves must be used to detect it. Experiments to be described in subsequent chapters have shown that composite fermions indeed experience the magnetic field B^* rather than B, a fact that has numerous striking observable consequences. It has also been confirmed extensively (Chapter 6) that the low-energy spectrum of interacting electrons in the lowest Landau level resembles, in a model independent manner, the low-energy spectrum of noninteracting fermions at B^*.

5.13 Reality of the Λ levels

Intimately connected to the reality of B^* is the formation of Λ levels. How do we know that the lowest Landau level really splits into Λ levels of composite fermions, which are analogous to the electronic Landau levels in the effective magnetic field?

We stress a fundamental theoretical distinction between Λ levels and Landau levels. Landau levels are the solutions of the single electron problem, starting from which we build many-particle states. The Λ levels are more subtle and complex because they are an emergent structure, and provide a single particle-like interpretation for the inherently many-particle states in the lowest Landau level. A derivation of Λ levels for a single composite fermion is not possible (a single composite fermion is itself a nonsensical notion); the only way of convincing ourselves of the existence of Λ levels is to postulate them, and then to put their consequences to the test.

Subsequent chapters show that, in spite of the "theoretical" difference, Λ levels manifest themselves in experiments with amazing similarity to Landau levels. The FQHE is understood as a direct consequence of the formation of Λ levels just as the IQHE is of Landau levels. Various excitations across Λ levels have been observed. Computer experiments show higher LL-like structures appearing within the lowest Landau level, and also establish a direct connection between the LLL wave functions at ν with wave functions involving many Landau levels at ν^*.

5.14 Lowest Landau level projection

This section has three parts. It begins with a brief discussion of the need for LLL projection. The subsequent part addresses in what sense the CF physics survives projection into the lowest Landau level. Then we learn how the projection is performed.

5.14.1 Why projection?

In the limit of $B \rightarrow \infty$, only the lowest Landau level is relevant. However, the FQHE can, and indeed does, occur at finite B. Since some LL mixing is always present at a finite B, this proves that the quantization of Hall resistance is invariant under at least a small amount of admixture with higher Landau levels.

Although not necessary for the FQHE, the limit $B \to \infty$ is very convenient for several reasons. First, it enormously simplifies computer experimentation; the dimension of the relevant Fock space is finite because of the LLL constraint, allowing us to obtain *exact* results, which, in turn, enable a rigorous and unbiased testing of any theoretical ideas. Second, a reliable treatment of LL mixing is an extra complication that distracts from the physics of the FQHE, and is, therefore, best dispensed with when the primary concern is to establish the basic principle of the phenomenon. Stated differently, the $B \to \infty$ limit simplifies the problem by eliminating one parameter (the electron band mass) from the problem. Finally, nonuniversal quantities (e.g., the excitation gaps) calculated in the limit $B \to \infty$ often provide decent approximations for the experimentally measured quantities.

As seen above, the unprojected wave functions of Eq. (5.31) have some amplitude in higher Landau levels. We wish to adiabatically deform them to produce wave functions that reside entirely in the lowest electronic Landau level. How can that be accomplished? Let us note that these wave functions are already predominantly in the lowest Landau level. A measure of LL mixing in the unprojected wave function is the kinetic energy per particle, which is in general small, as shown in Table 5.1. Eliminating the part that does not reside in the lowest electronic Landau level is called "LLL projection." Two methods for LLL projection are outlined below, with further details in Appendix J.

The unprojected wave functions of Eq. (5.31) do not describe the physics of LL mixing. They have a fixed amount of LL mixing, whereas the LL mixing of the actual state is a function of the magnetic field. Even for an investigation of the effects of LL mixing, it is appropriate to begin with the projected wave functions and then incorporate admixture with higher Landau levels through an additional Jastrow factor [219, 434, 435, 529, 567]. Melik-Alaverdian and Bonesteel [434] have considered a mixture of projected and unprojected wave functions to treat LL mixing, which amounts to a partial projection.[20] Güçlü *et al.* [218, 219] have shown that a moderate amount of hybridization with higher Landau levels leaves the phases of the LLL wave function unchanged to a very good approximation; therefore, the fixed phase diffusion Monte Carlo offers a reliable method for dealing with LL mixing, provided we have in our possession an accurate LLL wave function.

5.14.2 Composite fermions in the lowest Landau level

The physics of the formation of composite fermions and the effective magnetic field is most transparent in the unprojected wave function $\Phi_{\nu*}\Phi_1^{2p}$, but becomes obscure after LLL projection. This is best illustrated by specializing to the filling factor range $\nu > 1/3$. To satisfy antisymmetry, a wave function confined to the lowest Landau level must have the form (apart from the Gaussian) $\prod_{j<k}(z_j - z_k)^{2p+1}F_S[\{z_i\}]$, where the polynomial $F_S[\{z_i\}]$ is symmetric under an exchange of any two coordinates. This wave function has $\nu \leq 1/(2p+1)$. Therefore, for $\nu > 1/3$, the wave function contains only a single power of

---

[20] This is akin to the well-known Gutzwiller projection in studies of the Hubbard model.

the factor $\prod_{j<k}(z_j - z_k)$, implying that each electron binds *one and only one* vortex. That seems, naively, to exclude the formation of composite fermions in the lowest Landau level in this filling factor range.

Yet, the CF physics undoubtedly survives in the lowest Landau level. The defining property of composite fermions, namely that they experience an effective magnetic field, has been established within the lowest Landau level, both in real and computer experiments (subsequent chapters), and forms the basis for the explanation of many phenomena.

How do we reconcile the two preceding paragraphs? By appealing to the powerful principle of adiabatic continuity of physics. Let us ask what happens to the vortices of the unprojected wave function when it is projected into the lowest Landau level. The *two* essential pieces in the unprojected wave function, the Jastrow factor and Φ_{ν^*}, are, *together*, interpreted as composite fermions at filling ν^*. It should be stressed, however, that the factor Φ_{ν^*} itself contains many vortices *and* antivortices. When the full wave function is projected, the vortices from the Jastrow factor and the vortices and the antivortices from Φ_{ν^*} get all mixed up (with all antivortices annihilated by vortices). The two pieces of physics thus get entangled beyond recognition. Nonetheless, unless a phase transition (or a level crossing) takes place during the act of projection, the CF physics is guaranteed to survive projection into the lowest Landau level.

This, in fact, brings out the beauty of the CF theory. The LLL wave functions of the FQHE states are, in general, so complicated that understanding their physics is a formidable challenge. The CF theory recognizes that they are LLL projections of certain simpler wave functions, from which the CF physics can be "read off." While composite fermions become effectively disguised when we project the wave functions into the lowest Landau level, the qualitative consequences of composite fermions' existence remain sharply recognizable.

It may help the reader to recall other, well-known examples where the physics of a complex system is understood by its adiabatic continuity to a simpler system. When the interaction between fermions is switched on in a Landau Fermi liquid, free fermions are renormalized into quasiparticles that are complex, many-body objects; their mass can sometimes be modified by two orders of magnitude, as in heavy fermion systems. Nonetheless, so long as no phase boundary is crossed, the physics of the interacting system is qualitatively similar to that of the free system, apart from a renormalization of parameters. Another example is that of superconductivity. In the limit of strong attractive interaction, electrons form tightly bound Cooper pairs, which are viewed as bosons, and superconductivity as their Bose–Einstein condensation. As the interaction strength is slowly reduced, the BEC state evolves into the Bardeen–Cooper–Schrieffer state in the other limit of weak attractive interactions. Spatially bound pairs of electrons are no longer identifiable, but the qualitative physics of the strong coupling limit continues to hold.

5.14.3 Use of higher Landau levels in the LLL physics

We should not be distressed by the use, at intermediate steps, of higher Landau levels, even if we were only interested in the $B = \infty$ physics. An accepted strategy in condensed matter physics is to enlarge the Hilbert space (sometimes into unphysical directions, as done in slave-boson, large-N, large-S and large-d approaches), hope that the physics is more transparent there, obtain an approximate solution, project back into the original space, and pray that the essential physics survives. To be sure, there is nothing wrong in using higher Landau levels – LL mixing is at least not unphysical. We can also directly ascertain the validity of the final solution against exact results; after all, that is what ultimately counts.

Most importantly, the use of higher Landau levels is not just a technical tool but reveals the emergence of Λ level structure within the lowest Landau level, which has numerous experimental manifestations.

5.14.4 Projection methods

The term "LLL projection" refers to the step which produces a LLL wave function starting from an unprojected wave function of the form $\Psi_\nu = \Phi_{\nu*} \Phi_1^{2p}$. The factor $\Phi_{\nu*}$ is in general a linear superposition of Slater determinants. The following theorems are useful for LLL projection.

Theorem 5.1 The product of two LLL wave functions is also in the lowest Landau level (although at a different magnetic field).

The reader is asked to verify this statement using explicit wave functions for the disk and spherical geometries. A simple corollary is that Φ_1^{2p} lies in the lowest Landau level. Therefore, the basic term that we need to project into the lowest Landau level is the product of a higher LL single particle wave function (from the expansion of $\Phi_{\nu*}$) and a LLL wave function (from Φ_1^{2p}).

Theorem 5.2 In spherical geometry, the LLL projection of the product of a LLL wave function and a higher LL wave function can be written as

$$\mathcal{P}_{\text{LLL}} Y_{Qlm}(\mathbf{\Omega}) \Psi_{Q'Q'm'}(\mathbf{\Omega}) = \hat{Y}_{Qlm}^{Q'}(\mathbf{\Omega}) \Psi_{Q'Q'm'}(\mathbf{\Omega}), \tag{5.86}$$

where $\Psi_{Q'Q'm'}$ is an arbitrary LLL wave function at monopole strength Q' and $\hat{Y}_{Qlm}^{Q'}(\mathbf{\Omega})$ is an m' independent operator. Similarly, in the disk geometry, the LLL projection is given by

$$\mathcal{P}_{\text{LLL}} \eta_{nm}(r) \eta_{0,m'} = \hat{\eta}_{nm}(r) \eta_{0,m'}, \tag{5.87}$$

where $\eta_{0,m'}$ is an arbitrary LLL wave function.

We can, of course, project any single particle wave function into the lowest Landau level. The point of this theorem is that the operator $\hat{Y}_{Qlm}^{Q'}$ or $\hat{\eta}_{nm}$ accomplishes LLL projection independent of the LLL state it acts upon.[21]

---

[21] $\hat{Y}_{Qlm}^{Q'}$ depends on Q' but that is not a problem because, for the LLL projections of our interest, it acts upon Φ_1^{2p}, which has a fixed $Q' = p(N - 1)$. The superscript is often omitted.

The explicit expression for $\hat{Y}_{Qlm}$ is given in Appendix J; for now it will be sufficient to know that it exists. The operator $\hat{\eta}$ for disk geometry is simpler, and we show here that it is given by

$$\hat{\eta}_{nm}(\bar{z}, z) =: \eta_{nm}\left(\bar{z} \rightarrow 2\frac{\partial}{\partial z}, z\right) :, \tag{5.88}$$

where the normal ordering of a function (denoted by two colons ::) amounts to moving $\bar{z}$ to the left of z. (We also follow the convention that the derivatives do not act upon the Gaussian factor.) This follows because

$$\mathcal{P}_{\mathrm{LLL}}\bar{z}^n z^m \mathrm{e}^{-\frac{1}{4}z\bar{z}} = \mathrm{e}^{-\frac{1}{4}z\bar{z}}\left(2\frac{\partial}{\partial z}\right)^n z^m. \tag{5.89}$$

Proof One way to proceed is to write

$$\bar{z}^n z^m \,\mathrm{e}^{-\frac{1}{4}z\bar{z}} = a_0 z^{m-n} \,\mathrm{e}^{-\frac{1}{4}z\bar{z}} + \text{higher LL terms}, \tag{5.90}$$

and determine a_0 by multiplying both sides by $\bar{z}^{m-n}\mathrm{e}^{-\frac{1}{4}z\bar{z}}$ and integrating over the position. The answer can be obtained without performing Gaussian integrals:

$$\int \mathrm{d}^2\boldsymbol{r}\, \bar{z}^{m-n}\bar{z}^n z^m \mathrm{e}^{-\frac{1}{2}z\bar{z}} = \int \mathrm{d}^2\boldsymbol{r}\, \bar{z}^{m-n}z^m \left(-2\frac{\partial}{\partial z}\right)^n \mathrm{e}^{-\frac{1}{2}z\bar{z}} \tag{5.91}$$

$$= \int \mathrm{d}^2\boldsymbol{r}\, \bar{z}^{m-n}\left[\left(2\frac{\partial}{\partial z}\right)^n z^m\right]\mathrm{e}^{-\frac{1}{2}z\bar{z}}, \tag{5.92}$$

which demonstrates that the projection of $z^{m-n}\,\mathrm{e}^{-\frac{1}{4}z\bar{z}}$ on $\bar{z}^n z^m \,\mathrm{e}^{-\frac{1}{4}z\bar{z}}$ is the same as its projection on $\mathrm{e}^{-\frac{1}{4}z\bar{z}}\left(2\frac{\partial}{\partial z}\right)^n z^m$. □

A wave function strictly in the lowest Landau level has no powers of $\bar{z}$ in the polynomial part. The preceding discussion clarifies that even when the wave function $\Phi \prod_{j<k}(z_j - z_k)^{2p}$ contains powers of $\bar{z}$, it has a nonzero amplitude in the lowest Landau level. In fact, it may be predominantly in the lowest Landau level [272, 275]. To project $\Phi \prod_{j<k}(z_j - z_k)^{2p}$, let us consider a typical term in its expansion, which contains factors of the type

$$\phi = \bar{z}^n z^{n+m}\,\mathrm{e}^{-\frac{1}{4}z\bar{z}} \tag{5.93}$$

for each particle. Typically, n is of order one (bounded by the index of the highest occupied Landau level in Φ), and m is a large number (because, for large N, terms with the highest weights in the expansion of $\prod_{j<k}(z_j - z_k)^{2p}$ contain large powers of z_j's). Thus, the typical factor has $m \gg n$. Some algebra (Exercise 5.3) shows that

$$\frac{\langle \phi_p | \phi \rangle}{\langle \phi | \phi \rangle} = \frac{(m+n)!(m+n)!}{m!(m+2n)!}, \tag{5.94}$$

where ϕ_p is the LLL projection of ϕ. For $m \gg n$, the right hand side differs from unity only by terms of order $1/m$, demonstrating that ϕ is almost the same as its LLL projection ϕ_p.

In other words, ϕ has only a small amplitude outside the lowest Landau level. This single particle argument suggests that the many-body wave function $\Phi \prod_{j<k}(z_j - z_k)^{2p}$ might be mostly in the lowest Landau level even without projection. In fact, one might wonder if the amplitude in higher Landau levels vanishes in the thermodynamic limit. Explicit calculations show that the mixing with higher Landau levels in these wave functions is small but nonzero (See Table 5.1), and an explicit projection is required to obtain strictly LLL wave functions.

Two methods have been used for projecting $\Phi_{\nu^*} \Phi_1^{2p}$ into the lowest Landau level.

5.14.5 Method I

The most natural method for projecting any given wave function into the lowest Landau level is to express it as a linear superposition of Slater determinant basis states (involving higher Landau level orbitals), and then retain only the part that fully resides in the lowest Landau level. Formally, we can express the projected wave function in the spherical geometry as

$$\Psi = \mathcal{P}_{\text{LLL}} \Phi(Y) \Phi_1^{2p} = \hat{\Phi}(\hat{Y}) \Phi_1^{2p}, \tag{5.95}$$

where $\hat{\Phi}(\hat{Y})$ is an operator obtained by replacing monopole harmonics in $\Phi(Y)$ by the corresponding operators. In the disk geometry, we have

$$\mathcal{P}_{\text{LLL}} e^{-\sum_j |z|_j^2/4} \Phi(\eta) \prod_{j<k}(z_j - z_k)^{2p} = e^{-\sum_j |z|_j^2/4} \hat{\Phi}(\hat{\eta}) \prod_{j<k}(z_j - z_k)^{2p}. \tag{5.96}$$

LLL projection through this procedure is rather cumbersome, as each term of $\hat{\Phi}(\hat{\eta})$ involves a large number of derivatives. The projection has been implemented for up to 10 particles by an expansion into Slater determinant basis states [96, 115, 116, 549, 703], but becomes impractical for larger systems, because the number of LLL basis states quickly grows too large to be stored on a computer.

5.14.6 Method II

An alternative projection method (Jain and Kamilla [280, 281]) allows calculations for much larger systems (systems with up to 100 composite fermions have been studied [571]; larger systems are treatable if needed) and has proved crucial for obtaining quantitative information from the CF theory. The key is to write the projected wave function in a way that can be evaluated without expansion into Slater determinant basis functions.

We first consider the disk geometry, while assuming that Φ is a single Slater determinant. We write (modulo an overall sign)

$$\prod_{j<k}(z_j - z_k)^{2p} = \prod_{j \neq k}(z_j - z_k)^p \equiv \prod_j J_j^p, \tag{5.97}$$

where

$$\mathcal{J}_j = \prod_k{}' (z_j - z_k), \tag{5.98}$$

and the prime denotes the condition $k \neq j$. The Jastrow factor can be subsumed into the Slater determinant:

$$\Psi = \begin{vmatrix} \eta_1(\mathbf{r}_1) & \eta_1(\mathbf{r}_2) & \cdot & \cdot & \cdot \\ \eta_2(\mathbf{r}_1) & \eta_2(\mathbf{r}_2) & \cdot & \cdot & \cdot \\ \cdot & & \cdot & \cdot & \\ & \cdot & & & \cdot \\ \cdot & \cdot & & \cdot & \end{vmatrix} \prod_{j<k} (z_j - z_k)^{2p}$$

$$= \begin{vmatrix} \eta_1(\mathbf{r}_1)\mathcal{J}_1^p & \eta_1(\mathbf{r}_2)\mathcal{J}_2^p & \cdot & \cdot & \cdot \\ \eta_2(\mathbf{r}_1)\mathcal{J}_1^p & \eta_2(\mathbf{r}_2)\mathcal{J}_2^p & \cdot & \cdot & \cdot \\ \cdot & & \cdot & \cdot & \\ & \cdot & & & \cdot \\ \cdot & \cdot & & \cdot & \end{vmatrix} \tag{5.99}$$

We now project each element individually to *define* the LLL projection as

$$\mathcal{P}_{\text{LLL}}\Psi = \begin{vmatrix} \hat{\eta}_1(\mathbf{r}_1)\mathcal{J}_1^p & \hat{\eta}_1(\mathbf{r}_2)\mathcal{J}_2^p & \cdot & \cdot & \cdot \\ \hat{\eta}_2(\mathbf{r}_1)\mathcal{J}_1^p & \hat{\eta}_2(\mathbf{r}_2)\mathcal{J}_2^p & \cdot & \cdot & \cdot \\ \cdot & & \cdot & \cdot & \\ & \cdot & & & \cdot \\ \cdot & \cdot & & \cdot & \end{vmatrix} \tag{5.100}$$

An explicit expression for each matrix element can be obtained by the projection method I. The largest power of $\bar{z}$ in $\eta_{n,m}$ is $\bar{z}^n$, so we need to evaluate the nth derivative of $\mathcal{J}_j^p$. Explicit expressions for derivatives can be obtained by Mathematica when n is not too large. Further details can be found in Appendix J.

When Φ is a linear superposition of many Slater determinants, the corresponding wave function Ψ is obtained as

$$\Psi = \mathcal{P}_{\text{LLL}}\Phi\Phi_1^{2p}$$
$$= \Phi[\eta_\alpha(\mathbf{r}_j) \rightarrow \hat{\eta}_\alpha(\mathbf{r}_j)\mathcal{J}_j^p], \tag{5.101}$$

where α denotes the quantum numbers n, m. The projected wave function Ψ is thus obtained from Φ by replacing each single particle state by a complicated expression. The coefficients of superposition remain unchanged.

This trick also works for the spherical geometry, with

$$\prod_{j<k}(u_j v_k - v_j u_k)^{2p} = \prod_{j\neq k}(u_j v_k - v_j u_k)^p \equiv \prod_j \mathcal{J}_j^p, \qquad (5.102)$$

$$\mathcal{J}_j = \prod_k{}'(u_j v_k - v_j u_k). \qquad (5.103)$$

When Φ is a single Slater determinant,

$$\mathcal{P}_{\text{LLL}}\text{Det}[Y_\alpha(\mathbf{\Omega}_j)] \prod_{k<l}(u_k v_l - v_k u_l)^{2p} = \mathcal{P}_{\text{LLL}}\text{Det}[Y_\alpha(\mathbf{\Omega}_j)] \prod_k \mathcal{J}_k^p$$

$$= \mathcal{P}_{\text{LLL}}\text{Det}[Y_\alpha(\mathbf{\Omega}_j)\mathcal{J}_j^p]$$

$$\equiv \text{Det}[\mathcal{P}_{\text{LLL}}Y_\alpha(\mathbf{\Omega}_j)\mathcal{J}_j^p]$$

$$= \text{Det}[\hat{Y}_\alpha(\mathbf{\Omega}_j)\mathcal{J}_j^p]. \qquad (5.104)$$

For an arbitrary Φ, which is a linear superposition of Slater determinants, the corresponding wave function Ψ is obtained as

$$\Psi = \mathcal{P}_{\text{LLL}}\Phi\Phi_1^{2p} = \Phi[Y_\alpha(\mathbf{\Omega}_j) \rightarrow \hat{Y}_\alpha(\mathbf{\Omega}_j)\mathcal{J}_j^p]. \qquad (5.105)$$

The coefficients of superposition remain unchanged in going from Φ to Ψ.

The LLL projected wave functions from methods I and II are not identical. We should remember, however, that our aim is to obtain a LLL wave function starting from certain unprojected wave functions, and, therefore, we have no a-priori preference between different schemes that accomplish that. Each scheme must be tested individually against exact results to establish the level of its accuracy. We should not be surprised if more than one projection method turns out be valid; because the unprojected wave functions are mostly in the lowest Landau level to begin with, it should be unimportant how the higher LL part is eliminated.

5.15 Need for other formulations

So far, we have dealt with the physics of composite fermions and its quantitative implementation through microscopic wave functions. An advantage of this implementation is that its statements can be tested rigorously against exact results known for systems with up to 12–15 particles (Chapter 6). Extracting numbers from wave functions requires evaluation of $2N$-dimensional integrals, which can be accomplished numerically for fairly large N with the help of the Metropolis Monte Carlo method. The wave functions are suitable for the calculation of energy gaps and equal-time, zero-temperature correlation functions (which require only the ground state wave function). CF diagonalization can also tell us, in principle, if a state is not incompressible. Microscopic wave functions, however, are not useful for compressible states, because a proper account of the thermodynamic limit requires systems that are larger than what can currently be handled, and because the absence of a gap precludes a systematic truncation of the basis for CF diagonalization.[22] Wave functions are also impractical for time, wave vector, frequency, or temperature dependent response functions.

That constitutes part of the motivation for seeking other theoretical formulations of composite fermions, to which the remainder of the chapter is devoted. We note that none of the approaches discussed below *derives* composite fermions. They all assume composite fermions at the very outset, sometimes through an exact transformation. Also, resorting to uncontrolled approximations is necessary in all of the formulations, just as it was in writing wave functions. This fact, although hardly surprising in view of the complexity of the problem and the lack of a small parameter, necessitates independent tests of the "final" statements of each approach. Because the qualitative phenomenology of the FQHE and the CF Fermi sea follows from the CF principle without recourse to any specific theoretical scheme, the validity of any formulation is to be ascertained through detailed comparisons of its *quantitative* consequences against theoretical and experimental facts.

5.16 Composite fermion Chern–Simons theory

The Chern–Simons field theory of FQHE was pioneered by Zhang, Hansson, and Kilvelson [743], who mapped the problem at Laughlin fractions into a boson superfluid, following ideas of Girvin and MacDonald [193]. The composite fermion Chern–Simons (CFCS) theory (also called the "fermion CS theory") was introduced by Lopez and Fradkin [401], and further extended by Halperin, Lee, and Read [231], Murthy and Shankar [581], and several other groups. A thorough treatment of the CFCS approach is outside the scope of this book. We present here only a basic introduction, and refer the reader to several excellent review articles (Halperin [232, 233]; Lopez and Fradkin [404]; Murthy and Shankar [468]; and Simon [594]) for further details and successes of this approach, as well as for many relevant references.

5.16.1 Why field theory?

Some condensed matter physicists hold the view (subconsciously) that a field theory is more fundamental than a theory based on wave functions. This notion perhaps has its roots either in the success of field theory in high-energy physics[23], or, more likely, in the tendency to equate the term "theory" with complicated mathematical and technical formalism.

---

[22] Certain quantities for the compressible states at $\nu = 1/2p$ have been estimated by calculating them for incompressible states belonging to the sequence $\nu = n/(2pn \pm 1)$, followed by an extrapolation to the limit $n \to \infty$. Such a procedure, however, does not test the validity of the CF description at $\nu = 1/2p$; rather, it makes the assumptions that (i) composite fermions remain relevant in that limit, and (ii) the residual interaction between them does not cause an instability with increasing n. Both these assumptions are known to fail in special cases: In higher Landau levels, the CF description is invalid altogether, whereas in the second Landau level composite fermions exist at half filling but, apparently, form a paired FQHE state rather than a CF Fermi sea.

[23] Ironically, attempts are underway to move away from field theory in high-energy physics; the string theory program is based on quantum mechanics rather than field theory.

The possibility of creation and annihilation of particles necessitates field theory in high-energy physics. In nonrelativistic quantum many-body physics, with a fixed number of particles, the wave function and field theoretical approaches are merely different calculational tools for deducing quantitative information from the Schrödinger equation. The question is not which is more fundamental, but which is more reliable, which takes us further, and which offers new insights. *In principle*, they are equivalent, in the sense that the starting point of a field theory in condensed matter physics is an exact reformulation of the many-particle Schrödinger equation. In practice, however, the two approaches can be remarkably different, and offer distinct intuitions. To be able to identify what an approximation in one language means in the other is rare. Thinking of wave functions has facilitated initial breakthroughs in several important problems, for example, the BCS theory for superconductivity (wave functions for the ground and excited states) and Feynman's theory of superfluidity (wave function for the collective mode). On the other hand, field theory is an essential tool for a description of quantum or classical phase transitions.

Why, then, field theory? On general grounds, having in our possession as many theoretical formulations as possible of a new physical concept is desirable. While one may be more useful for one question, another may take us further in a different context. The power of a field theory lies in its treatment of spontaneous symmetry breaking and order parameter, and in organizing a perturbation theory through Feynman diagrams. The theory of superconductivity provides an ideal example. The Gorkov–Eliashberg field theoretical formulation obtains the Bardeen–Cooper–Schrieffer results in one limit, but takes us farther by enabling a more complete treatment of several features (electron–phonon coupling, effect of disorder), and even allows a first-principles determination of quantities such as the critical temperature.

It is worth stressing that the essential physics must be already known before a perturbation program can be implemented through a field theory. By definition, perturbation theory keeps us within the phase we started with (phase transitions involve nonperturbative reorganizations), and is of no use unless we begin in the correct phase. For example, with the noninteracting Fermi sea as the starting point, only the properties of a Landau Fermi liquid will be accessible in perturbative calculations; the theory of superconductivity must incorporate, at the very outset, the physics of pairing, which was discovered by other means.

5.16.2 Lopez–Fradkin theory

We started in Section 5.8.2 with electrons at B^*, added $2p$ flux quanta to each electron, and then made a mean-field approximation to end up with fermions at B. The Lopez–Fradkin [401] construction begins with electrons at B, attaches $2p$ flux quanta pointing in the direction opposite to B, and then performs a mean-field approximation to cancel part of the external field, producing fermions at B^* in the end. The basic idea is shown in Fig. 5.12.

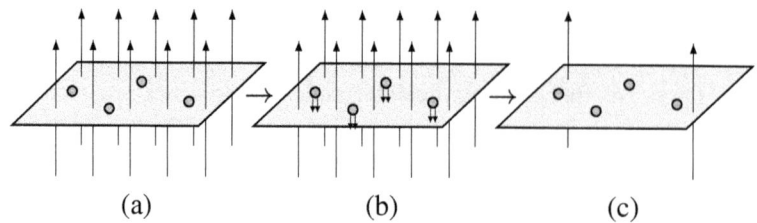

Fig. 5.12. The physical idea of the Lopez–Fradkin theory. The FQHE system is shown in (a). Two magnetic flux quanta are attached to each electron pointing in the downward direction to convert electrons into composite fermions (b). Composite fermions respond to a combination of the real and attached fields, and, consequently, behave as though they were in a smaller magnetic field (c).

We consider the Schrödinger equation

$$\left[\frac{1}{2m_{\mathrm{b}}} \sum_i \left(p_i + \frac{e}{c} A(r_i) \right)^2 + V \right] \Psi = E\Psi, \tag{5.106}$$

where V is the interaction. Through an exact singular gauge transformation defined by

$$\Psi = \prod_{j<k} \left(\frac{z_j - z_k}{|z_j - z_k|} \right)^{2p} \Psi_{\mathrm{CS}}, \tag{5.107}$$

known as the CFCS transformation, the eigenvalue problem can be expressed as

$$H'\Psi_{\mathrm{CS}} = E\Psi_{\mathrm{CS}}, \tag{5.108}$$

$$H' = \left[\frac{1}{2m_{\mathrm{b}}} \sum_i \left(p_i + \frac{e}{c} A(r_i) - \frac{e}{c} a(r_i) \right)^2 + V \right], \tag{5.109}$$

$$a(r_i) = \frac{2p}{2\pi} \phi_0 \sum_j{}' \nabla_i \theta_{ij}, \tag{5.110}$$

following algebra similar to that in Section 5.8.2. The vector potential $-a$ amounts to attaching a point flux of strength $-2p\phi_0$ to each electron. This is the starting model for a composite fermion in the Lopez–Fradkin theory.

Further progress is not possible without making approximations. The usual approach is to make a "mean-field" approximation, which amounts to spreading the CS flux on each composite fermion into a uniform CS magnetic field, and assuming that, to zeroth order, the particles respond to the sum of CS and external fields. Formally, we write

$$A - a \equiv A^* + \delta A, \tag{5.111}$$

$$\nabla \times A^* = B^* \hat{z}, \tag{5.112}$$

$$B^* = B - 2p\rho\phi_0, \tag{5.113}$$

where B^* is the effective magnetic field experienced by composite fermions. The transformed Hamiltonian can now be written as

$$H' = \frac{1}{2m_b} \sum_i \left(p_i + \frac{e}{c} A^*(r_i) \right)^2 + V + V' = H_0' + V + V', \qquad (5.114)$$

where V' contains terms proportional to δA. The solution to H_0' is trivial, describing free fermions in an effective magnetic field B^*. We have thus decomposed the Hamiltonian into two parts: H_0' can be solved exactly and the remainder, $V+V'$, is to be treated perturbatively.

The mean-field approximation of expanding around H_0' is a crucial, nonperturbative step. A mathematically exact CS transformation can be made to attach any amount of flux, integral or nonintegral, to each electron. But because solving the transformed problem exactly is no more possible than it was to solve the original problem, the mathematical exactness of the CS transformation is of little value. The question is what one can do with it. With the mean-field approximation, CS transformations with different values of attached flux become nonequivalent, producing distinct physics. The CS transformation must be physically motivated. The goal is to transform into particles that are weakly interacting, which can only be ascertained by comparing the consequences of the noninteracting model of the CS-transformed particles (H_0') with the observed phenomenology.

Lopez and Fradkin recast the problem in the language of functional integrals, which is suitable for studying corrections to the mean-field theory. For readers who are familiar with this method, the zero-temperature quantum partition function is written as

$$\mathcal{Z} = \int \mathcal{D}\psi \mathcal{D}\psi^* \mathcal{D}a \, \exp\left(\frac{i}{\hbar} \mathcal{S} \right), \qquad (5.115)$$

$$\mathcal{S} = \int d^2r \int dt \, \mathcal{L}, \qquad (5.116)$$

$$\mathcal{L} = \psi^*(i\partial_t - a_0)\psi + \frac{1}{2m_b} \left| \left(-i\hbar\nabla + \frac{e}{c}A - \frac{e}{c}a \right) \psi \right|^2$$

$$+ \frac{1}{2p\phi_0} a_0 \nabla \times a + \int d^2r'(\rho(r) - \bar{\rho})V(r - r')(\rho(r') - \bar{\rho}), \qquad (5.117)$$

where ψ and ψ^* are anticommuting Grassmann variables, $\bar{\rho}$ is the average density, and $V(r)$ is the interaction between electrons. Flux attachment is introduced through a Lagrange multiplier a_0. Because a_0 enters linearly in the action, it can be integrated out to produce a delta function that imposes the constraint

$$\nabla \times a(r) = 2p\phi_0\rho(r) = 2p\phi_0\psi^*(r)\psi(r). \qquad (5.118)$$

What is the connection with the topological Chern–Simons theory familiar to particle theorists? The Lagrangian for the electromagnetic field is given by the Lorentz scalar $\mathcal{L} = (1/4)F^{\mu\nu}F_{\mu\nu}$, with

$$F_{\mu\nu} = \partial_\mu A_\nu - \partial_\nu A_\mu, \qquad (5.119)$$

which is invariant under the gauge transformation $A_\mu \to A_\mu + \partial_\mu \Lambda(\mathbf{r}, t)$. (The standard summation convention for repeated indices is assumed.) The Chern–Simons Langrangian has the form

$$\mathcal{L}_{CS} \sim \epsilon^{\mu\nu\lambda} A_\mu F_{\nu\lambda} = 2\epsilon^{\mu\nu\lambda} A_\mu \partial_\nu A_\lambda. \tag{5.120}$$

Here $\epsilon^{\mu\nu\lambda}$ is the antisymmetric Levy–Civita tensor, with $\epsilon^{012} = 1$. The index μ takes values $0, 1, 2$, the first being the time component and the rest space components. The CS action is invariant, up to surface terms, under a gauge transformation, because the functional variation $\delta A_\mu = \partial_\mu \Lambda$ causes a change in $\mathcal{L}_{CS}$ that is a total derivative:

$$\delta\mathcal{L}_{CS} = \epsilon^{\mu\nu\lambda} \delta A_\mu \partial_\nu A_\lambda + \epsilon^{\mu\nu\lambda} A_\mu \partial_\nu \delta A_\lambda$$

$$= 2\epsilon^{\mu\nu\lambda} \delta A_\mu \partial_\nu A_\lambda$$

$$= -2\epsilon^{\mu\nu\lambda} \partial_\nu (A_\lambda \partial_\mu \Lambda). \tag{5.121}$$

Zhang, Hansson, and Kivelson [743, 744] note that the term proportional to $a_0 \nabla \times \mathbf{a}$ in Eq. (5.117), which enforces flux attachment, is precisely equal to the Chern–Simons Langrangian in the Coulomb gauge. This is clarified by writing

$$\mathcal{L}_{CS} = \frac{1}{4p\phi_0} \epsilon^{\mu\nu\lambda} a_\mu \partial_\nu a_\lambda$$

$$= \frac{1}{2p\phi_0} \epsilon^{ij} a_0 \partial_i a_j - \frac{1}{4p\phi_0} \epsilon^{ij} a_i \partial_0 a_j, \tag{5.122}$$

where i, j represent the spatial components ($i, j = 1, 2$); the time components have been displayed explicitly in the second step ($\partial_0 = \partial_t$). The first term on the right hand side of Eq. (5.122) is identical to the third term on the right hand side of Eq. (5.117). What about the last term in Eq. (5.122)? In Fourier space it is proportional to

$$\epsilon^{ij} a_i(\mathbf{q}, \omega)(-i\omega) a_j(-\mathbf{q}, -\omega). \tag{5.123}$$

By choosing the x-axis along $\mathbf{q}$, the Coulomb gauge condition $\mathbf{q} \cdot \mathbf{a} = 0$ implies $a_2(\mathbf{q}, \omega) = 0$, guaranteeing that the last term in Eq. (5.122) is identically zero.

The constraint of Eq. (5.118) is used to eliminate two factors of density in the last term of Eq. (5.117) in favor of $(2p\phi_0)^{-1}\nabla \times \mathbf{a}$. The action is then quadratic in the fermion field, which can be integrated out. Various response functions can be expressed as correlation functions of the vector potential field and their averages over the CS field configurations are evaluated perturbatively by standard diagrammatic methods.

The solutions of the unperturbed Hamiltonian H_0' are far from the actual solution, as can be seen both from the energy eigenvalues and the eigenfunctions. Specializing to $\nu = n/(2pn + 1)$, the "unperturbed" energy gap is given by

$$\Delta = \hbar\omega_c^* = \hbar\frac{eB^*}{m_b c} = \frac{\hbar}{(2pn+1)}\frac{eB}{m_b c}, \tag{5.124}$$

with $B^* = B/(2pn + 1)$. The actual energy gap in the lowest Landau level problem, in contrast, is proportional to $e^2/\epsilon\ell \sim \sqrt{B}$, the only energy scale in the LLL problem, and is independent of the electron mass. Coming to the "unperturbed" eigenfunctions, let us denote by $\Phi_n^\alpha(B^*)$ the eigenfunctions of H_0' at magnetic field B^*. Undoing the CFCS transformation gives the mean-field electron wave functions (using Eq. 5.107):

$$\Psi_\nu^\alpha(B) = \prod_{j<k}\left(\frac{z_j - z_k}{|z_j - z_k|}\right)^{2p}\Phi_n^\alpha(B^*), \qquad (5.125)$$

which are nothing but the wave functions encountered previously in Eq. (5.30). As explained in the discussion following Eq. (5.30), they are not satisfactory (also see Sitko [595, 597]; Ciftja and Wexler [86, 88]). The perturbation theory program faces the formidable challenge of producing repulsive correlations and getting rid of the electron mass.[24] What classes of Feynman diagrams will accomplish that is not known, precluding, so far, a first principles determination of various quantities in the CFCS approach. The difficulties confronted here are related to what would be faced in attempting a derivation of the quantized Landau levels in a perturbative treatment of the magnetic field.

Some comments are in order.

(i) Exorcising the electron mass from the problem requires a proper treatment of the LLL projection. For electrons, LLL wave functions have the form

$$\Psi = \prod_{j<k}(z_j - z_k)F_S[\{z_i\}]\exp\left[-\frac{1}{4}\sum_l |z_l|^2\right], \qquad (5.126)$$

where $F_S[\{z_i\}]$ is a symmetric polynomial. These wave functions translate into CS wave functions with the form:

$$\Psi_{CS} = \prod_{j<k}\frac{|z_j - z_k|^{2p}}{(z_j - z_k)^{2p-1}}F_S[\{z_i\}]\exp\left[-\frac{1}{4}\sum_l |z_l|^2\right]. \qquad (5.127)$$

Restricting solutions of the Chern–Simons problem to the space defined by such wave functions is difficult to implement, given that the constraint cannot be imposed on single particle wave functions.

(ii) The mean-field theory does not describe all qualitative features correctly, as illustrated by two examples. (a) The Hall resistance of the mean-field theory is $R_H = h/ne^2$. We see below how the correct value, $R_H = h/[n/(2pn\pm1)]e^2$, is recovered. (b) The "local charge" of the quasiparticle (defined in Section 9.3) is equal to the electron charge and not a fraction of it. This follows because, at the mean-field level, the charge excess associated with a quasiparticle at $\nu = n/(2pn\pm1)$ is the same as that at $\nu^* = n$. (The "correct" fractional charge is produced by the random phase approximation [401].)

(iii) One may encounter phrases to the effect: "composite fermions interact through the Chern–Simons gauge potential." That is indeed the way it appears in the CFCS formulation, but we prefer not to

---

[24] A renormalization of the the electron mass is not sufficient – it must be eliminated entirely.

use such language, because it does not give a quantitative account of the interaction between the "physical" composite fermions. The CFCS field theory is designed to capture the long-distance physics, and does not account for the short-distance part of the interaction between composite fermions. At the present, the most accurate determination of the inter-CF interaction is based on a theory that makes no reference to the CS gauge potential. Section 6.7 contains further discussion on this issue, as well as the actual form of the inter-CF interaction.

5.16.3 Halperin–Lee–Read theory

Halperin, Lee, and Read [231] (also Kalmeyer and Zhang [309]) make the remarkable prediction that composite fermions form a Fermi sea at $\nu = 1/2$, motivated by the observation that composite fermions, should they exist at $\nu = 1/2$, experience no effective magnetic field here (i.e., $B^* = 0$). The CF Fermi sea has been beautifully confirmed experimentally (see Chapter 10); the explanation of the physics of the compressible half-filled Landau level state has been one of the triumphs of the CF approach.

Halperin, Lee, and Read treat the Lopez–Fradkin theory as a starting point to develop an effective theory of the CF Fermi sea, following the spirit of the Landau theory of ordinary Fermi liquids, which expresses various observables in the low-temperature limit in terms of a finite number of parameters. The "energy scale problem" is avoided by replacing the electron mass m_b with a free parameter m^*, interpreted as the mass of composite fermions, to be fixed empirically. We list here the principal results of this approach, often without proof, closely following the reviews by Simon [594] and Halperin [232, 233].

Magnetoresistances   With composite fermions viewed as electrons bound to CS flux quanta, their charge current has associated with it a CS flux current. That, in turn, induces, through the Faraday effect, a CS electric field, given by (Exercise 5.4)

$$e_{CS} = 2p\frac{h}{e^2}(\hat{z} \times j),$$ (5.128)

where $j = -\rho e v$ is the current density, v being the local *average* drift velocity. The effect of e_{CS} must be included in the calculation of various magnetoresistances. Just as with the real and CS magnetic fields, composite fermions respond to the combination of real and induced CS electric fields. The current density is given by (switching to matrix notation)

$$j = \sigma_{CF}(E + e_{CS}(j))$$ (5.129)

which defines the CF conductivity. Here, j, E and e_{CS} represent two-dimensional vectors as column vectors, and σ_{CF} is a 2×2 matrix. Experiments do not measure σ_{CF}, however. Because the CS electric field is "internal" to the CF system, it is felt by composite fermions but not detected by a voltmeter in the laboratory. The physical resistivity matrix ρ (not to be confused with the density) is given by

$$E = \rho j,$$ (5.130)

which can be evaluated with the help of above equations. Equation (5.128) is conveniently rewritten as

$$e_{CS} = -\rho_{CS} j, \tag{5.131}$$

with

$$\rho_{CS} = \frac{h}{e^2} \begin{bmatrix} 0 & 2p \\ -2p & 0 \end{bmatrix}. \tag{5.132}$$

Equation (5.129) then becomes

$$E + e_{CS} = \rho_{CF} j, \tag{5.133}$$

with

$$\rho_{CF} = (\sigma_{CF})^{-1}. \tag{5.134}$$

With the help of

$$\rho = \rho_{CS} + \rho_{CF}, \tag{5.135}$$

the physical resistivity matrix can be obtained from the CF resistivity (ρ_{CS} is a fixed matrix).

The random-phase approximation (RPA) of the CS theory amounts to using the mean-field result for ρ_{CF}. Physical resistivities can thus be obtained from those of noninteracting fermions at an effective magnetic field. Let us see how this yields the correct Hall resistance at $\nu = n/(2pn \pm 1)$. In the absence of disorder, we have (at the mean-field level)

$$\rho_{CF} = \frac{h}{e^2} \begin{bmatrix} 0 & \pm\frac{1}{n} \\ \mp\frac{1}{n} & 0 \end{bmatrix}, \tag{5.136}$$

which produces, using Eqs. (5.132) and (5.135), the physical resistivity matrix

$$\rho = \frac{h}{e^2} \begin{bmatrix} 0 & \frac{2pn\pm1}{n} \\ -\frac{2pn\pm1}{n} & 0 \end{bmatrix}. \tag{5.137}$$

The point is this: At the mean-field level, the Hall resistance appears to be $R_H = h/\nu^* e^2$. The correct value of Hall resistance, $R_H = h/\nu e^2$, is obtained, however, once it is recognized that while composite fermions sense the aggregate of the real and (fictitious) CS electric fields, only the former is measured by a voltmeter. Consideration of the special case of $\nu = 1/2$, where $B^* = 0$, is instructive. A current of composite fermions also implies a CS flux flow, which induces a transverse CS electric field. However, for composite fermions to move straight ahead and carry current, they must not feel any transverse electric field. Therefore, a real transverse electric field must build up to cancel the effect of the internal Faraday electric field. While composite fermions sense a sum of the internal and external electric fields (which add to zero) only the latter is measured by the voltmeter, producing a nonzero Hall resistance.

Another way of understanding this result (Kirczenow and Johnson [338,341]) is to note that Eq. (5.128) can be expressed as

$$e_{CS} = -\frac{1}{c}v \times b,$$ (5.138)

in terms of the CS magnetic field $b = -2p\rho\phi_0\hat{z}$. This produces an *average* force of

$$-ee_{CS} = \frac{e}{c}v \times b,$$ (5.139)

which exactly cancels the average force due to the CS magnetic field, $-\frac{e}{c}v \times b$. Thus, the *average* Lorentz force experienced by composite fermions is identical to the average Lorentz force experienced by electrons, producing the Hall resistance $R_H = h/\nu e^2$ rather than $R_H = h/\nu^* e^2$.[25]

The longitudinal conductivity is obtained by inverting ρ to yield

$$\sigma_{xx} = \frac{\rho_{yy}^{CF}}{\text{Det}[\rho_{CF} + \rho_{CS}]} \approx \rho_{yy}^{CF}\left(2p\frac{h}{e^2}\right)^{-2},$$ (5.140)

where we have used, in the last step, the result that, for typical parameters, ρ_{CS} dominates in the denominator.

We next consider finite wave vector conductivity for the CF Fermi sea at $\nu = 1/2p$. For ρ_{xx}^{CF} we take the resistivity for noninteracting electrons at zero magnetic field, which, for low frequency, is given by [594]

$$\rho_{yy}^{CF}(q) = \frac{q}{k_F}\frac{h}{e^2} \quad \text{for} \quad q \gg \frac{2}{l^*}$$

$$= \frac{2}{k_F l}\frac{h}{e^2} \quad \text{for} \quad q < \frac{2}{l^*},$$ (5.141)

where $l^* = v_F^* \tau$ is the mean-free path due to scattering by disorder. This predicts the following wave vector dependence for the conductivity of the CF Fermi sea:

$$\sigma_{xx}(q) = \frac{q}{k_F}\frac{e^2}{(2p)^2 h} \quad \text{for} \quad q \gg \frac{2}{l^*}$$

$$= \frac{2}{k_F l}\frac{e^2}{(2p)^2 h} \quad \text{for} \quad q < \frac{2}{l^*}.$$ (5.142)

Electromagnetic response function For $\omega \ll q v_F$ the density response function $K_{00}(q, \omega)$ is given by

$$[K_{00}(q, \omega)]^{-1} = \frac{2\pi}{m^*}\left(1 + \frac{p^2}{3}\right) + V(q) + i\left(\frac{4\pi p}{q}\right)^2\left(\frac{2\hbar\rho}{m^*}\right)\frac{\omega}{q v_F},$$ (5.143)

[25] The cancellation applies [338,341] only to the slow guiding center drift of composite fermions, but not to the cyclotron orbit motion. Consequently, the latter is determined by the effective magnetic field, as confirmed experimentally (Section 10.1).

where $V(q)$ is the Fourier transform of the interaction. The static compressibility is given by $K_{00}(q \to 0, \omega = 0)$; it is nonzero for short-ranged interactions (e.g., for a delta function interaction, for which $V(q) = $ constant), but vanishes as q for the Coulomb interaction. For nonzero ω, $K_{00}(q, \omega)$ has a pole at an imaginary frequency

$$\omega \propto iq^3 V(q), \tag{5.144}$$

which is referred to as the "overdamped mode." For Coulomb interaction it behaves as $\omega \sim q^2$, i.e., is "diffusive" (the diffusion equation is given by $\partial \psi / \partial t \sim \nabla^2 \psi$).

CF mass divergence The mass of a quasiparticle in a condensed matter background is renormalized due to its interaction with other quasiparticles. Mass renormalization can be obtained from a calculation of its self energy. For composite fermions, the most dominant contribution at low energies comes from the transverse gauge field propagator, which is proportional to K_{00} and has a pole at the overdamped mode. That results in a logarithmic divergence to the CF mass at $\nu = 1/2p$:

$$m^* \sim \ln \omega. \tag{5.145}$$

Because the divergence is weak, one assumes that it does not invalidate the Fermi liquid starting point. The low-temperature specific heat of an ordinary Landau Fermi liquid is proportional to $m^* T$; for the CF Fermi sea, this gives

$$C_v \sim T \ln T. \tag{5.146}$$

Away from $\nu = 1/2$, say at $\nu^* = n$ for large n, one might expect that the singular behavior of the self energy would be cut off by the gap. Curone and Stamp [103] calculate the self energy perturbatively within the CFCS approach to find an unphysical vanishing of the quasiparticle spectral weight at the Λ level energies. They take this to mean either that a nonperturbative calculation is needed, or that the system is unstable.

Landau Fermi liquid modifications As a consequence of Galilean invariance, the density response function of a fully interacting system has a pole at the cyclotron energy $\hbar eB/m_b c$, which exhausts the f-sum rule at small wave vectors (Kohn [351]). In the CFCS-RPA response, the density response function has a pole at the cyclotron frequency with proper oscillator strength provided the electron band mass is retained. The $m_b \to m^*$ replacement results in a violation of Kohn's theorem [231]. Simon and Halperin [592] introduce a modified random phase approximation (MRPA), in the spirit of Silin's extension of the Landau Fermi liquid theory to deal with the long-range part of the interaction, to ensure that the high-frequency response is determined by m_b.

Simon, Stern, and Halperin [593] note that the static response to a spatially varying magnetic field is not correctly given within the MRPA framework. To see the problem, consider $B(r) = B_{1/2} + \delta B(r)$, where $B_{1/2}$ corresponds to $\nu = 1/2$. This reduces to noninteracting composite fermions with $B^*(r) = \delta B(r)$. From kinetic energy considerations, the CS theory would suggest density maxima at positions where $\delta B(r) = 0$, whereas we know from the original problem that density maxima occur at the minima of $\delta B(r)$. (A more complete treatment must solve for the density and the effective magnetic field self-consistently, because $B^*(r) = B(r) - 2\phi_0\rho(r)$.) Simon, Stern, and Halperin propose to fix the problem by attaching an orbital magnetization to each composite fermion (not related to its spin) which couples to the total magnetic field $B(r)$, resulting in the so-called magnetized MRPA, or M^2RPA.

The most important accomplishment of the CFCS theory is to provide description of the long-wavelength, low-energy behavior of incompressible and compressible states, especially the CF Fermi sea at $\nu = 1/2$, where its consequences are in general agreement with experiments. The CFCS approach, in particular, successfully predicts nontrivial structure in finite wave vector conductivity measured in a surface acoustic wave experiment (Fig. 10.4). It is, however, not expected to produce, from first principles, reliable numbers for various quantities (such as excitation gaps, CF masses, or the interaction between composite fermions) for incompressible FQHE states; the field theory focuses on the long-distance behavior, whereas such numbers are often governed by the short-distance physics. The Chern–Simon and the microscopic wave function approaches thus play complementary roles.

The Landau energy functional for the ordinary Landau Fermi liquids (at $B = 0$) has been justified from renormalization group considerations (Shankar [580]), which demonstrate that the terms left out are "irrelevant" in the low-energy sector. Some progress has been made toward a similar justification for the CFCS theory of the CF sea [474].

5.17 Other CF based approaches

5.17.1 Murthy–Shankar theory

There is a tension between the lowest Landau level physics, which has no knowledge of the kinetic energy for electrons, and CS flux attachment, which relies in a fundamental way on the kinetic energy term in the Hamiltonian. Therein lies the origin of the difficulty in imposing the LLL constraint and getting rid of the electron band mass in the CFCS formulation in a systematic manner.

Murthy and Shankar [462, 468, 581] propose a method to overcome this dilemma by separating inter-LL degrees of freedom that depend on the electron mass and the cyclotron energy, and intra-LL degrees of freedom that are determined by the interaction. They are motivated by the work of Bohm and Pines [32] on the electron gas in three dimensions. In that problem, the relevant objects are collective modes (plasmons) and particle–hole excitations. The plasmons are not independent objects; indeed the Hamiltonian contains no plasmons. Yet they are well-defined "particles" that can be seen as sharp resonances in scattering experiments, and to treat them as independent entities is tempting. Bohm and Pines showed that this can be done by enlarging the Hilbert space by introducing additional oscillator degrees of freedom, but at the cost of introducing certain constraints that eliminate double counting.

Murthy and Shankar (MS) treat "vortices" in an analogous manner. Like the plasmon of the three-dimensional electron gas, a vortex in the lowest Landau level is a collective excitation of the electron system. Murthy and Shankar introduce new, charge $2pv$, vortex operators to enlarge the Hilbert space, and then combine them with electrons to construct a Hamiltonian in terms of composite fermions coordinates. This is accomplished as follows:[26]

The kinetic energy Hamiltonian for a single electron is given by

$$H_0 = \frac{(\boldsymbol{p} + e\boldsymbol{A})^2}{2m_b} \equiv \frac{\pi^2}{2m_b} \tag{5.147}$$

where $\boldsymbol{B} = \nabla \times \boldsymbol{A} = -B\hat{z}$, and we have taken $\hbar = 1 = c$. Define the cyclotron coordinate

$$\boldsymbol{\eta} = \ell^2 \hat{z} \times \boldsymbol{\pi}, \tag{5.148}$$

and the guiding center coordinate

$$\boldsymbol{R} = \boldsymbol{r} - \boldsymbol{\eta}. \tag{5.149}$$

It can be shown that

$$[\eta_x, \eta_y] = i\ell^2 \qquad [R_x, R_y] = -i\ell^2, \tag{5.150}$$

$$H_0 = \frac{\eta^2}{2m_b \ell^2}, \tag{5.151}$$

and that $\boldsymbol{\eta}$ commutes with $\boldsymbol{R}$.

Let us consider now the filling $\nu = n/(2pn + 1)$ which maps into n filled Λ levels of composite fermions. For each electron coordinate Murthy and Shankar introduce a new guiding center coordinate for a vortex,[27] $\boldsymbol{R}_v$, defined by

$$[R_{vx}, R_{vy}] = \frac{i\ell^2}{g^2}, \tag{5.152}$$

which represents (compare Eqs. 5.150 and 5.152) an object of charge (in units of e)

$$e_v = g^2 \equiv 2p\nu = \frac{2pn}{2pn + 1}, \tag{5.153}$$

as appropriate for a vortex in the $\nu = n/(2pn + 1)$ state. We next combine $\boldsymbol{R}_v$ with $\boldsymbol{R}$ to define a new set of guiding center and cyclotron coordinates, which are identified with the CF coordinates:

$$\boldsymbol{R}^* = \frac{\boldsymbol{R} - g^2 \boldsymbol{R}_v}{1 - g^2} \tag{5.154}$$

and

$$\boldsymbol{\eta}^* = \frac{g}{1 - g^2} (\boldsymbol{R}_v - \boldsymbol{R}). \tag{5.155}$$

---

[26] We follow closely the development and conventions of Murthy and Shankar [468] in this subsection. The first version of their theory began with the CFCS Hamiltonian, followed by a canonical transformation to separate the magnetoplasmon degrees of freedom; it could be implemented only in the long-wavelength limit, however. The more powerful later version, the one described here, builds composite fermions directly in the lowest Landau level without going through the CS route.

[27] They call it a *pseudo*vortex to stress that it is an unphysical degree of freedom; however, the pseudovortex is expected to turn into a real vortex after projection into the physical sector.

(The knowledge of R^* and η^* is equivalent to knowing r^* and π^* for composite fermions.) These definitions imply the commutators

$$[\eta_x^*, \eta_y^*] = i\ell^{*2} = \frac{i\ell^2}{1 - g^2} = -[R_x^*, R_y^*], \tag{5.156}$$

which are the cyclotron and guiding center coordinates for an object of charge

$$e^* = -(1 - g^2) = -\frac{1}{2pn + 1}, \tag{5.157}$$

which is the "local charge" of a composite fermion (Section 9.3). Inverse operators are

$$R = R^* + \eta g = r^* - \frac{\ell^2}{(1 + g)} \hat{z} \times \pi^* \tag{5.158}$$

$$R_v = R^* + \frac{\eta}{g} = r^* + \frac{\ell^2}{g(1 + g)} \hat{z} \times \pi^*, \tag{5.159}$$

where r^* and π^* are composite fermion coordinate and velocity operators, respectively. The form of vortex and CF operators is uniquely determined by demanding the correct charge and commutators.

Let us now turn to the many-particle problem. The LLL projected Hamiltonian for many interacting electrons is given by

$$H = \sum_j \frac{\eta_j^2}{2m_b \ell^2} + \frac{1}{2} \sum_{j,k} V(q) e^{i\boldsymbol{q} \cdot (\boldsymbol{r}_j - \boldsymbol{r}_k)}$$

$$\rightarrow \frac{1}{2} \sum_{j,k} V(q) e^{-q^2 \ell^2 / 2} e^{i\boldsymbol{q} \cdot (\boldsymbol{R}_j - \boldsymbol{R}_k)}. \tag{5.160}$$

The constant kinetic energy has been suppressed. LLL projection of the interaction energy follows by substituting $r = R + \eta$ and then taking the expectation value within the lowest Landau level, which gives

$$\mathcal{P}_{\text{LLL}} e^{-i\boldsymbol{q} \cdot \boldsymbol{r}} = \langle e^{-i\boldsymbol{q} \cdot \boldsymbol{\eta}} \rangle_{\text{LLL}} e^{-i\boldsymbol{q} \cdot \boldsymbol{R}} = e^{-q^2 \ell^2 / 4} e^{-i\boldsymbol{q} \cdot \boldsymbol{R}}. \tag{5.161}$$

Expressed in terms of composite fermion variables, the Hamiltonian takes the form

$$H_{\text{MS}} = \frac{1}{2} \sum_{j,k} V(q) e^{-q^2 \ell^2 / 2} e^{i\boldsymbol{q} \cdot [(\boldsymbol{R}_j^* - \boldsymbol{R}_k^*) + g(\boldsymbol{\eta}_j^* - \boldsymbol{\eta}_k^*)]}, \tag{5.162}$$

which is the Murthy–Shankar Hamiltonian for composite fermions. No approximation has been made so far. The advantage of this formulation is that, because of the reduced degeneracy for composite fermions implied by their commutation relations, the IQHE wave function Φ_n, written in terms of CF coordinates r^*, is a natural trial wave function. This wave function can also be shown to be a valid Hartree–Fock solution for the MS Hamiltonian (i.e., H_{MS} does not mix different single particle excitations of a filled Landau level state [468]). That allows variational estimates for various kinds of gaps, which are seen to be in reasonable agreement with those obtained from microscopic wave functions.

The following subtlety deserves mention. H_{MS} depends only on $\boldsymbol{R} = \boldsymbol{R}^* + \eta g$ but not on $\boldsymbol{R}_v = \boldsymbol{R}^* + \eta/g$; it commutes with the latter. The eigenfunctions of H_{MS} are thus products of two factors, one dependent on $\boldsymbol{R}$ and the other on $\boldsymbol{R}_v$. Because $\boldsymbol{R}_v$ is unphysical, no observable may depend on it. This represents a gauge symmetry in the problem. Murthy and Shankar fix the gauge by demanding that the expectation value for "vortex density"

$$\chi(\boldsymbol{q}) = \sum_j e^{-i\boldsymbol{q}\cdot\boldsymbol{R}_{vj}}, \tag{5.163}$$

which commutes with the Hamiltonian of Eq. (5.162), vanish for all states. This constraint, which projects the enlarged space with unphysical degrees of freedom back onto the "physical" sector, is difficult to implement in practice, but Murthy and Shankar have demonstrated that satisfactory answers are obtained for various observables even in the unconstrained theory.[28]

The qualitative consequences of the Murthy–Shankar theory are generally consistent with those of the Halperin–Lee–Read theory, but provide alternative justifications. The Murthy–Shankar approach has been used to estimate gaps, collective mode dispersions, spin-related phase transitions, time and temperature dependent response of the CF liquid, as well as the effect of disorder on transport properties.

5.17.2 Modified Chern–Simons approaches

Ichinose and Matsui [262–265] use methods similar to those employed in treating the possibility of spin charge separation in the so-called t–J model to decompose the composite fermion into two slave fields, a fermionic "chargon" representing charge degrees of freedom, and a bosonic field "fluxon" representing the correlation hole. This generates a dynamical auxiliary gauge field (distinct from the Chern–Simons gauge field). They identify the low-temperature phase at $\nu = 1/2$ with the "deconfined" phase, in which the particle and flux separate, with the latter undergoing Bose–Einstein condensation.

Rajaraman [534] defines, through an exact nonunitary transformation, composite fermion operators which obey canonical anticommutation rules and carry unit charge. He shows how a mean-field approximation produces the unprojected wave functions of the type in Eq. (5.31) for single- and double-layer quantum Hall systems.

---

[28] This may seem surprising but has precedent. Already in 1961, Baym and Kadanoff [22] showed that theories violating gauge invariance can sometimes produce valid, gauge invariant response functions.

5.17.3 Semiclassical model for composite fermion dynamics

A semiclassical model for CF dynamics has been employed by several authors (Evers *et al.*
[144–146]; Fleischmann *et al.* [166]; Mirlin *et al.* [447–449]). The semiclassical equations
of motion for an electron at the Fermi energy are given by [14]

$$v(k) = \frac{1}{\hbar} \nabla_k \epsilon(k), \tag{5.164}$$

$$\hbar \dot{k} = -\frac{e}{c} v(k) \times B(r, t) - \nabla V(r), \tag{5.165}$$

where $\epsilon(k)$ is the energy dispersion for the electron and $V(r)$ is the potential sensed
by it. Analogous equations for a composite fermion at the Fermi energy are as follows
(Fleischmann *et al.* [166]):

$$v(k) = \frac{1}{\hbar} \nabla_k \epsilon^*(k), \tag{5.166}$$

$$\hbar \dot{k} = -\frac{e}{c} v(k) \times B^*(r, t) - \nabla V^*(r), \tag{5.167}$$

$$B^*(r) = B - 2p\phi_0 \rho(r), \tag{5.168}$$

$$V^*(r) = E_F^* - \epsilon^*(\rho(r)), \tag{5.169}$$

$$\rho = \frac{|k|^2}{4\pi}. \tag{5.170}$$

The replacement $B \to B^*$ naturally incorporates the effect of inhomogeneity in the effective
magnetic field induced by density variations. The naive expectation would be for composite
fermions to see the same potential as electrons. However, that would lead to a very different
density distribution for composite fermions from that for electrons at $B = 0$, as the two have
unrelated Fermi energies, and the density is determined by the condition $E_F = \epsilon(\rho(r)) +
V(r)$ (and a similar relation for composite fermions). A more reasonable assumption is that
the spatial density of composite fermions at $B^* \approx 0$ is the same as that of electrons at
$B = 0$, which implies that they see an effective potential V^*. By introducing a nonlinear
"time" variable $s(t)$ that satisfies $ds/dt = (1/\alpha)(d\epsilon^*/d\rho)|_{\rho=\rho(r(t))}$ (the arbitrary constant
α is introduced for dimensional reasons), and assuming that $v(k)$ and $\epsilon(k)$ depend only on
$|k|$ (hence are functions of ρ), we obtain

$$\frac{dr}{ds} = \frac{\alpha}{\hbar} \nabla_k \rho, \tag{5.171}$$

$$\hbar \frac{dk}{ds} = -\frac{e}{c} \frac{dr}{ds} \times B^* + \alpha \nabla \rho. \tag{5.172}$$

These equations are independent of the CF energy dispersion $\epsilon^*(k)$, implying that the
trajectories are independent of the CF mass. (We have already encountered that property
for the radius of the CF cyclotron orbit.) The semiclassical approximation has been used to
address nonlinear dynamics of composite fermions in nanostructures [166].

5.17.4 Boltzmann equation for composite fermions

The semiclassical Boltzmann transport equation is a powerful tool for analyzing transport phenomena. A Boltzmann equation can be written for composite fermions in close analogy to the standard Boltzmann equation for electrons. We begin by assuming a local distribution function $f(\mathbf{k}, \mathbf{r}, t)$, which measures the number of composite fermions in momentum state $\mathbf{k}$ at position $\mathbf{r}$ and time t. (The reason for the qualification "semiclassical" is that such a function is not strictly defined due to the Heisenberg uncertainty principle.) The current density is then given by

$$j(\mathbf{r}, t) = e \int d^2 k \frac{\mathbf{k}}{m^*} f(\mathbf{k}, \mathbf{r}, t) \tag{5.173}$$

where $\mathbf{v} = \mathbf{k}/m^*$ is the velocity. We need to determine how the distribution function is affected by the presence of applied electric field and temperature gradients in the presence of various types of scatterings. Besides using m^* and B^* for mass and magnetic field, a new aspect of the CF physics is that any external potential, e.g., created by random ionized impurities or by periodic modulation in a lateral superlattice, translates into a spatially varying $B^*(\mathbf{r})$. The Boltzmann equation for composite fermions (Mirlin and collaborators [447, 448]; Simon and Halperin [591]; Stern and Halperin [613]; von Oppen, Stern and Halperin [662]) is given by [447]

$$i(\omega - \mathbf{v} \cdot \mathbf{q}) f_1 + \frac{e}{c} (\mathbf{v} \times \mathbf{B}_0^*) \cdot \nabla_{\mathbf{k}} f_1 + e\mathbf{E} \cdot \nabla_{\mathbf{k}} f_0 = C\{f_1\}, \tag{5.174}$$

where f_1, the deviation of the distribution function from its equilibrium value f_0, is caused by the application of an electric field $\mathbf{E}(\mathbf{r}, t) = \mathbf{E} \exp(-i\omega t + i\mathbf{q} \cdot \mathbf{r})$. The collision integral, $C\{f_1\}$, includes the effects of scattering by random impurities as well as fluctuations in the effective magnetic field. The results for CF transport coefficients must be transformed back into electron transport coefficients before a contact with experiment may be made.

As for ordinary metals, the validity of the Boltzmann equation for composite fermions rests upon the existence of well-defined CF "quasiparticles" close to the Fermi energy. This requires a determination of the CF-quasiparticle width, which, in turn, is given by the imaginary part of the CF self energy. In the CFCS approach, the most singular correction to the CF self energy comes from transverse gauge field fluctuations, which gives (Halperin, Lee and Read [231]; Lee and Nagaosa [380])

$$\text{Re}\,\Sigma \sim \omega \ln \omega, \qquad \text{Im}\,\Sigma \sim \omega \tag{5.175}$$

for the Coulomb interaction. As the imaginary part of the self energy is of the same order as the energy itself, it does not satisfy the Landau criterion for the Landau Fermi liquid, namely that the quasiparticle width $(\text{Im}\,\Sigma)$ vanish faster than the energy; we have an example of what is called a "marginal" Landau Fermi liquid. Nonetheless, a Boltzmann equation description can be valid, as shown by Kim, Lee, and Wen [334]. They follow the approach of Prange and Kadanoff [525], who had derived transport equations for electrons coupled to phonons and found that they have the same form as from Landau's Fermi liquid theory even when

Landau quasiparticles are not well defined (for example, at temperatures large compared to the Debye temperature). Using a nonequilibrium Green's function approach, Kim, Lee, and Wen discover that the generalized Fermi surface displacement satisfies a closed equation of motion for composite fermions, called the quantum Boltzmann equation; the smooth Fermi surface fluctuations conform to the usual Landau Fermi liquid behavior, whereas the rough ones do not. In other words, the "quasiparticle approximation" is valid for quantities that are dominated by smooth fluctuations, which include the physically relevant density–density and current–current correlation functions, but not for all quantities (e.g., the energy gap away from $\nu = 1/2$).

Parenthetically, the quasiparticle approximation is not valid for interactions that have a shorter range than Coulomb, i.e., for which $V(q) \sim q^{-1+\eta}$ with $1 \geq \eta > 0$ ($\eta = 0$ corresponds to Coulomb). Such interactions are less effective in suppressing gauge field fluctuations, producing real and imaginary parts of the self energy that are proportional to $\omega^{2/(2+\eta)}$, implying a strong violation of the Landau criterion. For interactions shorter ranged than Coulomb, the quasiparticle approximation does not produce the same specific heat for the CF sea as that obtained from the free energy of the gauge field (Kim and Lee [335]).

The Kubo conductivity tensor can also be evaluated in the semiclassical approximation [220,552]. In a path integral representation, the Kubo expression for the conductivity can be written as a double sum over the semiclassical periodic orbits, one from each of two Green's functions. Quantum corrections to conductivity appear through the phase associated with each orbit, given by the classical action of the orbit, which can lead to quantum oscillations or other phase coherence effects as a function of the magnetic field. The semiclassical method is convenient for dealing with spatial fluctuations or modulations in the effective magnetic field, particularly for treating the "snake" trajectories of composite fermions along lines of zero effective magnetic field [75], and has been employed for calculating CF conductivity at a finite effective magnetic field, finite wave vector, and finite frequency [144–146, 166, 447–449].

5.17.5 Fermi hypernetted-chain approximation

The hypernetted-chain approximation is a widely used method for calculating the pair correlation function $g(r)$ for liquids in equilibrium [236]. The pair distribution function is expanded in powers of density to produce a cluster expansion (the Mayer–Montroll expansion), the terms of which are represented by linked irreducible diagrams. The sum of these diagrams can be converted into an integral equation. The exact equation is intractable, but can be solved self consistently by discarding a class of diagrams (those free of "nodal circles"). The resulting self-consistent integral equation is called the hypernetted-chain equation, which is accurate in the limit of low densities. It has been used to deduce $g(r)$ for Jastrow correlated wave functions for boson liquids. Ciftja and collaborators [82–87] have developed a Fermi hypernetted-chain method to carry out approximate computations of the pair correlation function and energies (which can be obtained from $g(r)$) for the unprojected wave functions of Eq. (5.31). The method captures the behavior qualitatively and semiquantitatively.

5.17.6 Evolution from one dimension to two dimensions

Bergholtz and Karlhede [27, 28] study the evolution of a FQHE system on a cylinder as a function of the radius of the cylinder. When the radius is very small, the circular single particle orbitals of the Landau gauge are far separated along the length of the cylinder. In that limit, at any rational filling, a crystal state is obtained with a gap to excitations. Bergholtz and Karlhede argue that at $\nu = n/(2pn \pm 1)$ this crystal evolves continuously into the bulk incompressible FQHE state as the radius of the cylinder is increased. At $\nu = 1/2$, on the other hand, it undergoes a transition into a compressible state when the radius of the cylinder is approximately equal to one magnetic length; numerical calculations make a convincing case that the transition is into a finite-size version of the CF Fermi sea. Furthermore, the $\nu = 1/2$ system at small but nonzero radii is well described by a short-range interaction model with nearest neighbor hopping [27, 28], which can be solved exactly with the help of the well-known Jordan–Wigner transformation, and produces a non-interacting Fermi sea. Understanding how this 1D Fermi sea evolves into the CF Fermi sea in the 2D limit should prove interesting.

5.17.7 Conformal field theory

A correspondence has been made between certain FQHE wave functions and certain correlation functions of chiral conformal field theory (Cristofano *et al.* [100]; Fubini [176]; Moore and Read [454]), which is very briefly outlined in this section. We use some standard results from conformal field theory without derivation. (The derivations can be found, for example, in the textbook of Di Francesco, Mathieu, and Senechal [118].)

Consider a free bosonic field in $1+1$-dimensional Euclidean spacetime, with its correlator given by

$$\langle \phi(z)\phi(z') \rangle = -\ln(z - z').$$ (5.176)

The so-called vertex operators are defined by

$$\mathcal{V}_\alpha(z) = e^{i\alpha\phi(z)}.$$ (5.177)

With the help of Wick's theorem, their correlators can be shown to be given by the expression

$$\left\langle \prod_i \mathcal{V}_{\alpha_i}(z_i) \right\rangle = \exp\left[-\sum_{j<k} \alpha_j\alpha_k \langle \phi(z_j)\phi(z_k) \rangle \right] = \prod_{j<k} (z_j - z_k)^{\alpha_j\alpha_k},$$ (5.178)

provided the neutrality condition,

$$\sum_i \alpha_i = 0,$$ (5.179)

is satisfied; otherwise the correlator vanishes identically. (The vertex operators are assumed to be normal ordered, so only Wick contractions between boson fields at different z_i are considered.)

Disregarding the neutrality condition, the choice $\alpha_j = \sqrt{m}$ gives precisely the Laughlin wave function on the right hand side of Eq. (5.178). (The neutrality condition can be taken care of by introducing a uniform background charge through a term $\exp[-i\sqrt{m} \int d^2z' \rho \phi(z')]$, where ρ is the charge density. This term produces $\exp(-\sum_j |z_j|^2/4)$, and an additional singular phase factor that is neglected.) Inserting the vertex operator $\mathcal{V}_{1/\sqrt{m}}(\eta)$ produces a Laughlin quasihole (Eq. 12.29) at η.

There is no fundamental reason why the correlation functions (or conformal blocks) of vertex operators in a two-dimensional Euclidean conformal field theory should bear any relation to the quantum mechanical wave functions of electrons in the lowest Landau level interacting via the Coulomb potential. Nonetheless, one can ask if some other correlation functions in conformal field theory may also qualify as legitimate FQHE wave functions. Of course, conformal field theory cannot *derive* FQHE wave functions; any ansatz wave function must be tested and confirmed in a quantum mechanical calculation.

Moore and Read [454] interpret the Pfaffian wave function of Eq. (7.11), originally obtained as a p-wave generalization of the Haldane–Rezayi wave function (Eq. E7.6), using conformal field theory. They introduce Majorana fermions (massless real fermions) χ, which satisfy $\langle \chi(z)\chi(z') \rangle = (z-z')^{-1}$. The Pfaffian wave function can then be expressed as (again suppressing the background term)

$$\left\langle \prod_j \chi(z_j) e^{i\sqrt{m}\phi(z_j)} \right\rangle. \tag{5.180}$$

Conformal field theory has also been extended to other FQHE states of the series $\nu = n/(2pn \pm 1)$ (Flohr [167]; Flohr and Osterloh [168]; Hansson et al. [239]). Hansson et al. have constructed conformal field theory representations for the quasiparticle of $\nu = 1/m$ (Eq. 12.37), as well as for the FQHE states at $\nu = n/(2pn + 1)$ (Eq. 5.32). For example, defining

$$\mathcal{V}^{(0)}(z) = e^{i\sqrt{3}\phi_0(z)} \quad \text{and} \quad \mathcal{V}^{(1)}(z) = \partial_z e^{i\frac{2}{\sqrt{3}}\phi_0(z)} e^{i\sqrt{\frac{5}{3}}\phi_1(z)}, \tag{5.181}$$

where ϕ_0 and ϕ_1 are two free scalar fields, the wave function for the 2/5 state is given by

$$\Psi_{2/5} = \sum_{i_1 < i_2 \cdots < i_n} (-1)^{\sum_k^n i_k} \left\langle \mathcal{V}^{(1)}(z_{i_1}) \ldots \mathcal{V}^{(1)}(z_{i_n}) \prod_{j \notin \{i_1 \ldots i_n\}}^N \mathcal{V}^{(0)}(z_j) \right\rangle \tag{5.182}$$

with $n = N/2$. This exactly reproduces the projected wave function of Eq. (5.32) with n composite fermions in the second Λ level and the remaining in the lowest [239]. The vertex operators $\mathcal{V}^{(0)}$ and $\mathcal{V}^{(1)}$ create composite fermions in the lowest two Λ levels. Vertex operators corresponding to composite fermions in higher Λ levels can be constructed by introducing one additional bosonic field for each level [239].

Exercises

5.1 Going from the mean-field wave function (Eq. 5.30) to the unprojected wave function (Eq. 5.31) lowers both the interaction and the kinetic energies. When the latter is projected into the lowest Landau level however, the kinetic energy is lowered at the cost of an increase in the interaction energy. The unprojected wave function thus has lower energy than the projected one at sufficiently small cyclotron energies. Taking the numbers in Table 5.1, and parameters appropriate for GaAs, determine the magnetic fields, and also the corresponding densities, below which the unprojected wave function has lower energy than the projected one at $v = 2/5$, $3/7$ and $4/9$. What does the result mean? Does it imply a phase transition as a function of the cyclotron energy?

5.2 Show that the wave functions $\Phi_{\pm v^*}$ and Ψ_v in Eq. (5.42) are confined to a disk of the same radius. (Hint: Determine the outermost occupied orbital.)

5.3 Derive Eq. (5.94).

5.4 This exercise concerns a derivation of Eq. (5.128) for the induced Chern–Simons electric field e_{CS}. This electric field is related, through Faraday's law, to the Chern–Simons magnetic field b as

$$\nabla \times e_{CS}(r, t) = -\frac{1}{c}\frac{\partial}{\partial t}b(r, t), \tag{E5.1}$$

where

$$b(r, t) = -2p\phi_0\rho(r, t)\hat{z}. \tag{E5.2}$$

Transform these relations, as well as the continuity equation

$$\nabla \cdot j + (-e)\frac{\partial\rho}{\partial t} = 0, \tag{E5.3}$$

to Fourier space by substituting $e_{CS}(r, t) = e_{CS}^0 e^{-i(qx-\omega t)}$ and analogous equations for other space and time dependent variables (the wave vector is chosen along the x direction without any loss of generality). Obtain Eq. (5.128) by eliminating ρ in favor of j. Note that Eq. (5.128) is independent of q and ω, and remains valid in the static limit.

5.5 This exercise follows the mean-field physics described in Sections 5.16.3 and 7.1 to obtain the correct Hall resistance for composite fermions at $v^* = n$. The apparent Hall resistance is $V_{apparent}/I = \pm h/ne^2$, where $V_{apparent} = V_{external} + V_{induced}$. Show that if one assumes that each electron carries $2p$ flux quanta, then Faraday's law implies that $V_{induced} = -2p\phi_0 I/ec$. Calculate $R_H = V_{external}/I$.

6

Microscopic verifications

Whenever new emergent particles are postulated, one must ask: "So what? What do these particles do for us?" The significance of such particles is determined by the scope and the importance of experimental phenomena they produce. For particles advertised as truly exotic to show up in but minor experimental details would be hardly satisfying.

The remainder of this book concerns experimental manifestations of composite fermions. Before proceeding to laboratory experiments, we ask, in this chapter, if composite fermions can be "seen" in computer experiments.

6.1 Computer experiments

Let us recall how quantum mechanics was verified. It began by explaining the spectral lines of the hydrogen atom. With further work, its predictions were seen to be in exquisite agreement with the excitation spectra of helium and other atoms in the periodic table, and also of molecules. This vast body of exceedingly detailed comparisons, now known as quantum chemistry, is what led to the establishment of quantum mechanics as the quantitative theory of matter. Quantum mechanics was then used with great effect in systems containing many electrons, for example metals and superconductors, and now the fractional quantum Hall effect.

This chapter is devoted to the quantum chemistry of composite fermions. Nature does not produce any *real* "FQHE atoms," but, fortunately, they can be created on the computer. These atoms are characterized by two parameters: the number of electrons and the amount of magnetic flux to which they are exposed. Theorists can perform computer experiments to determine their *exact* spectra and wave functions. Exact solutions are possible because, in the limit of $B \to \infty$, when the Hilbert space is restricted to the lowest Landau level, finite systems have only a finite number of many-body basis states, which allows a brute force numerical diagonalization of the Hamiltonian for the Coulomb interaction. That is why these studies are dignified as "experiments," as opposed to, say, "simulations."

In a completely independent calculation, the eigenenergies and eigenfunctions for the low-energy states can be obtained from the CF theory, by appealing to the analogy with the IQHE. We ask in this chapter how well the results from the two sets of calculations match.

While computer experiments cannot replace real laboratory ones, they offer certain complementary advantages. They produce a tremendous amount of – in fact, *all* possible – information, which allows for a much more extensive, rigorous and direct testing of a theory than can ever be possible with laboratory experiments. Suppose we wish to convince ourselves that our favorite wave function for the ground state is "correct." Because laboratory experiments do not tell us what the ground state wave function is, we must deduce the validity of our wave function on the basis of testing a finite number of its observable consequences against experiment. Computer experiments, on the other hand, enable a direct comparison of our wave function with the exact one. A single successful comparison of the wave function is equivalent to verifying a large number of observables at once.

The second advantage, in the context of the FQHE, is that computer experiments produce identical numbers no matter where, when or by whom they are performed, and therefore offer theory a single set of numbers to explain. The numbers produced in laboratory experiments are affected by several features extraneous to the physics of the FQHE, for example, the type of confinement potential (triangular, square well, or parabolic), the amount of Landau level mixing (which is determined by the electron density, the electron band mass, and the dielectric constant), and the ever-present disorder. As a result, different laboratory experiments produce different numbers. That is an unfortunate feature of man-made structures as opposed to naturally occurring systems like atoms or molecules. We could carry out a separate calculation for each sample, but even then the additional features mentioned above would be dealt with only approximately, making it difficult to determine what contributes how much to the discrepancy between theory and experiment. In computer experiments, on the other hand, it is customary to consider an idealized model in which all extrinsic features – disorder, Landau level mixing, and thickness – are set to zero, enabling a determination of the *intrinsic* accuracy of the theory of the FQHE. It is further assumed, in this chapter, that the electron system is fully spin polarized; the spin part of the wave function is completely symmetric and will not be considered explicitly.

6.2 Relevance to laboratory experiments

Computer experiments can only be performed on systems consisting typically of fewer than 12–15 electrons in the region of interest. What relevance do they have to the phenomenon of the FQHE which occurs for systems with many more electrons?

Even though computer systems are finite, they are sufficiently large to be nontrivial. If these can be understood in detail, it would not be surprising should the understanding extend to much larger systems, just as the principles tested in detail for atoms with only a few electrons carry over to the physics of metals and superconductors. The size of the largest computer samples significantly exceeds the typical length in the problem, namely the magnetic length, so it is expected that they reflect the physics of the thermodynamic limit, at least for the more robust features. Nevertheless, the appearance of some subtle problem on the way to the thermodynamic limit cannot be excluded. Also, certain weak states might not become manifest in system sizes amenable to computer experiments. It is therefore crucial to compare the consequences of theory also with laboratory experiments, where nature is diagonalizing the Hamiltonian for us in the thermodynamic limit. Such comparisons are presented in detail in the following chapters. If theory is able to explain both ends, we can feel secure in our understanding.

6.3 A caveat regarding variational approach

We compare below the wave functions of Eqs. (5.32) and (5.36) with exact wave functions. In that context, a caveat is in order. The variational theorem must be treated with great care in condensed matter physics, because it is entirely possible – and is often the case – that the energy is not particularly sensitive to the essential physics. The physics of interest in condensed matter systems has often to do with the long-distance behavior of the wave function, whereas the energy is dominated by short-distance correlations. A low energy, by itself, is not sufficient for the validity of a variational wave function. The overlap of a variational wave function with the exact one can also be misleading at times. Examples are known when a wave function with a manifestly wrong symmetry has a pretty high overlap with the exact wave function for small systems; on the flip side, a wave function with the correct physics built in may have a rather poor overlap with the exact one.

The CF theory, however, is not merely a calculational scheme. It makes precise qualitative statements, independent of which formulation is used to describe quantitative details, regarding the reorganization of the low-energy Hilbert space. We must first verify these qualitative features to make sure that the CF theory captures the correct physics. Only then will testing the wave functions be meaningful.

6.4 Qualitative tests

A system of electrons on a sphere is labeled $(N, 2Q)$, N being the number of electrons and $2Q$ the number of flux quanta passing through the surface of the sphere. Figures 6.1, 6.2 and 6.3 show a number of exact spectra (dashes). The energies here and below include contributions from electron–background as well as background–background interactions (Appendix I). The total orbital angular momentum L is a good quantum number, used to label eigenstates; each dash in the figure represents a multiplet of $(2L+1)$ degenerate states.

We begin by recalling that all states are degenerate in the absence of the Coulomb interaction, which therefore is solely responsible for the structure seen in these spectra. Ordinarily, interaction is expected to cause a featureless broadening of the density of many-particle states. As evident in the spectra, however, a nontrivial rearrangement of the low-energy Hilbert space takes place. Some states break off from the rest to form a low-energy band, which is separated from the rest by a gap. Formation of such a low-energy band occurs no matter how weak the interaction (which only alters the energy scale on the y-axis), and thus is a nonperturbative effect. The structure of the low-energy band changes in an apparently haphazard manner with the flux $2Q$. In certain cases, e.g., those shown in Fig. 6.1, the lowest band contains a single state. Here the interaction kills the enormous

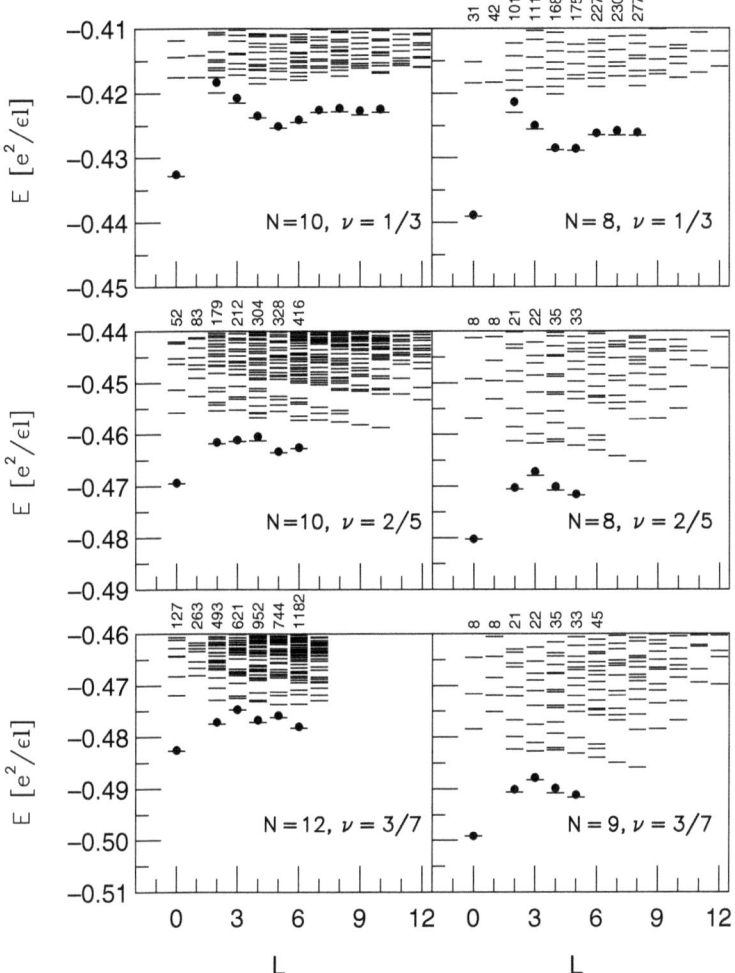

Fig. 6.1. Comparison of spectra obtained from exact diagonalization (dashes) and CF theory (dots). Top two spectra correspond to $(N, 2Q) = (10, 27)$ and $(8, 21)$; middle two spectra correspond to $(N, 2Q) = (10, 21)$ and $(8, 16)$; and bottom two spectra correspond to $(N, 2Q) = (12, 23)$ and $(9, 16)$. These are finite size representations of $\nu = 1/3$, $2/5$ and $3/7$, respectively. Each dash represents a multiplet of $2L + 1$ degenerate states. The CF energies are obtained with no adjustable parameters, without CF diagonalization. The $L = 0$ ground state is interpreted as an integral number of filled Λ levels, and the branch of low-energy excitations as an exciton of composite fermions; their wave functions are given in Eqs. (5.32) and (5.62). Exact eigenenergies are taken from He *et al.* (Ref. [244, 246–248], and private communication), and CF energies from Ref. [282]. The statistical uncertainty in CF results, obtained by Monte Carlo, is an order of magnitude smaller than the size of the dots. Source: J. K. Jain and R. K. Kamilla, Chapter 1 in *Composite Fermions*, ed. O. Heinonen (New York: World Scientific, 1998). (Reprinted with permission.)

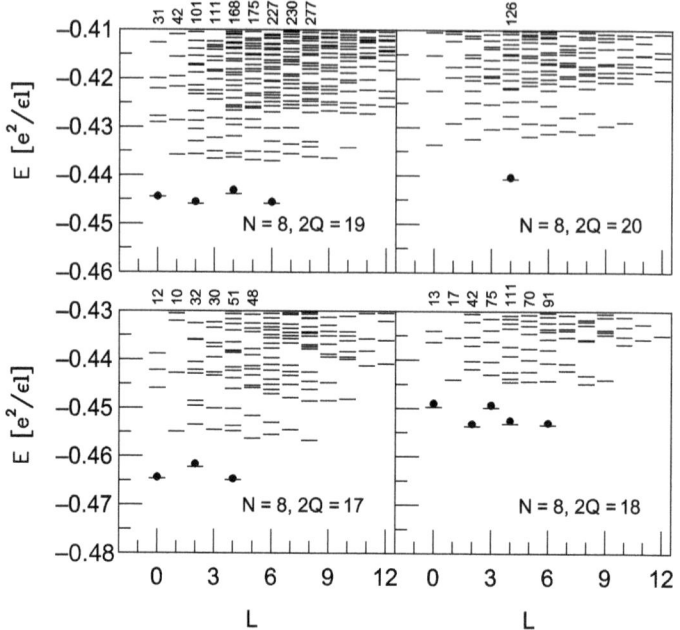

Fig. 6.2. Exact (dashes) and CF (dots) spectra for $(N, 2Q) = (8, 17)$, $(8, 18)$, $(8, 19)$ and $(8, 20)$. These describe the system in the filling factor range $1/3 < \nu < 2/5$, with the lowest Λ level fully occupied while the second one is partially occupied. Different states in the lowest band correspond to distinct configurations of composite fermions in the second Λ level. Exact energies are taken from He, Xie, and Zhang [244]. CF energies are obtained from Eq. (5.36) without any fitting parameters. Source: J. K. Jain and R. K. Kamilla, Chapter 1 in *Composite Fermions*, ed. O. Heinonen (New York: World Scientific, 1998). (Reprinted with permission.)

degeneracy of the noninteracting problem to produce a gap. An excited band can also be identified in some cases, although it is, in general, less well defined than the lowest energy band.

Let us now put to the test the prediction of the CF theory that the low-energy spectrum of interacting electrons at Q is analogous to that of weakly interacting fermions at

$$Q^* = Q - N + 1, \tag{6.1}$$

where we have set $p = 1$. The free fermion system at Q^* has bands of quasi-degenerate states in the *many*-body spectrum at Q^*, as a result of the quantization of the kinetic energy of a single fermion into Landau levels. The quantum numbers of states in various bands can be enumerated straightforwardly, given that the particles are noninteracting.

6.4.1 Lowest band

Let us consider $(N, 2Q) = (12, 23)$, which maps into a system of weakly interacting composite fermions at $(N, 2Q^*) = (12, 1)$. Here, the lowest kinetic energy state has the

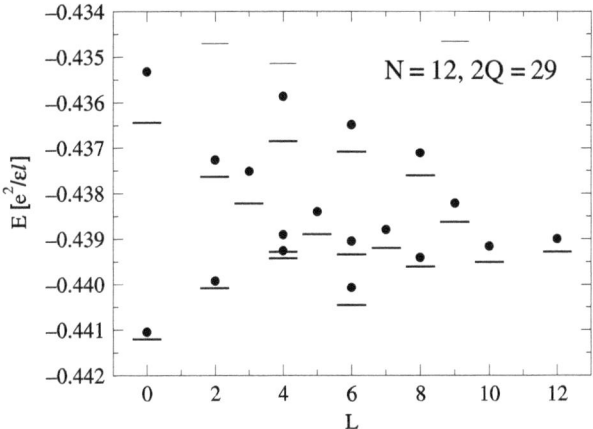

Fig. 6.3. Comparison between the exact spectrum (dashes) with the spectrum obtained by CF diagonalization (dots) for $(N, 2Q) = (12, 29)$. Only the very low-energy part of the exact spectrum is shown. Darker dashes form a low-energy band of states, which have a one-to-one correspondence with states predicted by the CF theory. Exact eigenenergies are taken from E. H. Rezayi (unpublished) and the CF energies from Refs. [67,424]. Source: J. K. Jain, C.-C. Chang, G.-S. Jeon, and M. R. Peterson, *Solid State Commun.* **127**, 809 (2003). (Reprinted with permission.)

lowest three Landau levels completely full, with 2, 4, and 6 electrons in them. Thus, the filling factor is $\nu^* = 3$, and the ground state is a filled shell state with $L = 0$, separated from other states by a gap. That is consistent with the exact spectrum.

When the system maps into a partially filled Λ level of composite fermions, many degenerate ground states result, the angular momentum quantum numbers of which can be obtained straightforwardly. To illustrate the procedure, let us consider the example of $N = 3$ fermions in an angular momentum shell $l = 3$. To obtain the allowed angular momenta within the Pauli constraint we make Table 6.1, which lists all possible occupations (m_1, m_2, m_3), with no two m_j's being the same, according to their total $L_z = m_1 + m_2 + m_3$ quantum number, where m_j is the z-component of the angular momentum of the jth electrons. Negative values of L_z are not shown because they can be obtained from the positive values by a sign change everywhere. Clearly, a multiplet with $L = 6$ is allowed. This contains 13 states with $L_z = 6, 5, \ldots, -5, -6$. The only state at $L_z = 5$ belongs to the $L = 6$ multiplet, and, therefore, there is no multiplet with $L = 5$. One linear combination of the two states at $L_z = 4$ belongs to $L = 6$, but the other to a new multiplet at $L = 4$. Proceeding in this manner, we determine that the allowed angular momentum multiplets occur at $L = 6, 4, 3, 2$, and 0. We have derived the relation (for fermions):

$$3 \otimes 3 \otimes 3 = 6 \oplus 4 \oplus 3 \oplus 2 \oplus 0. \tag{6.2}$$

Let us make sure we are not missing anything. The total number of states is $\binom{7}{3} = 35$. On the other hand, the total number of states for each multiplet is $2L + 1$, which, altogether,

Table 6.1. *Classification of all Fock states for three electrons at*
$Q = 3$.

L_z		States			
6	$(3,2,1)$				
5	$(3,2,0)$				
4	$(3,2,-1)$	$(3,1,0)$			
3	$(3,2,-2)$	$(3,1,-1)$	$(2,1,0)$		
2	$(3,2,-3)$	$(3,1,-2)$	$(3,0,-1)$	$(2,1,-1)$	
1	$(3,1,-3)$	$(3,0,-2)$	$(2,1,-2)$	$(2,0,-1)$	
0	$(3,0,-3)$	$(3,-1,-2)$	$(2,1,-3)$	$(2,0,-2)$	$(1,0,-1)$

Note: All possible (m_1, m_2, m_3) are listed for each L_z.

also gives $13 + 9 + 7 + 5 + 1 = 35$. The presence of a single state at $L = 0$ implies that if we can construct *a* wave function with $L = 0$, we have found the exact ground state.

The angular momentum quantum numbers predicted by the CF theory are in complete agreement with the actual angular momenta of the quasi-degenerate states of the lowest energy bands of the exact spectra shown in Figs. 6.1, 6.2, and 6.3 (Exercise 6.3).

6.4.2 Exciton branch

Let us consider again $(N, 2Q) = (12, 23)$, which corresponds to $(N, 2Q^*) = (12, 1)$ with three filled shells. The lowest-energy excitation is obtained by promoting a composite fermion from the third Λ level shell into the fourth Λ level shell. The angular momentum of the lowest Λ level is $l = Q^*$ and increases by one unit in each successive Λ level. Thus, the angular momentum of the excited composite fermion is $7/2$ and of the hole left behind is $5/2$. From angular momentum addition, we have

$$\frac{7}{2} \otimes \frac{5}{2} = 1 \oplus 2 \oplus 3 \oplus 4 \oplus 5 \oplus 6.$$

These values match the actual quantum numbers of states in the exciton branch in Fig. 6.1, with one exception: an $L = 1$ state is predicted for the CF-exciton for this as well as other incompressible states (Exercise 6.4), but is missing in the actual spectrum. This discrepancy is resolved when the wave function for the $L = 1$ CF-exciton is constructed according to the standard prescription; the $L = 1$ CF-exciton wave function is annihilated by the LLL projection [115] for all incompressible states.

6.4.3 Higher-energy bands

At higher energies, in general, no bands are clearly identifiable in exact spectra. Even when they are, discrepancies exist between the number of states in these bands and the number of states in corresponding bands at Q^*. The CF mean-field theory predicts many spurious states. (We already saw an example of that for the exciton band, which is the first excited band for incompressible states.) A breakdown of the correspondence is to be expected, given that the system $(N, 2Q)$, confined to the lowest Landau level, has a finite number of states, whereas the system $(N, 2Q^*)$, where we allow higher Landau levels, has an infinite number of states. Remarkably, in all cases studied so far [705], the LLL projection annihilates the spurious states in each band, bringing the prediction of the CF theory into accordance with actual spectra. How many states are annihilated is not generally understood.

6.4.4 Negative B^* versus particle–hole symmetry

For certain $(N, 2Q)$ values, an application of the CF transformation according to $Q^* = Q - p(N - 1)$ produces a negative value of Q^*, implying that the sign of magnetic field has switched in going from electrons to composite fermions. (The spectrum is independent of the direction of magnetic field, i.e., the sign of Q^*.) We can alternatively proceed by first making an exact particle–hole transformation into $(N_h, 2Q)$, where $N_h = 2Q + 1 - N$ is the number of holes in the lowest Landau level. Now, we turn the *holes* into composite fermions at $Q^* = Q - p(N_h - 1)$, which is positive. Which one is the right approach? While the physical descriptions may appear rather different, the two methods produce identical answers insofar as the energy level counting is concerned (Exercise 6.6).

6.5 Quantitative tests

Chapter 5 explains how the CF physics enables a construction of correlated wave functions for the lowest band states, as well as for excited states, from the knowledge of wave functions of noninteracting electrons at the effective magnetic field. If the wave function is unique, its Coulomb expectation value can be evaluated straightaway. When the basis contains more than one state, CF diagonalization is necessary.

Predicted energies are shown by dots in Figs. 6.1, 6.2, and 6.3. The exact eigenenergies and the energies predicted by the CF theory are explicitly given in Table 6.2, for both the ground state and the excitation with the particle–hole pair at maximum separation (on north and south poles), and in Table 6.3 for the system corresponding to Fig. 6.3. Table 6.4 gives overlaps of the wave functions of Eq. (5.32) with the exact ground states for several incompressible states. (All states are properly normalized for this calculation, so two identical states have an overlap of 1.0.)

A remark on overlaps: Overlaps decrease with increasing size. In fact, they eventually vanish exponentially with N. For illustration, let us consider a problem of bosons, and compare a trial wave function for a Bose condensate, $\prod_j \Phi_0(r_j)$, with the exact wave function, $\prod_j \Phi_{\text{exact}}(r_j)$. Further, we assume that Φ_0 and Φ_{exact} have an overlap close to unity:

$$\langle \Phi_0 | \Phi_{\text{exact}} \rangle = e^{-\epsilon}. \tag{6.3}$$

This implies that the N-particle wave function has an overlap $e^{-\epsilon N}$, which may be close to unity for small N but will eventually vanish exponentially with N. That, however, should

Table 6.2. *Comparison between exact and CF energies for*
incompressible states

ν	N	Ground state		Excited state	
		CF	Exact	CF	Exact
$\frac{1}{3}$	8	$-0.438\,86(1)$	$-0.439\,10$	$-0.426\,07(3)$	$-0.426\,55$
	10	$-0.432\,58(0)$	$-0.432\,84$	$-0.422\,42(3)$	$-0.422\,91$
$\frac{2}{5}$	8	$-0.480\,22(3)$	$-0.480\,24$	$-0.471\,44(8)$	$-0.471\,73$
	10	$-0.469\,34(7)$	$-0.469\,45$	$-0.462\,54(5)$	$-0.462\,74$
$\frac{3}{7}$	9	$-0.499\,14(7)$	$-0.499\,18$	$-0.491\,46(8)$	$-0.491\,62$
	12	$-0.482\,51(5)$	$-0.482\,64$	$-0.478\,19(5)$	$-0.478\,26$

Notes: Exact and CF energies per particle for the ground and excited
states for several systems at $\nu = 1/3, 2/5$ and $3/7$.

The excited state considered here is obtained by removing a single
composite fermion from the south pole in the highest occupied Λ level
and placing it on the north pole in the lowest unoccupied Λ level.

Energies are in units of $e^2/\epsilon\ell$, and include interaction with the
uniform positively charged background.

Statistical uncertainty in the last digit(s) of the CF energy, computed
by Monte Carlo techniques, is shown in brackets.

Source: Exact results are taken from He, Simon, and Halperin [247]
and Fano, Ortolani, and Colombo [151]. CF results are from Jain and
Kamilla [280, 281].

not be taken to mean that the wave function becomes invalid for large N, or that the large
overlaps for small N are meaningless. The wave function describes the correct physics
(all bosons in the same state), and continues to predict accurately many observables in
the $N \to \infty$ limit, for example, the energy per particle, energy differences, pair correlation
function, etc. How do we decide when overlaps are meaningful? Coming back to the FQHE
problem, we can make the following statements:

(i) The quantity $\epsilon = -\ln(\text{overlap})/N$ is very small.
(ii) A random wave function with the correct symmetry (with given L and L_z quantum numbers in
the spherical geometry) will have an overlap of $1/\sqrt{D}$, where D is the number of basis states in
the relevant sector. Near unity overlaps are, thus, not coincidental.
(iii) Because energies are largely determined by the short-distance behavior, it is plausible that the
energies from the CF theory will remain accurate in the limit $N \to \infty$.
(iv) The qualitative correctness of the CF physics has been confirmed by other means.

Figure 6.4 shows the exact densities at several N and Q, shown on the figures, as a
function of the distance from the north pole. The ground state multiplets occur at $L = 4$

Table 6.3. *Comparison between exact and CF
energies for the lowest band*

L	E^{exact}	E_{CF}
0	−0.441 214	−0.441 051(89)
	−0.436 440	−0.435 323(69)
2	−0.440 457	−0.439 922(86)
	−0.437 646	−0.437 268(59)
3	−0.438 226	−0.437 516(72)
4	−0.439 422	−0.439 260(46)
	−0.439 280	−0.438 904(58)
	−0.436 844	−0.435 864(49)
5	−0.438 904	−0.438 400(63)
6	−0.440 547	−0.440 072(75)
	−0.439 337	−0.439 050(42)
	−0.437 093	−0.436 488(80)
7	−0.439 190	−0.438 794(79)
8	−0.439 613	−0.439 404(98)
	−0.437 615	−0.437 108(88)
9	−0.438 632	−0.438 215(10)
10	−0.439 507	−0.439 160(50)
12	−0.439 287	−0.439 000(11)

Notes: Energies for lowest energy band at $2Q = 29$ for $N = 12$
particles.
　　CF energy obtained by CF diagonalization; exact energy
taken from E. H. Rezayi (unpublished).
　　The spectra are shown in Fig. 6.3.
　　The number of multiplets at each L is obtained correctly
by the CF theory.
Source: Jain, Chang, Jeon, Peterson [285].

and 3 for the left panels, and $L = 4$ and 2.5 for the right panels. (The displayed densities
are for the $L_z = L$ member of the multiplet.) These systems represent CF-quasiparticles and
CF-quasiholes (Exercise 6.4; see Fig. 5.4 for a depiction of the CF-quasiparticle and CF-
quasihole of $\nu = 3/7$). Also shown in Fig. 6.4 are the density profiles calculated from the
unique wave functions for the CF-quasiparticle and the CF-quasihole given in Eqs. (5.60)
and (5.61). (For $\nu = 1/3$, the CF-quasihole wave function is the same as the one proposed
earlier by Laughlin.)

Table 6.4. *Overlaps between exact and CF wave*
functions for several incompressible states

ν	n	N	Overlap
$\frac{1}{3}$	1	7	0.9964
		8	0.9954
		9	0.9941
$\frac{2}{5}$	2	6	0.9998
		8	0.9996
$\frac{3}{7}$	3	9	0.9994
$\frac{2}{3}$	-2	6	0.9965
		8	0.9982
		10	0.9940

Sources: Results are taken from Fano, Ortolani, and Colombo
[151] for $\nu = 1/3$; Dev and Jain [116] for $\nu = 2/5$; and Wu,
Dev, and Jain [703] for $\nu = 3/7$ and $2/3$.

6.5.1 Gaps

Since the gap is an $O(1)$ energy obtained as the difference between two large $O(N)$ energies,
its accuracy is expected to be less than that of either the ground state or the excited state
energy. A comparison of gaps predicted by the CF theory with the corresponding exact gaps
(Table 6.5) shows that the discrepancy is on the order of a few percent. The full dispersion
of the CF-exciton branch has a similar level of accuracy, as seen in Fig. 6.1.

6.5.2 Composite fermions in negative B^*

A filling factor of the type $f = n/(2n - 1)$ can be understood either as an effective filling
$\nu^* = n$ of composite fermions, but with the effective magnetic field *anti*parallel to the
external magnetic field, or as the hole analog of $f' = (n - 1)/[2(n - 1) + 1]$, which
maps into $\nu'^* = n - 1$ of composite fermions. The results shown above, combined with
the exactness of the particle–hole transformation in the lowest Landau level, guarantee
that the latter interpretation is correct. Table 6.6 and Fig. 6.5 (also the results for 2/3 in
Table 6.4) demonstrate the validity of the idea of composite fermions in a negative B^*,
which maps 2/3, 3/5 and 4/7 into 2, 3 and 4 filled Λ levels, for both the ground state and
the CF-exciton. Thus, both interpretations are equivalent. The fact that $\mathcal{P}_{\text{LLL}} \Phi_1^2 \Phi_{n+1}^*$ and
$\mathcal{P}_{\text{LLL}} \Phi_1^2 \Phi_n$ are accurately related by particle–hole symmetry is nontrivial, because there is
no a-priori reason for it. As seen in Chapter 10, several experiments have directly confirmed
that the cyclotron orbit of current carriers for $\nu > 1/2$ is consistent with a negative B^*.

Table 6.5. *Comparing exact gaps with CF predictions*

ν	N	Gap	
		CF	Exact
$\frac{1}{3}$	8	0.1023(3)	0.1004
	10	0.1016(3)	0.0993
$\frac{2}{5}$	8	0.0702(9)	0.0681
	10	0.0681(12)	0.0671
$\frac{3}{7}$	9	0.0691(14)	0.0681
	12	0.0518(12)	0.0525

Notes: Exact and CF gaps for several systems at $\nu = 1/3, 2/5$ and $3/7$.

The excited state considered here is the largest CF exciton, with the CF-quasiparticle and CF-quasihole on opposite poles.

Source: Exact results are taken from He, Simon, and Halperin [247] and Fano, Ortolani, and Colombo [151]. CF results are from Jain and Kamilla [280, 281].

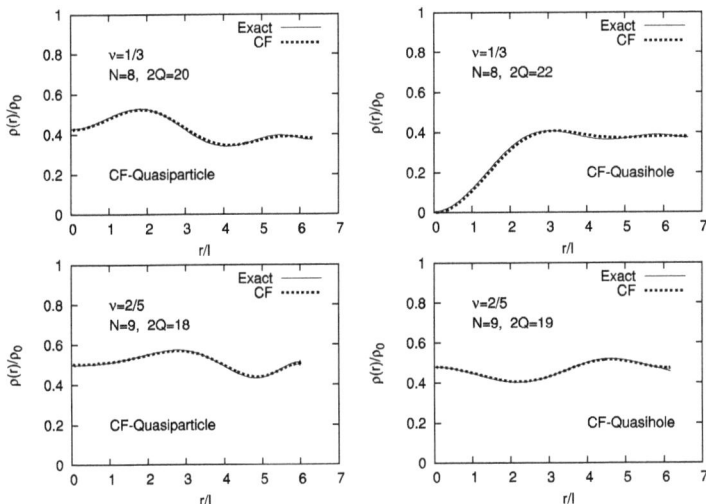

Fig. 6.4. Comparison of the densities calculated from the unique wave functions of the CF theory (Eqs. 5.60 and 5.61) with the exact densities for the ground states of certain eight and nine particle systems at Q values shown on the figure. These states represent a CF-quasiparticle or a CF-quasihole of the 1/3 and 2/5 states (Section 5.9.4). The densities are quoted in units of $\rho_0 = (2\pi l^2)^{-1}$, where l is the magnetic length. The overlaps with the exact wave functions are 0.990 (CF-quasiparticle at 1/3), 0.993 (CF-quasiparticle at 2/5), 0.988 (CF-quasihole at 1/3), and 0.992 (CF-quasihole at 2/5); the $L_z = 0$ Fock space dimensions for these systems are 5529, 2860, 11 975, and 4890, respectively. Source: C.-C. Chang and J. K. Jain, unpublished (2006).

Table 6.6. *Testing CF theory for negative B^**

v	$(N, 2Q)$	overlap	E_{CF}	E_{exact}
	(6,9)	0.996 53	−0.539 39	−0.539 64
2/3	(8,12)	0.998 20	−0.534 02	−0.534 15
	(10,15)	0.994 03	−0.530 80	−0.531 02
3/5	(9,16)	0.999 38	−0.499 15	−0.499 18

Notes: Comparison of the wave functions of Eq. (5.32) for the ground states at $v = 2/3$ and $v = 3/5$, modeled as composite fermions at $v^* = 2$ and $v^* = 3$ in negative B^*, with the exact Coulomb ground states.

E_{CF} and E_{exact} are energies of the trial and the true ground states (including the background contribution).

Source: Wu, Dev, and Jain [703].

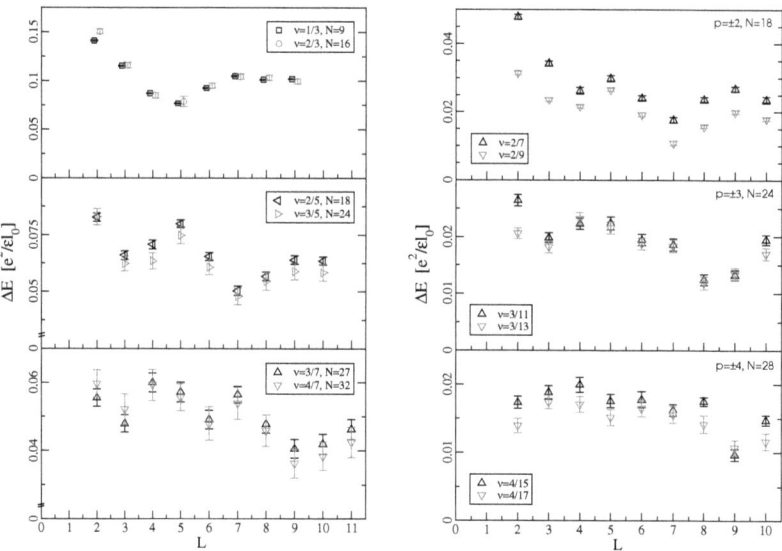

Fig. 6.5. Left column: CF-exciton energies for composite fermions at $v = n/(2n-1)$ and $v = n/(2n+1)$, modeled as filling factor n of composite fermions in negative and positive B^*, respectively. Right column: Same as in the left column, but for $v = n/(4n-1)$ and $v = n/(4n+1)$. The quantity p in this figure denotes the number of filled Λ levels, i.e., it is the same as our n. Source: G. Möller and S. H. Simon, *Phys. Rev. B* **72**, 045344 (2005). (Reprinted with permission.)

Möller and Simon [451] compute the CF-exciton dispersion for the FQHE states at $n/(4n-1)$, also shown in Fig. 6.5. The dispersion shares its qualitative features with that at $n/(4n+1)$, and the two dispersions become quantitatively the same with increasing n. There would have been no reason to suspect this in the absence of the CF theory. Möller and Simon further calculate the ground state energy and the CF-exciton dispersions at $\nu = 1/3$ by modeling it as filling factor one of ^{4}CFs, but in a negative effective magnetic field, i.e., with wave functions of the form $\mathcal{P}_{\mathrm{LLL}} \Phi_1^4 \Phi_{-1}^\alpha$. This is an alternative to the usual description as filling factor one of ^{2}CFs. They find that the ground state energies from the two methods are extremely close. The excitation gaps at the roton minimum and the large wave vector limit are quite similar, though not identical, for finite N, but seem to extrapolate to the same value at infinite N.

6.6 What computer experiments prove

Numerous such tests of the CF theory against computer experiments have been carried out by many groups (Bonesteel [33]; Dev and Jain [115, 116]; Girlich and Hellmund [188]; Kasner [328]; Möller and Simon [451]; Rezayi and MacDonald [548]; Rezayi and Read [549]; Rezayi and Haldane [550]; Wu *et al.* [703]; Yang [714]) and the results presented above are typical. More comparisons are shown below for the disk geometry. The following observations help to appreciate the full significance of the results:

- The results from the computer experiments are, of course, exact. The CF predictions are obtained from the CF theory, as defined in the previous chapter, without making any approximations. A comparison between the two provides a rigorous, detailed, nontrivial, and unbiased test of the CF theory.
- The wave functions and their Coulomb energies for the incompressible ground states (Eq. 5.32), CF-exciton (Eq. 5.62), CF-quasiparticle (Eq. 5.60), and CF-quasihole (Eq. 5.61) contain no free parameters.
- The number of distinct multiplets in an L sector is, in general, quite large. Therefore, it would be improbable for a random state with the correct symmetry to have an energy close to the exact value. To take an example, the $L = 6$ eigenstate of the twelve-particle $\nu = 3/7$ system is defined by 1181 coefficients, all of which must be predicted accurately by the CF theory in order for it to produce an almost exact energy. The quantitative agreement between the CF theory and the exact result is thus nontrivial for even a single case.

The computer experiments provide a clear and definitive confirmation of the CF theory. The following aspects are unambiguously established:

- The qualitative comparisons demonstrate, in a model-independent manner, the fundamental proposition of the CF theory, that the nonperturbative structures appearing in the problem of interacting electrons in the lowest Landau level can be explained in terms of weakly interacting fermions in an effective magnetic field.
- The quantitative comparisons confirm the correspondence between the FQHE and the IQHE at a microscopic level, and also establish the degree of quantitative accuracy of the CF theory.

A prediction of the CF theory is the splitting of the lowest Landau level into Λ levels of composite fermions, which resemble Landau levels of electrons at an effective magnetic field. As stressed in the preceding chapter, the existence of Λ levels cannot be established at the single particle level but must be deduced from the many-body solutions. The studies presented above provide a direct confirmation of the emergence of higher-LL-like physics within the lowest Landau level.

In short, the zero-parameter CF theory successfully predicts, with atomic-physics-like precision,[1] the structure, the quantum numbers, the wave functions, and the energies of the low-lying states for a range of relevant filling factors. The following chapters of the book give an account of a vast body of experimental phenomena that are explained by composite fermions.

6.7 Inter-composite fermion interaction

The interaction between bound states, for example atoms, molecules, or protons, is always more complex than that between the constituent particles. Because composite fermions are complex bound states of electrons, the interaction between them, a remnant of the Coulomb interaction between electrons, is also expected to be complicated. The following observations can be made, however:

- Composite fermions do feel some residual interaction. Had they been noninteracting, all states in the lowest band of the exact spectrum would be degenerate, which is not the case.
- The residual interaction is weak. This can be seen in a model-independent manner directly from the exact spectra. The splitting of the quasi-degenerate states in a given band is a measure of the residual interaction between composite fermions. Composite fermions are weakly interacting in the sense that this splitting for the states in the ground band is small compared to the gap separating them from higher-energy bands. Such a separation of energy scales is indicated by the very presence of bands. As a concrete example, let us consider the 12-particle system of Fig. 6.3. The width of the low-energy band (dark dashes) in this figure is $\sim 0.07\, e^2/\epsilon\ell$ (the energies shown are *single* particle energies), which is small compared to the energy gap at $\nu = 1/3$.
- At $\nu = n/(2pn \pm 1)$, a neglect of the inter-CF interaction is qualitatively valid because of the presence of a gap, in analogy to an explanation of the integral quantum Hall effect in terms of free electrons. For $\nu \neq n/(2pn \pm 1)$, however, the inter-CF interaction determines the physics, and can produce qualitatively new states, for example, new FQHE. The physics of such states is governed by much smaller energies, and is described in terms of composite fermions rather than electrons.
- Even though the wave functions for composite fermions are obtained from the corresponding wave functions of *noninteracting* electrons at an effective magnetic field, they are practically exact, and thus describe fully interacting electrons in the FQHE regime. The energies calculated from them capture the actual energy splitting of states in a given band, demonstrating that these wave functions go beyond the mean-field picture of noninteracting composite fermions, and provide an account of the residual interaction between composite fermions. In particular, the CF diagonalization automatically includes the effect of interactions between composite fermions. The

---

[1] In quantum chemistry studies of *atoms*, a few percent accuracy in a variational approach requires introduction of several fitting parameters through two- and three-body Jastrow factors and backflow corrections [478].

effective interaction between composite fermions can thus be determined from the wave function approach.

- Composite fermions appear to be strongly interacting in the CF–Chern–Simons formulation. It ought to be understood, however, that the strongly interacting objects, in that formulation, are the "bare" composite fermions, i.e, electron–*flux* composites. A perturbation theory of interaction should, in principle, take us to weakly interacting "dressed" composite fermions, in the same way as interacting electrons evolve into weakly interacting quasiparticles in Landau's Fermi liquid theory. Exact diagonalization studies allow us to make statements directly about the dressed composite fermions.

- A more quantitative account of the inter-CF interaction is possible, as seen in the following subsection.

6.7.1 CF pseudopotentials

Landau's theory of ordinary Fermi liquids deals with the effective interaction between quasiparticles near the Fermi energy, characterized by what are known as Landau parameters. Is it possible similarly to define and calculate the effective interaction between composite fermions near the CF Fermi level? And if so, what is it good for?

The Haldane pseudopotential V_m for electrons in a given Landau level is defined as the energy of two electrons in the relative angular momentum m state. The pseudopotentials for composite fermions, V_m^{CF}, are defined analogously by Sitko *et al.* [596] as the energy of two composite fermions in relative angular momentum m state in a given Λ level. An evaluation of the CF pseudopotentials requires a true many-body calculation (as is also the case for Landau parameters). Sitko *et al.* [596] evaluate V_m^{CF} from exact diagonalization on small systems. Lee *et al.* [382, 383] obtain the thermodynamic limits by computing the Coulomb expectation values for the appropriate wave functions (which have an integral number of Λ levels fully occupied and two composite fermions in the lowest unoccupied Λ level). The results are shown in Fig. 6.6. The left panel shows pseudopotentials for two composite fermions in the second, third and fourth Λ levels. The right panel shows pseudopotentials for two CF-holes in various Λ levels. The distance between two composite fermions increases with m, so Fig. 6.6 tells us how the interaction between two composite fermions (particles or holes) depends on their separation. A real-space interaction between composite fermions can be constructed [382, 383], which reproduces CF pseudopotentials in a given Λ level. As explained in Section 3.11.2, several very different looking real-space interactions can produce the same pseudopotentials, and, therefore, the real-space form of the interaction does not give any more insight than the pseudopotentials. The CF-exciton dispersion, discussed earlier, gives pseudopotentials for interaction between a CF particle and hole. Pseudopotentials for CF particles and holes in a given Λ level look similar, suggesting approximate particle–hole symmetry, even though very different wave functions are used for their evaluation.

The CF pseudopotentials in Fig. 6.6 are approximately an order of magnitude smaller than the electron pseudopotentials, again confirming the feebleness of the inter-CF interaction.

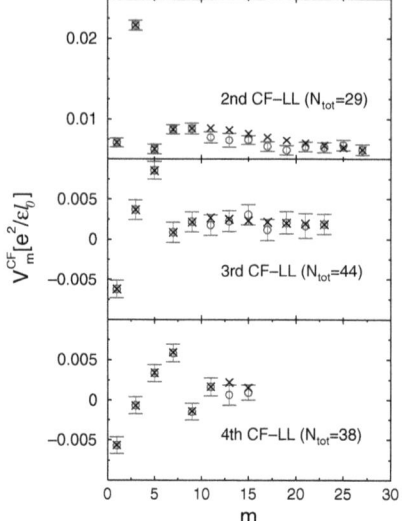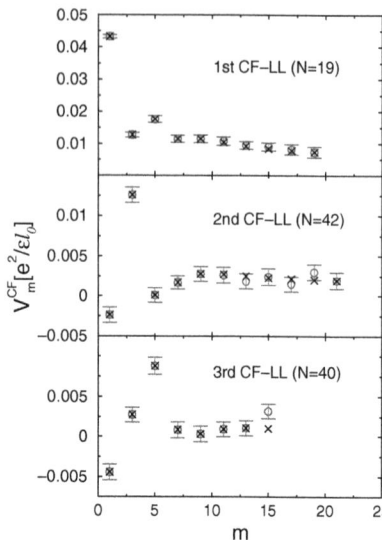

Fig. 6.6. Interaction pseudopotentials for composite fermion particles (left panel) and holes (right panel) in various Λ levels (called CF-LLs in this figure). Energies are defined up to an overall additive constant; only the energy differences are physically relevant. Circles are pseudopotentials computed with the help of many-body wave functions of the type shown in Eqs. (5.60) and (5.61); crosses are pseudopotentials of an approximate real-space interaction that can be found in the references given below. N is the total number of particles used for the calculation; the results are believed to represent the thermodynamic limit. Source: S.-Y. Lee, V. W. Scarola, and J. K. Jain, *Phys. Rev. Lett.* **87**, 256803 (2001); *Phys. Rev. B* **66**, 085336 (2002). (Reprinted with permission.)

A striking aspect is the absence of a hard-core interaction for composite fermions in *higher* Λ levels. Except for the lowest Λ level, the inter-CF interaction is actually *attractive* at short distances, as indicated by the fact that the minimum value of V_m^{CF} occurs at a small m. Composite fermions in higher Λ levels thus not only do not repel one another but form bound states. This reflects in the possibility of phase separation into stripes or bubble crystals, which is further discussed in Section 15.5. The binding energy is not large enough to destabilize the underlying FQHE state, however.

The CF pseudopotentials allow an approximate treatment of situations with $\nu^* \neq n$ by defining an effective problem in terms of *only* composite fermions in the topmost partially filled Λ level, which leads to a significant technical simplification due to the reduction in the number of particles that need to be considered. (Analogous studies at $\nu = n + \bar{\nu}$ treat the lowest n Landau levels as inert and consider only the electrons in the partially occupied Landau level.) This is the most reliable theoretical method available for some situations of interest, and has been employed in several calculations, described elsewhere in this book [382, 383, 531, 568, 596, 693, 694, 697, 698, 724]. For example, Fig. 6.7 shows the exact diagonalization spectrum for seven and eight electrons (pluses) containing three CF-holes in the lowest Λ level; circles show exact diagonalization for

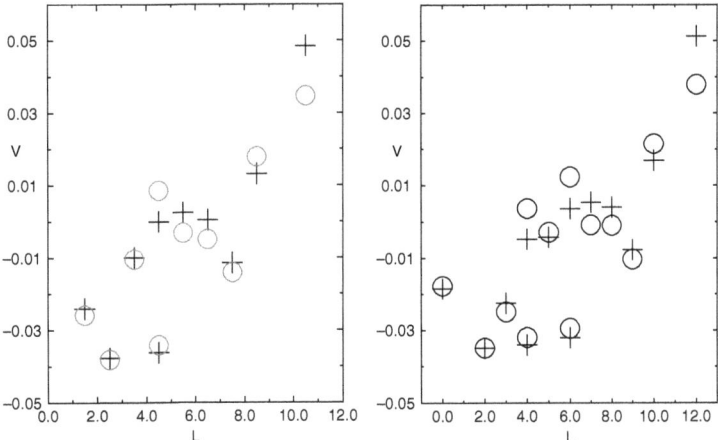

Fig. 6.7. Crosses show the exact low-energy spectrum for three quasiholes of the $\nu = 1/3$ state, for seven (left panel) and eight (right panel) electrons. Circles are obtained from the diagonalization of a three-fermion problem, with their interaction defined by CF pseudopotentials. Source: P. Sitko, S. N. Yi, K. S. Yi, and J. J. Quinn, *Phys. Rev. Lett.* **76**, 3396–3399 (1996). (Reprinted with permission.)

three particles with pseudopotentials obtained as above. The method, however, is not as accurate as CF diagonalization of Fig. 5.10 (which deals with *all* composite fermions), because (i) it neglects possible contributions from three and higher body interaction terms in the effective Hamiltonian for composite fermions, and (ii) it assumes that two-body pseudopotentials are independent of the filling factor of composite fermions in the Λ level of interest.

6.7.2 Conditions for composite fermions

Electrons do not *always* capture vortices to transform into composite fermions. The agreement between the CF theory and exact results is not completely satisfactory in the second Landau level, and the CF description almost completely fails in still higher Landau levels. That is consistent with the FQHE being most pronounced in the lowest Landau level, much less so in the second, and almost nonexistent in the third or higher Landau levels.

Landau levels differ through their Coulomb pseudopotentials. One may wonder what conditions the pseudopotentials of the electron–electron interaction need to satisfy to produce composite fermions. Composite fermions with two bound vortices are formed when the interaction is sufficiently strongly repulsive at short distances, i.e., has a large V_1 pseudopotential. This condition is well satisfied in the lowest Landau level but not in higher Landau levels, as can be understood intuitively by noting that the size of the maximally localizable wave packet increases with the LL index. Further work by Quinn, Wójs, and collaborators [530, 531, 690–692, 694] has defined the conditions more precisely.

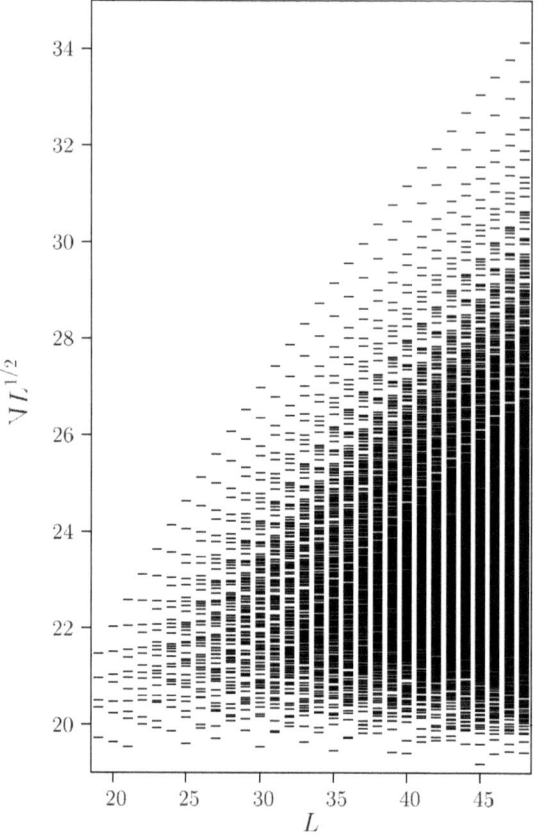

Fig. 6.8. Exact interaction energy V (dashes) as a function of the angular momentum L for $N = 6$ particles. Energies are quoted in units of $e^2/\epsilon\ell$; they have been multiplied by $\sqrt{L}$ to make the ground state energy approximately L independent. Source: G.-S. Jeon, C.-C. Chang and J. K. Jain [299].

6.8 Disk geometry

The spherical geometry resembles a three-dimensional atom, except that the electron radius is fixed and the nucleus has both electric and magnetic charges, the former producing a uniform neutralizing background and the latter a radial magnetic field. This section concerns FQHE atoms in a plane. The confinement for such 2D atoms is produced by a parabolic potential, and electrons form a droplet characterized by their total angular momentum L (which now is the z component of the angular momentum). We set the strength of the confinement potential to zero below; it can be included straightforwardly at the end, as shown in Section 3.12. Confinement is produced by fixing the total angular momentum.

Many exact diagonalization studies of lowest LL electrons on a disk have been conducted, beginning with the work of Laughlin [369], Yoshioka, Halperin and Lee [728] and Girvin and Jach [189]. Figure 6.8 shows the full spectrum for six electrons over a range of L values.

An inspection raises many questions. What is the physics of the low-energy states? How do we describe them qualitatively and quantitatively? We note that the ground state energy does not vary smoothly as a function of L, but has structure. Of special importance are the "cusp" states, where the ground state energy has a downward cusp, because, when the confinement energy is added to the interaction energy, one of these cusp states becomes the ground state. As a parameter like the magnetic field or the confinement strength is varied, the quantum dot jumps from one cusp state to another. The angular momenta where cusps appear are called "magic" angular momenta. Why do cusps appear? What are these magic angular momenta, and what is special about them?

6.8.1 Composite fermions in a quantum dot: the disk geometry

The CF theory can be generalized to electrons in the disk geometry (Beenakker and Rejaei [26]; Cappelli *et al.* [50]; Dev and Jain [116]; Han and Yang [235]; Jain and Kawamura [279]; Jain and Kamilla [281]; Jeon, Chang, and Jain [295, 296]). We begin, as usual, by constructing a correlated basis $\{\Psi_L^\alpha\}$, labeled conveniently by the L quantum number, for low-energy states:

$$\Psi_L^\alpha = \mathcal{P}_{\mathrm{LLL}} \Phi_{L^*}^\alpha \prod_{j<k} (z_j - z_k)^{2p}, \tag{6.4}$$

where

$$L = L^* + pN(N - 1). \tag{6.5}$$

The CF physics is included through the standard factor $\prod_{j<k}(z_j - z_k)^{2p}$, which contributes $pN(N - 1)$ to the angular momentum. The system of interacting electrons at L thus maps into that of weakly interacting composite fermions at L^*. Here, $\{\Phi_{L^*}^\alpha\}$ is a basis for *non-interacting* electrons with total angular momentum L^*; all *distinct* states with the *smallest* kinetic energy are kept in the simplest approximation. The dimension of the resulting basis $\{\Psi_L^\alpha\}$ is denoted by $D_{\mathrm{CF}}^{(0)}$. (In general, $D_{\mathrm{CF}}^{(0)}$ is not equal to the number of independent basis states of noninteracting electrons at L^*, because the composite-fermionization of the latter does not necessarily produce as many linearly independent states.) The CF vorticity $(2p)$ is chosen to minimize the dimension of this basis, to produce maximum simplification.

As seen in Table 6.7, the dimension of the CF basis is, in general, much smaller than that of the full basis D_{ex}, and does not increase with L as rapidly. Furthermore, its growth can be curtailed by increasing the CF vorticity $2p$. Figure 6.9 compares the spectrum obtained by CF diagonalization in the L range described by composite fermions with vorticity $2p = 2$. The CF theory gives an excellent qualitative and quantitative account of low-energy states, predicting the positions of all cusp correctly. It is notable that for many values of L, the CF basis contains a single state.

Table 6.7. Dimensions of exact and CF bases

L	D_{ex}	L	D_{ex}	L	D_{ex}	$D_{CF}^{(0)}$	$D_{CF}^{(1)}$
19	5	49	1 945	79	26 207	1	3
20	7	50	2 172	80	28 009	1	3
21	11	51	2 432	81	29 941	1	4
22	14	52	2 702	82	31 943	3	7
23	20	53	3 009	83	34 085	2	8
24	26	54	3 331	84	36 308	1	6
25	35	55	3 692	85	38 677	1	5
26	44	56	4 070	86	41 134	3	11
27	58	57	4 494	87	43 752	2	10
28	71	58	4 935	88	46 461	5	18
29	90	59	5 427	89	49 342	2	13
30	110	60	5 942	90	52 327	1	6
31	136	61	6 510	91	55 491	3	14
32	163	62	7 104	92	58 767	7	25
33	199	63	7 760	93	62 239	2	16
34	235	64	8 442	94	65 827	4	27
35	282	65	9 192	95	69 624	1	10
36	331	66	9 975	96	73 551	2	18
37	391	67	10 829	97	77 695	5	31
38	454	68	11 720	98	81 979	9	47
39	532	69	12 692	99	86 499	1	17
40	612	70	13 702	100	91 164	2	26
41	709	71	14 800	101	96 079	4	41
42	811	72	15 944	102	101 155	7	58
43	931	73	17 180	103	106 491	12	83
44	1057	74	18 467	104	111 999	18	111
45	1206	75	19 858	105	117 788	1	28
46	1360	76	21 301	106	123 755	1	39
47	1540	77	22 856	107	130 019	2	55
48	1729	78	24 473	108	136 479	3	74

Notes: Dimensions of various bases as a function of L for six electrons in a disk. D_{ex} is the dimension of the Hamiltonian matrix in the exact diagonalization study.

The correlated CF basis has dimension $D_{CF}^{(0)}$ when only CF configurations with the smallest "kinetic" energy are kept, and $D_{CF}^{(1)}$ when states with one higher unit of "kinetic" energy are also included.

The dimensions of the CF bases at $L + pN(N-1)$ is the same as at L, with the basis wave functions at $L + pN(N-1)$ obtained by multiplying the basis at L by an appropriate Jastrow factor.

Source: Jeon, Chang and Jain [299].

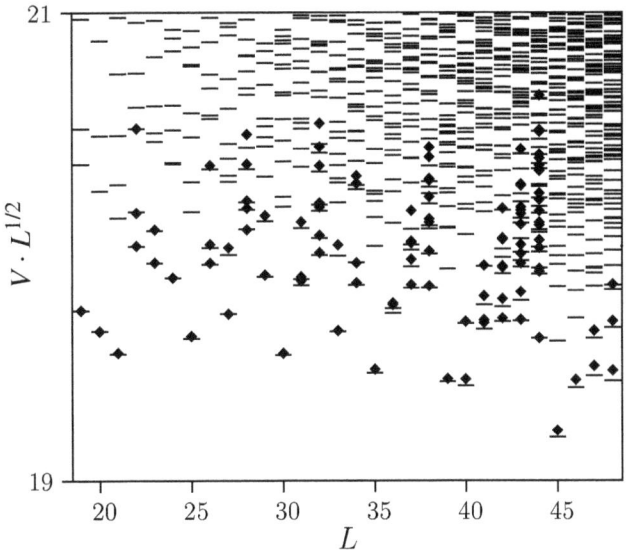

Fig. 6.9. Comparing the CF prediction (dots) with the exact spectrum (dashes) for $N = 6$ particles on a disk. Composite fermions carrying two vortices (^{2}CFs) are used for CF diagonalization. Source: G.-S. Jeon, C.-C. Chang and J. K. Jain, cond-mat/0611309 (2006). (Reprinted with permission.).

6.8.2 Mean-field composite fermion model

Insight into the above results can be gained by back-of-the-envelope calculations of the mean-field model, which treats composite fermions as noninteracting fermions at L^* with an effective cyclotron energy. In this model, the ground state energy is the *kinetic* energy of these fermions. Figure 6.10 compares the L dependence predicted by the mean-field theory with the exact results, showing that the mean-field theory correctly obtains the cusp positions. (The effective cyclotron energy is chosen conveniently, and an overall constant has been added to facilitate comparison.)

The following mean-field physical picture explains the origin of cusps. Let us begin with a sufficiently large L^* so that all composite fermions of a given vorticity (say ^{2}CFs) can be accommodated in the lowest Λ level. Initially, L^* can be decreased by moving composite fermions to lower m states within the lowest Λ level. When we reach $L^* = N(N-1)/2$, each orbital from $m = 0$ to $m = N - 1$ is occupied in the lowest Λ level. Further reduction in L^* is possible only by pushing a composite fermion to the second Λ level, thus producing a cusp in energy at $L^* = N(N-1)/2$. The angular momentum L^* can now be reduced without any change in kinetic energy until we reach the configuration in which each orbital from $m = 0$ to $m = N - 2$ is occupied in the lowest Λ level, and the $m = -1$ orbital is occupied in the second Λ level. Further reduction in L^* is again associated with a jump in the kinetic energy, producing a cusp at $L^* = \frac{1}{2}(N-1)(N-2) - 1$. Cusps in energy thus appear whenever further squeezing of the state is not possible without pushing a composite fermion into a higher Λ level. The cusp states are examples of "compact states" [279].

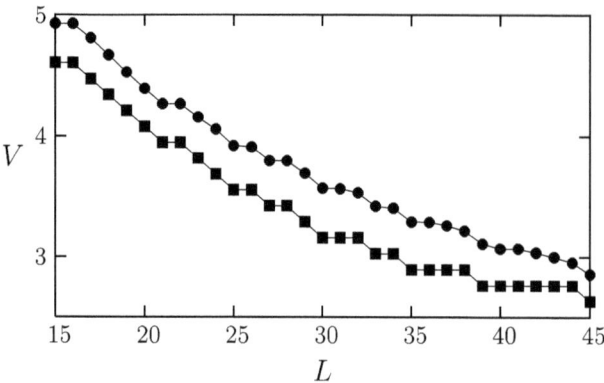

Fig. 6.10. The exact interaction energy V (circles) as a function of L for $N = 6$ particles. Such a plot of the ground state energy is often called the "yrast line," a terminology borrowed from nuclear physics. The squares are predictions of the mean-field model, where the energy is identified with the "kinetic" energy at L^*, with the Λ level separation fixed empirically. The CF mean-field curve is shifted vertically for clarity. Source: G. S. Jeon, C.-C. Chang, and J. K. Jain, cond-mat/0611309 (2006). (Reprinted with permission.)

Compact states   A compact state $[N_0, N_1, \ldots, N_k]$ is defined by the following conditions:

- N_n is the number of composite fermions in the nth Λ level. The Λ levels with $n > k$ are unoccupied.
- Composite fermions occupy each Λ level compactly, without leaving any hole; i.e., the innermost ($m = -n, -n + 1, \ldots, -n + N_n - 1$) single-particle orbitals of the nth Λ level are occupied.
- The occupation numbers satisfy the condition $N_{n+1} \leq N_n + 1$.

It follows that: (a) reducing the angular momentum of a compact state requires increasing the Λ level index of at least one composite fermion; (b) the CF kinetic energy cannot be lowered without increasing the angular momentum.

Two compact states, $[N]$ and $[N - 1, 1]$, were considered earlier in this subsection. Figure 6.11 shows the compact states $[3, 3]$ and $[4, 1, 1]$ for ^{2}CFs. All cusp states are easily seen to be compact. (The converse is not true. Why?) Otherwise: (i) If one of the Λ levels is not compactly occupied, the total angular momentum can be reduced without increasing the kinetic energy, by simply shifting a fermion to a lower m state within the same Landau level. (ii) For $N_{n+1} \geq N_n + 2$, the outermost composite fermion in the $(n+1)$th Λ level can be demoted to the nth Λ level without changing the total angular momentum, demonstrating that such a state is not the lowest energy state in the relevant L sector. The compact states of the disk geometry are the analogs of filled Λ level states in the spherical geometry, and their wave functions do not contain any free parameters.

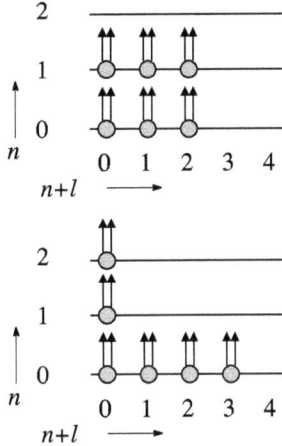

Fig. 6.11. Schematic depiction of two compact CF states, $[3, 3]$ and $[4, 1, 1]$ for $N = 6$ electrons at $L = 33$.

6.9 A small parameter and perturbation theory

Accurate as the wave functions of composite fermions already are, they can be further improved in a systematic perturbative approach. Perturbation theory requires a small parameter, and we have stressed the lack of a small parameter for the FQHE problem. That was for electrons, however. A small "parameter" can be identified for composite fermions: Λ level mixing.

Λ level mixing for composite fermions in the wave function $\Psi = \mathcal{P}_{\mathrm{LLL}} \Phi \Phi_1^{2p}$ is *defined* through Landau level mixing in Φ. The wave functions Ψ discussed so far have been derived from the lowest kinetic energy space for Φ; in other words, Λ level mixing has been neglected. The closeness of these wave functions to the exact wave functions indicates that Λ level mixing is small.

This understanding suggests a systematic way of improving the ground state wave function. We enlarge the basis $\{\Phi\}$ by including, successively, states with higher and higher kinetic energy, and obtain at each level the ground state by CF diagonalization of the corresponding basis $\{\Psi\}$. The dimension of the enlarged basis space is still much smaller than the dimension of the full LLL basis, at least at low orders of the perturbation theory. Figure 6.12 shows the result of CF diagonalization for $\nu = 1/3$ on a sphere in a basis containing zero, one, and two CF-excitons, with the energy of the lowest state in each L sector listed in Table 6.8. Figure 6.13 shows the spectrum in the disk geometry recalculated by including states with one higher unit of kinetic energy of composite fermions; the dimension of the enlarged basis is given in Table 6.7 along with the dimension of the full basis.

As more and more Λ level mixing is incorporated, the size of the CF basis will eventually become equal to the full basis and one would recover the exact result.[2] What is the advantage

---

[2] That would not be the most efficient way of calculating the exact spectrum.

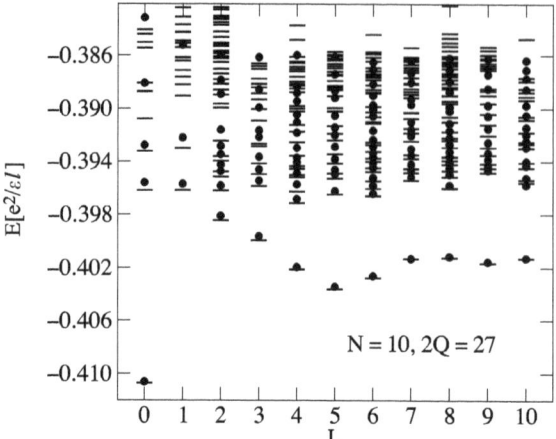

Fig. 6.12. Low-energy spectrum of ten electrons at $\nu = 1/3$ predicted by the CF theory (dots). States with 0, 1, and 2 CF excitons are considered in CF diagonalization. Dashes are the exact eigenenergies obtained by numerical diagonalization. E is the energy per particle. The statistical uncertainty in the energy, coming from Monte Carlo sampling, is much smaller than the size of the dots. The lowest energy in each L sector is given in Table 6.8, labeled $E^{(2)}$. Source: M. R. Peterson and J. K. Jain, *Phys. Rev. B* **68**, 195310 (2003). (Reprinted with permission.)

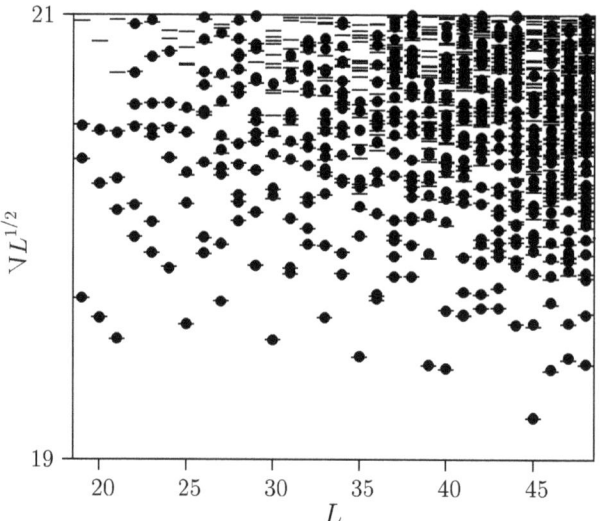

Fig. 6.13. Comparing the CF and the exact spectra for $N = 6$ particles on a disk. All ^{2}CF configurations with lowest and first excited "kinetic" energies are used in the CF diagonalization. This figure shows how results of Fig. 6.9 improve by including Λ level mixing at the lowest order. The dimension of the CF basis, $D_{CF}^{(1)}$, is given in Table 6.7. Source: G.-S. Jeon, C.-C. Chang and J. K. Jain, cond-mat/0611309 (2006). (Reprinted with permission.).

Table 6.8. *Improving CF energies by allowing Λ level mixing*

L	E^{exact}	$E_{\text{CF}}^{(1)}$	$E_{\text{CF}}^{(2)}$	error (%)
0	−0.410 628 97	−0.410 39(2)	−0.410 62(3)	–
1	−0.396 090 58	–	−0.395 7(1)	0.1
2	−0.398 368 47	−0.397 2(1)	−0.398 1(1)	0.07
3	−0.399 854 58	−0.399 16(8)	−0.399 6(1)	0.06
4	−0.402 038 92	−0.401 7(1)	−0.401 94(5)	0.02
5	−0.403 540 97	−0.403 24(8)	−0.403 43(4)	0.03
6	−0.402 684 31	−0.402 3(1)	−0.402 61(5)	0.02
7	−0.401 235 1	−0.400 94(8)	−0.401 3(2)	–
8	−0.401 153 14	−0.400 81(6)	−0.401 15(5)	–
9	−0.401 578 15	−0.401 27(5)	−0.401 5(1)	–
10	−0.401 203 51	−0.400 98(8)	−0.401 3(1)	–

Notes: Second column gives exact energy per particle for the lowest energy states at orbital angular momenta $L = 0, 1, \ldots, 10$ for 10 particles at $\nu = 1/3$.

Energies $E_{\text{CF}}^{(1)}$ and $E_{\text{CF}}^{(2)}$ are obtained from CF diagonalization including Λ level mixing at the first and the second orders. (The former considers states with at most one CF exciton, whereas the latter also includes two CF excitons.)

Last column gives the % error for cases where $E_{\text{CF}}^{(2)}$ differs significantly from E^{exact}.

Source: Peterson and Jain [512].

of organizing the perturbation theory in this manner? As always, the question boils down to what to include first and what to postpone for later (or forever). With composite fermions, accurate numbers are produced already at the zeroth order, and only in exceptional cases is there any need for appealing to the first or the second order perturbation theory.

For the most prominent FQHE states, Λ level mixing does not cause any phase transition; the energies are shifted somewhat, but the qualitative structure and the gap remain intact (e.g. Fig. 6.12). Λ level mixing *can* destabilize the FQHE, however. That would be the case, for example, at very low fillings, where the FQHE is unstable to a crystal state.

Exercises

6.1 Verify that the ground states of the systems shown in Fig. 6.1 contain an integral number of filled Λ levels of composite fermions.

6.2 Show that there exists only one $L = 0$ state for $(N, 2Q) = (3, 6)$ and $(N, 2Q) = (4, 6)$, which correspond to $\nu = 1/3$ and $\nu = 2/5$. Hence, any trial wave function with $L = 0$ is trivially exact for these systems.

6.3 Obtain CF predictions for angular momenta of the lowest band states for systems of Figs. 6.2 and 6.3. You will need to prove that:

$$\frac{7}{2} \otimes \frac{7}{2} = 0 \oplus 2 \oplus 4 \oplus 6,$$

$$3 \otimes 3 \otimes 3 = 0 \oplus 2 \oplus 3 \oplus 4 \oplus 6,$$

$$\frac{5}{2} \otimes \frac{5}{2} \otimes \frac{5}{2} \otimes \frac{5}{2} = 0 \oplus 2 \oplus 4,$$

$$\frac{9}{2} \otimes \frac{9}{2} \otimes \frac{9}{2} \otimes \frac{9}{2} = 0^2 \oplus 2^2 \oplus 3 \oplus 4^3 \oplus 5 \oplus 6^3 \oplus 7 \oplus 8^2 \oplus 9 \oplus 10 \oplus 12.$$

6.4 Calculate the mean-field prediction for the quantum numbers of the single CF exciton for the systems shown in Fig. 6.1.

6.5 Show that the systems of Fig. 6.4 represent a single CF-quasiparticle or a CF-quasihole. Obtain the CF prediction for the angular momentum of the ground state for each system.

6.6 Analyze the system $(N, 2Q) = (12, 19)$ in both ways described in Section 6.4.4 (either as composite fermions in a negative effective field, or by relating it by particle–hole symmetry to a system of composite fermions in a positive effective field) to predict the angular momenta of states in the lowest band.

7

Theory of the FQHE

The central postulate of composite fermion theory is that the liquid of strongly interacting electrons in a high magnetic field B is equivalent to an assemblage of weakly interacting composite fermions in an effective magnetic field B^*. The effective magnetic field is such a direct, fundamental, and dramatic consequence of the formation of composite fermions that it is taken as *the* defining property of composite fermions, and its observation is tantamount to an observation of composite fermions themselves. Computer experiments on small systems, described in Chapter 6, demonstrate that the dynamics of interacting electrons in the lowest Landau level resembles that of weakly interacting fermions in an effective magnetic field. The present and the subsequent chapters analyze numerous experimental facts within the CF theory. We begin with the explanation of the FQHE.

7.1 Comparing the IQHE and the FQHE

Compelling evidence for composite fermions comes from the simple act of plotting the FQHE data as a function of the effective magnetic field B^*, which amounts to rigidly shifting the B-axis by a constant $(B - B^* = 2p\rho\phi_0)$, and then comparing it to the IQHE of electrons, which is a system of weakly interacting fermions. To facilitate comparison for samples with different densities, we plot the resistance trace in the IQHE as a function of $1/\nu$, which is proportional to B, and compare it with the resistance trace in high magnetic field plotted as a function of $1/\nu^*$, which is proportional to B^*. Plotted this way, a close correspondence between the major features of two traces becomes apparent (Fig. 7.1). The FQHE of electrons is indistinguishable from the IQHE of certain fermionic particles. The IQHE of electrons is better developed than the IQHE of composite fermions, but the difference between the two traces is no more significant than that between IQHE traces for samples with different mobilities, or for the same sample but at different temperatures.[1] Such

---

[1] We attempted to find in the literature a *pure* IQHE trace, i.e., one which shows either all integers, or all even integers, but no fractions. Figure 7.1, upper panel [89], was the best we could find. It has weak FQHE structure on it, arising from inter-electron interactions. A similar structure in the lower panel occurs due to residual interactions between composite fermions.

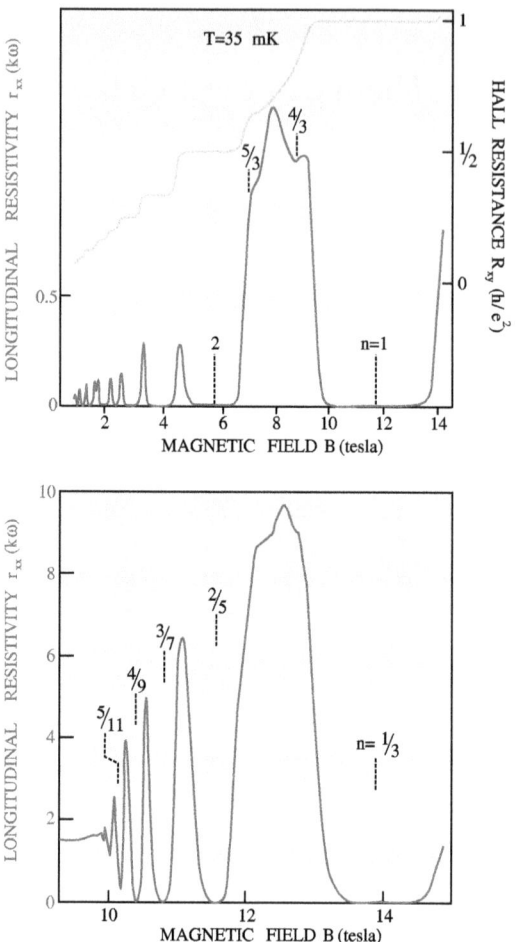

Fig. 7.1. Comparing the FQHE and the IQHE. The top panel shows the IQHE of electrons (from Clark, *et al.* [89]) and bottom panel shows the FQHE of electrons starting from $\nu = 1/2$ (from Stormer and Tsui [625]). A close correspondence between prominent features in the longitudinal resistance is manifest. Source: J. K. Jain, *Physics Today* **53** (4), 39 (2000). (Reprinted with permission.)

a comparison demonstrates, unambiguously, that the strongly correlated liquid of interacting electrons at B indeed behaves like a weakly interacting gas of composite fermions at B^*. While striking in retrospect, the close similarity between the IQHE and the FQHE had not been noted prior to the CF theory.

How about the Hall resistance? The measured Hall resistances at ν^* and ν are not equal, for reasons explained in Section 5.16.3. If one, however, uses Eq. (5.135) to obtain ρ_{CF} from the measured ρ in the FQHE regime, which amounts to shifting down the Hall resistance in the FQHE regime vertically by an amount $2ph/e^2$, then a close correspondence with the IQHE Hall resistance is obtained.[2]

The trace in the lower panel starts from $B^* = 0$, which coincides with $\nu = 1/2$. One could have also included in this figure negative values of B^*, corresponding to $\nu > 1/2$. The magnetoresistance trace here shows a correspondence with the IQHE trace with negative magnetic field B. The magnetoresistances in the filling factor range $1/3 > \nu \geq 1/5$ also resemble IQHE magnetoresistances when plotted as a function of $1/\nu^*$, but now for ^{4}CFs. Comparisons between the FQHE and the IQHE are also shown in the articles by Smet [603] and Stormer [627].

7.2 Explanation of the FQHE

In Chapter 9 we learn that incompressibility (i.e., the presence of a gap) at a fractional filling $\nu = f$ in a disorder-free system produces, in the presence of a weak disorder, a plateau quantized at $R_H = h/fe^2$. Here, we assume that to be the case, and ask how the fractions obtained by the CF theory compare with those observed experimentally.

Implicit in the comparison between the FQHE and the IQHE in the previous section is the explanation for the FQHE. The FQHE of electrons is a manifestation of the integral QHE for composite fermions. The latter occurs because a gap opens when composite fermions fill an integral number of Λ levels, i.e., when $\nu^* = n$. These fillings correspond to electron filling factors given by the sequences

$$\nu^* = n \iff \nu = \frac{n}{2pn \pm 1}. \tag{7.1}$$

A gap here results in a FQHE plateau at $R_H = h/fe^2$, with

$$f = \frac{n}{2pn \pm 1}. \tag{7.2}$$

FQHE at f also implies a FQHE at the hole partner

$$f = 1 - \frac{n}{2pn \pm 1}. \tag{7.3}$$

These fractions can be obtained by defining the original problem in terms of holes – rather than electrons – in the lowest Landau level, and making composite fermions by binding vortices to holes.

An analogy of the 2D electron system in a magnetic field with a giant atom containing N ($\sim 10^9$) electrons is instructive. The Landau levels are the quantized kinetic energy shells of this atom. The atoms with an integral number of filled shells are especially simple, akin to noble gas atoms: the ground state wave function is completely known (a Slater determinant), and excitations cost a nonzero energy. (See Fig. 4.1.) This uniqueness of the ground state

---

[2] The resistance would be ρ_{CF} for composite fermions for a hypothetical voltmeter that could measure both the real and the CS electric fields.

Table 7.1. *The fractional table*

$|n|$	1	2	3	4	5	6	7	8	9	10	11
f	1/3	2/5	3/7	4/9	5/11	6/13	7/15	8/17	9/19	10/21	
state	$^2\mathrm{CF}_1$	$^2\mathrm{CF}_2$	$^2\mathrm{CF}_3$	$^2\mathrm{CF}_4$	$^2\mathrm{CF}_5$	$^2\mathrm{CF}_6$	$^2\mathrm{CF}_7$	$^2\mathrm{CF}_8$	$^2\mathrm{CF}_9$	$^2\mathrm{CF}_{10}$	
f	1/5	2/9	3/13	4/17	5/21	6/25					
state	$^4\mathrm{CF}_1$	$^4\mathrm{CF}_2$	$^4\mathrm{CF}_3$	$^4\mathrm{CF}_4$	$^4\mathrm{CF}_5$	$^4\mathrm{CF}_6$					
f	1/7	2/13	3/19								
state	$^6\mathrm{CF}_1$	$^6\mathrm{CF}_2$	$^6\mathrm{CF}_3$								
f	1/9	2/17									
state	$^8\mathrm{CF}_1$	$^8\mathrm{CF}_2$									
f	1	2/3	3/5	4/7	5/9	6/11	7/13	8/15	9/17	10/19	
state	$^2\mathrm{CF}_{-1}$	$^2\mathrm{CF}_{-2}$	$^2\mathrm{CF}_{-3}$	$^2\mathrm{CF}_{-4}$	$^2\mathrm{CF}_{-5}$	$^2\mathrm{CF}_{-6}$	$^2\mathrm{CF}_{-7}$	$^2\mathrm{CF}_{-8}$	$^2\mathrm{CF}_{-9}$	$^2\mathrm{CF}_{-10}$	
f	1/3	2/7	3/11	4/15	5/19	6/23					
state	$^4\mathrm{CF}_{-1}$	$^4\mathrm{CF}_{-2}$	$^4\mathrm{CF}_{-3}$	$^4\mathrm{CF}_{-4}$	$^4\mathrm{CF}_{-5}$	$^4\mathrm{CF}_{-6}$					
f	1/5	2/11	3/17								
state	$^6\mathrm{CF}_{-1}$	$^6\mathrm{CF}_{-2}$	$^6\mathrm{CF}_{-3}$								
f	1/7	2/15									
state	$^8\mathrm{CF}_{-1}$	$^8\mathrm{CF}_{-2}$									

Notes: Prominently observed fractions in the lowest Landau level (for fully spin-polarized electrons), which correspond to the IQHE of composite fermions. These are giant, filled-shell atoms of composite fermions.

Rows and columns are arranged according to the vorticity ($2p$) and the number of filled Λ level shells ($|n|$).

Filled shell states at $f = n/(2pn+1)$ are denoted by $^{2p}\mathrm{CF}_n$; negative values of n, which produce the fractions $f = |n|/(2p|n|-1)$, refer to the situation in which B^* is negative.

A prediction of the CF theory is that the table gets filled without vacancies. When a fraction appears more than once, the description in terms of the simpler composite fermions is better and more natural, although both are valid.

at integral fillings lies at the heart of the IQHE. In the lowest Landau level, electrons transform into composite fermions by capturing vortices, and the lowest electronic Landau level splits into Λ levels of composite fermions. The quantum fluid at large B is a giant atom of composite fermions, and the FQHE states are the noble gas atoms of composite fermions. See Fig. 5.3 for some examples.

We note:

- The fractions f given by Eqs. (7.2) and (7.3) are precisely the prominently observed fractions.
- They have odd denominators because the vortex quantum number $2p$ is an *even* integer.
- The fractions appear in sequences because they are all derived from the integer sequence of the IQHE.

A physically meaningful way of displaying the observed fractions is shown in Table 7.1. (For several fractions shown in this table, the evidence is seen only in ρ_{xx}, and, therefore, their existence is not yet definitive.) Many of the fractions in this table were observed subsequent to the introduction of the CF theory, and have conformed with the CF theory prediction. Many more fractions, not shown for simplicity, follow straightforwardly from those given in this table. For example, particle–hole symmetry in the lowest Landau level implies fractions $f = 1 - n/(2pn \pm 1)$ (not shown). Also, a given fraction f in the lowest Landau level implies the *possibility* of FQHE at $k + f$ in the kth Landau level.

Different flavors of composite fermions occur in different filling factor regions. The filling factor range $1/2 \geq \nu \geq 1/3$ is described in terms of composite fermions carrying two vortices; the range $1/3 > \nu \geq 1/5$ in terms of composite fermions carrying four vortices (with B^* antiparallel to B for $1/3 > \nu > 1/4$), and so on. The region $1 > \nu \geq 1/2$ of electrons maps into $0 < \nu \leq 1/2$ of holes as a result of particle–hole symmetry in the lowest Landau level, and can be understood in terms of composite fermions made of holes.

Apparently dissimilar interpretations are possible at some filling factors. For example, the region $2/3 \geq \nu \geq 1/2$ can be described either in terms of composite fermions of electrons carrying two vortices in a negative B^* (i.e., B^* pointing opposite to B), or in terms of particle–hole symmetric states derived from the region $1/2 \geq \nu \geq 1/3$. As shown in Section 6.5.2, the two descriptions are equivalent.[3] In some cases, one interpretation is more natural. For example, $\nu = 1/3$ can be thought of either as one filled Λ level of ^{2}CFs ($\mathcal{P}_{LLL}\Phi_1\Phi_1^2$) or as "$-1$" filled Λ level of ^{4}CFs ($\mathcal{P}_{LLL}\Phi_{-1}\Phi_1^4$). The former description is preferred, as it does not require LLL projection, and is slightly more accurate; it is also more natural because the gap at $1/3$ is comparable to the gaps at $2/5$, $3/7$, etc., rather than the gaps at $2/7$, $3/11$, etc. An extreme example is the $\nu = 1$ state, which obviously ought to be bunched with the IQHE states, even though it can be obtained as the ^{2}CF-FQHE state at $\nu = \nu^*/(2\nu^* - 1)$ with $\nu^* = 1$.

---

[3] When the spin degree of freedom is relevant, however, a consideration of composite fermions in negative B^* is necessary for explaining non-fully spin-polarized FQHE in the region $3/2 > \nu > 1/2$, e.g., the spin singlet FQHE at $\nu = 2/3$ [703].

7.3 Absence of FQHE at $v = 1/2$

An intriguing experimental "observation" is the absence of FQHE at even-denominator fractions, most strikingly at $v = 1/2$, the simplest fraction. The CF theory suggests an explanation (Halperin, Lee and Read [231]; Kalmeyer and Zhang [309]). The fraction $f = 1/2$ is obtained as the $n \to \infty$ limit of the sequence $f = n/(2n + 1)$. If the model of noninteracting composite fermions continues to be valid in this limit, their effective magnetic field B^* vanishes and they form a Fermi sea, called the CF Fermi sea. The lack of FQHE follows because the Fermi sea has no gap to excitations. Chapter 10 details several experimental observations of the CF Fermi sea at $v = 1/2$.

No general principle excludes FQHE at even-denominator fractions. Such FQHE has been observed, for example, at $f = 5/2$. The CF theory provides a natural explanation for why even-denominator FQHE is rare. The model of noninteracting composite fermions produces only odd-denominator fractions. Any even-denominator fraction must necessarily owe its existence to weak residual interactions between composite fermions, and, therefore, can be expected to be much weaker. This is analogous to the FQHE of electrons being more delicate than their IQHE. Other even denominators may be observed in the future, and the possibility cannot be ruled out that even the 1/2 state in the lowest Landau level will eventually turn out be a FQHE state.

7.4 Interacting composite fermions: new fractions

Were composite fermions completely noninteracting, the sequences of Eq. (5.20) would exhaust the possible fractions. The inter-CF interaction can produce delicate *new* FQHE.

7.4.1 Next generation fractions: FQHE of composite fermions

Pan *et al.* [494, 496] have discovered (Fig. 7.2), in very high quality samples and at very low temperatures, new fractions in the filling factor range $2/7 < v < 2/5$, for example, $v = 4/11$, which do not belong to the sequences of Eq. (5.20). These survive to very high magnetic fields and have little dependence on the Zeeman energy, indicating that the states are fully spin polarized. A hint of $f = 4/11$ and $7/11$ was seen earlier by Goldman and Shayegan [206] (these experiments showed substantial dependence on the Zeeman energy, suggesting a role of the spin degree of freedom).

While the fractions $v = n/(2pn \pm 1)$ are obtained most immediately in the CF theory, other fractions are not ruled out. An intuitively appealing scenario for the "next generation" fractions is as follows: Let us consider electrons in the filling factor range

$$1/3 < v < 2/5 , \tag{7.4}$$

which map into ^{2}CFs in the range

$$1 < v^* < 2 . \tag{7.5}$$

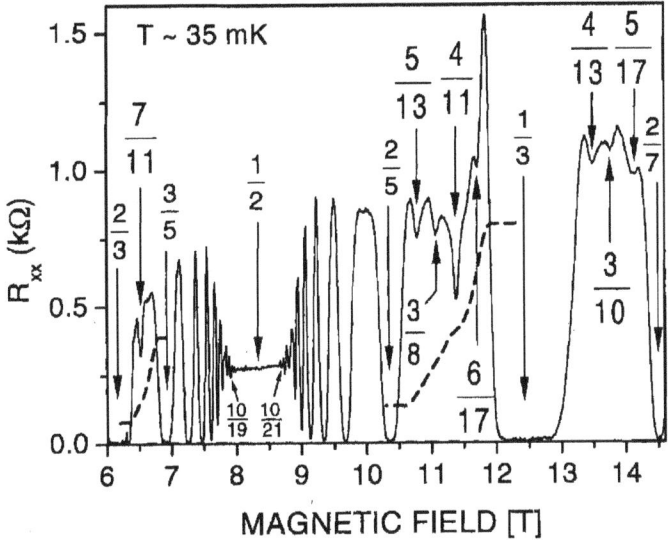

Fig. 7.2. Evidence for new fractions between 1/3 and 2/5, and also between 1/3 and 2/7. Solid trace is the longitudinal resistance and the dashed one shows the Hall resistance. Source: W. Pan, H. L. Stormer, D. C. Tsui, L. N. Pfeiffer, K. W. Baldwin, and K. W. West, *Phys. Rev. Lett.* **90**, 016801 (2003). (Reprinted with permission.)

The lowest ^{2}CF Λ level is fully occupied and the second one partially occupied. (Because we are interested here in fully spin-polarized states, we take composite fermions to be fully spin polarized.) No FQHE would result in this region for noninteracting composite fermions, just as noninteracting electrons in the partially filled second Landau level do not exhibit any FQHE. The weak residual interaction between composite fermions, however, can possibly cause a gap to open at certain filling factors. A natural conjecture is that the strongest new CF fractions are

$$f^* = 1 + \frac{1}{3} \text{ and } f^* = 1 + \frac{2}{3}, \tag{7.6}$$

which produce new electron fractions

$$f = \frac{4}{11} \text{ and } f = \frac{5}{13} \tag{7.7}$$

between the familiar fractions 1/3 and 2/5, consistent with the experimental observations of Pan *et al.* [494, 496]. Here, ^{2}CFs in the partially filled second Λ level capture two more vortices (as a result of the residual ^{2}CF–^{2}CF interaction) to turn into higher-order composite fermions (^{4}CFs), which condense into their own Λ levels, thereby opening a gap and producing the quantum Hall effect. Many more FQHE states can similarly be constructed (Exercise 7.1).

Table 7.2.   *Testing trial wave functions for $\nu = 4/11$*

E^{ex}	E^{tr}	$E^{(0)}$	overlap
$-0.441\ 214$	$-0.440\ 88(4)$	$-0.441\ 05(9)$	$0.993(2)$

Notes: Exact Coulomb state $\Psi^{\mathrm{ex}}_{4/11}$; trial wave function $\Psi^{\mathrm{tr}}_{4/11} = \mathcal{P}_{\mathrm{LLL}} \Psi_{4/3} \Phi_1^2$; and ground state $\Psi^{(0)}_{4/11}$ obtained by CF diagonalization as in Fig. 6.3.

Results are for $N = 12$ particles. The overlap is defined as $\langle \Psi^{\mathrm{tr}} | \Psi^{(0)} \rangle / \sqrt{\langle \Psi^{(0)} | \Psi^{(0)} \rangle \langle \Psi^{\mathrm{tr}} | \Psi^{\mathrm{tr}} \rangle}$.

Source: Chang and Jain [67].

A remark on the nomenclature is in order. In principle, the ^{4}CF state at $\nu = n/(4n \pm 1)$ can be viewed as the $\nu^* = n/(2n \pm 1)$ *fractional* QHE of ^{2}CFs. However, the interpretation as $\nu^* = n$ IQHE of ^{4}CFs is simpler. The standard practice is to classify only those states as the *fractional* QHE of *composite fermions* that do not entertain an IQHE interpretation.

The physical picture that the $\nu = 4/11$ FQHE as the $\nu^* = 4/3$ FQHE of composite fermions can be tested quantitatively. We have already seen in Chapter 6 that the state in the region $1/3 > \nu > 2/5$ is well described in terms of composite fermions. The $\nu = 4/11$ state occurs at

$$2Q = \frac{11}{4} N - 4 , \tag{7.8}$$

with $\bar{N} = (N+4)/4$ composite fermions in the second Λ level. (A derivation of this relation is left to Exercise 7.3.) The spectrum obtained from CF diagonalization for 12 particles at $2Q = 29$, shown in Fig. 6.3, is a finite size representation of $\nu = 4/11$. In the spirit of the CF theory, a trial wave function for the ground state is written as

$$\Psi^{\mathrm{tr}}_{4/11} = \mathcal{P}_{\mathrm{LLL}} \Psi_{4/3} \Phi_1^2 , \tag{7.9}$$

which requires as input the wave function $\Psi_{4/3}$ for the FQHE state of electrons at $\nu = 4/3$. Substituting for $\Psi_{4/3}$ the exact (fully spin-polarized) Coulomb ground state at $\nu = 4/3$ produces a 4/11 wave function with energy within 0.1% of the exact energy for $N = 12$ electrons (Table 7.2). The overlap of $\Psi^{\mathrm{tr}}_{4/11}$ with $\Psi^{(0)}_{4/11}$, the ground state obtained from CF diagonalization (which is essentially exact, as seen in Table 7.2) is 99.3%. This level of accuracy is comparable to that of the accepted wave functions for other fractions, thus confirming the link between the physics of the FQHE at 4/11 and 4/3 at a microscopic level.

An alternative trial wave function for 4/11 can be constructed by taking for $\Psi_{4/3}$ in Eq. (7.9) the state in which the lowest Landau level is full and electrons in the second Landau level form the Laughlin 1/3 wave function. That produces a poor variational state, with an energy $-0.43670(4)$ $e^2/\epsilon\ell$ and an overlap of $0.51(1)$ with $\Psi^{(0)}_{4/11}$ for $N = 12$ particles [67]. That is not surprising, because the Laughlin 1/3 wave function is an inadequate

Table 7.3.   *Testing trial wave functions in the second Landau level*

ν	$\bar{N}$	3	4	5	6	7	8	9
4/3	N	8	12	16	20	24	28	32
	gap	"C"	0.035	"C"	0.024	0.000 64	–	–
	overlap	0	0.4765	0	0.528 5	0.607 1	0.572 0	0.4794
7/5	N	–	9	–	16	–	23	–
	gap	–	"C"	–	0.015 85	–	0.001 829	–
	overlap	–	0	–	0.008(1)	–	0.193(1)	–

Notes: Overlaps of exact wave functions at $\bar{\nu} = 1/3$ and $\bar{\nu} = 2/5$ in the second Landau level (producing a total filling of $\nu = 4/3$ and $\nu = 7/5$, respectively) with wave functions of Eqs. (5.16) and (5.32).

$\bar{N}$ is the number of electrons in the second Landau level and N is the total number of electrons.

Spherical geometry is used; mixing between Landau levels is neglected; and electrons are taken as "spinless."

The symbol "C" indicates that the ground state is compressible (i.e., does not have $L = 0$). The $\nu = 7/5$ state is defined only for even $\bar{N}$.

Source: d'Ambrumenil and Reynolds [105] for $\nu = 4/3$ and Chang and Jain [68] for $\nu = 7/5$.

representation of the exact 1/3 wave function in the second Landau level, as shown by d'Ambrumenil and Reynolds [105]. (See Table 7.3.)

There is an additional complication. A combination of exact and CF diagonalization (He, Xie, and Zhang [244]; Mandal and Jain [424]; Wójs and Quinn [693]) shows that while the 4/11 state is incompressible for $N = 12$ and 20 (i.e., has an $L = 0$ ground state with a gap to excitations), it is compressible for $N = 8$, 18, and 24 with an $L \neq 0$ ground state. That appears troublesome [424], until one realizes that very similar finite size effects are seen for the FQHE at $\nu = 4/3$ (Table 7.3), where the state is incompressible for $N = 12$ and 20, compressible for $N = 8$ and 16, and barely incompressible for $N = 24$ (the gap is extremely small for $N = 24$). The 4/3 state becomes incompressible in the thermodynamic limit [105], and the same can be expected of 4/11.

The wave function $\Psi^{tr}_{4/11}$ describes complex "nested" correlations (see "Wheels within wheels" by Smet [605]). To begin with, $\Psi_{4/3}$ itself is quite complex, with the lowest Landau level of electrons completely full and a ^{2}CF state in the second Landau level. Its composite-fermionization creates a structure with two flavors of composite fermions. It would be difficult to take this wave function seriously if it did not turn out to be so accurate.

The explanation of the next generation FQHE states is yet another, nontrivial illustration of the analogy between the physics at ν and ν^*, and of higher-LL-like physics appearing within the lowest Landau level, exhibiting the same subtleties as the electronic FQHE in higher electronic Landau levels. Just as the 4/3 state is much weaker than the IQHE states at 1 and 2, so is the 4/11 state compared to 1/3 and 2/5.

As discussed in Section 7.7, the FQHE is scarce in higher Landau levels. That implies a scarcity of the FQHE of composite fermions in higher Λ levels, i.e., of FQHE of electrons *in between* the fractions $f = n/(2pn \pm 1)$. The lack of FQHE in fourth or higher Landau levels strongly suggests that no fractions other than $n/(2n \pm 1)$ are possible in the range $4/7 > \nu > 4/9$.

The experimental observation of new FQHE states between 1/3 and 2/5 has motivated several studies of the FQHE of composite fermions using other approaches (Flohr and Osterloh [168]; Goerbig, Lederer, and Morais Smith [197, 198]; Lopez and Fradkin [405]; Merlo *et al.* [442]). Wójs, Yi, and Quinn [532, 533, 697] propose an alternative mechanism for the new FQHE states, in terms of pairing of composite fermions in the second Λ level. Wójs, Wodziński, and Quinn [698] compute the pair correlation function for composite fermions in the second Λ level, interacting through CF pseudopotentials (Section 6.7.1), and find a shoulder at small separations, which they take as evidence of cluster formation. Sitko [598] considers a condensation of higher Λ level composite fermions into a Pfaffian state (next subsection), and deduces the charge and statistics of the excitations of this state. Goerbig, Lederer, and Morais Smith [198] calculate energies of various solid and liquid phases of composite fermions in the second Λ level, for a fully spin-polarized state, in a Hartree–Fock approximation of the Murthy–Shankar theory, and argue that the liquid phases have lower energies at several filling factors, including at $\nu = 4/11$.

7.4.2 $\nu = 5/2$: pairing of composite fermions

Willett *et al.* [680] have seen evidence for a FQHE plateau at $R_H = h/(5/2)e^2$. More recent measurements (Pan *et al.* [489]; Xia *et al.* [706]) in extremely high mobility samples (17 million cm^2/V s) at ultra-low temperatures (4 mK) confirm the accuracy of quantization to better than 2×10^{-6}, establishing the 5/2 state as a true FQHE state. A well-developed 5/2 plateau can be seen in Fig. 4.16.

Writing

$$\frac{5}{2} = 2 + \frac{1}{2}, \tag{7.10}$$

and treating the lowest filled Landau level as inert (which, counting the spin degree of freedom, accounts for the 2 on the right hand side), shows that 5/2 corresponds to a filling of 1/2 in the second Landau level. Although the 5/2 FQHE was initially thought to involve the electron spin in an essential manner (Haldane and Rezayi [224]), subsequent studies supported a fully spin-polarized ground state of the half-filled second Landau level. One indication was that the 5/2 FQHE is seen at fairly high magnetic fields. Furthermore, in an exact diagonalization study with up to 12 electrons, Morf [457] finds that, even with zero Zeeman splitting, the spin singlet state at the half-filled second Landau level has higher energy than the fully polarized state.

The problem of electrons at $\nu = 1/2$ in the second Landau level can be simulated by electrons at $\nu = 1/2$ in the lowest Landau level subject to an effective interaction. The first instinct would be to expect a CF Fermi sea, by analogy to the physics of the half-filled lowest Landau level. Experiments show dramatically different behaviors at 1/2 and 5/2, however.

An attractive proposal for the explanation of the 5/2 FQHE (Greiter, Wen, and Wilczek [213,215]; Moore and Read [454]) is that composite fermions form a p-wave paired state, described by a Pfaffian[4] wave function, written by Moore and Read [454]:

$$\Psi_{1/2}^{\text{Pf}} = \text{Pf}\left(\frac{1}{z_i - z_j}\right) \prod_{i<j}(z_i - z_j)^2 \exp\left[-\frac{1}{4}\sum_k |z_k|^2\right], \tag{7.11}$$

where "Pf" stands for Pfaffian. Without the Pfaffian factor, the wave function still describes a system at $\nu = 1/2$, but has a wrong exchange symmetry. The Pfaffian of an antisymmetric matrix M is defined, apart from an overall normalization factor, as [410]

$$\text{Pf } M_{ij} = A(M_{12}M_{34}\ldots M_{N-1,N}), \tag{7.12}$$

where A is the antisymmetrization operator. The Pfaffian factor makes $\Psi_{1/2}^{\text{Pf}}$ properly antisymmetric without altering the filling factor. In the spherical geometry, the Moore–Read wave function generalizes to

$$\Psi_{1/2}^{\text{Pf}} = \Phi_1^2 \text{ Pf } M, \tag{7.13}$$

with $M_{ij} = (u_i v_j - v_i u_j)^{-1}$.

While Pf M_{ij} is a complicated function, its square is relatively simple [410]:

$$[\text{Pf } M]^2 = \det M. \tag{7.14}$$

This property is useful in the calculation of the energy of the Moore–Read wave function, for which only the modulus of the squared wave function is needed.

The usual Bardeen–Cooper–Schrieffer wave function for fully polarized electrons can be written as [110]

$$\Psi_{\text{BCS}} = A[\phi_0(r_1, r_2)\phi_0(r_3, r_4)\ldots\phi_0(r_{N-1}, r_N)], \tag{7.15}$$

which is a Pfaffian. (The fully symmetric spin part is not shown explicitly.) Analogously, $\text{Pf}\left(\frac{1}{z_i - z_j}\right)$ describes a p-wave pairing of electrons (p-wave because the system is fully spin polarized), and $\Psi_{1/2}^{\text{Pf}}$ is interpreted as the p-wave paired state of composite fermions carrying two vortices.

The feasibility of the concept of CF pairing has been investigated in several quantitative studies. Park *et al.* [501] show that the Moore–Read wave function has a substantially lower energy than the CF Fermi sea wave function in the second Landau level (with reverse

---

[4] Named after Johann Friedrich Pfaff

Table 7.4.   *Testing the Pfaffian wave function at* $\nu = 5/2$

$\bar{N}$	8	10	14	16
overlap	0.87	0.84	0.69	0.78

Notes: $\bar{N}$ is the number of electrons in the second Landau level. Spherical geometry is used, and mixing between Landau levels is neglected.
Source: Scarola, Jain, and Rezayi [570].

energy ordering in the lowest Landau level). Several studies compare the Moore–Read wave function with the exact Coulomb ground state wave function at $\nu = 1/2$ in the second Landau level (Greiter, Wen, and Wilczek [213,215]; Morf [457]; Rezayi and Haldane [550]; Scarola, Jain, and Rezayi [570]). The overlaps of the Moore–Read wave function with the exact Coulomb ground state are given in Table 7.4 for several N. Morf [457] and Rezayi and Haldane [550] have shown that the overlaps can be improved by tweaking the form of the interaction for the exact state, and also by particle–hole symmetrizing the Moore–Read wave function. The overlaps are decent by the standards of second LL FQHE, and taken together, the numerical studies strongly support the interpretation of the 5/2 state as a paired state of composite fermions. Unlike the BCS wave function for superconductors, the Moore–Read wave function does not have any variational freedom, and how one can be introduced is not known; such freedom should prove useful in improving the wave function, as well as for determining if the pairing is in the strong or the weak coupling limit.

Pairing usually results from attractive interactions. In the present case, the original Hamiltonian only consists of the repulsive Coulomb interaction between electrons. How might pairing occur in spite of the strong repulsion? The point is that the objects that pair up are not electrons but composite fermions, the interaction between which is different from that between electrons. Whereas electrons repel one another strongly, the interaction between composite fermions is weak, and there is no reason why it can sometimes not be weakly *attractive*. Essentially, the Coulomb interaction is screened through the binding of vortices to electrons, which creates a correlation hole around each electron. Since the number of bound vortices cannot change continuously, such a binding can possibly *over*screen the Coulomb interaction for certain parameters. Further insight into the origin of pairing can be gained by choosing the CF Fermi sea as the starting point at 5/2, and asking if it is unstable to a pairing of composite fermions through a Cooper instability. There is numerical evidence (Scarola, Park, and Jain [568]) that the interaction between two composite fermions on top of the CF Fermi sea is repulsive at 1/2 but attractive at 5/2. Such a picture is further supported by the experiment of Willett, West, and Pfeiffer [687], wherein they detect CF Fermi surface properties at 5/2 after eliminating the FQHE by raising the temperature.

It must be stressed that the 5/2 state has no off-diagonal long-range order. The 5/2 state is not a superconductor. Pairing of composite fermions simply produces a gapped FQHE state for electrons. It is not known in what way, operationally, the 5/2 FQHE differs from other FQHE, and how experiments can verify that composite fermions are paired in this state.

7.5 FQHE and spin

For small Zeeman energies, the spin degree of freedom adds to the richness of FQHE. While the spin physics does not produce any new fractions, it creates many new FQHE states at the old fractions. This physics is discussed in Chapter 11.

7.6 FQHE at low fillings

As the filling factor is reduced, composite fermions with increasingly greater numbers of vortices are expected to form, eventually giving way to a crystal phase. Many fractions from the ^{2}CFs and ^{4}CFs sequences, $f = n/(2n \pm 1)$ and $f = n/(4n \pm 1)$, have been seen (Table 7.1). Pan *et al.* [495] have seen transport evidence, through weak minima in R_{xx}, for several fractions belonging to ^{6}CF and ^{8}CF sequences: $n/(6n + 1)$ (1/7, 2/13, 3/19); $n/(6n - 1)$ (2/11, 3/17); $n/(8n + 1)$ (1/9, 2/17); and $n/(8n - 1)$ (2/15). These FQHE states appear only above certain filling-factor dependent temperatures; Pan *et al.* conclude that the zero temperature phase in this filling factor range is a crystal, which melts into a CF liquid upon raising temperature. Earlier magneto-optical studies of Buhmann *et al.* [41] also report evidence for FQHE down to 1/9 in the luminescence spectrum, where they identify the FQHE liquid through a weakening of the intensity of the luminescence line associated with the pinned crystal.

7.7 FQHE in higher Landau levels

Most observed fractions belong to the lowest Landau level. A few fractions have been seen in the second Landau level. For example, $2 + 1/3$, $2 + 2/5$, $2 + 2/3$, $2 + 3/5$, $2 + 1/5$, $2 + 2/7$, and $2 + 4/5$ can be seen in Fig. 4.16, in addition to the even-denominator fraction $2 + 1/2$. Gervais *et al.* [184] have seen evidence for $4 + 1/5$ and $4 + 4/5$ in the third Landau level.

Composite fermions carrying two vortices are the most robust flavor of composite fermions in the lowest Landau level, and their states at $f = n/(2n \pm 1)$ are the most accurately understood FQHE states. The wave functions that are successful in the lowest Landau level are poor quantitative representations of the second LL ground states, as seen in Table 7.3. (That underscores the nontriviality of the excellent accuracy of the CF theory in the lowest Landau level.) Nonetheless, because the observed fractions in the second Landau level also belong to the standard sequences (1/3, 2/5, 2/3, 3/5), the expectation is natural that they have the same physical origin as the LLL FQHE; i.e., it should be possible to vary the interaction pseudopotentials from their LLL values to the second LL values without crossing any phase boundary. (See Wójs [695] for a contrary opinion.) Nonetheless, ^{2}CFs form only with difficulty in the second Landau level.

Table 7.5.   *Overlaps for the $\nu = 1/5$ wave function in the second Landau level*

$\bar{N}$	3	4	5	6	7
overlap	1.000	0.9893	0.9982	0.9590	0.9818

Notes: $\bar{N}$ is the number of electrons in the second Landau level. Spherical geometry is used, and mixing between Landau levels is neglected.

Source: d'Ambrumenil and Reynolds [105].

Somewhat surprisingly, the $2 + 1/5$ state, the first member of the ^{4}CF series $2 + n/(4n + 1)$, is well described by the Laughlin wave function (d'Ambrumenil and Reynolds [105]; MacDonald and Girvin [411]), as seen in Table 7.5. That is consistent with the observation of several ^{4}CF states (1/5, 4/5, 2/7) in the second Landau level.

7.8 Fractions *ad infinitum*?

The reader may encounter in literature the statement that *all* odd-denominator fractions will eventually be observed as the system is made sufficiently pure. That is incorrect. While one can make a *scenario* for FQHE at any odd-denominator fraction (in fact, also at any even denominator fraction), e.g. $100\,000/200\,000\,01$, that does not mean that it would necessarily occur for the Coulomb interaction. Some well-known counter-examples illustrate this point. A FQHE at $f = 16 + n/(2n + 1)$ could, in principle, result for the same reason as at $f = n/(2n + 1)$ in the lowest Landau level, but it is certain that the Haldane pseudopotentials of the ninth Landau level do not stabilize a FQHE state; the actual state is probably a simpler, Hartree–Fock charge density wave state. Similarly, at sufficiently low fillings in the lowest Landau level, the ground state surely is a crystal. It is possible that FQHE along the sequence $f = n/(2n + 1)$ will cease beyond a certain maximum n. A time may come in the future when we know all of the fractions that nature supports. That will not be the end of the field though. There is more to FQHE than a collection of fractions.

Exercises

7.1 List all electron fractions that correspond to FQHE of composite fermions (of vorticity $2p$) in the nth Λ level. What is their order of stability?

7.2 Show that any odd denominator fraction can be reduced to an integer by repeated applications of three operations: composite fermionization ($\nu \rightarrow \nu^*$, with $\nu^{*-1} = \nu^{-1} - 2p$); particle–hole symmetry ($\nu \rightarrow 1 - \nu$); and LL subtraction ($\nu \rightarrow \nu - 1$). This implies that, in principle, FQHE is possible at any odd-denominator fraction (see, however, Section 7.8). Hint: Consider p_0/q_0 where p_0 and q_0 are relatively prime and

q_0 is an odd integer. If $p_0/q_0 > 1/2$, use LL subtraction or particle–hole symmetry to relate it to $p_1/q_1 < 1/2$. Now composite-fermionize to obtain $p_2/q_2 > 1/2$, with $q_2 < q_1$. Repeat the process until $q_m = 1$. (Source: Jain [275].)

7.3 Show that the FQHE state at

$$\nu = \frac{3n+1}{8n+3} \tag{E7.1}$$

occurs at

$$2Q = \frac{8n+3}{3n+1}N - \frac{12n+n^2+3}{3n+1}, \tag{E7.2}$$

and the number of composite fermions in the second Λ level is

$$\bar{N} = \frac{n(N+n+3)}{3n+1}. \tag{E7.3}$$

Hint: First determine the flux $2Q^*$ at which the state $\nu^* = 1 + n/(2n+1)$ occurs. For this purpose, make use of the fact that the $\bar{\nu} = n/(2n+1)$ state of $\bar{N}$ electrons in the second Landau level occurs at the same LL *degeneracy* for the second Landau level as it does in the lowest Landau level. (Then one can upgrade the LLL wave function to the second Landau level straightforwardly.) Adding the lowest Landau level gives the total number of electrons. Composite-fermionize this state to obtain the desired result. (Source: Mandal and Jain [424].)

7.4 Derive the relation $2Q = 2N - 3$ for the Moore–Read wave function of Eq. (7.11) in the spherical geometry. Show that for $N = 6$ and 12, omitted in Table 7.4, the state at $2Q = 2N - 3$ has other interpretations.

7.5 Consider the BCS wave function for a spin-singlet state, given by

$$\Psi_{\mathrm{BCS}} = A[\phi_0(r_1, r_{[1]}) \ldots \phi_0(r_{N/2}, r_{[N/2]}) u_1 \ldots u_{N/2} d_{[1]} \ldots d_{[N/2]}], \tag{E7.4}$$

where $[j] \equiv j + N/2$, $\phi_0(r_i, r_{[j]})$ is symmetric under label exchange, and u_j and $d_{[j]}$ represent up and down spinors. Show that, apart from an overall normalization factor, this wave function can be written as

$$\Psi_{\mathrm{BCS}} = A[(\det M) u_1 \ldots u_{N/2} d_{[1]} \ldots d_{[N/2]}], \tag{E7.5}$$

where $M_{ij} = \phi_0(r_i, r_{[j]})$ is a symmetric $\frac{N}{2} \times \frac{N}{2}$ matrix. Det M is the wave function for the configuration in which the first $N/2$ electrons have spin up, and the remainder have spin down. Hint: The coefficient of the term $u_1 \ldots u_{N/2} d_{[1]} \ldots d_{[N/2]}$ in Eq. (E7.4) is obtained by antisymmetrizing the factor $\phi_0(r_1, r_{[1]}) \ldots \phi_0(r_{N/2}, r_{[N/2]})$ with respect to an exchange of any two like-spin coordinates. (Source: Bouchaud, Georges, and Lhuillier [36].)

7.6 The Haldane–Rezayi wave function is given by

$$\Psi_{\mathrm{HR}} = \det\left(\frac{1}{(z_j^\uparrow - z_k^\downarrow)^2}\right) \prod_{j<k}(z_j - z_k)^2 \exp\left[-\frac{1}{4}\sum_i |z_i|^2\right]. \tag{E7.6}$$

Show that it describes a state at $\nu = 1/2$. Also show that it represents a spin-singlet pairing of composite fermions (Hint: use results from the preceding exercise). This wave function is the exact ground state for a "hollow-core" interaction model (see Section 12.8.4), but is not known to be relevant for any realistic interaction. (Source: Haldane and Rezayi [224, 225].)

8

Incompressible ground states and their excitations

This chapter is devoted to the properties of incompressible ground states and their excitations. Ground state properties, such as pair correlation function and static structure factor, as well as energies of several kinds of charged and neutral excitations are evaluated using the CF theory. Quantities in the thermodynamic limit are obtained by an extrapolation of finite system results. Experimental studies of the FQHE state through transport, inelastic light scattering, and scattering by ballistic phonons have identified a variety of excitations which can be compared to theory. As we have seen in Chapter 6, the CF theory predicts excitation energies with an accuracy of a few % for $N \leq 12$. A reasonable expectation is that the thermodynamic limits obtained from the CF theory are also accurate to within a small percentage. Unfortunately, a comparison with real life experiments also necessitates an inclusion of the effects of nonzero thickness of the electron wave function, Landau level mixing, and disorder, which are not as well understood as the FQHE, and the accuracy of quantitative comparisons between theory and experiment is determined largely by the accuracy with which these other effects can be incorporated into theory. At present, the quantitative agreement between theory and laboratory experiment is roughly within a factor of two, although a 10–20% agreement has been achieved in some cases. Fully spin-polarized FQHE is assumed throughout this chapter; the spin physics is the topic of Chapter 11.

8.1 One-particle reduced density matrix

The one-particle reduced density matrix $\rho_1(r, r')$, which is related to the equal-time Green's function, is defined as (Appendix H)

$$\rho_1(r', r) = \left\langle \hat{\psi}^\dagger(r')\hat{\psi}(r) \right\rangle, \tag{8.1}$$

where the field operators are given by $\hat{\psi}(r) = \sum_m \eta_m(r)c_m$ and $\hat{\psi}^\dagger(r) = \sum_m \bar{\eta}_m(r)c_m^\dagger$. In terms of the ground state wave function, we have (Section F.1.3)

$$\rho_1(r, r') = \frac{N \int d^2r_2 \ldots d^2r_N \, \Psi^*(r', r_2, \ldots, r_N)\Psi(r, r_2, \ldots, r_N)}{\int d^2r_1 \ldots d^2r_N \, \Psi^*(r_1, r_2, \ldots, r_N)\Psi(r_1, r_2, \ldots, r_N)}. \tag{8.2}$$

We now show that for a uniform density, isotropic liquid, $\rho_1(r,r')$ can be evaluated by inspection, and is completely specified by the filling factor. Its analytic structure indicates that it must have the form (using the symmetric gauge in the disk geometry)

$$\rho_1(r,r') = \frac{\nu}{2\pi \ell^2} \exp\left[f(\bar{z}',z) - \frac{1}{4}\bar{z}'z' - \frac{1}{4}\bar{z}z \right],$$
$$(8.3)$$

where the function $f(\bar{z}',z)$ is analytic in $\bar{z}'$ and z. The function $f(\bar{z}',z)$ can be determined from the knowledge that the diagonal element of $\rho_1(r,r')$ is equal to the expectation value of $\rho(r) = \sum_i \delta^{(2)}(r_i - r)$, which is simply the density, i.e., $\rho_1(z,z) = \rho = \nu/(2\pi \ell^2)$. We therefore must have

$$\lim_{z' \to z} f(\bar{z}',z) - \frac{1}{4}\bar{z}'z' - \frac{1}{4}\bar{z}z \to 0.$$
$$(8.4)$$

That fixes $f(\bar{z}',z) = \frac{1}{2}\bar{z}'z$, to give

$$\rho_1(r,r') = \frac{\nu}{2\pi \ell^2} \exp\left[\frac{1}{2}\bar{z}'z - \frac{1}{4}\bar{z}'z' - \frac{1}{4}\bar{z}z \right]$$

$$= \frac{\nu}{2\pi \ell^2} \exp\left[-\frac{|z - z'|^2}{4} + \frac{z\bar{z}' - z'\bar{z}}{4} \right]$$
$$(8.5)$$

for any uniform density, isotropic liquid in the lowest Landau level. This result was first derived by Girvin and MacDonald [193] directly from Eq. (8.1) (Exercise 8.1).

Because the last term inside square brackets of Eq. (8.5) is a pure phase, the one-body reduced density matrix has a Gaussian fall off. Thus, we have $\rho_1(r,r') \to 0$ in the limit $|r - r'| \to \infty$ for any FQHE state. This implies an absence of off-diagonal long-range order (ODLRO) in the FQHE state in the one-particle reduced density matrix. (See Appendix H for the definition of ODLRO.)

8.2 Pair correlation function

An important quantity for a liquid is its pair distribution function $g(r)$, proportional to the probability of finding a particle at a distance r from a given particle in the ground state. The liquid phase is defined by the property that this probability approaches a constant at large r. The pair correlation function contains in it information about the short-distance correlations in the liquid.

The expression for $g(r)$ is given by

$$g(r) = \frac{1}{\rho} \left\langle \sum_i{}' \delta^{(2)}(r - r_i + r_j) \right\rangle$$

$$= \frac{1}{\rho N} \left\langle \sum_{i \neq j} \delta^{(2)}(r - r_i + r_j) \right\rangle,$$
$$(8.6)$$

where the prime denotes the condition $i \neq j$, the expectation value is evaluated in the ground state, ρ is the electron density, $\mathbf{r}_j$ is the position of the jth electron, N is the total number of electrons, and translational symmetry is assumed. The function $g(r)$ is normalized so that it approaches unity at large r. More explicitly, it is given by

$$g(r) = \frac{1}{\rho N} \int d^2 r_1 \ldots d^2 r_N \sum_{i \neq j} \delta^{(2)}(r - r_i + r_j) |\Psi(r_1, r_2, \ldots, r_N)|^2$$

$$= \frac{(N-1)}{\rho} \int d^2 r_1 \ldots d^2 r_N \delta^{(2)}(r - r_1 + r_2) |\Psi(r_1, r_2, \ldots, r_N)|^2$$

$$= \frac{(N-1)}{\rho} \int d^2 r_2 \ldots d^2 r_N |\Psi(r_2 + r, r_2, \ldots, r_N)|^2$$

$$= \frac{N(N-1)}{\rho^2} \int d^2 r_3 \ldots d^2 r_N |\Psi(r, 0, r_3, \ldots, r_N)|^2, \tag{8.7}$$

where translational invariance has been used in the last step. Another equivalent form, useful for a Monte Carlo evaluation, is

$$g(|r_1 - r_2|) = \frac{N(N-1)}{\rho^2} \int d^2 r_3 \ldots d^2 r_N |\Psi(r_1, r_2, \ldots, r_N)|^2, \tag{8.8}$$

where the wave function Ψ is assumed to be normalized to unity.

Using Eq. (F.30) the pair correlation function can be related to the equal-time density–density correlation function

$$g(r) = \rho^{-2} \langle \hat{\psi}^\dagger(0) \hat{\psi}^\dagger(r) \hat{\psi}(r) \hat{\psi}(0) \rangle$$

$$= \rho^{-2} \langle \hat{\rho}(0) \hat{\rho}(r) \rangle, \tag{8.9}$$

where $\hat{\rho}(r) = \hat{\psi}^\dagger(r) \hat{\psi}(r)$ and $r \neq 0$ has been assumed.

The pair correlation function can be obtained for any given wave function by evaluating an integral. For Hartree–Fock states, which are described by a single Slater determinant, it may be calculated without difficulty. Substituting the N-particle state

$$\Psi(r_1, \ldots, r_N) = \frac{1}{\sqrt{N}} \begin{vmatrix} \eta_0(r_1) & \eta_1(r_1) & \eta_2(r_1) & \cdot & \cdot \\ \eta_0(r_2) & \eta_1(r_2) & \eta_2(r_2) & \cdot & \cdot \\ \eta_0(r_3) & \eta_1(r_3) & \eta_2(r_3) & \cdot & \cdot \\ \cdot & \cdot & \cdot & \cdot & \cdot \\ \cdot & \cdot & \cdot & \cdot & \cdot \end{vmatrix} \tag{8.10}$$

into Eq. (8.8) yields (the $(N-2)!$ terms make equal contribution, with the factor $(N-2)!/N!$ canceling the factor $N(N-1)$ in Eq. (8.8))

$$g(|r_1 - r_2|) = \frac{1}{2\rho^2} \sum_{\alpha,\beta} |\eta_\alpha(r_1)\eta_\beta(r_2) - \eta_\beta(r_1)\eta_\alpha(r_2)|^2, \qquad (8.11)$$

where the sum is over all occupied single particle orbitals. This allows, in particular, the determination of $g(r)$ for IQHE states.

A good approximation for the pair correlation function of a FQHE state can be obtained with the help of the wave functions of Eq. (5.32), with multi-dimensional integrals evaluated numerically by Monte Carlo methods. The results for several fractions of the form $\nu = n/(2n+1)$ are shown in Fig. 8.1 [315]. The systems are sufficiently large that the results are essentially N independent, and can be considered to represent the thermodynamic limit. In all cases, $g(r)$ becomes constant for large r/ℓ, confirming that the wave functions describe a liquid. For a crystal with long-range order, the pair correlation function would oscillate all the way to infinity.[1]

8.3 Static structure factor

Another ground state property of interest is the static structure factor $S(k)$, defined as

$$S(k) = \frac{1}{N}\langle\rho_k\rho_{-k}\rangle - N\delta_{k0}, \qquad (8.12)$$

where

$$\rho_k = \sum_{j=1}^N e^{-ik\cdot r_j} \qquad (8.13)$$

is the Fourier transform of the electron density:

$$\rho_k = \int d^2r\, e^{-ik\cdot r}\rho(r)$$

$$= \int d^2r\, e^{-ik\cdot r} \sum_{j=1}^N \delta^{(2)}(r - r_j)$$

$$= \sum_{j=1}^N e^{-ik\cdot r_j}. \qquad (8.14)$$

The last term in Eq. (8.12) cancels the $k=0$ divergence in the first term on the right hand side.

---

[1] Translational invariance, or rotational invariance in the spherical geometry, does not necessarily imply a liquid. A Hartree–Fock crystal state projected on to a definite angular momentum produces a *rotating* crystal, which has uniform density. An example is the Laughlin wave function $\Psi_{1/m}$ for $m > 72$, which explicitly has rotational symmetry on the sphere, but describes a crystal, as revealed by pair correlation function.

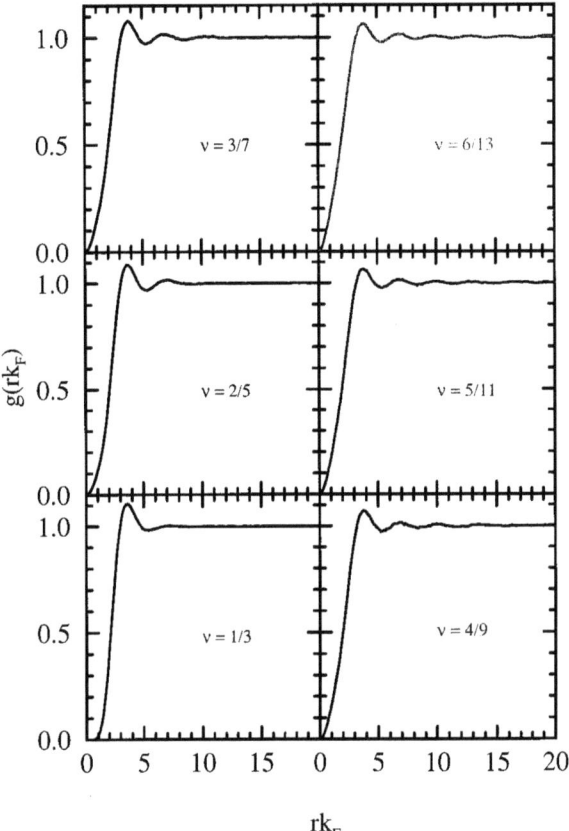

Fig. 8.1. Pair distribution functions for several FQHE states at $\nu = n/(2n + 1)$, modeled as n filled Landau levels of composite fermions (Eq. (5.32)). $N = 50, 54, 60, 60, 60,$ and 54 particles have been used for $\nu = 1/3, 2/5, 3/7, 4/9, 5/11,$ and 6/13, respectively. The quantity k_F is defined as $k_F = \sqrt{4\pi\rho}$. Source: R. K. Kamilla, J. K. Jain, and S. M. Girvin, *Phys. Rev. B* **56**, 12411 (1997). (Reprinted with permission.)

The static structure factor is related to the pair correlation function through Fourier transformation. Let us consider $k \neq 0$ first:

$$S(k) = \frac{1}{N} \int d^2 r_1 \ldots d^2 r_N \left[N + \sum_{j \neq k} e^{-i k \cdot (r_j - r_k)} \right] |\Psi(r_1, r_2, \ldots, r_N)|^2$$

$$= 1 + (N - 1) \int d^2 r_1 \ldots d^2 r_N \, e^{-i k \cdot (r_1 - r_2)} |\Psi(r_1, r_2, \ldots, r_N)|^2$$

$$= 1 + (N - 1) \int d^2 r \, e^{-i k \cdot r} \int d^2 r_2 \ldots d^2 r_N |\Psi(r + r_2, r_2, \ldots, r_N)|^2$$

$$= 1 + \rho \int d^2 r \, e^{-i k \cdot r} g(r). \tag{8.15}$$

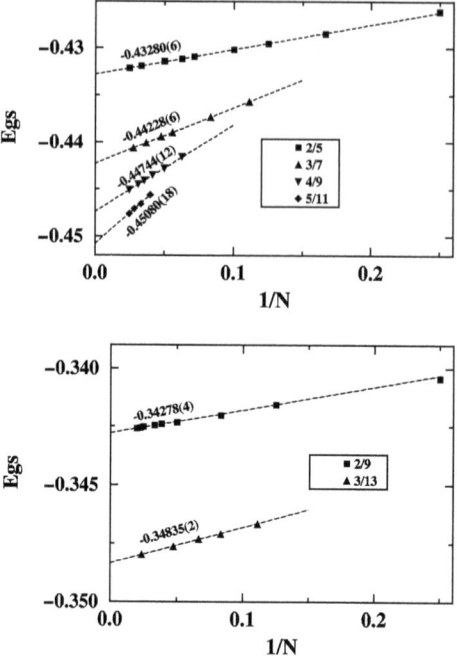

Fig. 8.2. Ground state energy as a function of $1/N$ for several incompressible states. The thermodynamic values are shown on the figure. Source: J. K. Jain and R. K. Kamilla, *Int. J. Mod. Phys. B* **11**, 2621 (1997). (Reprinted with permission.)

For all k, $g(r)$ and $S(k)$ are related by

$$S(k) - 1 = \rho \int d^2r \, e^{ik.r}(g(r) - 1).$$
(8.16)

8.4 Ground state energy

The ground state energy of incompressible states is of interest, even though it cannot be measured directly, because it serves as a criterion when comparing different theories of the FQHE, and also in investigating the stability of a FQHE state relative to another state with a different symmetry, e.g., the Wigner crystal (WC) [675].

Figure 8.2 shows the N dependence of the ground state energy per particle for certain fractions. The energy has been corrected for a finite size deviation of the electron density from its thermodynamic value through a factor $\sqrt{\rho/\rho_N} = \sqrt{2Q\nu/N}$, which reduces the N dependence and leads to a better extrapolation to the thermodynamic limit (Appendix I). Thermodynamic estimates are given in Table 8.1.

Figure 8.3 shows the ground state energy of the CF liquid along with the best estimate for the energy of the Wigner crystal (Lam and Girvin [363]). It illustrates why the Wigner crystal

Table 8.1. *Ground state energies and gaps for FQHE states*

ν	Ground state energy	Gap
1/3	−0.409 828(27)	0.106(3)
2/5	−0.432 804(62)	0.058(5)
3/7	−0.442 281(62)	0.047(4)
4/9	−0.447 442(115)	0.035(6)
5/11	−0.450 797(175)	0.023(10)
1/2	−0.465 3(2)	−
1/5	−0.327 499(5)	0.025(3)
2/9	−0.342 782(35)	0.016(2)
3/13	−0.348 349(19)	0.014(2)
4/17	−0.351 189(39)	0.011(3)

Source: Jain and Kamilla [292]

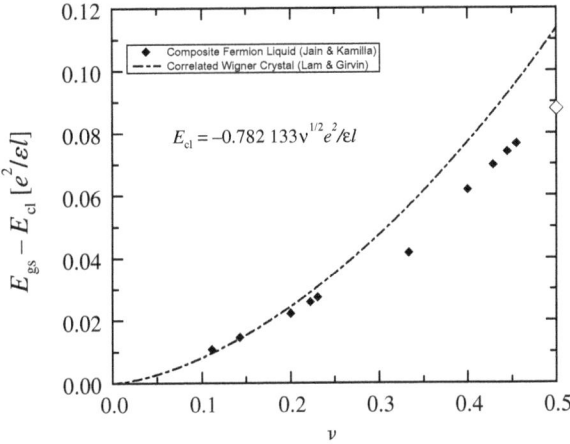

Fig. 8.3. Energies per particle for the CF liquid and the WC as a function of the filling factor. The WC energy is taken from Lam and Girvin [363]. Energies are shown relative to $E_{cl} = -0.782\,133\sqrt{\nu}e^2/\epsilon\ell$, the energy of a classical two-dimensional WC with triangular symmetry. The open diamond on the right vertical axis is the estimate of the CF Fermi sea energy at $\nu = 1/2$ $(E_{gs,1/2} - E_{cl} = 0.0877(2)e^2/\epsilon\ell)$ obtained by an extrapolation of the solid diamonds. The energy of the CF liquid is shown only at the special $n/(2pn + 1)$ filling factors; the full curve has cusps at these points. Source: J. K. Jain and R. K. Kamilla, Int. J. Mod. Phys. B **11**, 2621 (1997). (Reprinted with permission.)

is not observed as soon as the kinetic energy is suppressed by forcing all electrons into the lowest Landau level. The Wigner crystal is preempted by the CF liquid for a range of filling factors in the lowest Landau level. As the system becomes more classical with decreasing ν, a crystal is eventually expected. More information on this topic can be found in Chapter 15.

8.5 CF-quasiparticle and CF-quasihole

CF-quasiparticle and CF-quasihole are introduced in Section 5.9.4. They are images of IQHE states obtained by adding an electron to or removing an electron from the state with n full Landau levels. Figure 6.4 contains a comparison of the densities predicted by these wave functions with exact densities for the CF-quasiparticles and CF-quasiholes at $\nu = 1/3$ and $2/5$. Density profiles for larger N are shown for several FQHE states in Figs. 8.4, 8.5, and 8.6; although evaluated from the CF theory, they should be viewed as being practically exact.

The following comments are in order.

- CF-quasiparticle and CF-quasihole are sometimes referred to simply as "quasiparticle" and "quasihole."
- CF-quasiparticles and CF-quasiholes have the topological properties of fractional "local charge" and fractional "braiding statistics," discussed in Chapter 9.
- In the spherical geometry, a CF-quasiparticle or a CF-quasihole is obtained with a proper choice of the flux $2Q$ and the number of particles N (Exercise 8.7).
- A CF-quasiparticle or a CF-quasihole is actually the *ground* state at the proper $(N, 2Q)$ values (Exercise 8.7). It is thought of as an excitation for reasons explained in the next section.
- While excitations have a simple interpretation in terms of composite fermions, they are complex states of electrons.

Quasiparticle–quasihole asymmetry   One might have expected the density profiles of the quasihole and the quasiparticle of a given FQHE state to be mirror images, but that is not the case. For example, as seen in Fig. 8.6, the quasihole at $\nu = 1/3$ has a deep minimum, whereas the quasiparticle has a ring-like shape, with maximum density somewhat away

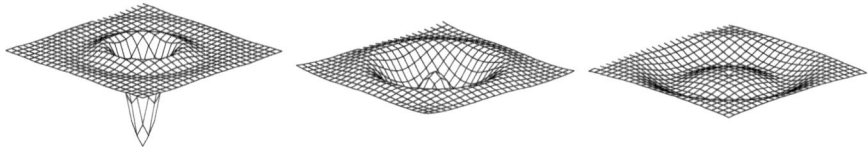

Fig. 8.4. Density profiles $(\rho(r) - \rho_0)/\rho_0$ for the CF-quasiholes at $\nu = 1/3, 2/5$, and $3/7$. ρ_0 is the average electron density. Source: K. Park, Ph.D. thesis (Stony Brook, State University of New York, 2000). (Reprinted with permission.)

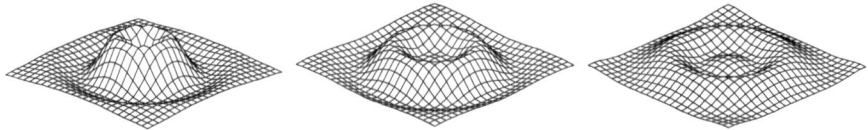

Fig. 8.5. Density profiles $(\rho(r) - \rho_0)/\rho_0$ for the CF-quasiparticles at $\nu = 1/3, 2/5$, and $3/7$. ρ_0 is the average electron density. Source: K. Park, Ph.D. thesis (Stony Brook, State University of New York, 2000). (Reprinted with permission.)

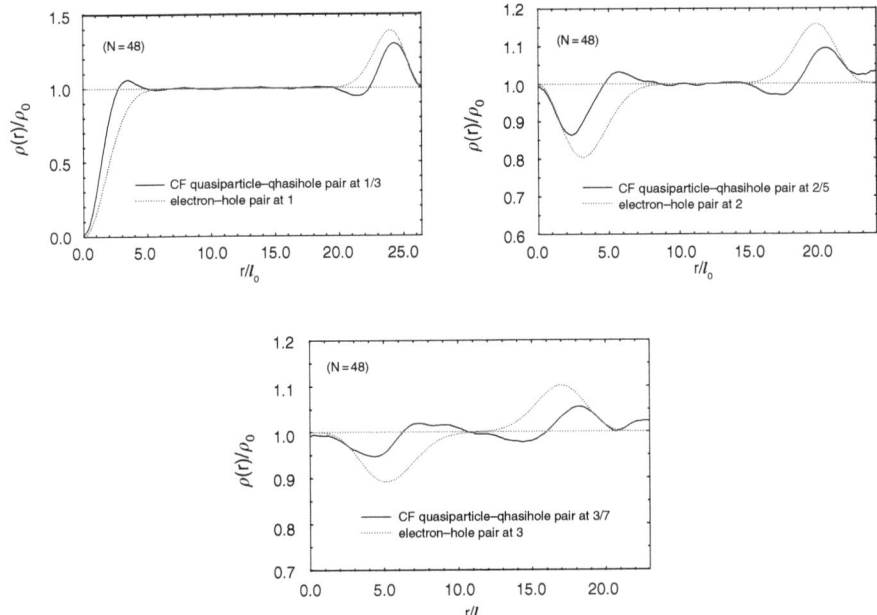

Fig. 8.6. The density profiles for a CF-quasihole and a CF-quasiparticle at $\nu = 1/3$, $\nu = 2/5$, and $\nu = 3/7$. The CF-quasihole is at the north pole (far left) and the CF-quasiparticle is located at the south pole (far right). The distance (r) is measured along the arc from the north pole to the south pole in units of the magnetic length l_0 at ν. Wave functions given in Eq. (5.62) are used for the calculation, obtained from the IQHE wave function with a particle and a hole at opposite poles. For reference, density profiles for a particle–hole excitation are also given for $\nu = 1$, $\nu = 2$, and $\nu = 3$, calculated at the effective magnetic field (dashed curve). Source: K. Park, Ph.D. thesis (Stony Brook, State University of New York, 2000). (Reprinted with permission.)

from its center. The energy required to create a quasiparticle is also larger than that for a quasihole.

The CF theory provides insight into these issues. The asymmetry occurs because, for $\nu = n/(2pn \pm 1)$, the CF-quasihole is a missing composite fermion in the nth Λ level, whereas the CF-quasiparticle is a composite fermion residing in the $(n+1)$th Λ level. This understanding suggests that the density profiles of the CF-quasiparticle of the $\nu^* = n$ state ought to be a mirror image of the density profile of the CF-quasihole of the $\nu^* = n + 1$ state, which is indeed very nearly the case; compare, for example, the CF-quasiparticle at 1/3 with the CF-quasihole at 2/5 in Fig. 8.6. The greater energy of the CF-quasiparticle is explained by noting that it involves excitation to a higher Λ level. (For noninteracting electrons at $\nu^* = n$, the hole costs zero energy, while the particle costs $\hbar\omega_c$.)

The analogy with the IQHE can be pursued further. Figure 8.6 shows a resemblance between the density profile of a CF-quasiparticle or a CF-quasihole of a FQHE state, and the density profile of a particle or a hole of the corresponding IQHE state. They share many qualitative features, and have approximately the same size. That is remarkable because the wave functions for the former are strictly in the lowest Landau level, whereas the density profiles of the latter are determined by the wave function of a single electron or a single hole in the appropriate Landau level. This is one of the signatures of the appearance of higher LL-like structure within the lowest Landau level.

8.6 Excitations

Experiments have demonstrated a variety of excitations of the FQHE state: independent quasiparticle–quasihole pairs, rotons, bi-rotons, trions, flavor altering excitons, spin waves, independent spin-flip quasiparticle–quasihole pairs, spin-flip rotons, and skyrmions. We consider spin conserving excitations here; those involving spin reversal are taken up in Chapter 11.

8.6.1 The CF-exciton

The principal excitation of an incompressible FQHE state is obtained by promoting a composite fermion to an unoccupied Λ level, as shown schematically in the inset of Fig. 8.7. This excitation amounts to creating a particle–hole pair of composite fermions, called a CF-exciton (Section 5.9.5). The energy of a CF-exciton has three contributions: the self energy of the CF-quasiparticle, the self energy of the CF-quasihole, and their interaction energy. When the CF-quasiparticle and CF-quasihole are far apart, the third term is negligible, and the energy of the CF-exciton is equal to the energy it takes to create an independent

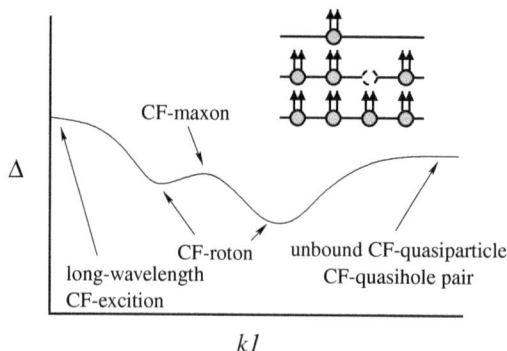

Fig. 8.7. Schematic dispersion of the CF-exciton. This excitation is obtained by promoting a composite fermion across one Λ level, as shown in the inset. The typical energy dispersion contains several minima (maxima), called rotons (maxons), and approaches a constant at large wave vectors, where the excitation contains a far separated pair of CF-quasiparticle and CF-quasihole that can move independently and contribute to transport.

CF-quasiparticle CF-quasihole pair. A somewhat counterintuitive feature is that the distance between the constituent particles of an exciton is proportional to its wave vector; i.e., the CF-quasiparticle and CF-quasihole are far apart in the limit of large wave vectors. This is most easily seen by considering a particle–hole excitation of an IQHE state in the Landau gauge, where the position of an eigenstate is $y = k\ell^2$; changing the wave vector by Δk amounts to a displacement of $\Delta y = \Delta k\ell^2$.

A comparison with superconductivity is interesting. The ground state of a superconductor is a Bose–Einstein condensate of Cooper pairs, and a low-energy excitation is obtained by breaking a Cooper pair into its constituents. In contrast, the lowest-energy excitation of a CF liquid is obtained by promoting a composite fermion as a whole into a higher Λ level. Section 8.6.6 considers excitations that involve dissociation of a composite fermion into its constituents.

The CF-exciton wave function is given in Eq. (5.62). A comparison of the CF-exciton dispersion with the exact dispersion is shown in Fig. 6.1 for finite systems at various fillings of the form $\nu = n/(2n+1)$. Using the CF theory, the exciton dispersion can be calculated for larger systems and for a wider range of filling factors than available in exact diagonalization. The results are conveniently presented as a function the linear wave vector k, related to the orbital angular momentum of the spherical geometry as $k = L/R$, where $R = \sqrt{Q}\ell$ is the radius of the sphere. Several dispersions are shown in Fig. 8.8. The deviation in energy from the large $k\ell$ value is a measure of the interaction between the CF-quasiparticle and CF-quasihole, which is a nonmonotonic function of the separation because of the complicated density profiles.

Many studies estimate the quantitative effect of nonzero thickness on excitation energies (Morf [458]; Ortalano, He, and Das Sarma [488]; Park, Meskini, and Jain [502];

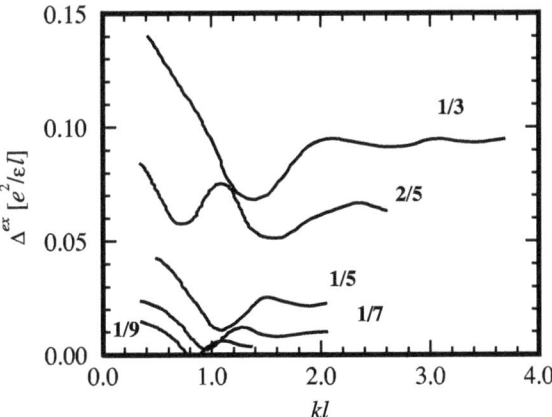

Fig. 8.8. Theoretical dispersions of CF-excitons at 1/3, 2/5, 1/5, 1/7 and 1/9 for 20, 30, 25, 25, and 20 particles, respectively. Source: J. K. Jain and R. K. Kamilla, Chapter 1 in *Composite Fermions*, ed. O. Heinonen (New York: World Scientific, 1998). (Reprinted with permission.)

Yoshioka [730]; Zhang and Das Sarma [742]) and conclude that it lowers the energies by approximately 10–20% for typical parameters. Corrections due to Landau level mixing (Melik-Alaverdian and Bonesteel [434, 435]; Price and Das Sarma [529]; Scarola, Park, and Jain [567]) are smaller still, on the order of 5–10% for typical densities. The effect of LL mixing is weaker for a nonzero width system than it is for a strictly two-dimensional system, because the short-distance part of the effective interaction, which is primarily responsible for causing LL mixing, becomes less effective with increasing thickness. Various approximations made in the determination of the effective interaction are the most significant sources of error in the theoretical predictions. Based on an estimation of the accuracy of this approach [488], the theoretical energies should not be trusted to better than 20%. The intrinsic error in wave functions is a relatively minor effect.

A composite fermion can be excited across a Λ level by a variety of experimental means, the principal ones being thermal excitation, inelastic Raman scattering [109, 129–131, 253, 254, 323, 324, 515–519], optical spectroscopy [47, 357–360], and phonon absorption [117, 439, 576, 577, 740]. We summarize results from these experiments.

8.6.2 Transport gaps

It was appreciated quite early on (e.g., Chang *et al.* [60]) that the longitudinal resistance exhibits Arrhenius behavior, $\rho_{xx} \sim \exp(-\Delta/2k_B T)$, in an intermediate range

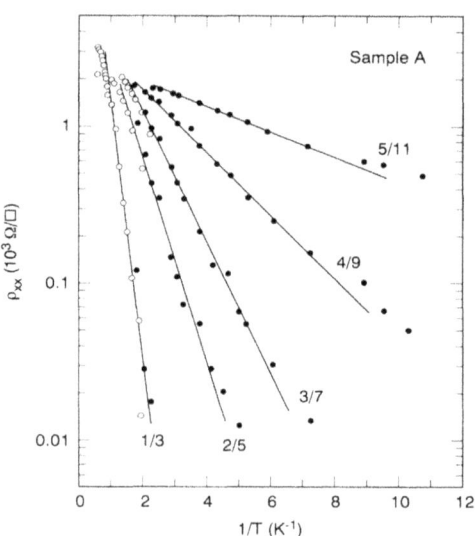

Fig. 8.9. Longitudinal resistance as a function of temperature at several filling factors of the sequence $\nu = n/(2n + 1)$. Source: R. R. Du, H. L. Stormer, D. C. Tsui, L. N. Pfeiffer, and K. W. West, *Phys. Rev. Lett.* **70**, 2944 (1993). (Reprinted with permission.)

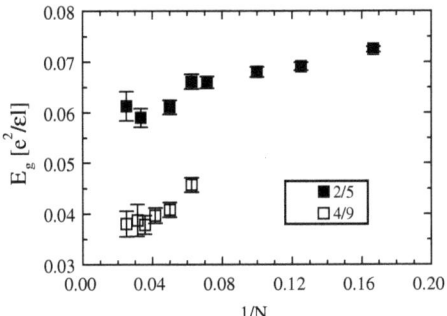

Fig. 8.10. *N* dependence of the gap for 2/5 and 4/9. Source: J. K. Jain and R. K. Kamilla, *Int. J. Mod. Phys. B* **11**, 2621 (1997). (Reprinted with permission.)

of temperatures,[2] as seen in Fig. 8.9. The activation energy Δ, referred to as the transport gap, is interpreted as the energy required to create a far-separated pair of CF-quasiparticle and CF-quasihole, each of which can move independently and thus contribute to transport. Theoretically, the energy of such an excitation can be obtained in two ways: by calculating the energies of the CF-quasiparticle (Fig. 5.4(b)) and the CF-quasihole (Fig. 5.4(c)) separately, or by determining the large wave vector limit of the CF-exciton dispersion (Fig. 8.7). Activation gaps for two states related by particle–hole symmetry are the same when measured in units of $e^2/\epsilon\ell$ (Exercise 8.5). (It should be remembered, however, that the particle–hole symmetry is exact only in the absence of LL mixing.) Figure 8.10 shows typical dependence of gaps on the number of particles N. For $\nu = 1/(2p+1)$, the transport gap has been computed by Bonesteel [33].

Fig. 8.11 shows transport gaps for two high-quality samples with different densities for the sequence $\nu = n/(2n+1)$ (Du *et al.* [123]). Also shown are gaps predicted by the CF theory (Park, Meskini, and Jain [502]), including modification of the interaction nonzero thickness in an LDA scheme. The nonzero thickness reduces gaps from their pure 2D values, bringing them into better agreement with experiment, but a discrepancy of ~50% remains. As mentioned earlier, LL mixing cannot account for the discrepancy. The ever-present disorder is suspected to cause a substantial suppression of the gap, but no reliable theoretical method can currently confirm this quantitatively.

8.6.3 CF-rotons

The dispersion of the CF-exciton has several minima, named "rotons" (or magnetorotons) by Girvin, MacDonald, and Platzman [191], drawing on the analogy to the "roton" minimum in the dispersion of the phonon excitation of superfluid ^{4}He. Extending the analogy, maxima in the dispersion can be called "maxons." In the case of superfluid ^{4}He, the phonon, the maxon, and the roton are part of a single excitation branch (as confirmed by neutron

---

[2] At high temperatures the transport is not activated, whereas at very low temperatures, it is dominated by variable range hopping [523].

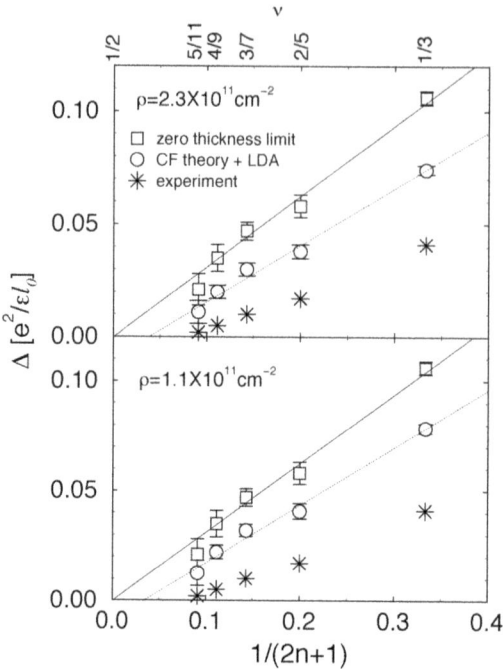

Fig. 8.11. Comparison of theoretical and experimental gaps for two different densities shown on the figure. Squares are for a pure two-dimensional system, and circles for a heterojunction (with the nonzero width correction to interaction evaluated in a local density approximation). Stars are gaps from the experiment of Du *et al.*[123]. Source: K. Park, N. Meskini, and J. K. Jain, *J. Phys. Condens. Matter* **11**, 7283 (1999). (Reprinted with permission.)

scattering), with no conceptual distinction between them.[3] Rotons and maxons in the FQHE are similarly special cases of the CF-exciton. They occur simply because the internal structure in the density profiles of the CF-quasiparticle and the CF-quasihole makes their interaction nonmonotonic at short distances, i.e., at small wave vectors. Girvin, MacDonald, and Platzman [191] have developed a density-wave picture for neutral excitations of the FQHE states, known as the single mode approximation, which successfully predicts the roton minimum at $\nu = 1/3$ (Section 12.6.) Two differences from ^{4}He ought to be noted: (i) There are in general several rotons for a given FQHE state. (ii) Because of the absence of a massless phonon-like mode at small wave vectors, *the* lowest energy excitation is a roton.

One reason why the minima (and maxima) in the CF-exciton dispersion deserve a new name is that they produce van Hove singularities in the density of states for the CF-exciton, which makes them easier to detect experimentally. Raman scattering experiments (Davies *et al.* [109]; Dujovne *et al.* [129, 131]; Gallais *et al.* [177, 178]; Hirjibedin *et al.* [253, 254]; Kang *et al.* [324]; Pinczuk *et al.* [515, 517]) have given a remarkable amount of information

---

[3] The name roton was originally coined by Landau [365], who envisioned it as an excitation characterized by a rotational velocity flow, distinct from the phonon. It has been described colorfully as a "ghost of a vanishing vortex ring" (Feynman [161]) or a "ghost of a Bragg spot" (Nozières [482]).

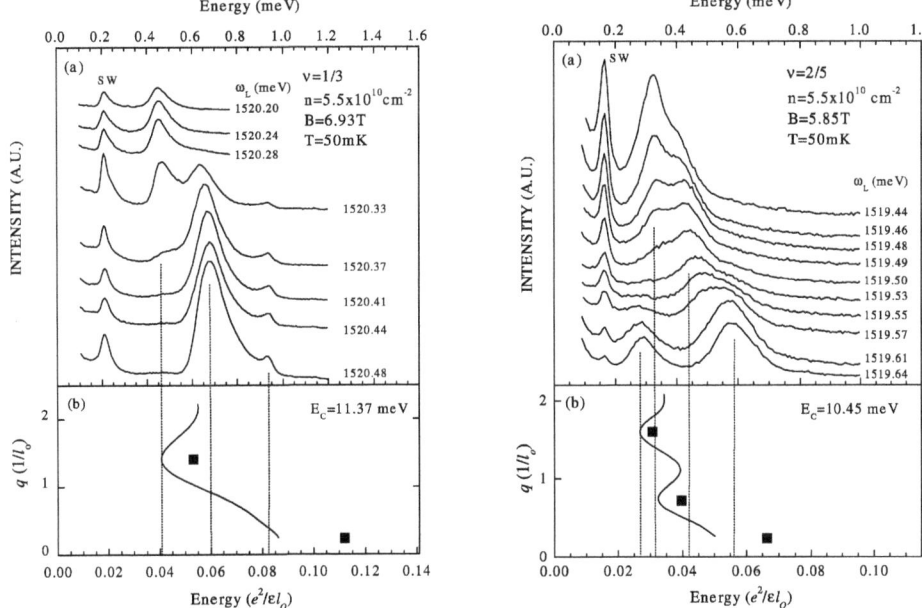

Fig. 8.12. Left panel: Resonant Raman spectra at $\nu = 1/3$ (top panel). Vertical dashed lines mark, from left to right, the roton, the short-wavelength mode, and the long-wavelength mode. The bottom panel shows the calculated dispersion from Ref. [567] (assuming zero width), scaled down by a constant factor. Solid squares show calculated energies at the critical points in the dispersion, but including the effect of nonzero thickness. Right panel: Same as in left panel, but for $\nu = 2/5$. Vertical dashed lines mark, from left to right, a roton, the short-wavelength mode, another roton, and the long-wavelength mode. Source: M. Kang, A. Pinczuk, B. S. Dennis, L. N. Pfeiffer, and K. W. West *Phys. Rev. Lett.* **86**, 2637 (2001). (Reprinted with permission.)

about various kinds of excitations in the FQHE regime, including rotons. Ideally, rotons are inaccessible to light scattering, due to their large wave vector, but disorder-induced breakdown of wave vector conservation, combined with the singularity in the density of states at the roton energy, has made it possible for inelastic Raman scattering to observe them. Figure 8.12 shows the Raman spectra at $\nu = 1/3$ and $2/5$ obtained by Kang *et al.* [324]. Also shown are theoretical CF-exciton dispersions [567], scaled down by a constant factor to facilitate assignment of the observed modes. Besides Raman scattering, rotons have also been probed by phonon absorption (Mellor *et al.* [439]), and by ballistic phonon scattering (Zeitler *et al.* [740]).

Roton energies have been determined theoretically (Scarola, Park, and Jain [567]) for many fractions by an extrapolation to the thermodynamic limit (studying systems with as many as ~60 composite fermions). Energies for a purely two-dimensional system are given in Table 8.2. To compare theory with experiment, Ref. [567] also incorporates the effect of nonzero thickness in the local-density approximation. A comparison between theory and experiment is given in Table 8.3, where each theoretical energy is calculated for the density

Table 8.2. *Energies of CF rotons and long-wavelength excitations*

Mode	ν	Energy	α_R
$k\ell = 0$	1/3	0.15	–
	2/5	0.087(1)	–
	3/7	0.068(5)	–
Roton	1/3	0.066(1)	0.0079(3)
	2/5	0.037(1)	0.0090(11)
	3/7	0.027(3)	0.0095(32)

Notes: Neutral excitations at 1/3, 2/5, and 3/7, for strictly two-dimensional system, in units of $e^2/\epsilon\ell$.

Energy of $k\ell = 0$ excitation at $\nu = 1/3$ agrees with energy obtained by single-mode-approximation scheme of Girvin, MacDonald, and Platzman [191].

Estimate for roton "mass," $m_R^* \equiv \alpha_R m_e \sqrt{B[T]}$, defined in Eq. (8.17), where m_e is the electron mass in vacuum. Statistical uncertainty in last digit(s) is shown in parentheses.

Source: Scarola, Park, and Jain [567].

Table 8.3. *Comparison between experiment and CF theory prediction for the roton and exciton energies*

	$k\ell = 0$		Roton		Experimental
ν	Experiment	Theory	Experiment	Theory	reference
1/3	0.082	0.104(1)	0.044	0.050 (1)	[517]
	0.084	0.113(1)	–	0.052(1)	[515]
	–	0.09(2)	0.041(2)	0.045(1)	[439]
	0.074	0.095(1)	0.047	0.047(1)	[109]
	–	0.092(1)	0.036(5)	0.045(1)	[740]
2/5	–	0.054(1)	0.021(2)	0.026(1)	[439]
	–	0.055(1)	0.025(3)	0.027(1)	[117]
3/7	–	0.044(2)	0.014(2)	0.017(2)	[439]

Notes: Units of $e^2/\epsilon\ell$.

In some cases, we have assumed particle–hole symmetry within the lowest Landau level, which implies that the energy gaps at ν and $1 - \nu$ are identical when measured in units of $e^2/\epsilon\ell$.

Source: Scarola, Park, and Jain [567].

and geometry of the corresponding experimental sample. Theoretical energies are obtained with no adjustable parameters; they do not include correction due to LL mixing, which have been estimated to be on the order of 5% for typical densities [567].

The agreement between theory and experiment is better for rotons than for transport gaps, presumably because rotons are less sensitive to disorder. Due to their overall charge neutrality, rotons have a weaker dipolar coupling to the disorder potential created by charged donor atoms. The coupling is further diminished because the disorder potential in modulation doped samples is typically smooth on the scale of the size of a roton (a few magnetic lengths). Further experimental evidence for the insensitivity of the roton energy to disorder is that the same roton energy is found for samples for which transport gaps differ by a factor of two (Mellor *et al.* [439]). That is also in line with attributing the discrepancy between theory and experiment for transport gaps to disorder.

Near a roton, the energy dispersion of a CF-exciton has a parabolic form, which allows a definition of the "roton mass" m_R^* (Platzman and Price [520]) through

$$\Delta_k^{ex} = \Delta + \frac{\hbar^2 (k - k_0)^2}{2m_R^*}. \qquad (8.17)$$

Theoretical values of m_R^* for several fillings are quoted in Table 8.2. The lowest-energy roton governs the thermodynamics of the FQHE state at low temperatures. Platzman and Price [520] show that it determines the stability of the CF liquid against Wigner crystallization.

8.6.4 Long-wave-length excitation: the bi-roton

The long-wavelength mode of $v = 1/3$ was observed by Pinczuk *et al.* [515] in 1993 by Raman scattering; see, for example, Fig. 8.12. As seen in Table 8.3, the calculated energy at $1/3$ is off by $\sim$30% for $k\ell = 0$. Girvin, MacDonald, and Platzmann [192] have suggested that the lowest energy excitation at $k\ell = 0$ may be composed of *two* rotons with opposite wave vectors at this filling factor. That is surely the case for $v = 1/5$ and $v = 1/7$, where, as seen in Fig. 8.8, twice the roton energy is significantly smaller than the energy of a single exciton. At $v = 1/3$, the two energies happen to be very close. Exact diagonalization studies indicate that the single CF-exciton may not be very accurate at long wavelengths. The agreement between the exact eigenenergy and the single CF-exciton energy worsens somewhat at small $k\ell$ (Fig. 6.1). Also, $\mathcal{P}_{LLL}$ projects away an increasingly bigger fraction of the single-exciton wave function as $k\ell \to 0$ [312], eventually annihilating it completely [115].

Figure 6.12 demonstrates that the predicted dispersion at small wave vectors improves significantly upon inclusion of Λ level mixing at the lowest order, indicating that the actual excitation here hybridizes with bi-excitons, i.e., states in which *two* composite fermions have been excited from the lowest Λ level to the second. The natural expectation is that the lowest energy is produced when the bi-exciton is a bi-roton. A wave function can be

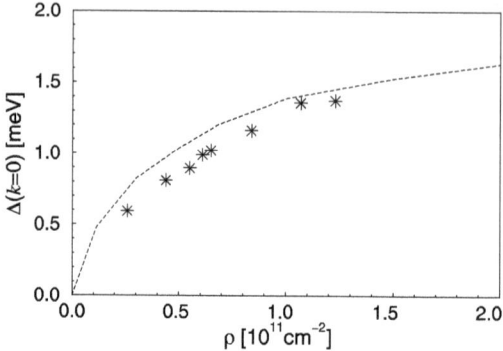

Fig. 8.13. Comparison between theory (dashed line) and experiment (stars) for the energy of the long-wavelength mode. Theoretical energy for the CF bi-roton includes effects of LL mixing (5% correction) and nonzero sample thickness (33 nm wide quantum well). Experimental results are taken from Kang *et al.*[324]. Source: K. Park and J. K. Jain, *Phys. Rev. Lett.* **84**, 5576 (2000). (Reprinted with permission.)

constructed for the bi-roton by creating two rotons of opposite wave vector, by analogy to the two-exciton state at $\nu = n$ that contains two particle–hole pairs, and has $\sim 10\%$ lower energy than the single CF-exciton state in the long-wavelength limit (Park and Jain [504]). Ghosh and Baskaran [185] estimate the binding energy of the bi-roton in a variational scheme, exploiting the oriented-dipole character of the roton. Figure 8.13 shows that, after correcting for nonzero thickness and LL mixing, the theory is in satisfactory agreement (20% or better) with experiment. Hirjibehedin *et al.* [254] estimate the binding energy of the bi-roton mode to be approximately 10% of the bi-roton energy, in agreement with theoretical estimates, by comparing its energy with that of two unbound rotons (twice the energy of a single roton).

The bi-roton interpretation of the observed long-wavelength mode may be relevant for another issue. The oscillator strength of the single exciton mode, which coincides with the density wave mode in the $k\ell = 0$ limit, vanishes rapidly with wave vector $(\sim k^4)$ [191,192], because, according to Kohn's theorem, the $\hbar\omega_c$ mode exhausts the f-sum rule in the limit of $k\ell = 0$. There is no reason for the oscillator strength to vanish for the bi-roton mode, however. Also, while the two-exciton states form a continuum, the bi-*roton* states lead to a singularity in the density of states at the lower edge of the continuum. A quantitative understanding of the Raman scattering cross sections remains unsatisfactory, especially because of the complications introduced by the resonant nature of Raman scattering. For theoretical work along these lines, the reader is referred to Das Sarma and Wang [108] and Platzman and He [521,522].

Hirjibehedin *et al.* [254] investigate the wave vector dependence of the long-wavelength excitation in a low-density sample, which allows, by tilting the sample in the backscattering geometry, access to larger $k\ell$ than was possible previously. They observe the surprising result that the single peak splits into two at relatively larger wave vectors (Fig. 8.14), suggesting

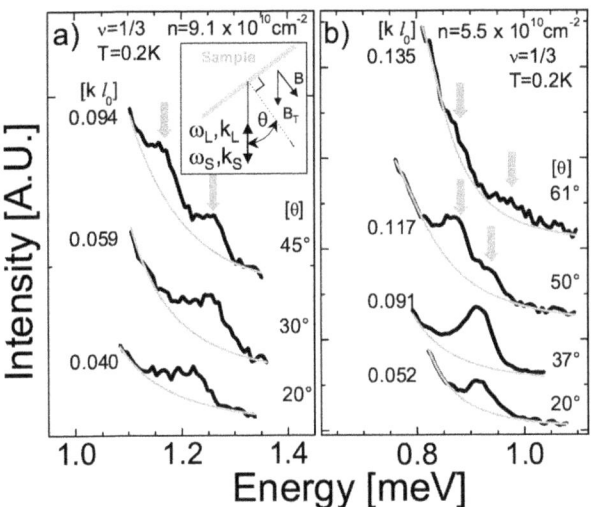

Fig. 8.14. Resonant Raman spectra of the low-lying mode at $\nu = 1/3$ from two samples as a function of the tilt angle in the backscattering geometry (see inset), which is related to the wave vector of the excitation k. (l_0 is the magnetic length.) The single peak splits into two as the wave vector is increased. Source: C. F. Hirjibehedin, I. Dujovne, A. Pinczuk, B. S. Dennis, L. N. Pfeiffer, and K. W. West, *Phys. Rev. Lett.* **95**, 066803 (2005). (Reprinted with permission.)

that the long-wavelength excitation is an unresolved doublet. The physical origin of this effect is not fully understood at present. Tokatly and Vignale [640] have argued, using continuum elasticity theory (Section 12.7), that such mode splitting is a generic consequence of classical hydrodynamic equations of motion for the incompressible FQHE liquid.

8.6.5 Charged trions

Park [507] has shown, theoretically, that the CF-roton can lower its energy by forming a bound state with an already existing CF-quasiparticle or CF-quasihole. This bound state is known as a CF-trion. He estimates the binding energy to be roughly $0.02e^2/\epsilon\ell$. CF-trions may be observable in resonant inelastic light scattering experiments, since some localized quasiparticles are likely to be always present in real experiments (induced by disorder). Hirjibehedin *et al.* [254] see additional excitations at energies 0.005–$0.01e^2/\epsilon\ell$ below the ordinary roton energy, which may be a signature of the CF-trion. (Theoretical estimate is expected to overestimate the binding energy.)

8.6.6 Flavor-altering excitations

Hirjibehedin *et al.* [253] investigate the evolution of various excitation energies as a function of ν in the filling factor range $1/3 \geq \nu \geq 1/5$. The relevant composite fermions carry four vortices (^{4}CFs) in this filling factor range. Experiments show that new low-energy excitations appear abruptly for $\nu < 1/3$, as expected, which are excitations of ^{4}CFs across

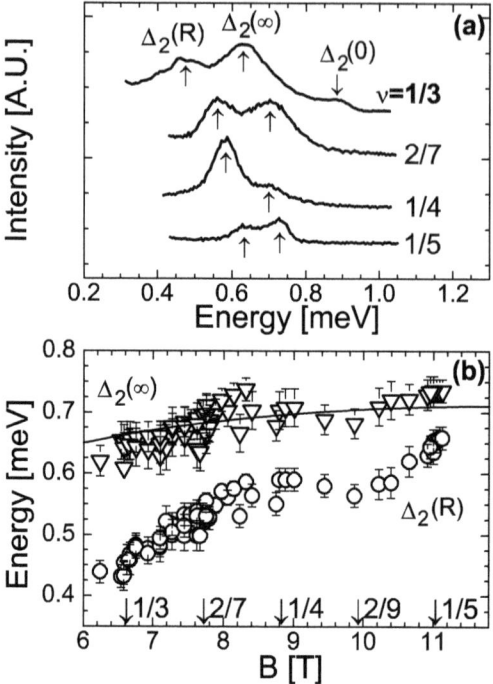

Fig. 8.15. Resonant Raman spectra in the filling factor range $1/5 \leq \nu \leq 1/3$. Lower energy part (not shown) exhibits *new* modes. Source: C. F. Hirjibehedin, A. Pinczuk, B. S. Dennis, L. N. Pfeiffer, and K. W. West, *Phys. Rev. Lett.* **91**, 186802 (2003). (Reprinted with permission.)

their Λ levels; their energies are in reasonable agreement with those predicted theoretically [313, 567]. Surprisingly, however, excitations from $\nu > 1/3$ do not disappear, but evolve continuously as the filling factor changes across $\nu = 1/3$ toward $\nu \to 1/5$. This is shown in Fig. 8.15; the modes $\Delta_2(R)$, $\Delta_2(\infty)$, and $\Delta_2(0)$ are the continuous evolutions of the roton, $k\ell = \infty$, and the long-wavelength modes, respectively, of composite fermions carrying *two* vortices (^{2}CFs). (The new Δ_4 modes, corresponding to the excitons of ^{4}CFs, occur at lower energies and are not shown on this figure. These modes disappear more easily upon raising temperature than do the Δ_2 modes.)

At first sight, the existence of ^{2}CF modes in the region $1/3 \geq \nu \geq 1/5$ appears inconsistent with the understanding that the relevant composite fermions in the filling factor region are ^{4}CFs. The ^{2}CF modes at $\nu < 1/3$, however, represent a new class of excitations. Previous sections have described excitations wherein the integrity of the composite fermion remains intact, i.e., the CF flavor does not change. In the new excitations, two of the vortices attached to a composite fermion are stripped away:

$$^4\text{CF} \;\to\; ^2\text{CF} + \text{two vortices}. \tag{8.18}$$

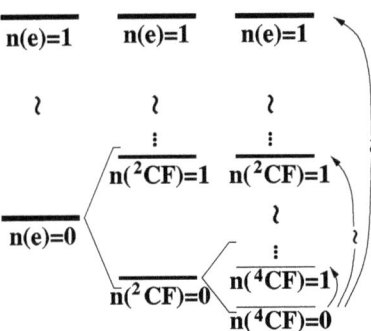

Fig. 8.16. Flavor-altering excitations. The lowest electronic Landau level splits into Λ levels of ^{2}CFs, and the lowest ^{2}CF-Λ level further splits into ^{4}CF-Λ levels. The Λ level index is denoted by n(P) where "P" is the type of relevant particle: "e" (electron), "^{2}CF," or "^{4}CF." Three ladders of excitations are indicated at the far right, two of them being flavor-changing excitations. Source: M. R. Peterson and J. K. Jain, *Phys. Rev. Lett.* **93**, 046402 (2004). (Reprinted with permission.)

The basic idea is explained in Fig. 8.16, which illustrates a ladder of such flavor-altering excitations. This interpretation is supported by finite system calculations (Peterson and Jain [513]) that produce energies consistent with experiment, as well as by the calculations of Wójs and Quinn [530,531,694], which show that bands in the energy spectrum (obtained in exact diagonalization studies) are characterized by the number of vortices bound to electrons. The Kohn mode of a FQHE state is also an example of a flavor-altering excitation, because it involves an excitation from a Λ level to the second (electronic) Landau level.

8.6.7 Fractionally charged excitons

Byszewski *et al.* [47] have seen evidence for fractionally charged excitons in optical spectroscopy. Such excitations are relevant because an electron–hole exciton (with hole in the valence band) can lower its energy by forming a bound state with a fractionally charged CF-quasiparticle (Wójs, Gladysiewicz, and Quinn [699]). Byszewski *et al.* [47] observe a new low-energy peak in the emission spectrum in the region between the quantized plateaus, where the CF-quasiparticles are mobile, and interpret it as a fractionally charged exciton.

8.7 CF masses

The understanding of incompressible ground states at $\nu = n/(2pn \pm 1)$ as n filled Λ levels of composite fermions, and of excited states in terms of excitations of composite fermions across Λ levels, invites an interpretation of the excitation gap as the cyclotron energy of composite fermions. That, in turn, suggests the following definition for an effective mass m^* for composite fermion (Du *et al.* [126]; Halperin *et al.* [231]):

$$\Delta = \hbar\omega_c^* \equiv \hbar \frac{eB^*}{m^*c} = \frac{m_e}{m^*} \frac{\hbar\Omega_c}{(2pn \pm 1)},\tag{8.19}$$

where

$$\hbar\Omega_c = \hbar\frac{eB}{m_e c} \tag{8.20}$$

is the cyclotron energy of an electron in vacuum. A problem with this definition is that a constant value for m^* would imply a wrong parametric dependence for the energy gap at a fixed filling factor, namely $\Delta \propto B$. (Because the Coulomb energy is the only energy scale in the LLL problem, the actual gap must behave as $\Delta \propto e^2/\epsilon\ell \sim \sqrt{B}$.) A constant, parameter-independent mass for composite fermions is thus impossible. Rather than giving up, let us ask what is the best we can do within this constraint. We write

$$\Delta = \frac{1}{\alpha(2pn \pm 1)} \frac{e^2}{\epsilon\ell}, \tag{8.21}$$

which gives

$$\frac{m^*}{m_e} = \alpha\frac{\hbar\Omega_c}{V_C} = 0.0263\alpha\sqrt{B[T]}, \tag{8.22}$$

where we have used $\hbar\Omega_c = \hbar eB/m_e c \approx 1.34B[T]$ K and $e^2/\epsilon\ell \approx 51\sqrt{B[T]}$ K (the last terms quote the energy in kelvin for parameters appropriate for GaAs, with $B[T]$ in tesla). Because the magnetic field does not change much across a given sequence (only by a factor of 1.5 across the sequence $\nu = n/(2n+1)$) m^* is approximately constant along a sequence. The dominant variation of gap then comes from the factor $(2pn \pm 1)$ in the denominator.

Several comments are in order.

- The effective mass of composite fermion is defined in units of the electron mass in vacuum to underscore that it bears no relation to the electron band mass in the host semiconductor. (It is also independent of the electron mass in vacuum, which only provides a convenient mass scale.) Experimental systems with different band masses produce composite fermions with the same mass (to the extent the effects of LL mixing and nonzero width can be neglected, which all renormalize the mass).
- All gaps can be calculated from the microscopic theory without defining any effective mass, and a filling-factor-dependent effective mass can always be defined as above. The concept of effective mass is useful only if the mass turns out to be approximately independent of the filling factor, modulo the $\sqrt{B}$ dependence imposed by general considerations. The validity of CF theory is not predicated upon the composite fermion having a parameter-independent mass.
- In condensed matter systems, the quasiparticle mass is, in general, a less fundamental quantity than, say, the quasiparticle charge, because the mass is renormalized by interactions. The problem is especially severe for composite fermions. While in many situations (e.g., Landau Fermi liquid theory), we begin with a bare mass and study how it is affected by the interaction, the CF mass is generated entirely from interactions. The composite fermion does not have a bare mass. Not surprisingly, the CF mass depends sensitively on various parameters, for example, the magnetic field or disorder. This feature is manifest through the fact, as seen below, that the value of mass differs significantly from one experiment to another. Care must be exercised not to overinterpret the CF mass in excessive quantitative detail. It provides an attractive, but approximate, way of parametrizing the energy scales involved in the problem.

- A crucial difference from Landau's approach to the ordinary Fermi liquids is implicit in the above discussion, and we stress it yet again. In the Landau philosophy one derives renormalizations of various parameters, from which physical observables are obtained. For the CF liquid, theory can directly calculate the observables, which are then interpreted in terms of renormalized parameters.
- To speak of *the* mass for composite fermions (as done in preceding paragraphs) is incorrect. We have purposely avoided specifying the energy gap Δ so far, because different masses can be defined by equating Δ to different energies governed by different physics. There is nothing strange about it; the Landau quasiparticle of an ordinary Fermi liquid also has many context-specific masses.

8.7.1 Activation mass

The "activation mass" of composite fermions, m_a^*, is defined by identifying the gap in Eq. (8.19) with the energy of a far separated quasiparticle–quasihole pair of composite fermions. Theoretical values of activation gaps for the sequence $\nu = n/(2n + 1)$ have been obtained for up to $\nu = 7/15$, and are found to be approximately proportional to $1/(2n + 1)$, which is a nontrivial affirmation of the usefulness of the concept of CF mass. It has been estimated [231, 281, 571] that the activation gaps are well approximated by $\Delta = V_C/[3.0(2n + 1)]$ (Halperin *et al.* [231]), which gives

$$\frac{m_a^*}{m_e} = \alpha \frac{\hbar\omega}{V_C} \approx 0.079\sqrt{B[T]}. \tag{8.23}$$

Measured gaps from two experiments are shown in Figs. 8.17 and 8.18. If we used Eq. (8.21) directly to determine the mass, it would diverge beyond 5/11 or 6/13, where the gap vanishes. Du *et al.* [123] argue that this divergence is not real but an artifact of the disorder-induced Λ level broadening. They note that the activation gaps do open linearly from $\nu = 1/2$, consistent with the effective mass interpretation. The intercept on the y-axis has a negative value, interpreted as a measure of the Λ level broadening, Γ. Taking Γ to be Λ level index independent, the expression

$$\Delta = \frac{1}{\alpha(2n + 1)} \frac{e^2}{\epsilon\ell} - \Gamma \tag{8.24}$$

provides a good fit to experimental data. The activation mass, determined from the slope, is shown on Figs. 8.17 and 8.18; the experimental values for m_a^* are approximately a factor of two larger than those predicted theoretically.

8.7.2 Shubnikov–de Haas mass

As the filling factor is varied away from $\nu = 1/2$, oscillations are observed in ρ_{xx}, which develop into fully formed FQHE states for larger $|B^*|$. Leadley *et al.* [374] and Du *et al.* [124] have demonstrated that the temperature dependence of the unquantized oscillations is well described in terms of Shubnikov–de Haas(SdH) oscillations of noninteracting composite fermions of mass m^*_{SdH} at B^*. The analogous oscillations for noninteracting electrons are described by the Ando–Lifschitz–Kosevich formula [9, 573]

$$\Delta\rho_{xx} \propto \frac{X}{\sinh X} \exp\left(-\frac{\pi}{\omega_c \tau}\right) \cos 2\pi(\nu - 1/2), \qquad (8.25)$$

where $X = 2\pi^2 k_B T / \hbar\omega_c$ and τ is the quantum lifetime. In the simplest approximation, which turns out to be reasonable, this expression is modified for composite fermions by making the substitutions $B \rightarrow B^*$, $\nu \rightarrow \nu^*$, $m_e \rightarrow m^*_{SdH}$, and $\tau \rightarrow \tau^*$ (τ^* is assumed to be temperature independent). This equation provides an excellent fit to the experimental temperature and magnetic field dependences of the resistance oscillations (with the exception of the region near $\nu = 2/3$ in Ref. [374]), as seen in Fig. 8.19. Du et al. [124] find similarly good fits. For small numbers of occupied Λ levels, the CF masses obtained phenomenologically from the best fits are all roughly of the same order, $m^*_{SdH}/m_e \approx 0.5$–1.0, which is also close to the activation mass. Several experiments (Coleridge et al. [92]; Du et al. [125]; Manoharan et al. [428]) notice a strong enhancement in the SdH mass as the Λ level index n increases beyond 5 along the sequence $\nu = n/(2n \pm 1)$; see Fig. 8.20 for measurements from several samples. (For large n, the FQHE states at $\nu = n/(2n \pm 1)$ are not sufficiently well developed for a determination of the activation mass.)

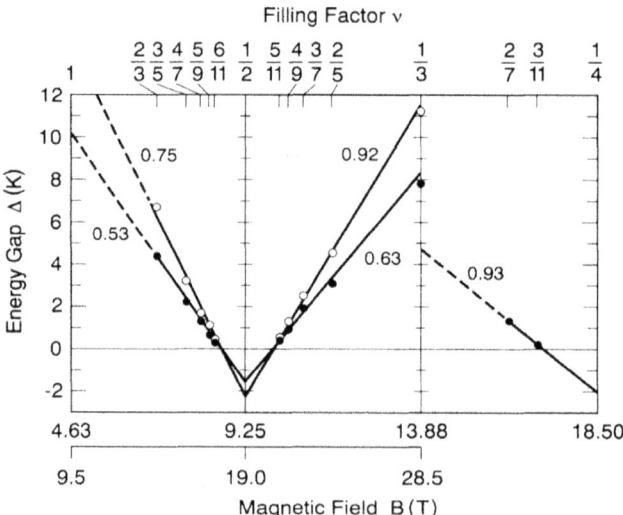

Fig. 8.17. Activation gaps for the sequences $\nu = n/(2n \pm 1)$, and also for some members of $\nu = n/(4n - 1)$. Straight lines are the best fits for a constant m^*_a; numbers near the lines give the ratio m^*_a/m_e. (The masses do not change appreciably for a $\sqrt{B}$ dependent fit [625].) Source: R. R. Du, H. L. Stormer, D. C. Tsui, L. N. Pfeiffer, and K. W. West, *Phys. Rev. Lett.* **70**, 2944 (1993). (Reprinted with permission.)

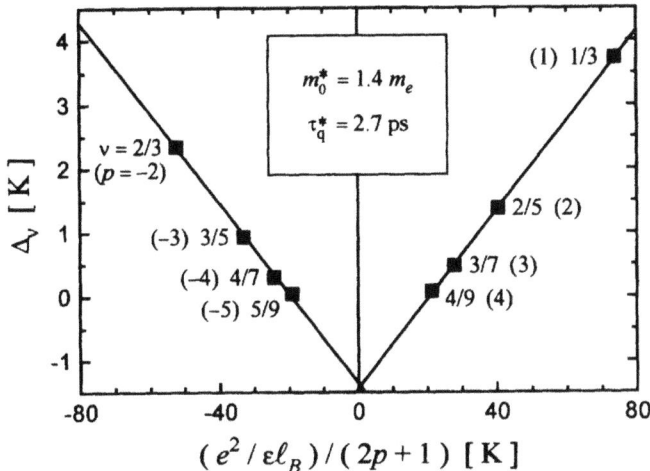

Fig. 8.18. Activation gaps for the sequence $v = n/(2n+1)$ in a hole-doped sample. The symbol m_0^* denotes the activation mass. Source: H. C. Manoharan, M. Shayegan, and S. J. Klepper, *Phys. Rev. Lett.* **73**, 3270 (1994). (Reprinted with permission.)

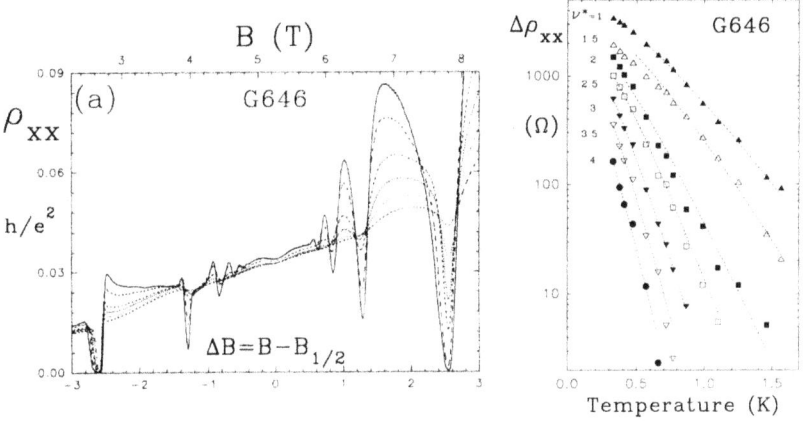

Fig. 8.19. Temperature and magnetic field dependences of SdH oscillations of composite fermions (left panel). The right panel shows the fit to Eq. (8.25), suitably modified for composite fermions, at the resistance maxima (open symbols; $v^* = n + 1/2$) and minima (filled symbols; $v^* = n$). Source: D. R. Leadley, R. J. Nicholas, C. T. Foxon, and J. J. Harris, *Phys. Rev. Lett.* **72**, 1906 (1994). (Reprinted with permission.)

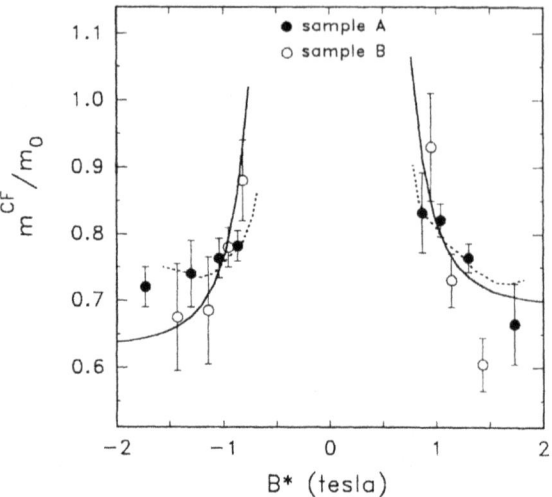

Fig. 8.20. SdH mass of composite fermions (denoted by m^{CF} on this figure) in units of the free electron mass (m_0). Solid and open circles are from two samples of Coleridge *et al.* [92]; dotted curve is from Du *et al.* [123] and solid curve is from Du *et al.* [125]. Masses from different samples are scaled by a factor of $\sqrt{B}$ for a proper comparison. Source: P. T. Coleridge, Z. W. Wasilewski, P. Zawadzki, A. S. Sachrajda, and H. A. Carmona, *Phys. Rev. B* **52**, R11603 (1995). (Reprinted with permission.)

The SdH oscillations have been studied with the help of the Boltzmann equation for composite fermion transport, with spatial fluctuations in the effective magnetic field incorporated at a semiclassical level (Aronov *et al.* [12] and Mirlin *et al.* [446, 449]). Mirlin, Polyakov, the and Wölfle [449] attribute quantum oscillations to the quantum Hall effect as opposed to the conventional SdH oscillations that occur due to density of states oscillations; that, they argue, explains the fast damping of oscillations with decreasing effective magnetic field.

Leadley *et al.* [375, 376, 477] have extended their experiments to higher values of $|B^*|$ with the use of a 50 tesla pulsed magnetic field (where samples with higher density could be studied). They find that the SdH mass follows, approximately, a sample-independent behavior

$$\frac{m_{SdH}^*}{m_e} = 0.5 + 0.08|B^*[\text{T}]| \tag{8.26}$$

on either side of $\nu = 1/2$, covering more than a factor of 25 variation in B^*. The mass itself changes from 0.5 at low B^* to 1.6 at the highest B^* in their experiment. In an ideal system, one expects the mass to be proportional to $\sqrt{B}$, and, presumably, corrections due to nonzero width will be important for a theoretical understanding of this result. The effect of tilting (or, application of a parallel magnetic field) has been studied by Gee and collaborators [179, 180]; while the mass at each tilt angle is well defined, its tilt dependence requires a consideration of the variation of the transverse component of the wave function.

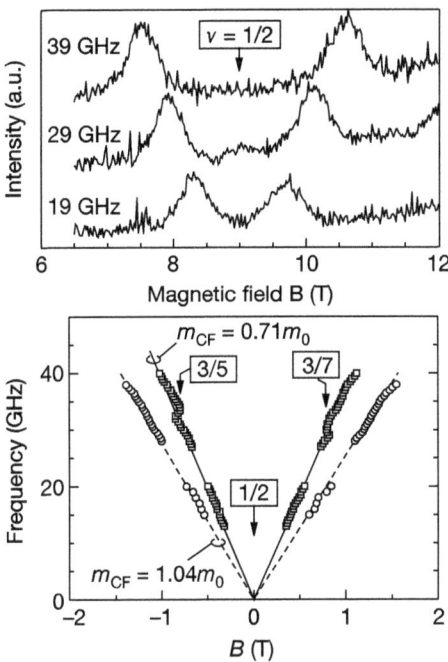

Fig. 8.21. Top panel depicts the microwave absorption amplitude in the vicinity of $\nu = 1/2$ as a function of the frequency of the incident radiation. Bottom panel shows positions of the CF cyclotron mode as a function of the effective magnetic field for two different densities. The measured CF cyclotron mass at $\nu = 1/2$ is well described by the form in Eq. (8.23), but with a four times larger coefficient on the right hand side. In this figure, m_{CF} denotes the CF cyclotron mass and m_0 the electron mass in vacuum (for which we have used the notations m_c^* and m_e, respectively). Source: I. V. Kukushkin, J. H. Smet, K. von Klitzing, and W. Wegscheider, *Nature* **415**, 409 (2002).

8.7.3 Cyclotron mass

Kukushkin *et al.* [360, 361] have observed the cyclotron resonance of composite fermions, which causes an enhanced absorption of microwave radiation at certain magnetic-field-dependent frequencies, as seen in Fig. 8.21. The cyclotron mass of composite fermions, m_c^*, is found to be

$$m_c^* = 0.32 m_e \sqrt{B[\text{T}]}. \tag{8.27}$$

It is approximately a factor of four larger than the theoretical value for the activation mass in Eq. (8.23).

8.7.4 Polarization mass

Yet another mass, called the "polarization mass" (m_p^*), is defined in connection with the spin polarizatio of various FQHE states and of the CF Fermi sea. This mass is discussed in Chapter 11. It has been measured directly at $\nu = 1/2$ and $3/2$, agrees well with theory, and exhibits no divergence. Being related to a thermodynamic quantity, it is less affected by disorder than the activation and the SdH masses.

8.7.5 Flavor dependence of CF mass

While most work has focused on the mass of ^{2}CFs along the fractions $n/(2n \pm 1)$, the activation and SdH masses of ^{4}CFs have also been measured. Yeh *et al.* [723] investigate the FQHE states near $\nu = 3/4$ in tilted field experiments, and conclude that the data can be explained very well within the CF theory. They further note that the effective mass and the g factor for ^{4}CFs are very similar to those for ^{2}CFs (with $\sqrt{B}$ correction for the mass). Pan *et al.* [491, 492] study the region near $\nu = 1/4$ by both activation gap and Shubnikov-de Haas measurements. The resolution of FQHE states at $n/(4n \pm 1)$ requires very high quality samples and high B fields. They find that the mass exhibits an asymmetry around $\nu = 1/4$. For $\nu > 1/4$ the behavior of the mass is similar, both qualitatively and quantitatively, to that observed for ^{2}CFs around $\nu = 1/2$. However, for $\nu < 1/4$, the mass increases linearly with $|B^*|$. They attribute this anomalous dependence on $|B^*|$ to proximity to the insulating phase in the vicinity of $\nu = 1/5$.

These observations are not understood theoretically at present. Onoda *et al.* [484, 485, 487] have found, in exact diagonalization studies with up to 18 particles (on sphere), that the excitation energy of composite fermions at $B^* = 0$ extrapolates to zero linearly in the thermodynamic limit. They deduce the mass from the slope and conclude that composite fermions become heavier as they swallow more vortices. Their calculated masses, including masses for incompressible states, are shown in Fig. 8.22. The mass varies continuously within the region defined by a single flavor of composite fermions, but changes abruptly

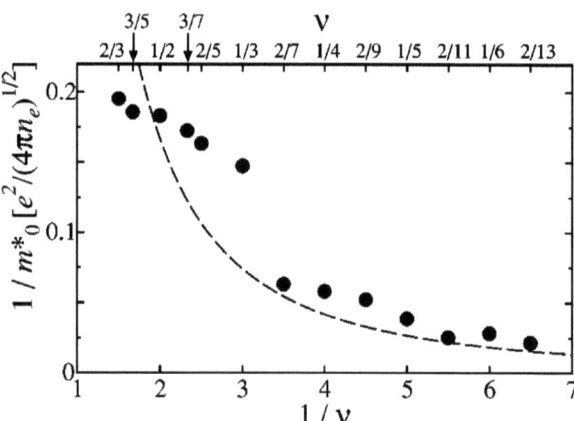

Fig. 8.22. Inverse effective mass of composite fermions (dots) as a function of the inverse filling factor. The quantity n_e is the electron density, and the CF activation mass is denoted m_0^* on this figure. The dashed line is a mean-field result from Ref. [462]. Source: M. Onoda, T. Mizusaki, and H. Aoki, *Phys. Rev. B* **64**, 235315 (2001). (Reprinted with permission.)

from one such region to another (for example, at $\nu = 1/3$). Möller and Simon [451] calculate gaps for several fractions of the form $\nu = n/(4pn \pm 1)$ using the CF theory; they find that the masses and gaps are more or less symmetric around $\nu = 1/4$.

8.8 CFCS theory of excitations

The collective modes of composite fermions have been studied by Lopez and Fradkin [402], He, Simon and Halperin [247, 592], and Xie [709] in the standard random phase approximation (called the "semiclassical approximation" in Ref. [402]) of the CFCS theory. This approach captures certain qualitative aspects of various kinds of excitations and their dispersions. The Murthy–Shankar approach has also been used to compute transport gaps as well as dispersions of the magnetoexcitons for several incompressible CF states [463,464].

The gaps or masses depend strongly on the short-distance part of the interelectron interaction. An exception is for the FQHE states with many filled Λ levels; the CF-quasiparticles and CF-quasiholes of such states have a large size, so their energies have a substantial dependence on the long-range part of the interaction. These may be amenable to a field theoretical treatment, which focuses on the long-range part of the interaction. Halperin, Lee, and Read [231] obtain the following form for the CF mass at $\nu = 1/2$:

$$m^*(\omega) = \frac{2p^2}{\pi} \frac{\epsilon k_F}{e^2} |\ln \omega|, \qquad (8.28)$$

where $k_F = \sqrt{4\pi\rho}$ and ω is the energy measured from the Fermi energy, predicting a logarithmic divergence for the mass close to the Fermi energy (Section 5.16.3). They show that this implies a transport gap of the form [231,613]

$$\Delta \approx \frac{\pi e^2}{2\epsilon\ell} \frac{1}{(2n+1)[\ln(2n+1)+C]} \qquad (8.29)$$

for the $f = n/(2n \pm 1)$ fractions asymptotically close to $\nu = 1/2$. The constant C cannot be determined within their approach, due to its dependence on the short-range part of the interaction, but the coefficient of the $\ln(2n+1)$ term has been argued to be exact [613] in the $n \to \infty$ limit through the use of Ward identities that determine the leading term in the divergence of m^*. Verification of the logarithmic divergence by exact diagonalization is difficult because such studies are unable to obtain gaps for states with many filled Λ levels where this effect would be sizable. The wave functions of Eq. (5.32) for the FQHE states close to $\nu = 1/2$ may not have sufficient accuracy to capture subtle logarithmic corrections. The mass enhancement observed in Shubnikov–de Haas experiments is much stronger than logarithmic.

8.9 Tunneling into the CF liquid: the electron spectral function

The electron spectral function, defined in Appendix G, is of conceptual importance in the Landau Fermi liquid theory of ordinary metals, because it tells us whether *electron-like* quasiparticles are well defined or not. They are well defined if the electron spectral function has a sharp (ideally a delta function, but typically a broadened Lorentzian) peak at some energy, i.e., has the form

$$A(\alpha, E) = \frac{Z_\alpha}{\pi} \frac{\Gamma_\alpha}{(E - E_\alpha)^2 + \Gamma_\alpha^2} + \text{a broad incoherent part.} \qquad (8.30)$$

The position of the peak, E_α, gives us the dispersion, and hence the mass, of the quasiparticle. The quantity $\Gamma_\alpha = \hbar/\tau_\alpha$ is the quasiparticle broadening, or the inverse lifetime, due to disorder. The weight under the peak, Z_α, is called the quasiparticle renormalization factor. Only when electron-like quasiparticles are well defined is the Landau mapping into a weakly interacting fermion system with renormalized parameters valid.[4] The electron spectral function has been of great interest in the context of recent searches for systems in which strong interactions cause a breakdown of the Landau Fermi liquid paradigm. It can not only signal such a breakdown (at least in principle – interpretation of experimental data is not always straightforward), but also shed light on the nature of the non-Landau–Fermi liquid state.

As discussed in Appendix G, the spectral function can be measured by tunneling of an external electron into the system. The electron spectral function for a superconductor was measured by Giaever [186], tunneling electrons from a normal metal into a superconductor through a thin potential barrier, providing a spectacular confirmation of the Bardeen–Cooper–Schrieffer theory of superconductivity. The low-energy spectral function for 2D electrons has been measured by Murphy *et al.* [461], who study the tunnel conductance (I/V) for transport across two parallel 2D layers (called a bilayer system). The width of the I/V resonance peak as a function of V is inversely proportional to the quasiparticle lifetime at the Fermi energy. At low temperatures, the peak width gets contributions from electron–electron scattering as well as disorder; the latter is expected to be more or less temperature independent. The temperature-dependent part of the peak width is found to behave approximately as T^2, which is consistent with the Landau Fermi liquid theory prediction $\hbar(\tau_{ee})^{-1} \sim T^2 \ln(T_F/T)$ (Giuliani and Quinn [195] and Hodges, Smith, Wilkins [256]), where τ_{ee} is the electron–electron scattering contribution to the electron lifetime.

The CF liquid is a non-Landau Fermi liquid.[5] A breakdown of the Landau paradigm is signaled by the emergence of new quasiparticles – composite fermions – which are qualitatively distinct from the original fermions (electrons). Nonetheless, one can ask if the *electron* spectral function will teach us something about this state.

The tunneling conductance has been measured in the FQHE regime for bilayer systems (Ashoori *et al.* [15]; Eisenstein, Pfeiffer, and West [138]) with each layer contacted separately. As seen in Fig. 8.23, I–V curves show a "Coulomb gap," with the current

---

[4] We have stressed the *electron-like* nature of the quasiparticles of a Landau Fermi liquid, a qualification that is left implicit in many discussions. There can be perfectly well-defined quasiparticles in liquids where Landau's theory fails, except that they are qualitatively distinct from electrons (or whatever fermions we started out with).

[5] We make a distinction between "Fermi liquids" and "*Landau* Fermi liquids." Any liquid of fermions is a Fermi liquid, including the liquid of composite fermions. *Landau* Fermi liquids, on the other hand, are those Fermi liquids for which the original fermions *adiabatically* evolve into weakly interacting quasiparticles as the interaction is turned on.

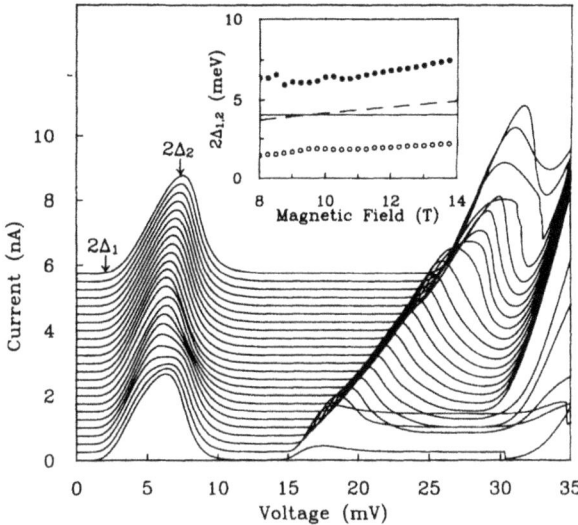

Fig. 8.23. I–V characteristics for tunneling in a bilayer system at several magnetic fields between 8 T (lowest trace) and 13.75 T (highest trace). (The traces are offset for clarity.) The suppression of current at small biases, followed by a peak at voltage $2\Delta_2$, indicates a Coulomb gap to tunneling. $2\Delta_1$ is the onset voltage for the tunneling peak. The dashed line in the inset is $0.3e^2/\epsilon\ell$ and the solid line is $e^2/\epsilon a$, where a is the average interparticle separation. Source: J. P. Eisenstein, L. N. Pfeiffer, and K. W. West, *Phys. Rev. Lett.* **69** 3804 (1992). (Reprinted with permission.)

exponentially suppressed at low bias, and a peak at approximately $0.5e^2/\epsilon\ell$. Such a gap signifies strong Coulomb correlations of the lowest Landau level state. A number of calculations of the electron spectral function by exact diagonalization (Hatsugai, Bares, and Wen [240]; He, Platzman, and Halperin [246]; Rezayi [545]) and also by the CFCS approach (Hatsugai, Bares, and Wen [240]; He, Platzmann, and Halperin [246]), obtain a Coulomb gap of the order of $e^2/\epsilon\ell$. Johansson and Kinaret [305] explain the Coulomb gap by modeling the state as a Wigner crystal (the gap is essentially the energy of an interstitial), while Efros and Pikus [134] accomplish the same within a model of a classical electron liquid. That is not surprising, as the experiments do not show any special structures related to the fractional quantum Hall effect. (The irrelevance of FQHE to this effect is also indicated by the observation of a Coulomb gap for the compressible state at $\nu = 1/2$.) Leonard and Johnson [387, 388] calculate the dynamical structure factor within the "modified random phase approximation" of the CFCS approach, and argue that the CF liquid at $\nu = 1/2$ undergoes a stronger shake-up upon the sudden introduction of a charged impurity than the electron system at zero magnetic field;[6] as a result, the authors predict, the CF Fermi sea exhibits a more severe orthogonality catastrophe (also see Ref. [246]).

---

[6] This is known as the X-ray edge singularity problem, the physics of which was clarified by the classic works of Mahan [415,416] and Nozières and DeDominicis [481].

An investigation of the problem using the microscopic wave functions of the CF theory predicts that the spectral function for an electron in the $m = 0$ state has a delta function peak (which would be broadened by disorder), followed by a broad incoherent part at higher energies (Jain and Peterson [287]). In other words, a sharply defined electron-like quasiparticle exists even in the FQHE state. The sharp peak of the "electron-quasiparticle"[7] of the $\nu = n/(2pn \pm 1)$ FQHE state is actually a complex bound "atom" of $(2pn \pm 1)$ excited CF-quasiparticles (Exercise 8.8), which should reveal itself through a sharp resonance in the tunnel conductance into the zero angular momentum state. (The electron tunneling in the experiment of Ref. [138] is believed to be into extended states with well-defined wave vector. Tunneling into the zero angular momentum state can, in principle, be achieved through an STM tip.[8]) The calculated renormalization factor Z decreases rapidly with increasing Λ level filling n, vanishing for the CF Fermi sea, which can be understood by noting that increasingly more complicated CF bound states represent the coherent part of the electron-quasiparticle.

How can the Landau paradigm break down in spite of the existence a long-lived electron-quasiparticle? Because other states appear at *lower* energies that are not describable in terms of electrons. The electron-quasiparticle is long lived (in the absence of disorder) because of its inability to decay into the lower-energy states as a result of angular momentum selection rules. The FQHE system thus exemplifies a new paradigm for the breakdown of the Landau Fermi liquid theory.

It is worth reminding ourselves how remarkable this structure is. First, composite fermions are formed, which are collective bound states of electrons. This may be thought of as a "fractionalization" of electrons (Section 9.3.9). Now we discover (yet to be confirmed experimentally) that an odd number of excited composite fermions form a complex bound state to give back an electron-quasiparticle.[9]

Further insight into this result comes from the Conti–Vignale continuum elasticity theory, discussed in Section 12.7, which treats the FQHE liquid as a continuous elastic medium. The dynamics is described in terms of the displacement field, which, when quantized, behaves as a collection of independent bosons (Eq. 12.76), which are the phonons of the elastic medium. To compute the spectral function for an electron in the zero angular momentum state, Conti and Vignale write the following effective Hamiltonian:

$$H = \frac{\hbar \omega_c}{2} a_0^\dagger a_0 + \sum_q \hbar \omega_q b_q^\dagger b_q + \sum_q M_q \left(b_q + b_{-q}^\dagger \right) a_0^\dagger a_0, \qquad (8.31)$$

where a_0 and $a_0^\dagger$ are, respectively, the destruction and creation operators for the test electron in the zero angular momentum sate, $\hbar \omega_q$ is the dispersion of the phonon, and the last term describes the coupling between the electron and the bath of phonons, with the coupling constant M_q being a function of the dynamical shear modulus. This Hamiltonian models

---

[7] The *true* "quasielectron" of the FQHE.
[8] This was pointed out to the author by Professors. E. Demler and J. P. Eisenstein.
[9] This is not equal to the original electron in its full glory but only captures a small piece of it. It is not possible to construct composite fermions out of the electron-quasiparticles to establish a perpetual "do loop."

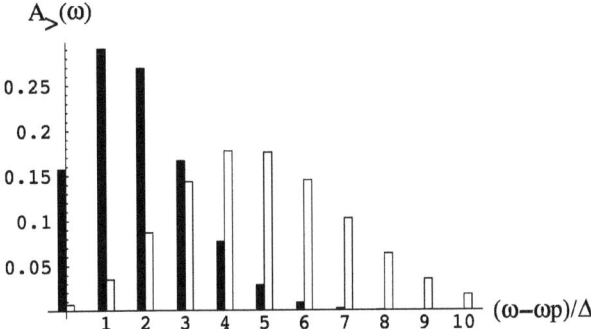

Fig. 8.24. Electron spectral function at $\nu = 1/3$ and 2/5 (black and white bars, respectively, representing delta function peaks) calculated from the Conti–Vignale continuum elasticity theory, assuming a dispersionless "phonon" at energy Δ. The energy $\hbar\omega_p$ is the "addition energy" (or the Coulomb gap) for a test electron added into the zero angular momentum state. Source: G. Vignale, *Phys. Rev. B* **73**, 073306 (2006). (Reprinted with permission.)

the renormalization of the test electron by the incompressible liquid through an electron–phonon interaction term. It is similar to the Hamiltonian for the "polaron" problem (an electron coupled to phonons), except simpler, in that it does not allow virtual transitions of the test electron out of the zero angular momentum state. This simplification enables an exact analytic calculation of the spectral function for this Hamiltonian.

Using this model, Vignale [657] obtains the spectral function for an electron in the zero angular momentum state for several filling factors, shown in Fig. 8.24. The first, zero-phonon delta function peak at the Coulomb gap energy is a consequence of the gap in the phonon dispersion, and the weight under it, namely Z, is shown to be $Z = e^{-|\hbar\omega_p|/\Delta}$ where $\hbar\omega_p$ is the Coulomb addition energy for an electron (the "polaron shift"), and Δ is the energy of the phonon, assumed to be dispersionless in the simplest approximation. The two energies, expressed in units of $e^2/\epsilon\ell$, are related by the expression

$$\hbar\omega_p = -\frac{1-\nu}{16\Delta}. \tag{8.32}$$

With Δ taken as the $q = 0$ limit of the CF-exciton energy, a roughly ν independent value is obtained for the Coulomb gap $\hbar\omega_p$, which is consistent with the value obtained from experiment or other calculations. The calculated Z decreases rapidly with the number of filled Λ levels ($Z = 0.156, 0.007, 0.0004$, for 1/3, 2/5, and 3/7, respectively). The subsequent delta function peaks in Fig. 8.24 are an artifact of the neglect of dispersion for the phonon, and would be replaced by broadened structures in a more realistic calculation. The nonzero value of Z for incompressible states and its decline along a FQHE sequence are qualitatively consistent with the results obtained in Ref. [287] from a microscopic calculation.

Exercises

8.1 Derive Eq. (8.5) for the one-particle reduced density matrix from Eq. (8.1), by directly substituting expressions for the field operators and using the property

$$\left\langle c_{m'}^{\dagger} c_m \right\rangle = \nu \delta_{m,m'},\tag{E8.1}$$

as expected for an isotropic and uniform ground state. (Source: Girvin and MacDonald [193].)

8.2 Derive the relation

$$G(r, 0^+; r', 0) = -\frac{i}{2\pi} \exp\left[\frac{\bar{z}z'}{2} - \frac{|z|^2}{4} - \frac{|z'|^2}{4}\right] + i\rho_1(r - r')\tag{E8.2}$$

in the LLL subspace.

8.3 Show that, for a uniform density state, the interaction energy per particle is given by

$$E = \frac{1}{N}\left(\left\langle \sum_{j<k} V(r_j - r_k) \right\rangle + V_{e-b} + V_{b-b}\right)$$
$$= \frac{\rho}{2} \int d^2r\, V(r)[g(r) - 1],\tag{E8.3}$$

where $V(r)$ is the interaction between electrons, and V_{e-b} and V_{b-b} are electron–background and background–background interaction terms. Using Eq. (8.9) this can be expressed as

$$E = (2\rho)^{-1} \int d^2r\, V(r)\langle \delta\hat{\rho}(r)\delta\hat{\rho}(0)\rangle,\tag{E8.4}$$

where $\delta\hat{\rho}(r) = \hat{\rho}(r) - \rho$, and the point $r = 0$ is excluded.

8.4 Several quantities discussed in this chapter can be obtained for integral quantum Hall states straightforwardly, and this problem asks you to do that for the one filled Landau level state at $\nu = 1$.

(i) Derive

$$g(r) = 1 - \exp\left[-\frac{r^2}{2\ell^2}\right].\tag{E8.5}$$

Hint: Use Eq. (8.11). The algebra can be simplified by exploiting translational invariance to choose $r_1 = 0$ and $r_2 = r$, and noting that $\eta_m(0) = \frac{1}{\sqrt{2\pi}}\delta_{m,0}$. Remember that $\rho = 1/(2\pi \ell^2)$ for $\nu = 1$.

(ii) Show that the static structure factor is given by

$$S(q) = 1 - \exp\left[-\frac{q^2\ell^2}{2}\right].\tag{E8.6}$$

(iii) Obtain the interaction energy per particle, assuming Coulomb interaction $V(r) = e^2/r$, with the help of Eq. (E8.3). Answer [151]:

$$E_1 = -\sqrt{\frac{\pi}{8}} \frac{e^2}{\ell}. \tag{E8.7}$$

8.5 Using the particle–hole symmetry in the lowest Landau level, the eigenenergies at $1 - \nu$ can be related to those at ν; in particular, the energy spectra at ν and $1 - \nu$ are seen to be identical, apart from an overall additive constant. An explicit relation between the eigenenergies is derived in this exercise for the spherical geometry, for N electrons at Q. Consider electrons in the lowest Landau level with interaction (cf. Appendix F)

$$H = \sum_{m_1,m_2,m_3,m_4} V_{m_1 m_2, m_3 m_4} c_{m_1}^\dagger c_{m_2}^\dagger c_{m_4} c_{m_3}. \tag{E8.8}$$

(i) Show that the interaction matrix elements can be chosen to satisfy the following symmetries:

$$V_{m_1 m_2, m_3 m_4} = -V_{m_1 m_2, m_4 m_3} = -V_{m_2 m_1, m_3 m_4} = V_{m_3 m_4, m_1 m_2}^*. \tag{E8.9}$$

(ii) Introduce creation and annihilation operators for holes, $d_m^\dagger = c_m$ and $d_m = c_m^\dagger$, which obey the canonical anticommutation relations. Show that

$$H = 2 \sum_m U_m (1 - 2d_m^\dagger d_m) + H_d, \tag{E8.10}$$

where $U_m = \sum_{m_1} V_{mm_1, mm_1}$ and

$$H_d = \sum_{m_1,m_2,m_3,m_4} V_{m_1 m_2, m_3 m_4} d_{m_1}^\dagger d_{m_2}^\dagger d_{m_4} d_{m_3}. \tag{E8.11}$$

(iii) Show that the energy of a completely full Landau level (at the same Q, and for "spinless" electrons) is given by $E_1 = 2 \sum_m U_m$.

(iv) For uniform density states (which have $L = 0$), derive the relation

$$H = \left(1 - \frac{2N_h}{2Q + 1}\right) E_1 + H_d, \tag{E8.12}$$

where $N_h = 2Q + 1 - N$ is the number of holes. This relation is valid for all eigenstates because, for a rotationally invariant interaction, U_m can be shown [451] to be independent of m.

(v) Show that the energy of an eigenstate, E_ν, is related to the energy of its particle–hole conjugate state, $E_{1-\nu}$, by the relation

$$E_\nu = \left(1 - \frac{2N_h}{2Q + 1}\right) E_1 + E_{1-\nu}. \tag{E8.13}$$

The energies E_ν, $E_{1-\nu}$ and E_1 are defined for different numbers of particles but for the same magnetic field (i.e., same Q). Show that the incorporation of electron–background and background–background terms does not alter this relation.

(vi) Until now, the E's are the total energies. Show that, in terms of energies *per particle*, the relation reduces, in the thermodynamic limit, to

$$\nu E_\nu = -(1 - 2\nu)\,E_1 + (1 - \nu)E_{1-\nu}. \tag{E8.14}$$

(Source: Möller and Simon [451].)

8.6 This exercise alerts the reader to certain pitfalls in assigning filling factors to finite systems.

(i) The $\nu = 4/9$ state occurs for $2Q = \frac{9}{4}N - 6$. The system $(N, 2Q) = (8, 12)$ satisfies this relation. Analyze it in terms of composite fermions to determine which incompressible state it *really* describes.

(ii) Consider the system $(N = n^2, 2Q = 2pN - 2p)$, which maps into $Q^* = 0$. Show that it satisfies the $2Q$–N relation for both $\nu = n/(2pn + 1)$ and $\nu = n/(2pn - 1)$. Furthermore, since $Q^* = 0$, this state may also be interpreted as a finite size representation of the CF Fermi sea [549]. For incompressible states, systems with $N > n^2$ lend themselves to an unambiguous interpretation.

8.7 Show that a single CF-quasiparticle or a single CF-quasihole of $\nu = n/(2pn+1)$ occurs at

$$Q = \left(p + \frac{1}{2n}\right)N - \left(p + \frac{n}{2} \pm \frac{1}{n}\right). \tag{E8.15}$$

This state corresponds to the IQHE state at Q^* which contains n full Landau levels plus an additional electron (hole) in the $(n + 1)$th (nth) Landau level.

8.8 Consider the FQHE state at $\nu = 1/3$ on a sphere. Add an electron at the north pole, which produces a state with $L = Q$, $L_z = Q$. Thus we have gone from the state $(N, Q, L, L_z) = (N, Q, 0, 0)$ to $(N + 1, Q, Q, Q)$. Analyze the new state in terms of composite fermions. What is the lowest energy Λ level occupation with the indicated quantum numbers? Repeat the exercise for $\nu = 2/5$ and $3/7$, as well as for a hole with quantum numbers $(N - 1, Q, Q, -Q)$ at $\nu = 1/3, 2/5$, and $3/7$. (Source: Jain and Peterson [287].)

9

Topology and quantizations

Topology is the study of properties of geometric configurations that remain invariant under continuous deformations. It is of relevance in graph theory, network theory, and knot theory. Many physical phenomena have topological content. Perhaps the best known is the Aharonov–Bohm phase for a closed path of a charged particle around a localized magnetic flux, which is independent of the shape and the length of the path so long as it is outside the region containing the magnetic flux. It is a special case of a more general concept, known as the Berry phase. The latter is the phase acquired when the parameters of a Hamiltonian execute a closed loop in the parameter space; the Berry phase does not depend on the rate at which the loop is executed, provided it is sufficiently slow. (There is no speed limit for the Aharonov–Bohm phase. In this case the adiabatic approximation of Berry becomes exact.)

The most dramatic examples of topology lead to macroscopic quantizations, such as the quantization of flux (in units of $h/2e$; $2e$ because of pairing) through a superconducting loop or through a vortex in a type-II superconductor. The quantization of flux results from the topological property that the phase of the order parameter can only change by an integral number times 2π around the flux.

Topology enters into the physics of the fractional quantum Hall effect through the formation of composite fermions, one constituent of which, namely the vortex, is topological. The vorticity of a composite fermion ($2p$) is quantized to be an even integer due to single-valuedness and antisymmetry properties of the wave function. Such a quantization is implicit in that we *count* the number of vortices bound to the composite fermion. Because the composite fermion is a topological particle, all phenomenology of the CF quantum fluid has a topological origin. The most direct, robust and well-confirmed manifestation of the topological order of the CF state is the effective magnetic field, numerous consequences of which are discussed elsewhere in the book. This chapter focuses on three other effects arising from it: fractional "local charge" and fractional "braiding statistics" of the FQHE quasiparticles, and the fractional quantization of Hall resistance.

9.1 Charge charge, statistics statistics

Some confusion exists in the literature regarding the issues of charge and statistics in the FQHE. For example, the oft-heard phrase, "FQHE quasiparticles have fractional charge and obey fractional statistics," appears, on the face of it, inconsistent with the fact that they are excited composite *fermions* (Section 8.5). The confusion arises because, in reality, *two* distinct kinds of charges ("intrinsic" and "local") and statistics ("exchange" and "braiding") have been considered in the literature, but often without explicit qualification. Which one is relevant depends on the experimental measurement in question.

One aspect in which the two charges and statistics differ is in the choice of "vacuum," i.e., the reference state, relative to which they are defined. The two natural choices for vacuum are:

- the "null" vacuum: the state containing *no* particles,
- the FQHE vacuum: a (nearby) FQHE state with uniform density.

The charge depends on which vacuum is chosen as the reference state. For example, let us ask how many particles we have in the incompressible ground state at $v = n/(2pn + 1)$. From the second perspective the state has no particles, being itself the vacuum. But from the perspective of the null vacuum, it is teeming with composite fermions. A state at a filling factor v in the range

$$\frac{n+1}{2n+3} > v > \frac{n}{2n+1} , \qquad \text{i.e., } n+1 > v^* > n \qquad (9.1)$$

can be viewed with reference to (i) the null vacuum, (ii) the $v = n/(2n + 1)$ FQHE state, or (iii) the $v = (n + 1)/(2n + 3)$ FQHE state, which all result in apparently different descriptions.

Furthermore, the concept of "braiding statistics" is distinct from the ordinary exchange statistics that we encounter in our elementary quantum mechanics course, as explained in more detail in Section 9.8.1.

9.2 Intrinsic charge and exchange statistics of composite fermions

"Intrinsic charge" and the ordinary "exchange statistics" are defined relative to the null vacuum containing no particles. The number of composite fermions is equal to the number of electrons. Increasing the number of composite fermions by one unit amounts to adding a net charge $-e$ to the system. This gives the "intrinsic charge" of composite fermions as $-e$. Furthermore, an exchange of two composite fermion coordinates in the quantum mechanical wave function produces a negative sign, because the CF coordinates are the same as the electron coordinates. Therefore, the exchange statistics of composite fermions is fermionic. That is the reason for calling them *fermions*.

This definition treats *all* composite fermions equivalently. Because a CF-quasiparticle is also a composite fermion, its intrinsic charge is $-e$ and exchange statistics fermionic. The intrinsic charge and the exchange statistics remain sharply defined quantities for composite fermions independent of whether they are in a compressible or an incompressible state. (In contrast, as seen below, the values of "local charge" and "braiding statistics" for

CF-quasiparticles or CF-quasiholes depend on the reference FQHE state, and cannot be defined for compressible states.)

9.3 Local charge

Next we take a uniform FQHE state as the vacuum and define various quantities through their deviations from it. (This is in the spirit of Laudau's Fermi liquid theory.) The deviations from the FQHE vacuum occur in the form of CF-quasiparticles or CF-quasiholes. As seen in Chapter 8, a CF-quasiparticle or a CF-quasihole is a localized defect on top of the uniform density incompressible state. We define the charge *excess* or *deficiency* associated with this defect as the "local charge" of the CF-quasiparticle or the CF-quasihole. If we draw a circle large enough to enclose fully a single CF-quasiparticle or CF-quasihole, then the local charge is equal to the total charge inside the circle minus what the charge would have been without the CF-quasiparticle or CF-quasihole inside it.

The local charge is clearly not $-e$. When an electron gets dressed into a composite fermion through capture of vortices, the vortices push nearby particles outward to create a correlation hole. The local charge is equal to the intrinsic charge of the composite fermion plus the charge deficiency associated with the correlation hole. In other words, the "vacuum" partially screens the charge. Such screening is a necessary consequence of the physics that each composite fermion sees $2p$ vortices on every other composite fermion, including the ones that we have decided to count as belonging to the vacuum.

Remarkably, the local charge of a CF-quasiparticle is a precisely quantized fraction of the electron charge, the value of which depends only on the filling factor of the background FQHE state. Several derivations of fractional charge are given below. With one exception, they use only the general principles of the CF theory but no microscopic details, as appropriate for a universal quantity.

Various authors do not always qualify the charge as "intrinsic" or "local." Which charge is being discussed should be clear from the context. Very simply: A fractional charge always refers to the local charge, and the intrinsic charge is always $-e$.

9.3.1 An electron equals 2pn + 1 CF-quasiparticles

The clearest proof of fractional charge is based on a simple counting argument. It proceeds by asking what happens to an incompressible FQHE system when an *electron* is added to it. Let us consider the ground state at $\nu = n/(2pn + 1)$ in the spherical geometry, which maps into n filled Λ levels of composite fermions. That is, the effective flux

$$Q^* = Q - p(N - 1) \tag{9.2}$$

is such that the total number of single particle states in the lowest n Λ levels is precisely N. Now add one electron to this state *without changing the real external flux Q*. The new state of $N' = N + 1$ particles corresponds to a modified effective flux

$$Q'^* = Q - p(N' - 1) \tag{9.3}$$
$$= Q^* - p \, .$$

At Q'^* the degeneracy of each Λ level is reduced by $2p$ compared to Q^*. Thus $2p$ composite fermions from each of the $n \, \Lambda$ levels must be pushed up to the $(n + 1)$th Λ level. Including the added particle, the $(n + 1)$th Λ level has $2pn + 1$ CF-quasiparticles. These are identical, can move independently, and can be removed far from one another. Since a net charge $-e$ was added, each CF-quasiparticle must carry an excess charge of precisely

$$-e^* = -\frac{e}{2pn + 1} \, . \tag{9.4}$$

In short, the conversion of the additional electron into a composite fermion creates a correlation hole around the electron, which forces $2pn$ additional CF-quasiparticles out of the FQHE vacuum (Fig. 9.1). When viewed in terms of electrons, the system rearranges itself in a fantastically complicated manner under the addition (or removal) of electrons.

Although we do not need to use the actual wave functions, the underlying microscopic theory gives credence to the derivation, because it tells us that the words and concepts we are using are not pure fiction. It must be stressed, however, that the accuracy of the charge is not related to the accuracy of the microscopic wave functions. So long as the number of CF-quasiparticles created upon the insertion of a single electron is $2pn + 1$, Eq. (9.4) follows exactly, independent of the accuracy of the wave function. The actual wave function depends on the interaction between electrons or the amount of LL mixing, but e^* remains unaffected as long as the CF physics is valid. That is the reason why this counting argument provides the most powerful and convincing derivation of the fractional charge. In particular, the result does not depend on whether or not the wave function is projected into the lowest Landau level, although the derivation is certainly valid for the projected wave function of Eq. (5.32).

The local charge of a CF-quasihole can be obtained similarly (Exercise 9.1). Alternatively, we may deduce its charge to be $+e/(2pn + 1)$ by noting that a pair of CF-quasiparticle and CF-quasihole has no net charge, because it can be created without changing the values of N and $2Q$.

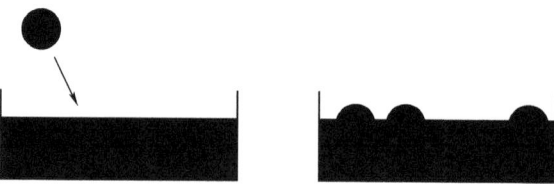

Fig. 9.1. Adding an electron to a FQHE state excites several CF-quasiparticles out of the FQHE state.

9.3.2 A CF-quasiparticle equals an electron + 2p vortices

A vortex at $\eta = Re^{-i\theta}$ is defined by the wave function

$$\Psi_\eta = N_R \prod_j (z_j - \eta)\Psi .$$ (9.5)

Avoidance of the position η creates a charge deficiency there. The amount of the charge deficiency can be determined following Arovas, Schrieffer, and Wilczek [13]. The Berry phase associated with a closed loop (for counterclockwise traversal) of the vortex is calculated in Section 5.11.4, where we see that each particle inside the loop contributes 2π to the Berry phase, giving a total phase of $2\pi N_{enc}$, where N_{enc} is the number of particles inside the loop. Equating it to the Aharonov–Bohm phase of a particle of charge e_V gives

$$2\pi N_{enc} = 2\pi \frac{BA\, e_V}{hc}$$

$$= 2\pi \frac{N_{enc}}{\nu} \frac{e_V}{e} ,$$ (9.6)

where we have used $\nu = N_{enc}\phi_0/BA$ (BA/ϕ_0 is the number of single particle states in area A, and neglected order-one corrections. The charge of the vortex therefore is

$$e_V = \nu e .$$ (9.7)

The local charge of a CF-quasiparticle at $\nu = n/(2pn+1)$ can now be ascertained by noting that it is the bound state of an electron and $2p$ vortices:

$$-e^* = -e + 2pe_V = -\frac{e}{2pn+1} .$$ (9.8)

The notion of binding of electrons and vortices is transparent in the unprojected wave functions of Eq. (5.31), but becomes obscure after projection into the lowest Landau level. It is expected, however, that while the act of LLL projection (or any other perturbation) causes a rearrangement of the charge within a CF-quasiparticle, it does not create any *new* CF-quasiparticles, and hence leaves the local charge intact.

9.3.3 A vortex equals n CF-quasiholes

Further insight into the meaning of a vortex can be gained through the CF theory. For illustration, consider a vortex at the origin in the $\nu = 2/5$ state. Its wave function, disregarding LLL projection for now, is given by

$$\Psi^{(2/5)}_{\eta=0} = N_0 \left(\prod_j z_j \right) \Psi_{2/5}$$

$$= N_0 \prod_{k<l} (z_j - z_l)^2 \left(\prod_j z_j \right) \Phi_2 ,$$ (9.9)

where Φ_2 is the wave function of two filled Landau levels, given explicitly in Eq. (3.51). The above equation indicates that the vortex at 2/5 is related to the vortex at filling factor two. Referring to Eq. (3.51), the latter has the wave function

$$\left(\prod_j z_j\right)\Phi_2 = \begin{vmatrix} z_1 & z_2 & z_3 & \cdot & \cdot \\ z_1^2 & z_2^2 & z_3^2 & \cdot & \cdot \\ \cdot & \cdot & \cdot & \cdot & \cdot \\ z_1^{N/2} & z_2^{N/2} & z_3^{N/2} & \cdot & \cdot \\ \bar{z}_1 z_1 & \bar{z}_2 z_2 & \bar{z}_3 z_3 & \cdot & \cdot \\ \bar{z}_1 z_1^2 & \bar{z}_2 z_2^2 & \bar{z}_3 z_3^2 & \cdot & \cdot \\ \cdot & \cdot & \cdot & \cdot & \cdot \\ \bar{z}_1 z_1^{N/2} & \bar{z}_2 z_2^{N/2} & \bar{z}_3 z_3^{N/2} & \cdot & \cdot \end{vmatrix} \exp\left[-\frac{1}{4}\sum_i |z_i|^2\right], \tag{9.10}$$

which has two holes at the origin, one in each of the two occupied Landau levels. The vortex $\Psi_{n=0}^{(2/5)}$ is thus not a fundamental excitation but a combination of two CF-quasiholes at the origin, one in each Λ level. Analogously, a vortex at η in the $n/(2pn \pm 1)$ state is a collection of n CF-quasiholes at η. The n CF-quasiholes, which happen to be coincident in the vortex, can split apart and move independently (their wave function is complicated but can, in principle, be written down with the help of the CF theory). The charge of a single CF-quasihole, therefore, is

$$e^* = \frac{e\nu}{n} = \frac{e}{2pn \pm 1} . \tag{9.11}$$

While this derivation is presented in terms of the unprojected wave function, the number of CF-quasiholes required to make one vortex, and hence also the charge of a single CF-quasihole, is expected to be robust to mild perturbations.

9.3.4 Adiabatic flux insertion

This subsection uses only one consequence of the CF theory, namely incompressibility at certain fractional fillings, to demonstrate fractionally charged excitations. This is how Laughlin first demonstrated fractional charge [369]. With certain additional plausible assumptions, the value of the *elementary* fractional charge is determined. This derivation has the advantage of illustrating that fractional charge is a general consequence of incompressibility, but it is too general to provide a microscopic or intuitive understanding of what the fractionally charged objects really are.

Let us assume that the state at ν is incompressible. An excitation of this state can be obtained by Laughlin's trick of adiabatic flux insertion [369] as follows. We consider, in addition to the uniform magnetic field, a point flux of strength $\alpha\phi_0$ threading the plane at the origin in the normal direction, produced by the vector potential (Appendix C)

$$A_\alpha = \frac{\alpha}{2\pi}\phi_0 \nabla\theta . \tag{9.12}$$

Beginning with the *exact* (though unknown) ground state at $\alpha\phi_0 = 0$, we change the flux slowly from 0 to ϕ_0 adiabatically. The ground state evolves into a new state during this process. At the end of the process, the flux is equal to ϕ_0, which can be gauged away to obtain a new exact eigenstate of the Hamiltonian. This new state has a charge deficiency or excess at the origin, which we denote by e_ϕ.

Keeping track of how the interacting ground state evolves under the above process is, in practice, impossible, because that would require the knowledge of the exact ground state at each possible value of α. Nevertheless, e_ϕ can be determined as follows: As shown in Appendix D, when the flux is adiabatically changed from 0 to ϕ_0, each single particle state moves to the next one. In the presence of a gap, each orbital must carry its charge with it. Far from the insertion point the behavior is trivial: the amount of charge that moves radially outward is precisely equal to the charge contained in one single particle orbital, i.e., νe for the incompressible state at ν. This is equal to the charge excess or deficiency at the origin:

$$e_\phi = \nu e \ . \tag{9.13}$$

The presence of a gap is crucial to this argument; while each single particle orbital always moves to the next one, only for gapped states does it carry its charge along.

A related derivation of the charge uses Faraday's law:

$$\oint E \cdot dl = -\frac{d\phi}{dt} \ , \tag{9.14}$$

which gives

$$E_\phi = -\frac{1}{2\pi r}\frac{d\phi}{dt} \ . \tag{9.15}$$

The current density is $j_r = \sigma_H E_\phi$, where $\sigma_H = \nu e^2/h$ is the Hall conductivity. The charge leaving the area defined by a circle of radius r per unit time is $2\pi r j_r$. The total charge leaving this area in the adiabatic process is

$$Q = \int 2\pi r j_r \, dt = -\sigma_H \phi_0 = -\nu e. \tag{9.16}$$

This yields for the charge of the excitation

$$e_\phi = \nu e. \tag{9.17}$$

Given that the object produced by adiabatic flux insertion has the same charge as a vortex, it is tempting to equate the two. This is precisely what was done by Laughlin [369]. The correspondence is not microscopically exact, however, and is not needed for the purpose of determining the charge.

The flux insertion process is guaranteed to produce an exact eigenstate. Because e_ϕ is fractional, the above argument establishes a fractionally charged excitation. However, this excitation can in general be a combination of several elementary excitations, called quasiholes (we pretend ignorance of composite fermions in this section). We obtain the charge e^* of the quasihole following an argument put forth by Su [629]. Let us specialize to $\nu = n/(2pn \pm 1)$ and assume that e_ϕ is a collection of an integral number (r_1) of elementary excitations. In other words, we have

$$e_\phi = \frac{n}{2pn \pm 1}e = r_1 e^*. \tag{9.18}$$

As an additional input, let us ask what happens if an electron is removed from the system. That would create, in general, several quasiholes. Assuming only one kind of quasihole, an integral number (r_2) of them must have the same total charge as one missing electron:

$$e = r_2 e^* . \tag{9.19}$$

That gives

$$\frac{r_1}{r_2} = \frac{n}{2pn \pm 1} . \tag{9.20}$$

Because the numerator and the denominator of the filling factor are relatively prime, the only solution is $r_1 = rn$ and $r_2 = r(2pn \pm 1)$, where r is an arbitrary integer, giving $e^* = e/[r(2pn \pm 1)]$. If we further assume that the elementary quasihole has the largest charge allowed by above constraints, we recover the familiar result $e^* = e/(2pn \pm 1)$.

Thus, the presence of a gap at a fractional filling factor not only implies fractional charge, but is a sufficiently powerful restriction to fix the value of the charge uniquely, provided certain plausible assumptions are made. This derivation emphasizes that the fractional charge has nothing to do with the microscopic origin of the FQHE, but is a consequence of incompressibility in a partially filled Landau level. It also provides additional insight into why the value of e^* is robust, insensitive to the detailed form of the wave function. Further, the value of charge is closely related to the filling factor, and it is no accident that $e^* = e/(2pn + 1)$ at $\nu = n/(2pn + 1)$.

The correctness of the above assumptions is established by the CF theory, which confirms that there is only one kind of elementary excitations (CF-quasiparticles), and that $e/(2pn + 1)$ is indeed the smallest charge (otherwise, the CF theory would not give a complete description of the low-energy spectrum). The CF theory further tells us that quasiparticles and quasiholes really are CF-quasiparticles (electrons dressed by vortices) and CF-quasiholes (their absence), and that the "vortex" state produced by adiabatic flux insertion is, in general, a highly excited state containing one CF-quasihole in each occupied Λ level.

9.3.5 Direct calculation from wave function

The excess charge can be obtained directly from the wave functions of Eqs. (5.60) and (5.61), by evaluating the integral

$$-e^* = -e \int d^2 r [\rho(r) - \rho] , \tag{9.21}$$

where $\rho(r)$ is the density in the presence of the CF-quasiparticle and ρ is the density of the incompressible state. Figures 8.4 and 8.5 show the charge densities associated with CF-quasiparticles and CF-quasiholes at $\nu = 1/3, 2/5$, and $3/7$. The above integral has been performed and gives the expected fractional charge within numerical error.

Strictly speaking, this is not a derivation of e^*, but rather a check on the consistency of the trial wave function. This method does not make it obvious that the local charge is robust against perturbations that modify the wave functions. The counting arguments presented in the previous subsections are more satisfactory for demonstrating the sharp fractional quantization of the local charge.

The above proofs are generally applicable to the quasiparticle or quasihole of any FQHE state at $\nu = n/(2pn \pm 1)$. The quasihole at a Laughlin fraction is a simple vortex [369], the charge of which was deduced by Laughlin by mapping it into a two-dimensional one-component plasma (Section 12.4), and by Arovas, Schrieffer, and Wilczek [13] by the Berry phase calculation described above. These methods are special to the quasihole at $\nu = 1/(2p + 1)$.

As seen in all of the derivations, the fractional quantization of the local charge is critically dependent on the incompressibility of the state. For compressible states of composite fermions, the local charge of composite fermions is not a meaningful quantum number, although their intrinsic charge continues to be sharply defined.

9.3.6 A quasihole ≠ a vortex; a quasiparticle ≠ an antivortex

It is sometimes stated, erroneously, that quasiholes and quasiparticles are vortices and antivortices (and the neutral excitations are vortex–antivortex pairs). The misconception arises because the first excitation to be understood (Laughlin [369]) was the quasihole at a Laughlin fraction, which happens to be a vortex (Section 12.5). The Laughlin quasihole is an exception, however. Other quasiholes and quasiparticles are not simple vortices and antivortices but have a much more intricate structure. They are to be understood in terms of composite fermions.

9.3.7 Quasiparticle size: a crude estimate

A reliable determination of the size of a CF-quasiparticle or a CF-quasihole requires a detailed calculation of its density profile. A crude estimate can be obtained by assuming that it spreads itself out into a disk of uniform charge density. Consider an island of $\nu_{n+1} = (n + 1)/(2pn + 2p + 1)$ state in the background of the $\nu_n = n/(2pn + 1)$ state. When we add a composite fermion to the island, the excess charge in the island increases by $e^*/e = 1/(2pn + 1)$, increasing the size of the ν_{n+1} island. From the result that the excess charge per flux quantum is $\nu_{n+1} - \nu_n$, it follows that the island size increases by an area that encloses

$$\frac{(e^*/e)}{\nu_{n+1} - \nu_n} = 2pn + 2p + 1 \tag{9.22}$$

flux quanta. Identifying that increase with the size of a single CF-quasiparticle, πa^2, its radius is given as

$$a = \sqrt{2(2pn + 2p + 1)}\ \ell .\qquad(9.23)$$

The area enclosing $2pn + 2p + 1$ flux quanta contains $(2pn + 2p + 1)\nu_{n+1} = n + 1$ electrons, which gives a rough estimate for the number of electrons in the region containing a single CF-quasiparticle. Detailed calculation shows that the quasiparticle size obtained from this simple argument is an underestimate.

9.3.8 Charge addition to an inhomogeneous FQHE state

We saw (Fig. 9.1) that the addition of an electron to a uniform density compact state with a fixed area (for example, a FQHE state on a sphere) results in the creation of many CF-quasiparticles. For an inhomogeneous state, consisting of regions of several fillings and edges, the situation is somewhat more complicated, because charge can also accumulate at the boundaries. From the null vacuum perspective, adding a unit charge is equivalent to adding a composite fermion, which, however, causes nonlocal changes in the electronic state; the nonlocality arises because the addition of a composite fermion also involves the addition of $2p$ vortices. We illustrate this with two examples.

We first consider a $1/3$ droplet with an edge, denoted by $[N_0]$. (We use the notation $[N_0, N_1, \ldots]$ where N_j denotes the number of composite fermions in the jth Λ level.) Adding a composite fermion to the edge of the lowest Λ level simply expands the size of the $1/3$ state by three flux quanta to give $[N_0 + 1]$. If the new composite fermion is placed in the second Λ level in the interior of the sample, the resulting state $[N_0, 1]$ has an excess charge of $1/3$ in the interior and $2/3$ at the edge. The latter arises because the addition of two vortices pushes the edge out by an area enclosing two flux quanta. The net additional charge, of course, is one unit.

Let us next take an island of $2/5$ state surrounded by the $1/3$ state, denoted by $[N_0, N_1]$. Adding a composite fermion at the boundary of the island produces $[N_0, N_1 + 1]$, which amounts to an excess charge of $1/3$ at the edge of the $2/5$ island (increasing its area by $5\phi_0$), and of $2/3$ to the outer edge of the $1/3$ island. What happens when the new composite fermion is placed in the third Λ level, producing the state $[N_0, N_1, 1]$? Naively, there seems to be an ambiguity. From the perspective of the $1/3$ vacuum, the excess charge of the CF-quasiparticle is $1/3$, but from the perspective of the $2/5$ vacuum the charge excess is $1/5$. Which one is correct? Both are. The excess charge sticking out of the $2/5$ state is indeed $1/5$. However, the $2/5$ island itself also expands by an area that encloses two more flux quanta, which produces an additional excess charge of $2/15$ at the edge of the island (using results from the previous subsection), giving a total excess charge of $1/3$ relative to the $1/3$ vacuum. Of course, the larger $1/3$ island also expands by two flux quanta, equivalent to a charge of $2/3$, consistent with a unit increase in the total number of electrons relative to the null vacuum. As shown in Fig. 9.2, in the transition $[N_0, N_1] \rightarrow [N_0, N_1, 1]$, which is rather simply stated in the CF description, the additional charge distributes itself in three

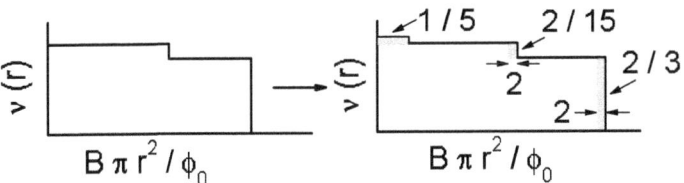

Fig. 9.2. The figure on the left shows a 2/5 island on top of a 1/3 state. This state is denoted as $[N_0, N_1]$, with the two entries being the number of composite fermions in the lowest two Λ levels. The figure on the right shows how the state changes when a composite fermion is added at the center of the 2/5 island (which corresponds to the state $[N_0, N_1, 1]$). The y-axis shows the local filling factor at a distance r from the center; the x-axis is the flux enclosed by the disk of radius r. The charge of the added composite fermion appears in three places, indicated by the shaded regions, with the charges in each region being 1/5, 2/15, and 2/3, as shown in the figure. Both the 2/5 and the 1/3 islands expand to enclose two more flux quanta. The charge density fluctuations at the edges as well as for the charge 1/5 quasiparticle have been suppressed for simplicity.

places. This example also demonstrates how the charge of a CF-quasiparticle depends on the choice of the vacuum state.

9.3.9 "Electron fractionalization"

Treating a uniform density FQHE state as the vacuum, it appears as though an added electron "fractionalizes" into several fractionally charged CF-quasiparticles, as shown in Figs. 9.1 and 9.2. The fractional quantum Hall liquid is the prototypical example of a "fractionalized" quantum fluid. Of course, the system is still made up of integrally charged objects (by shifting to the null vacuum, one can describe the physics perfectly well in terms of unit charge composite fermions plus a uniform neutralizing background); it is just that the charge density is nonuniform, and the nonuniformity manifests through several localized, fractionally charged, charge deficiencies or excesses relative to the uniform density ground state. While the fractional charge in the FQHE is not as revolutionary as a real breaking up of an electron would be, it does constitute an example where nature produces a sharply defined, fractional value for excess charge. (An individual atom inside a molecule also has a fraction of an electron charge on it, but the charge is not sharply quantized.)

9.4 Quantized screening

The binding of vortices to an electron creates a correlation hole around it, and therefore can be viewed as a screening of the electron (Goldhaber and Jain [201]). The composite fermion can thus be thought of as a screened electron. The word screening, however, is used here in a somewhat nonstandard sense. Many familiar examples of screening, such as that occurring in a conventional insulating medium, are perturbative, and often investigated theoretically in perturbative treatments. The screening of an electron into a composite fermion is a fundamentally nonperturbative effect. It occurs through the binding of *exactly* an even number of quantized vortices to each electron. Being quantized, vortices cannot be attached or removed adiabatically (akin to our inability to untie a knot while holding the ends of a rope fixed). That is the reason why the local charge of the CF-quasiparticle, the sum of the electron charge and the charge of the correlation hole, is precisely quantized, the value of which depends only on the filling factor, but not on details of interaction or wave function.

Quantization of screening is not unprecedented. The nonperturbative screening of a charge in a superconductor is also quantized, producing a local charge that is exactly zero. The quantization of screening in the composite fermion fluid, however, is the first example in physics where sharp, nonzero values for the local charge are produced.

9.5 Fractionally quantized Hall resistance

We saw in Chapter 4 that gaps at integral fillings combined with Anderson localization result in quantized Hall resistance plateaus at $R_H = h/ne^2$. The gap at $\nu = f$ in a pure system similarly produces quantized plateaus at $R_H = h/fe^2$ when a weak disorder potential is introduced. (A strong disorder will destroy the gap and the FQHE.) The explanations of Chapter 4 carry over to the FQHE problem almost verbatim, with "electron" replaced by "composite fermion" and "Landau level" by "Λ level." We go over the basic ideas briefly.

9.5.1 Plateaus as a consequence of incompressibility

Let us assume that we know, from other considerations, that the state at $\nu = f$ has a gap in the absence of disorder. We consider Fig. 4.6 in which ideal leads are attached on both ends of the disordered sample. The chemical potential is adjusted to lie in the bulk gap of the ideal regions, so the filling factor in these regions is $\nu = f$. The Hall resistance in the ideal regions is

$$R_H = \frac{h}{fe^2} \; . \tag{9.24}$$

In the disordered part the filling factor is different from $\nu = f$ due to the presence of disorder-induced localized defects. (We do not need to know that the defects are fractionally charged quasiparticles or quasiholes.) We assume that: (i) all defects are localized; (ii) the edge states are extended along the entire length of the sample; and (iii) backscattering is absent. These assumptions guarantee that the current is the same everywhere, as also is the Hall voltage. Consequently, the Hall resistance has the quantized value given in Eq. (9.24) also for the disordered region.

This explanation demonstrates that the fractional Hall quantization at $R_H = h/fe^2$ is a direct consequence of incompressibility at the fractional filling $\nu = f$, combined with disorder-induced Anderson localization. The only role of composite fermions is to produce incompressibility at a fractional filling.

No use is made above of the fractional local charge of a quasiparticle or quasihole. The quantized Hall resistance is a property of the incompressible "condensate"; the quasiparticles are localized and do not contribute to transport.

9.5.2 Laughlin's explanation

Referring to Fig. 4.12, let us extend the ideal region at the inner edge all the way to the origin. The azimuthal current is given by

$$I = c\frac{dU(\phi_t)}{d\phi_t} = c\frac{\Delta U}{\phi_0}, \qquad (9.25)$$

where ϕ_t is an additional test point flux piercing the origin, U is the energy of the system, and ΔU is the change in the energy for $\Delta\phi_t = \phi_0$.

It was shown in Section 9.3.4 that piercing an ideal FQHE state adiabatically by one flux quantum produces an object of charge $e_\phi = fe$, where f is the filling factor of the *ideal* region. This charge comes in from the edge of the system. Although now the ideal region is connected to a disordered sample, the extended states (that go around the origin) are everywhere gapped, and therefore the charge e_ϕ must still come from the outer edge. In other words, a charge e_ϕ has moved from the outer edge to the inner edge as the flux is varied adiabatically from 0 to ϕ_0, implying $\Delta U = e_\phi V_H$. Thus,

$$R_H = \frac{V_H}{I} = \frac{h}{e_\phi e} = \frac{h}{fe^2}. \qquad (9.26)$$

This argument again did not require anything beyond the incompressibility of the state at $v = f$. In particular, we did not need to know that for $v = n/(2pn \pm 1)$, n composite fermions, one in each Λ level, are transported from the inner edge to the outer edge under adiabatic insertion of one flux quantum, each carrying its local charge $e^* = e/(2pn \pm 1)$ with it.

9.5.3 Exactness of the Hall quantization

Why is the Hall resistance so accurately quantized? Its value is tied to the value of the filling factor where a gap appears in the absence of disorder. That, in turn, is precisely given by $f = n/(2pn \pm 1)$, with no correction, because the right hand side contains whole numbers, and, therefore, is not susceptible to small perturbations. The topological quantization of the vorticity of composite fermions $(2p)$ lies at the root of the exactness of the FQHE.

9.5.4 Anderson localization of composite fermions

Anderson localization of a single electron was considered in Section 4.6. The problem becomes notoriously complicated when both disorder and interaction are present, especially because interactions play a nonperturbative role in the FQHE problem. Some progress can be made within the CF theory. To the extent that composite fermions may be treated as noninteracting, the problem is identical to that of localization of free electrons, discussed in Section 4.6. That view [276] is consistent with experimental studies of scaling behavior in the FQHE regime. Engel *et al.* [142] ascertain that the width of the transition region for the $1/3 \to 2/5$ vanishes as $\Delta B^* \sim T^{\kappa^*}$ with $\kappa^* \approx 0.42$; Koch *et al.* [348] measure an exponent in the range $\kappa^* = 0.56\text{--}0.77$ for the $1 \to 2/3$ transition (which is the $\nu^* = 0$ to $\nu^* = 1$ transition of composite fermions made of holes in the lowest Landau level [277]). These exponents are of roughly the same order as those for the IQHE transitions (Section 4.6), and also nonuniversal, presumably as a result of the residual interaction between composite fermions.

9.6 Evidence for fractional local charge

The appearance of e_ϕ in Eq. (9.26) in Laughlin's explanation of Hall quantization might suggest that the fractional Hall plateau itself is a measurement of, or proof for, the fractional quantization of local charge. But the fractionally charged quasiparticles, being *localized* on the quantum Hall plateau, do not contribute to transport. The Hall current is carried by the background incompressible state containing no quasiparticles or quasiholes. Both the fractional local charge and the fractional quantization of the Hall resistance are consequences of incompressibility of the pure state at a fractional filling. The measurement of the fractional local charge of the quasiparticles must rely on the movement of these quasiparticles, i.e., on *deviations* from the perfect quantum Hall effect. A number of experiments have reported evidence for fractional charge [114, 172, 209, 432, 551, 565, 590].

Simmons *et al.* [590] observe quasiperiodic resistance fluctuations, near the resistance minima, in narrow Hall bars as a function of the magnetic field. These are naturally interpreted as arising from resonant tunneling through quasi-bound states localized on equipotential contours around potential hills or valleys along the Hall bar [270], created by unintentional impurity potentials. The observed quasi-period ΔB in the $\nu = 1/3$ minimum is approximately three times larger than that in the $\nu = 1, 2, 3, 4$ minima, which is taken as an indication of fractional charge. A detailed understanding of the experimental result is lacking. A ϕ_0 period follows generally from the Byers–Yang argument [46], which shows that, independent of the correlations in the electronic state, the ordering of various eigenstates does not change due to the addition of an integral number of flux quanta through a region devoid of electrons. Kivelson [343] argues that the period should be $\phi_0^* = hc/e^*$ for a fixed particle number (as opposed to a period of $\phi_0 = hc/e$ for a fixed chemical potential); the experiment can possibly be in that regime due to energetic considerations involving Coulomb blockade.

Goldman and Su [209] study resonant tunneling through a quantum antidot (a potential hill) etched into the 2D layer, both as a function of B, which alters the filling factor at a fixed density, and as a function of a backgate voltage, which changes the density at a fixed B through a capacitive coupling to the 2D electron system. The area A of the closed loop around the antidot is determined from the quasi-period ΔB, assuming that the two successive resonant tunneling peaks are separated by $A\Delta B = \phi_0$. From the knowledge of the dependence of the average charge density on the backgate voltage, it is surmised that the change in the charge in an area A between two resonant tunneling peaks (as a function of the backgate voltage) is $e^* \approx e/3$ for filling factor $\nu = 1/3$.

Franklin *et al.* [172] also observe quasi-periodic peaks associated with resonant tunneling through a quantum antidot. They conclude that the periods ΔB are the same at $\nu = 1/3$ and in the integral quantum Hall regime. This experiment, as the one in the previous paragraph, provides a demonstration of Byers and Yang theorem [46] in the FQHE regime.

Saminadayar *et al.* [565] and de Picciotto *et al.* [114] measure shot noise in the current due to tunneling from one edge to another in a narrow part of the sample (i.e., a quantum point contact), from which they deduce the charge of the tunneling objects. "Shot noise" refers to time-dependent fluctuations in the current arising from the property that the current flow is not continuous but occurs through discrete pulses in time, due to the discreteness of the particles carrying the current. (This is analogous to the noise created by raindrops falling on a tin roof.) The shot noise is conveniently characterized by the frequency-dependent noise power, $S(\omega)$, defined as $S = 2 \int dt \langle \delta I(0) \delta I(t) \rangle \cos(\omega t)$, which is the Fourier transform of the correlator of the time-dependent fluctuations in the current at a given voltage and temperature. When the tunneling of particles is completely random, the noise spectral density is given by $S = 2qI$, q being the charge of the object that is tunneling. Any correlations between successive tunnelings turn it into an inequality $S \leq 2qI$. In devices like tunnel junctions, the maximum value of the noise power indeed occurs when $q = e$.

For the FQHE experiments, in the limit of $k_B T / qV \rightarrow 0$ and weak coupling between edges (so the tunneling probability is small), one expects $S = 2qI$ for uncorrelated tunnelings of quasiparticles (i.e., composite fermions). More general expressions for S including the T and V dependences have been derived (Chamon, Freed, and Wen [59]; Fendley, Ludwig, and Saleur [153]; Kane and Fisher [318,319]) using Wen's chiral Tomonaga–Luttinger (TL) liquid model for the edge states [673] (Chapter 14). The value of effective shot-noise charge thus determined in the experiments of Saminadayar *et al.* [565] and de Picciotto *et al.* [114] is consistent with $q = e/3$ at $\nu = 1/3$. The local charge of quasiparticles is not quantized at the edge, due to the presence of massless excitations; it may be argued, however, that tunneling through the bulk filters out a sharp fractional charge.

Subsequent shot noise experiments have produced somewhat puzzling results. Chung *et al.* [80] find that at $\nu = 2/5$ and $3/7$, the shot-noise charge has the expected value of approximately 1/5 or 1/7 at certain filling-factor-dependent "high" temperatures (> 50 mK 2/5), but changes continuously to $q = \nu e$ at low temperatures (~ 9 mK). In the same experiment, the transmission through the quantum point contact is seen to increase with decreasing temperature for certain parameters, contrary to the expectation based on the chiral TL liquid model [319], which predicts the transmission to be completely suppressed at $T = 0$. Further observations contradicting the predictions of the chiral TL liquid model have been reported by Comforti *et al.* [93] and Chung *et al.* [81] for situations when the beam of "edge quasiparticles" impinging on the quantum point contact is very dilute, as can be achieved by passing the current through an additional weak backscatterer. For example, the former experiment [93] finds that in this situation a 1/3 charge can tunnel through a nearly opaque barrier. These observations are not understood. Wen's TL model used in the analysis of the shot-noise charge has been questioned by experiments probing tunneling from an ordinary Landau Fermi liquid into the FQHE edge (Chapter 14). These experiments determine the TL exponent (defined in Chapter 14), which governs the long-distance, low-energy behavior of various correlation functions, to be nonuniversal and filling-factor dependent. This may have implications for the shot-noise experiments, because a single parameter in the TL model determines both the TL exponent and the shot-noise charge. A measurement of the filling-factor dependence of the shot-noise charge across the 1/3 plateau should further confirm the interpretation of the experiments.

Martin *et al.* [432] investigate the microscopic nature of localization in the FQHE regime using a scanning single-electron transistor, which allows them to identify the charging of a localized state as a function of the overall density of the system. They observe approximately three times as many charging lines at $\nu = 1/3$ and $2/3$ as at $\nu = 1$ and 2, which is consistent with localization of fractionally charged quasiparticles.

9.7 Observations of the fermionic statistics of composite fermions

How does one measure the particle statistics? The proof of statistics stems from the inability to explain experimental results without recourse to the concept of particle statistics, both in condensed matter physics and in quantum chemistry. In the former, particle statistics announces itself loudly through the formation of collective quantum states, for example a Bose–Einstein condensate or a Fermi sea, the properties of which are impossible to understand without appealing to statistics. In the latter, the statistics plays a crucial role in quantitative predictions of the energy levels of atoms and molecules. An explanation of the energy levels of the H_2 molecule or the He atom would not be possible if we did not know that electrons are fermions, and the extremely precise agreement between theory and experiment on atomic and molecular spectra is a "quantum chemistry" proof of the fermionic statistics of electrons. It is only satisfying that a concept as novel as statistics should have decisive and dramatic manifestations.

The fermionic exchange statistics of composite fermions is firmly established through a variety of facts, both of quantum chemistry and condensed matter varieties. Listing the most prominent of these is worthwhile:

- The low-energy spectra exhibit a one-to-one correspondence with the energy levels of weakly interacting fermions. The description in terms of fermions neither misses any states at low energies nor predicts spurious states.
- The eigenenergies and eigenfunctions are predicted accurately by the CF theory.
- The FQHE is explained as the IQHE of composite fermions. The appearance of the sequences of fractions given in Eq. (5.20) is a direct consequence of the fermionic nature of composite fermions.

- The actual incompressible states are well described, quantitatively, as n filled Λ levels of composite fermions.
- The excited states are explained, in great detail, as excitations of composite fermions across Λ levels.
- The state at $\nu = 1/2$ is a Fermi sea of composite fermions.
- The FQHE state at $\nu = 5/2$ is currently believed to be a BCS-like paired state of composite fermions.

We note that the exchange statistics of neither ordinary particles nor composite fermions bears any relation to their charge. Should the reader be wondering why we bring up this rather obvious point at all, the reason is that the same is not true for the braiding statistics. The braiding statistics of the CF-quasiparticles is so intimately connected with their local charge that disentangling the two in a practical measurement is nontrivial.

9.8 Leinaas–Myrheim–Wilczek braiding statistics

Indistinguishability of particles imposes a fundamental constraint on the space of allowed quantum mechanical wave functions. They must be either even or odd under an exchange of two indistinguishable particles, giving bosonic or fermionic statistics. This property is independent of the Hamiltonian describing the particles, and the relevant symmetry group is the permutation group.

Leinaas and Myrheim [386] and Wilczek [677] (also see Goldin, Menikoff, and Sharp [202, 203]) introduced the fascinating theoretical concept of "braiding statistics" in which particle exchange is viewed as a continuous process. The corresponding symmetry group is Artin's braid group rather than the permutation group, and gives new possible structure in two space dimensions (but not in higher dimensions where the braidings can be disentangled). The Leinaas–Myrheim–Wilczek (LMW) braiding statistics is distinct from the ordinary exchange statistics in that it is an emergent consequence of interactions between the constituent particles, and it is defined through a dynamical Berry phase associated with the braiding of particles around one another. The exchange statistics of CF-quasiparticles is fermionic, but they are seen below to have a fractional braiding statistics.

Although it is not always specified in the literature which statistics – exchange or braiding – is being discussed, that should be clear from the context. Fractional statistics always refers to the braiding statistics. Particles obeying fractional braiding statistics are called "anyons" ("any" statistics).

9.8.1 General concept: ideal anyons

To illustrate the basic idea, we consider two indistinguishable particles in two space dimensions. Switching to the center of mass and relative coordinates, and denoting the latter by (r, θ), the part of the wave function depending on the azimuthal coordinate θ is

$$e^{im\theta} . \tag{9.27}$$

The parameter m must be an integer to ensure that the wave function is single valued. Furthermore, because an exchange of particles corresponds to $\theta \to \theta + \pi$, odd (even) integral values of m imply an antisymmetric (symmetric) wave function under exchange. The parameter m is the eigenvalue of the canonical angular momentum operator $L_\theta = -i\partial/\partial\theta$. Thus we see that the relative angular momentum quantum number is an an even integer for bosons, and an odd integer for fermions. (For simplicity, we assume that the particles are spinless; that leads to no inconsistency in nonrelativistic quantum mechanics.)

Now consider two particles for which the θ dependent part of the wave function is

$$e^{i(m+\alpha)\theta} , \tag{9.28}$$

where α is an arbitrary number and m is taken to be an even integer. An exchange of particles now produces a phase factor

$$e^{i\alpha\pi} , \tag{9.29}$$

which is neither $+1$ nor -1 but a complex quantity. Such particles are said to obey fractional statistics, i.e., are anyons. The relative angular momentum of anyons contains a fractional piece (α), although the different allowed values continue to be spaced by two units, as for bosons or fermions.

Consider a single particle on a ring. Its wave function is given by $e^{im\theta}$, where m is an integer to ensure continuity under a full traversal. Wave functions of Eq. (9.28) are solutions with "twisted" boundary conditions; they can be obtained by inserting a magnetic flux of appropriate strength through the ring. Analogously, anyons can be modeled as bosons (or fermions) each carrying a flux of an appropriate amount with it. To see how it works mathematically, consider the Schrödinger equation

$$H\Psi = E\Psi, \tag{9.30}$$

$$H = \sum_i \frac{p_i^2}{2m_b}, \tag{9.31}$$

for noninteracting anyons of mass m_b obeying fractional braiding statistics α. This can be seen to be equivalent to the bosonic problem:

$$H_B\Psi_B = E\Psi_B, \tag{9.32}$$

$$\Psi_B = e^{-i\alpha \sum_{j<k} \theta_{jk}} \Psi = \prod_{j<k} \left[\frac{z_j - z_k}{|z_j - z_k|} \right]^\alpha \Psi, \tag{9.33}$$

$$H_B = \sum_i \frac{1}{2m_b} \left(p_i + \frac{e}{c} a_i \right)^2, \tag{9.34}$$

$$a_i = \frac{\alpha\phi_0}{2\pi} \sum_j{}' \nabla_i \theta_{ij}, \tag{9.35}$$

$$\nabla_i \times \mathbf{a}_i = \alpha\phi_0 \sum_j {}' \delta^{(2)}(\mathbf{r}_j), \tag{9.36}$$

$$\theta_{ij} = \arg(\bar{z}_i - \bar{z}_j) = -i \ln \frac{\bar{z}_i - \bar{z}_j}{|\bar{z}_i - \bar{z}_j|}, \tag{9.37}$$

$$\theta_{ji} = \pi + \theta_{ij}. \tag{9.38}$$

Here, Ψ_B is a wave function for bosons, i.e., is single valued and symmetric under particle exchange. The wave function Ψ,

$$\Psi = e^{+i\alpha \sum_{j<k} \theta_{jk}} \Psi_B, \tag{9.39}$$

satisfies the desired statistics under the exchange of particles. Each boson sees a flux tube of strength $\alpha\phi_0$ on every other boson due to the vector potential $\mathbf{a}(\mathbf{r})$. The prime on the sum denotes the condition $j \neq i$. The angle θ_{ij} is the angle that the vector $\mathbf{r}_i - \mathbf{r}_j$ makes relative to the x-axis, and is only defined modulo 2π. Recall that $z = x - iy = re^{-i\theta}$. We have thus shown that the problem of anyons, with nonanalytic, non-single-valued wave functions, can be recast into a more familiar language of bosons, but at the cost of introducing a singular vector potential that ties to each boson a point flux tube of appropriate strength. (This is an example of how a singular gauge transformation can shift statistics from wave function to the Hamiltonian.) These flux tubes have no mass, no independent dynamics, and do not refer to a physical magnetic field that requires currents and stores energy. Their *raison d'être* is to impose appropriate boundary conditions.

We close this subsection with the following remarks.

- Equation (9.28) implies that a full circuit of one anyon around another produces a phase factor $e^{i\alpha}$ *independent* of the path. Fractional braiding statistics thus entails a topological distinction between the loops in which a particle goes around another and those in which it does not (paths labeled A and B in Fig. 9.3). It can be consistently defined only in two dimensions, because, in

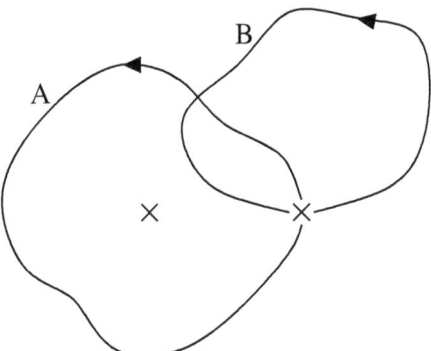

Fig. 9.3. Fractional braiding statistics implies a topological difference between classes of closed loops that enclose another particle and those that do not (marked A and B, respectively). Each cross represents a particle.

higher dimensions, one kind of loop can be continuously deformed into the other kind by peeling it off the plane of the page containing the two particles. Even in two dimensions, removal of the coincident points from the configuration space is necessary (for each particle the surface then has a complex topology with punctures at the positions of the other particles) to ensure that the loops enclosing different numbers of particles cannot be deformed into one another, i.e., are topologically distinct. This can be achieved by adding to the Hamiltonian an infinitely strong short-range repulsive interaction.

- Just because fractional braiding statistics can be consistently defined theoretically does not mean that it exists in the real world. It is an experimental fact that no elementary particles in nature have fractional braiding statistics. Why, then, do we talk of fractional braiding statistics at all? That would indeed be a valid question from an elementary particle physicist's perspective. No principle precludes, however, the possibility that certain *emergent* quasiparticles of a condensed matter system might obey fractional braiding statistics. It would obviously take a highly nontrivial state of matter to produce anyons, and the fractional quantum Hall state is currently the only viable candidate for such physics.

- A fundamental operational distinction between ordinary particles and anyons is that while we know the solution for noninteracting bosons or fermions, the problem of noninteracting anyons is nontrivial, analytically intractable, and remains unsolved. The bosonic reformulation of noninteracting anyons does not lend itself to an exact solution. The Hamiltonian H_B appears to be a sum of single particle terms due to bad notation, but it describes an interacting problem, because the vector potential depends on the coordinates of *all* particles. An expansion of the Hamiltonian

$$2m_b H_B = \sum_i \left[p_i^2 + \left(\frac{e}{c}a_i\right)^2 + \left(\frac{e}{c}\right) p_i \cdot a_i + \left(\frac{e}{c}\right) a_i \cdot p_i \right] \qquad (9.40)$$

shows that the Hamiltonian contains two- and three-body interactions. The flux attachment thus produces complex interactions that cannot be treated perturbatively. Because they cannot be considered weakly interacting, anyons are not true particles in the sense defined in Section 5.7.

- Unlike the ordinary statistics, which imposes a kinematic constraint on allowed wave functions, fractional braiding statistics is an intrinsically dynamical concept. That is perhaps seen most clearly when anyons are mapped into bosons; the fractional statistics is then equivalent to a long-range interaction mediated by the gauge potential.

- The many-body theory of bosons or fermions rests crucially on the fact that the many-particle wave functions can be built from single particle wave functions. That is not true for anyons.

- Another manner in which fractional braiding statistics differs from ordinary exchange statistics is that even distinguishable particles can have a well-defined relative braiding statistics. We see an example below.

9.8.2 FQHE quasiparticles: nonideal anyons

In view of Section 9.7, the reader may be wondering why we discuss fractional braiding statistics. Everything to date is well explained in terms of composite fermions, without any mention of fractional braiding statistics.

As a motivation, let us now consider the state in which an integral number of Λ levels are full, and the lowest unoccupied Λ level contains two additional composite fermions, i.e., CF-quasiparticles. Each CF-quasiparticle has an excess of precisely $1/(2pn+1)$ electrons associated with it. We know how to describe this state in great detail; we learned in Chapter 6 how to write explicit wave functions for states containing a single or many CF-quasiparticles.

Let us now switch our vantage point from the null vacuum to the FQHE vacuum. In other words, we only wish to keep track of the *deviations* from the uniform incompressible FQHE state. We then have only *two* quasiparticles, each with a fractional local charge. This seems a reasonable thing to do. We have gone from a large number of composite fermions to two quasiparticles, and can, therefore, hope for simplification.

We cannot simply forget about the composite fermions in the FQHE vacuum, however. They must be included indirectly through the effect they have on the two quasiparticles. Composite fermions in the filled Λ levels mediate complicated interactions between the two quasiparticles due to the strongly coupled nature of the state; after all, each composite fermion sees $2p$ vortices on all other composite fermions, independent of which Λ level they inhabit, and of whether or not they belong to what we have decided to call vacuum. We now demonstrate that the *long-distance* character of this induced interaction is precisely equivalent to fractional braiding statistics.

In our preceding discussion of braiding statistics, we assumed charge neutral particles. CF-quasiparticles are electrically charged objects in a magnetic field. As a result, a closed loop has two contributions to the phase: the Aharonov–Bohm phase (which occurs even when the loop does not enclose another quasiparticle), and the braiding phase (due to the enclosed quasiparticles). The latter is deduced by subtracting the former from the total phase (Fig 9.4). The braiding statistics is thus given by the *change* in the Berry phase when a new CF-quasiparticle is inserted inside a closed loop.

The appearance of fractional braiding statistics in the fractional quantum Hall effect was first suspected by Halperin [230], and demonstrated in a microscopic calculation for the

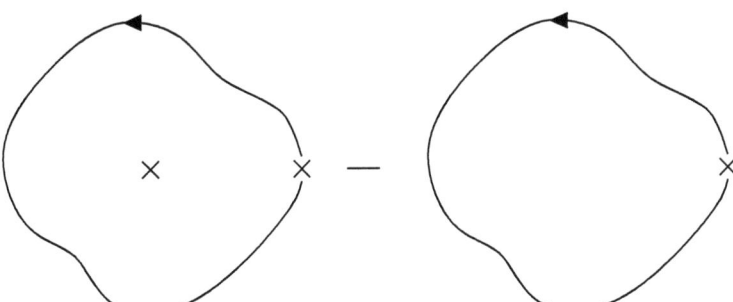

Fig. 9.4. The phase from braiding statistics is given by the change in the phase associated with a closed loop caused by the addition of a quasiparticle inside it.

quasiholes of $v = 1/m$ by Arovas, Schrieffer, and Wilczek [13] (ASW). It can be deduced as a general consequence of incompressibility at a fractional filling using the same kind of arguments as those leading to fractional charge [629].

The CF theory provides a simple and general derivation for fractional braiding statistics (Goldhaber and Jain [201]). Here, we go back to the "full" description, i.e., to the null vacuum, from where we see *all* composite fermions. The Berry phase associated with a closed loop of a CF-quasiparticle (or any other composite fermion for that matter) encircling an area A is given by Eq. (5.75), reproduced here for convenience:

$$\Phi^* = -2\pi \left(\frac{BA}{\phi_0} - 2pN_{enc} \right) . \tag{9.41}$$

The two terms on the right hand side are due to the two constituents of the composite fermion, charge $-e$ electron and $2p$ vortices, going around a loop enclosing BA flux and N_{enc} particles. The *average* change in the phase due to the insertion of another CF-quasiparticle inside the loop is given by

$$\Delta\Phi^* = 2\pi \, 2p\Delta\langle N_{enc}\rangle = 2\pi \cdot 2p \cdot \frac{e^*}{e} = 2\pi \frac{2p}{2pn+1} . \tag{9.42}$$

We have used that the change in the average number of electrons due to an extra CF-quasiparticle is $\Delta\langle N_{enc}\rangle = e^*/e = 1/(2pn+1)$, which is valid under the assumption that the two CF-quasiparticles do not overlap at any point of the closed trajectory; otherwise a part of the interior CF-quasiparticle would spill out of the loop. (The change in the Berry phase can also be evaluated when the two quasiparticles are overlapping, from the knowledge of the microscopic wave functions, but only with substantially greater effort.) With $\Delta\Phi^* = 2\pi\alpha$, the braiding statistics parameter is obtained as the product of the local charge and the number of vortices bound to a CF-quasiparticle:

$$\alpha_{qp} = 2p \cdot \frac{e^*}{e} = \frac{2p}{2pn+1} . \tag{9.43}$$

This completes the derivation of the braiding statistics of the CF-quasiparticles. It is a straightforward corollary of Eq. (5.75) or Eq. (9.41), which is also the equation that embodies the physics of the effective magnetic field. Essentially, even though each electron carries an even number of vortices, we have changed the charge inside the loop by a fraction of an electron charge, which amounts to the addition of a fractional number of vortices, thus producing a fractional braiding statistics. The fractional braiding statistics is a direct descendant of the fractional charge.

The long-distance limit of fractional braiding statistics of the excitations of the FQHE states at $v = n/(2pn \pm 1)$ can also be derived in the CFCS framework. Lopez and Fradkin [404] obtain it by considering fluctuations in the gauge field around the mean field at the Gaussian level, which, they argue, gives the exact long-distance behavior.

How about the LMW braiding statistics of the CF-quasihole? Let us begin by noting that the relative braiding statistics of the CF-quasiparticle and the CF-quasihole is given by

(reader, please derive this)

$$\alpha_{qh-qp} = -\frac{2p}{2pn + 1} .$$ (9.44)

The braiding statistics of a CF-exciton requires taking a quasiparticle quasihole pair going around another, which produces four terms:

$$\alpha_{qp} + 2\alpha_{qh-qp} + \alpha_{qh} = 0 .$$ (9.45)

The zero on the right hand side follows because the CF-exciton is a boson. This gives

$$\alpha_{qh} = \frac{2p}{2pn + 1} .$$ (9.46)

(Various braiding statistics parameters α are only defined mod 2.)

Equations (9.43) and (9.46) have been confirmed for $\nu = 1/3$ and $2/5$ in detailed microscopic calculations (Jeon, Graham, and Jain [293,294]; Kjønsberg and Myrheim [345]; Kjønsberg and Leinaas [346]), with the Berry phase integral evaluated by Monte Carlo for the appropriate wave function of composite fermions. Some results are shown in Figs. 9.5 and 9.6. The answer is independent of the path, provided CF-quasiparticles or CF-quasiholes are sufficiently far separated (typically more than 10 magnetic lengths), and also of whether the projected or unprojected wave function is used. The values for the braiding statistics parameter quoted in the literature (for example, Halperin [230]; Su [629]) differ from that in Eq. (9.43) by an integer, and sometimes by a sign; some of these differences can be traced to different conventions used in the definition of the braiding statistics.

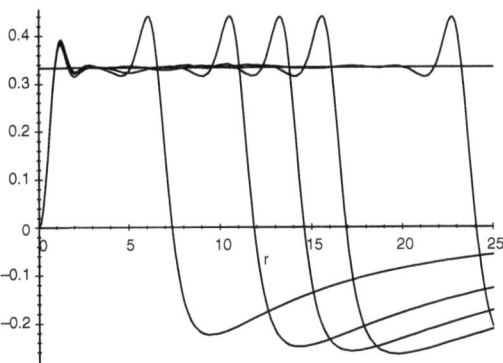

Fig. 9.5. The braiding statistics parameter (y-axis) for the quasihole at $\nu = 1/3$, determined from a Monte Carlo evaluation of the Berry phase for systems with up to 200 electrons. The parameter r is defined as $r = d/(2\sqrt{2})$, where d is the distance between the centers of the quasiholes in units of the magnetic length. The curves are, from left to right, for 20, 50, 75, 100, and 200 electrons, and the horizontal line marks $1/3$. Source: H. Kjønsberg and J. Myrheim, *Int. J. Mod. Phys. A* **14**, 537 (1999). (Reprinted with permission.)

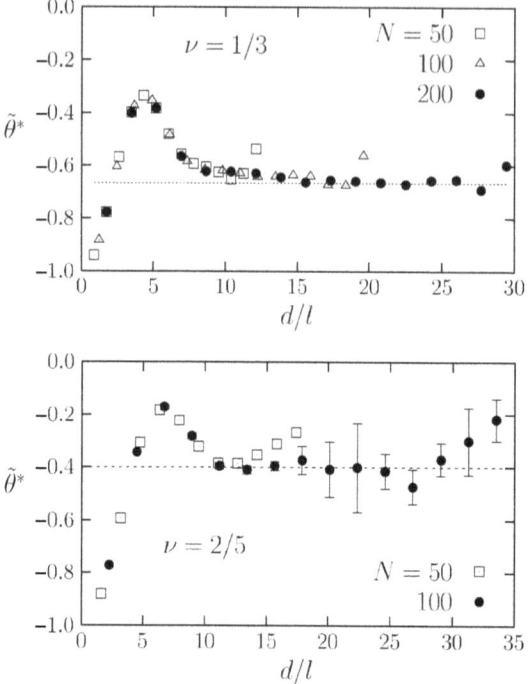

Fig. 9.6. The braiding statistics parameter (denoted $\tilde{\theta}^*$ in this figure) for the quasiparticles at $\nu = 1/3$ (upper panel) and $\nu = 2/5$ (lower panel) as a function of their separation d, calculated using the CF theory. N is the total number of composite fermions, and l is the magnetic length. The error bars from Monte Carlo sampling are not shown explicitly when they are smaller than the symbol size. The deviation at the largest d/ℓ for each N is due to proximity to the edge. The projected (unprojected) wave function is used for the CF-quasiparticles at $\nu = 1/3$ ($\nu = 2/5$). (The computation at $\nu = 2/5$ becomes prohibitively expensive if the projected wave function is used.) Source: G. S. Jeon, K. L. Graham, J. K. Jain, *Phys. Rev. Lett.* **91**, 036801 (2003). (Reprinted with permission.)

Two points need clarification. For the Laughlin fractions, Arovas, Schrieffer, and Wilczek considered the wave function

$$\Psi^{ASW}_{\eta,\eta'} = \prod_j (z_j - \eta)(z_j - \eta') \Psi_{\frac{1}{2p+1}} \tag{9.47}$$

for two quasiholes at η and η'. The quasiholes are vortices at these fractions. As shown in Section 5.11.4, the Berry phase associated with a closed loop of a vortex is $\Phi^* = 2\pi N_e$, with N_e being the number of electrons inside the loop. The difference in the Berry phase with or without the other quasihole is $\Delta\Phi^* = 2\pi \Delta N_e = -2\pi\nu$, equating which to $2\pi\alpha^{ASW}_{qh}$ gives the braiding statistics of the quasiholes at $\nu = 1/(2p+1)$:

$$\alpha^{ASW}_{qh} = -\frac{1}{2p+1}. \tag{9.48}$$

Equation (9.46), on the other hand, gives $\alpha_{qh} = 2p/(2p+1)$ for the quasiholes at $\nu = 1/(2p+1)$. The difference is merely that of convention. In the CF theory, the wave function that naturally appears is

$$\Psi_{\eta,\eta'} = (\eta - \eta') \prod_j (z_j - \eta)(z_j - \eta') \Psi_{\frac{1}{2p+1}} , \tag{9.49}$$

because the wave function for two holes at $\nu = 1$ is $(\eta - \eta') \prod_j (z_j - \eta)(z_j - \eta') \Psi_1$ – that is what we get by applying the destruction operators $\hat{\psi}(\eta)\hat{\psi}(\eta')$ to the $\nu = 1$ state, which amounts to replacing two of the particle coordinates in Φ_1 by η and η'. The reader can verify by an explicit ASW calculation that $\Psi_{\eta,\eta'}$ gives $\alpha_{qh} = -1/(2p+1) + 1$, the last term originating from the factor $(\eta - \eta')$, making the result consistent with Eq. (9.46).

Second, let us now check the result for consistency with the notion that $(2pn + 1)$ CF-quasiparticles (CF-quasiholes) make one electron (hole). The braiding statistics of a bound complex of $(2pn + 1)$ CF-quasiholes is given by

$$\alpha_h = (2pn + 1)^2 \alpha_{qh} = 2p(2pn + 1) \bmod 2 = 0 , \tag{9.50}$$

which is the result of $(2pn + 1)^2$ equal terms originating from $(2pn + 1)$ CF-quasiholes going around as many CF-quasiholes. The same is true of the braiding statistics of a bound complex of $(2pn + 1)$ CF-quasiparticles. This appears to contradict the expectation that the fermionic nature of the holes or electrons ought to produce an odd integer. Exercises 9.7 and 9.9 show, however, that the Berry phase calculation for two electrons or two holes in the lowest Landau level also produces $\alpha_h = \alpha_e = 0$. (In fact, the result was already derived in the previous paragraph for two holes.) The origin of the apparent inconsistency is explained in Exercise 9.7.

Several comments are in order.

- In essence: With the null vacuum as our reference, the state is described in terms of charge $-e$ fermions (in an effective magnetic field). On the other hand, relative to an incompressible FQHE state, the qualitative physics may be described, under appropriate conditions, in terms of a smaller number of quasiparticles, which are to be assigned fractional local charge and fractional braiding statistics.
- The braiding statistics of the FQHE quasiparticles is well defined only when their separation is large compared with their size. When they begin to overlap, the phases become path dependent and the topological notion of braiding statistics ceases to be meaningful. The FQHE quasiparticles are not ideal but "fuzzy anyons" because of their nonzero size, as depicted in Fig. 9.7. The statistical flux tube associated with them is not point-like but spread out.
- The composite fermion and anyon descriptions are not equivalent, the former being more general. (i) While the CF theory is valid for the low-energy physics at all fillings in the FQHE regime, the anyon description is applicable only in regions close to the fillings $\nu = n/(2pn \pm 1)$ where the quasiparticle density is dilute, i.e., when quasiparticles are far apart. (ii) Even when valid, the anyon language does not allow quantitative calculation. (iii) Anyons can be derived from composite fermions, but the reverse is not true.

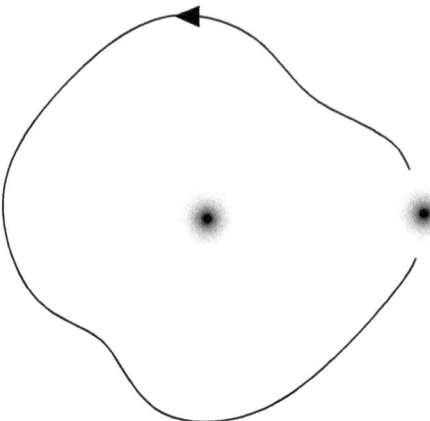

Fig. 9.7. The FQHE quasiparticles are not point objects but have a size of approximately $\geq 10\ell$ each.

- Fractional braiding statistics is not an additional concept but a property of composite fermions that follows from Eq. (9.41). It provides a natural interpretation for how the effective magnetic field changes upon a localized $O(1)$ variation in the particle density. A measurement of fractional braiding statistics will provide a more microscopic test of Eq. (9.41) than the measurement of the effective magnetic field, which is a more robust $O(N)$ quantity.
- The FQHE system is uncommon in one respect. It often pays to lighten the baggage that we must carry by working with only a few quasiparticles near the Fermi energy, because that usually results in a simplification. That is the Landau Fermi liquid theory philosophy. In the FQHE, however, the problem becomes more intractable when we shift our reference from the null vacuum (where we have almost free composite fermions) to the FQHE vacuum (where we end up with anyons). Nonetheless, the fractional braiding statistics is a remarkable manifestation of the correlations in the state, and we may ask how it can be measured.

9.8.3 Experimental situation

So far, neither a quantum chemistry nor a condensed matter type confirmation has been possible for the LMW braiding statistics of CF-quasiparticles. It has been asserted [372] that the non-Laughlin fractions of the type 2/5 and 2/7 are a measurable consequence, and prove the existence, of fractional braiding statistics. This claim is based on the so-called "hierarchy" approach for the FQHE (Section 12.1), which proposes to use the fractional braiding statistics of quasiparticles to explain the non-Laughlin fractions. If the explanation of 2/5 and 2/7 had indeed required fractional braiding statistics in the same essential manner as the bosonic statistics is required for BEC or the fermionic statistics for the Fermi sea, then such a claim would indeed be warranted. However: (i) the notion of fractional statistics becomes ambiguous at high quasiparticle or quasihole densities required for 2/5 and 2/7 (Section 12.1); and (ii) the CF theory explains all fractions (including 2/5 and 2/7) convincingly without any mention of fractional braiding statistics. A consideration of the totality of observed fractions indicates that the FQHE provides an evidence for the fermionic statistics of composite fermions.

The best chance for an observation of fractional braiding statistics may be through an experiment that implements the ASW geometry and measures the Berry phase induced by the insertion of a new quasiparticle inside the loop. The situation is somewhat complicated, in practice, by the fact that the order-one contribution originating from the braiding statistics sits atop a large Aharonov–Bohm phase (the AB phase is proportional to the area of the loop, which must be large enough to avoid overlap between the two quasiparticles in question). The conceptual question of disentangling the local charge, the braiding statistics, and the edge exponent is also relevant for any experimental situation, especially for the Laughlin fractions, for which, in a Wen-type approach for the edge state, all three arise out of a single parameter. In such situations, although interpretation in terms of one of them may seem more natural, it is not unique. Kim *et al.* [337] suggest that it would help to investigate other fractions of the sequences $\nu = n/(2pn \pm 1)$ for which the local charge, the braiding statistics, and the edge exponents have different values.

9.9 Non-Abelian braiding statistics

The excitations of certain FQHE states may show an even more complicated behavior under braiding, which goes under the name non-Abelian braiding statistics. The effective wave functions of such excitations transform as a non-Abelian representation of the braid group. We considered in Section 7.4.2 the Moore–Read Pfaffian wave function at $\nu = 1/2$, $\Psi^{\text{Pf}}_{1/2}$ (Eq. 7.11). Moore and Read [454] suggested that its quasiholes obey non-Abelian braiding statistics. An Arovas–Schrieffer–Wilczek-type adiabatic Berry phase calculation has not yet been possible, but we describe the essential physics. The following considerations are valid for a three-body short-range interaction (see Section 12.8.2 for further details) for which the Moore–Read wave function and its quasi-hole excitations are exact (zero energy) ground states.

The wave function for a vortex excitation at η is given by

$$\prod_j (z_j - \eta) \Psi^{\text{Pf}}_{1/2} = \prod_j (z_j - \eta) \text{Pf}\left(\frac{1}{z_i - z_j}\right) \Phi_1^2 , \qquad (9.51)$$

which has a local charge of $e/2$ associated with it (Section 9.3.2). However, as we saw in Section 9.3.3, the vortex is not necessarily an elementary excitation; for the FQHE states at $\nu = n/(2pn \pm 1)$ it is a collection of n quasiholes, due to one missing composite fermion in each Λ level. For the Moore–Read wave function also, the vortex can be split into two elementary excitations. This becomes clear by writing the following wave function for *two* quasiholes at η and η' (Moore and Read [454])

$$\Psi^{\text{Pf}}_{1/2}(\eta, \eta') = \text{Pf}\left(M_{ij}\right) \Phi_1^2 , \qquad (9.52)$$

where

$$M_{ij} = \frac{(z_i - \eta)(z_j - \eta') + (i \leftrightarrow j)}{(z_i - z_j)}. \tag{9.53}$$

For $\eta' = \eta$ it reduces to the charge $e/2$ vortex. Thus, pulling the vortex factor $\prod_j(z_j - \eta)$ into the Pfaffian shows that it actually represents *two* coincident quasiholes, each of which has a local charge $e/4$.

Moore and Read construct the wave function for $2n$ quasiholes at $\eta_1, \ldots, \eta_{2n}$ by generalizing Eq. (9.51) to n vortices, and then splitting them into $2n$ quasiholes by pulling the vortex factors into the Pfaffian. The wave function is given by Eq. (9.52), but with

$$M_{ij} = \frac{\prod_{\alpha=1}^{n}(z_i - \eta_\alpha)(z_j - \eta_{\alpha+n}) + (i \leftrightarrow j)}{(z_i - z_j)}. \tag{9.54}$$

Note that quasiholes can be created only in pairs.

To appreciate the origin of non-Abelian braiding statistics, let us first recall the case of several CF-quasiholes at $\nu = n/(2pn \pm 1)$. The lowest energy state is obtained when they are all in the topmost occupied Λ level. Because the wave function is specified uniquely by the positions of CF-quasiholes, when two CF-quasiholes adiabatically braid around one another, the final wave function may differ from the initial at most by a Berry phase factor. Two consecutive operations thus commute (the phases add), and the LMW braiding statistics is Abelian.

In contrast, the Moore–Read wave function for $2n$ quasiholes is not fully specified by their positions. In Eq. (9.54) we associated $\eta_1, \ldots, \eta_n$ with z_i and the remaining quasiholes with z_j, but we could equally well have chosen to associate different sets of η's with z_i and z_j to obtain different wave functions. This might suggest that the number of independent wave functions for a specified set of quasihole positions is equal to $(2n)!/[2(n!)^2]$, which is the number of ways $2n$ quasiholes could be arranged into two groups of n. But only 2^{n-1} of these wave functions turn out to be linearly independent, as can be seen from an explicit consideration of the $(2n)!/[2(n!)^2]$ polynomials (Moore and Read [454]; Nayak and Wilczek [475]). The initial wave function will, therefore, in general, be a linear superposition of 2^{n-1} orthogonal basis functions. An adiabatic interchange of two quasiholes will produce a new linear superposition which can be obtained from the initial one by application of a $2^{n-1} \times 2^{n-1}$ matrix. Matrices associated with consequent exchanges do not commute in general. Hence the name non-Abelian braiding statistics. The numerical Berry phase calculations by Tserkovnyak and Simon [648] support these ideas.

While non-Abelian braiding statistics has been established for a model interaction for which the Moore–Read wave functions are exact, its relevance for the excitations of the actual Coulomb 5/2 state is unclear. Ho has shown [255] that the Moore–Read wave function is adiabatically connected to the Halperin (331) wave function (Section 13.2), the excitations of which are known to obey Abelian braiding statistics. Exact diagonalization studies on finite systems [641, 642] indicate that the Moore–Read Pfaffian model for the quasiholes may not provide a very accurate representation of the Coulomb quasiholes of the 5/2 FQHE state. Further work is needed to ascertain the range of interactions for which the actual 5/2 FQHE state supports excitations with non-Abelian braiding statistics, and whether this range includes the Coulomb interaction.

9.10 Logical order

The logical order for the understanding of various concepts discussed in this chapter is shown in Fig. 9.8, which illustrates the cause-and-effect relation between them. The directions of the arrows are consistent with the following facts: (a) Incompressibility at certain fractions follows from the formation of composite fermions. Composite fermions, on the other hand, cannot be derived from incompressibility. Composite fermions form compressible states as well, and are thus more general than the FQHE. (b) FQHE, fractional local charge, and fractional LMW braiding statistics are fundamentally tied to, and a direct consequence of, incompressibility. The local charge or braiding statistics are not sharply defined concepts for compressible states. (c) The FQHE can be understood without appealing to the charge of the excitations. (d) Fractional local charge is a prerequisite for fractional braiding statistics. (e) Fractional local charge and fractional braiding statistics can be derived from composite fermions, but the reverse is not true.

Exercises

9.1 This exercise builds on the arguments of Section 9.3.1.

(i) Derive the charge of a CF-quasihole by asking what happens when an electron is removed from an incompressible state.

(ii) Show that the local charge of a CF-quasiparticle or a CF-quasihole is independent of its Λ level index.

(iii) Show that the charge of a CF-quasiparticle is given by $-e^* = -e/(2pn+1)$ even for filling factors away from the special filling factors, when many CF-quasiparticles or CF-quasiholes are already present (localized by disorder).

(iv) Show that, for the incompressible states at $n/(2pn-1)$, a composite fermion in the $(n+1)$th Λ level has an excess charge $+e/(2pn-1)$ (note the sign).

9.2 This exercise seeks to formulate the argument of Section 9.3.1 in the disk geometry. To avoid excitations at the edge, the size of the disk will be kept constant. Begin with the $n/(2pn+1)$ FQHE state, $\Psi = \prod_{j<k}(z_j - z_k)^{2p}\Phi_n$, confined to a disk of a given radius, which fixes the largest allowed power of z_j. Now add an *electron* to this system while insisting that the size of the system not change. The Jastrow factor increases the largest power of z_j, which requires taking some particles from the boundary of each Landau level in Φ and placing them in the $(n+1)$th Landau level. Show that the total

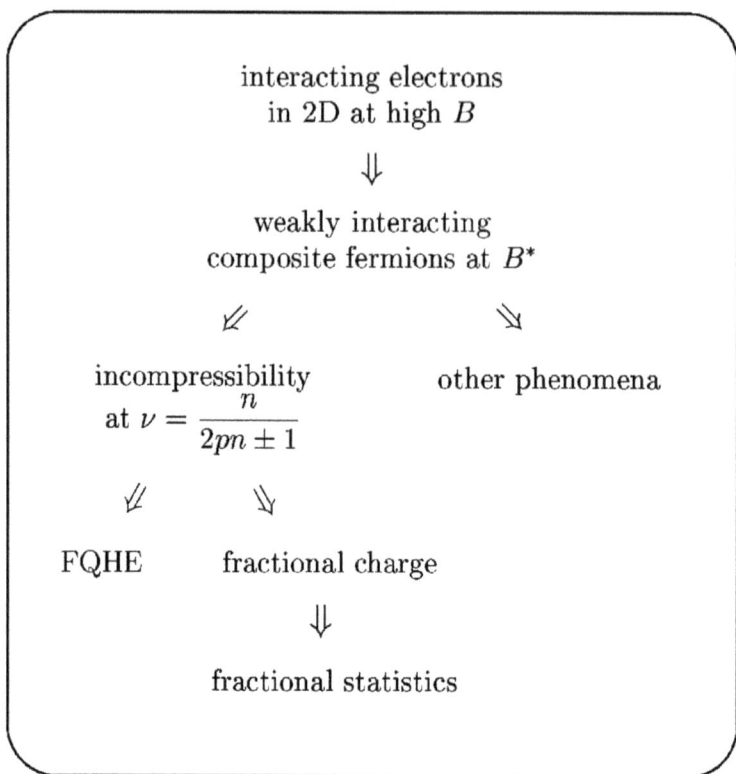

Fig. 9.8. The logical order of the topological quantities discussed in this chapter. Source: G. S. Jeon, K. L. Graham, J. K. Jain, *Phys. Rev. B* **70**, 125316 (2004). (Reprinted with permission.)

number of CF-quasiparticles in the $(n + 1)$th Λ level is $2pn + 1$, which produces the local charge $-e^* = -e/(2pn + 1)$.

9.3 Apply the arguments of Section 9.3.4 to an even denominator fraction. What can you conclude?

9.4 Following Section 9.3.7, estimate the size of a CF-quasihole.

9.5 Consider an island of the $\nu = (n+1)/(2n+3)$ state surrounded by the $\nu = n/(2n+1)$ state. Show schematically, as in Fig. 9.2 (but neglecting the outer edge of the $n/(2n+1)$ state), what happens when:
(i) a composite fermion is added to the boundary of the island in the nth Λ level;
(ii) a composite fermion is added to the center of the island in the $(n + 1)$th Λ level;
(iii) a composite fermion is moved from the edge of the nth Λ level to the center of the $(n + 1)$th Λ level. (Source: Jain and Shi [288].)

9.6 With the help of a singular gauge transformation, map (i) anyons into fermions; (ii) fermions into bosons.

9.7 The wave function for a localized wave packet centered at η is given by Eq. (E3.2)

$$\chi_\eta(z) = \frac{1}{\sqrt{2\pi}} \exp\left[\frac{1}{2}\bar{\eta}z - \frac{1}{4}|z|^2 - \frac{1}{4}|\eta|^2\right] \tag{E9.1}$$

where $\eta \equiv Re^{-i\theta}$. The Berry γ for a closed loop in which θ goes from 0 to 2π is shown in Appendix E to be $\gamma = -\pi R^2$. Now let us consider the problem of two electrons at η and η'. Antisymmetrizing and taking one particle centered at the origin ($\eta' = 0$) and the other at $\eta = Re^{-i\theta}$, the wave function is given by

$$\chi_{\eta,0}(\mathbf{r}_1,\mathbf{r}_2) = N \left(e^{\frac{\bar{\eta}z_1}{2}} - e^{\frac{\bar{\eta}z_2}{2}}\right) e^{-\frac{1}{4}(R^2+r_1^2+r_2^2)} . \tag{E9.2}$$

Show the following:
(i) The normalization constant is

$$N = \frac{1}{2\pi\sqrt{2[1 - \exp(-R^2/2)]}} . \tag{E9.3}$$

(ii) The Berry phase for the path $\theta = 0 \to 2\pi$ is

$$\gamma' = -\frac{\pi R^2}{1 - \exp(-R^2/2)} . \tag{E9.4}$$

(iii) The difference is given by

$$\Delta\gamma = \gamma' - \gamma = -\frac{\pi R^2}{\exp(R^2/2) - 1} . \tag{E9.5}$$

This gives the correction to the Berry phase due to the presence of another electron inside the loop.
(iv) The correction is exponentially small for $R/\ell \gg 1$. What does this imply for the braiding statistics? For what R does $\Delta\gamma$ become equal to 0.05? 0.01? In the other limit $R/\ell \ll 1$, we have

$$\Delta\gamma = -2\pi . \tag{E9.6}$$

That can be understood from the observation that for $R/\ell \ll 1$, the wave function reduces to

$$\chi_{\eta_1,\eta_2}(\mathbf{r}_1,\mathbf{r}_2) = \frac{1}{2}(z_1 - z_2)(\bar{\eta}_1 - \bar{\eta}_2) , \tag{E9.7}$$

that is, the particle $\mathbf{r}_1$ sees a simple vortex at $\mathbf{r}_2$, which has a phase of magnitude 2π associated with it.
Hint: Use the identity, derived in Appendix E:

$$\int \frac{d^2r}{2\pi} \frac{z\bar{\eta}}{2} \exp\left[\frac{1}{2}\left(\bar{\eta}z + \eta\bar{z} - r^2 - R^2\right)\right] = \frac{R^2}{2} . \tag{E9.8}$$

Comment: The above Berry phase calculation produces the statistics parameter $\alpha = 0$, contrary to the expectation of an odd-integer value of α for fermions. The reason is that the adiabatic process considered here amounts to an exchange of two "hydrogen atoms," with the η's playing the role of the proton coordinates. (The wave function is antisymmetric with respect to an exchange of both the electron coordinates and the η's.) The correct electron statistics is obtained after subtracting the statistics of the η's from the Berry phase result. Similar considerations are relevant to Exercise 9.9.

9.8 Show that, for charged *bosons* in the lowest Landau level, the change in the Berry phase of a closed loop due to the insertion of another boson is given by $\Delta\gamma = \pi R^2/(1 + e^{R^2/2})$.

9.9 This exercise repeats the calculation for one or two holes in the lowest Landau level. Denote the state in which the lowest Landau level is fully occupied by $|\Phi_1>$. A hole at η is created by the application of the field operator as follows:

$$|\chi_\eta\rangle = N_1 \Psi_\eta |\Phi_1\rangle$$

$$= N_1 \sum_{m=0}^{\infty} \frac{\eta^m}{\sqrt{2^m m!}} e^{-|\eta|^2/4} c_m |\Phi_1\rangle . \tag{E9.9}$$

The wave function for two holes, one at the origin and the other at η is similarly given by

$$|\chi_{\eta,0}\rangle = N_2 \sum_{m=1}^{\infty} \frac{\eta^m}{\sqrt{2^m m!}} e^{-|\eta|^2/4} c_m c_0 |\Phi_1\rangle . \tag{E9.10}$$

Here c_m is annihilates the electron in the angular momentum m orbital. Show the following:

(i) The normalization constants are given by

$$N_1 = 1, \qquad N_2 = [1 - \exp(-R^2/2)]^{-1/2} . \tag{E9.11}$$

(ii) For a single hole, the Berry phase for the closed loop defined by $\eta = R e^{-i\theta}$, $R = \text{constant}$, $\theta = 0 \to 2\pi$ is given by

$$\gamma = \pi R^2 = 2\pi \frac{\phi}{\phi_0} . \tag{E9.12}$$

(iii) When the loop encloses a hole at the origin, the Berry phase is given by

$$\gamma' = \frac{\pi R^2}{1 - \exp(-R^2/2)} . \tag{E9.13}$$

(iv) Discuss the behavior of $\gamma' - \gamma$ in the limiting cases $R/\ell \gg 1$ and $R/\ell \ll 1$.

9.10 We made several consistency checks in the text. Here is one more. A vortex at $v = n/(2pn + 1)$ is a collection of n CF-quasiholes. Show that the LMW braiding statistics of a vortex agrees with that of n CF-quasiholes (modulo an integer).

9.11 Another candidate for non-Abelian braiding statistics (Wen [672]) is the wave function $\Phi_1 \Phi_2^2$ (Jain [274]). Determine its filling factor, the charge of a vortex excitation, and the charge of an elementary excitation. (Hint: Into how many elementary excitations can a vortex be split?) The elementary excitations are CF-quasiholes or CF-quasiparticles in the Φ_2 factors. They obey non-Abelian braiding statistics, because specifying their positions does not uniquely determine the wave function; they can be arranged in many ways among the two Φ_2 factors. (The wave function, so far, is not known to be applicable to a realistic situation.)

10

Composite fermion Fermi sea

The physics of the state at filling factor $\nu = 1/2$ remained a puzzle for many years. Attention was redrawn to it in late 1989/early 1990 by the work of Jiang *et al.* [300], who reported a deep resistance minimum at $\nu = 1/2$ in certain high-quality samples, and by certain anomalies at $\nu = 1/2$ in surface acoustic wave absorption observed by Willett *et al.* [682]. This time, with composite fermions available, rapid progress was made.

The lowest Landau level problem has no kinetic energy. When electrons transmute into composite fermions, the interelectron interaction energy transforms, in the simplest approximation, into a "kinetic energy" of composite fermions. (In general, not all of the Coulomb interaction transforms into kinetic energy, which leaves behind a residual interaction between composite fermions.) The CF kinetic energy manifests dramatically through the quantized Λ levels and the FQHE at $\nu = n/(2pn \pm 1)$. These sequences terminate into $\nu = 1/2p$ in the limit of $n \to \infty$. Should composite fermions exist in this limit, the magnetic field experienced by them vanishes. Motivated by the experiments mentioned in the preceding paragraph, Halperin, Lee, and Read [231] (also see Kalmeyer and Zhang [309]) made the striking proposal that composite fermions form a Fermi sea here, called the CF Fermi sea:

$$\text{an infinite number of filled Landau levels} = \text{Fermi sea}, \tag{10.1}$$

$$\text{an infinite number of filled } \Lambda \text{ levels} = \text{CF Fermi sea}. \tag{10.2}$$

A strong, but indirect, clue for the presence of a CF Fermi sea at $\nu = 1/2$ comes from the observation of very many fractions along the sequence $f = n/(2n + 1)$; evidence exists [496], in ρ_{xx}, for fractions up to $f = 10/21$. Many properties of the FQHE states discussed in earlier chapters, for example the Shubnikov–de Haas oscillations of composite fermions and a "linear" opening of the gap, are also consistent with the notion of a CF Fermi sea at $\nu = 1/2$. Nonetheless, the CF Fermi sea must be verified by experimentation directly in the compressible region.

A comparison of the CF Fermi sea with the ordinary Fermi sea of electrons at $B = 0$ is instructive. For the latter we begin with a Fermi sea in the absence of interactions and argue that interactions do not destroy it.[1] At $\nu = 1/2p$, on the other hand, there was, a priori, no

---

[1] Strictly speaking, as shown by Kohn and Luttinger [352], the ordinary Fermi sea is ultimately unstable to a BCS pairing in some high angular momentum channel, but an instability of this type is believed to occur at such small temperatures that it can be neglected for all practical purposes.

reason to expect any kind of Fermi sea at all, and only when we think in terms of composite fermions does the possibility of a Fermi sea manifest itself. It will be hardly surprising, however, should the CF Fermi sea not turn out to be *completely* analogous to the ordinary Fermi sea of electrons at $B = 0$.

Recall how we "accessed" the FQHE from the IQHE in Chapter 5 by attaching point flux quanta to electrons and then adiabatically spreading them out. Crucial to the adiabaticity is the presence of a gap at integral fillings. The gapless nature of the Fermi sea at $B = 0$ makes it conceptually harder to connect it to a Fermi sea of composite fermions at $\nu = 1/2p$. Halperin, Lee and Read [231] note that the CF Fermi sea description at these fillings at least gives a unique ground state, and evaluate many of its properties using the Chern–Simons Landau Fermi liquid approach described in Chapter 5. We see below that many qualitative features of experiments can be well understood by viewing the 1/2 state as an ordinary Fermi sea of nearly independent composite fermions, and borrowing familiar expressions from the Fermi liquid theory of electrons.

While the CF Fermi sea is a distinct possibility in the CF theory, its existence is not necessary for either the FQHE at $\nu = n/(2pn \pm 1)$ or the validity of the CF theory. A non-CF physics cannot be excluded at $\nu = 1/2p$. Even if composite fermions *are* relevant at $\nu = 1/2p$, they need not necessarily assemble into a Fermi sea. They could form an Anderson insulator or a FQHE state. It is an experimental fact that the second-LL FQHE states at $\nu = 2 + n/(2n + 1)$ do not terminate into a compressible CF Fermi sea at $\nu = 2 + 1/2$, but instead into a FQHE state believed to be a paired state of composite fermions.

We have noted in Chapter 9 that, close to an incompressible FQHE state, an effective description in terms of quasiparticles with fractional local charge and fractional braiding statistics becomes possible (although not necessary), if one defines the uniform density FQHE state as the vacuum state. The sharpness of these fractional quantum numbers derives from the presence of a gap. Fractional local charge or fractional braiding statistics are of no relevance for the gapless CF Fermi sea. All discussion in this chapter is in terms of charge $-e$ composite fermions.

10.1 Geometric resonances

Many ingenious experiments have verified composite fermions and their CF Fermi sea by directly probing the compressible region near $\nu = 1/2$. Some of these measure the cyclotron radius of the charge carriers in the vicinity of the half-filled Landau level and find it to be consistent with the *effective field* rather than the external magnetic field, providing direct experimental proof for the occurrence of composite fermions in the compressible region near the half-filled Landau level. An advantage of these experiments is that the radius of the cyclotron orbit for composite fermions involves only experimentally determinable parameters, and, notably, does not depend on the CF mass. It is given by

$$R^*_c = \frac{\hbar k^*_F}{eB^*} , \tag{10.3}$$

with $k_F^* = \sqrt{4\pi\rho}$, as appropriate for a fully polarized CF Fermi sea. R_c^* is distinct from any length scale in the original problem; in the vicinity of $B^* = 0$, it can be orders of magnitude larger than the magnetic length, or the cyclotron radius for a "classical" electron at B. The observation of the cyclotron orbits of composite fermions establishes their validity at a semiclassical level. We refer the reader to the excellent review article by Smet [603] for further information on various experiments studying ballistic transport of composite fermions in semiconductor nanostructures.

10.1.1 Transport in an antidot superlattice

Weiss *et al.* [671] study electron transport in antidot superlattices, and find prominent new features in the data at low magnetic fields. They show that these can be understood in terms of classical orbits commensurate with the lattice, e.g., those shown in the inset of the top panel of Fig. 10.1. Naively, we can imagine that the cyclotron orbit gets pinned when it is commensurate with the lattice, producing peaks, or geometric resonances, in the resistance. (See Beenakker [25], and Fleischmann, Geisel, and Ketzmerick [165] for more detailed theoretical treatments.)

Kang *et al.* [320] observe that the presence of the antidot superlattice also strongly affects the resistance in the vicinity of $\nu = 1/2$. The lower panel of Fig. 10.1 compares the resistances near $B = 0$ (lower trace) and $B^* = 0$ (upper trace). Geometric resonances of composite fermions are clearly visible; their positions match with those of the strongest electron resonances, allowing their identification with the CF orbit encircling a single antidot. For the resonance at negative values of B^* ($\nu > 1/2$), composite fermions orbit in a direction opposite to the direction of classical electrons in the external magnetic field. The broader peakwidth for composite fermions and the absence of higher order peaks indicate a shorter mean free path for composite fermions than for electrons. The asymmetry of the peaks is attributed to the spatial variation in the effective magnetic field B^* near the walls of the antidots [166, 599].

Smet *et al.* [601] show that the geometric resonances in the resistance are absent in a random lattice of antidots, thus proving that they arise solely due to the presence of the periodic potential. Further, in periodic antidot lattices, in addition to the pinning resonances in the longitudinal resistance, they notice a quenching of the Hall resistance at $\nu = 1/2$ (it becomes flat), which they ascribe to channeling trajectories of composite fermions. This is similar to a quenching of the Hall resistance around zero magnetic field in antidot lattices [671] or in narrow junctions (Roukes *et al.* [558]). In Ref. [602], Smet *et al.* observe similar geometric resonances in a weak two-dimensional periodic modulation of the potential, demonstrating the robustness of the effect.

Nihey *et al.* [479] study transport in a triangular lattice of large diameter antidots, with diameter roughly equal to half of the lattice constant, and find a qualitatively different behavior. They do not see either commensurability peaks or FQHE structure in ρ_{xx}, presumably because the ballistic motion of electrons or composite fermions is blocked by the large antidots. They see, however, evidence for quasi-periodic oscillations in the

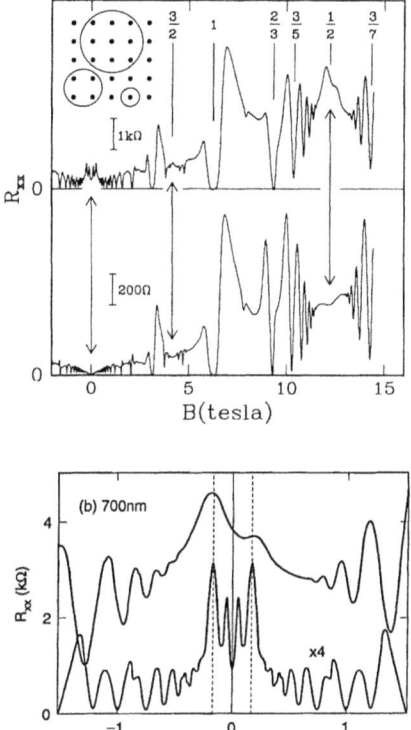

Fig. 10.1. Geometric resonances in an antidot superlattice. The upper panel shows the magnetoresistances with and without the superlattice (upper and lower traces, respectively), illustrating that differences appear near $B = 0$ and $\nu = 1/2$ and $3/2$. The superlattice is shown in the inset. The lower panel shows a comparison of magnetoresistances for electrons in the vicinity of $B = 0$ (lower trace) and composite fermions in the vicinity of $B^* = 0$ (upper trace). The latter has been scaled by a factor of $\sqrt{2}$ for comparison. Two resonances are seen for electrons, corresponding to the cyclotron orbits enclosing one and four lattice sites (inset). Only the principal resonance is seen for composite fermions, presumably because of their shorter mean free path. Source: W. Kang, H. L. Stormer, L. N. Pfeiffer, K. W. Baldwin, and K. W. West, *Phys. Rev. Lett.* **71**, 3850 (1993). (Reprinted with permission.)

resistance near both $B = 0$ and $B^* = 0$; they ascribe the latter to a quantization of the CF orbit around an antidot.

10.1.2 Magnetic focusing

Another method for determining the cyclotron radius is through transverse magnetic focusing, studied by van Houten and collaborators [655] for electrons in two dimensions at low magnetic fields. The geometry is shown in Fig. 10.2. A current is passed through the left (emitter) constriction and the voltage is measured across the right (collector) constriction.

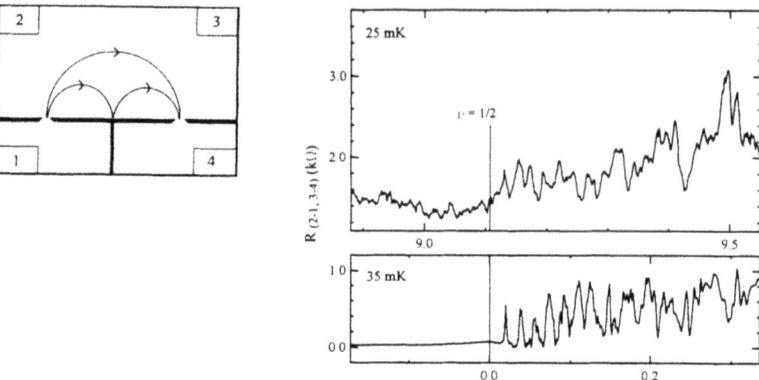

Fig. 10.2. Determination of the effective magnetic field by magnetic focusing of composite fermions. Composite fermions are injected into one constriction and collected into another (left panel), possibly after an integral number of specular bounces. The lower panel on the right shows the focusing peaks for *electrons* at $B \approx 0$, while the upper panel depicts the focusing peaks for composite fermions with $B^* \approx 0$. The two sets of focusing peaks (superimposed over mesoscopic resistance fluctuations due to disorder) align after scaling B^* by a factor of $\sqrt{2}$ to account for the difference that the electron Fermi sea is spin unpolarized whereas the CF Fermi sea is fully spin polarized. Source: V. J. Goldman, B. Su, and J. K. Jain, *Phys. Rev. Lett.* **72**, 2065 (1994). (Reprinted with permission.)

In the linear regime, the voltage is proportional to the applied current, and the ratio is the nonlocal "focusing resistance." This resistance displays peaks when electrons emerging from the left constriction are focused into the right one. The peaks occur at certain values of magnetic field only, and have information about the radius of the electron's cyclotron orbit. This is most easily seen in a classical picture, in which electrons come straight out of the left constriction with the Fermi velocity (in the inset of Fig. 10.2), with their trajectory bent due to the influence of the magnetic field. If the trajectory bends toward the right, it makes specular reflections at the boundary and, for certain values of the magnetic field, may hit the right constriction, charging the lower right part of the sample, thereby producing a peak in the focusing resistance. For other values of the magnetic field, electrons will miss the right constriction, and no voltage is induced, as is also the case when electrons are bent leftward upon emerging from the left constriction. The focusing into the right constriction occurs when B is such that the cyclotron diameter $2R$ is equal to the constriction separation L. Allowing for $j-1$ specular reflections, a series of focusing peaks occurs when $2jR = 2j\hbar k_F/eB = L$. The peaks occur periodically with B, spaced by $\Delta B = 2\hbar k_F/eL$. Such a classical picture captures the qualitative features of the more sophisticated, semiclassical treatments [655] that consider electrons coming out at an angle, as well quantum interference between various possible trajectories. Van Houten *et al.* observed such behavior (similar to that shown in the lower panel of Fig. 10.2), which is an observation of skipping orbits of electrons with specular scattering at the boundary. The electron-focusing peaks were suppressed in Ref. [655] at high magnetic fields in the QHE regime.

Goldman *et al.* [208] observe quasi-periodic focusing peaks in several double-constriction samples in the vicinity of $v = 1/2$, shown in Fig. 10.2 for one of the samples. A qualitative distinction can be seen between the behaviors for $v > 1/2$ (only noise) and $v < 1/2$ (noise plus peaks). The period in magnetic field, obtained from a Fourier analysis of the data, is consistent with the effective magnetic field for composite fermions, which can also be seen by noting that, after correcting B^* by $\sqrt{2}$, focusing peaks near $B = 0$ and $B^* = 0$ align to a good approximation. CF focusing peaks are delicate and washed out by the time the temperature is raised to 100 mK.

Smet *et al.* [599,600] investigate magnetic focusing of composite fermions through arrays of micrometer-size cavities (three arrays, each containing ten cavities with a period L). The advantage of using many cavities in a single experiment is that the resulting ensemble averaging suppresses conductance fluctuations. Figure 10.3 shows the measured resistances for transport through the region containing the cavities, both in the vicinity of $B = 0$ and $B^* = 0$. The peaks labeled F_{ij} for electrons would not be present without the antidot arrays, and are a result of focusing. The observation of peaks for composite fermions corresponding to the strong electron focusing peaks (F_{11}, F_{21}, F_{31}, etc.) makes a compelling case for CF focusing from one cavity into its nearest neighbor. Many of the observed focusing peaks correspond to situations for which the effective magnetic field points opposite to the external magnetic field. The lack of symmetry between positive and negative B^* is attributed to soft wall confinement, rationalized in Ref. [599] through simulations based on semiclassical dynamics of composite fermions.

10.1.3 Surface acoustic wave absorption

Wixforth, Kotthaus, and Weimann [689] probe the dynamical properties of a two-dimensional electron system in the integral quantum Hall regime through its interaction with surface acoustic waves. Surface acoustic waves (SAW) are emitted and collected by transducers at two ends of the two-dimensional system. Interaction with the electron system causes an attenuation of the sound wave, as well as a shift in its velocity, which can both be related to the conductivity of the electron system. The SAW experiments thus provide a means for measuring the dynamical conductivity of the two-dimensional electron system at the frequency and the wave vector of the SAW.

Willett and collaborators [682] extend the experiment to the fractional regime and note that the SAW attenuation and the velocity shift at $v = \frac{1}{2}$ behave, at high frequencies, in a manner contrary to that expected from the *dc* conductivity. Halperin, Lee, and Read [231] attribute this anomaly to interaction of small wave vector SAW with the gapless CF Fermi sea. Using the CFCS approach, they obtain the following expression for the zero frequency limit of the conductivity (with $\boldsymbol{q} = q\hat{\boldsymbol{x}}$)

$$\sigma_{xx}(q) = \frac{e^2 q}{8\pi \hbar k_{\mathrm{F}}},$$ (10.4)

which indicates a linear dependence on q. In the SAW experiments, the effect is washed out at very low frequencies because the SAW wavelength is long compared to the CF

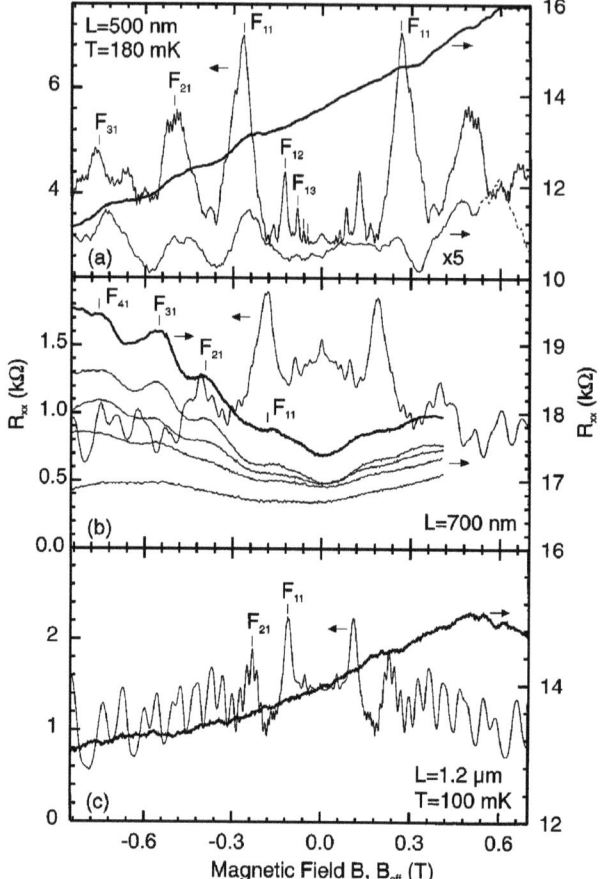

Fig. 10.3. Magnetic focusing spectra for electrons (left axis) and composite fermions (right axis) for transport through cavity arrays for three different periods L shown on the figure. B is the applied magnetic field and $B_{\text{eff}} = B^*$ is the magnetic field for composite fermions. The scale for the latter has been divided by a factor of $\sqrt{2}$ to facilitate comparison. The peaks labeled F_{11}, F_{21}, and F_{31} correspond to focusing from one cavity into its nearest neighbor after one, two, and three bounces, respectively. Those labeled F_{12} or F_{13} represent focusing from a given cavity into the second or the third nearest neighbors. In panel (a) a linear background has been subtracted from the CF trace to obtain the bottom curve, where the structure is more clearly visible. The panel (b) shows CF traces at different temperatures, with the top one at 100 mK and the bottom one at 830 mK. No CF focusing is observed for $L = 1.2\,\mu$m (c). Source: J. H. Smet, D. Weiss, R. H. Blick, G. Lütjering, K. von Klitzing, R. Fleischmann, R. Ketzmerick, T. Geisel, and G. Weimann, *Phys. Rev. Lett.* **77**, 2272 (1996). (Reprinted with permission.)

mean free path. For larger frequencies, when the SAW wavelength is shorter than the CF mean free path, the conductivity at $\nu = 1/2$ is indeed found to be proportional to q [683, 684]. Halperin, Lee, and Read also extend their calculation to $B^* \neq 0$ and predict commensurability minima in sound velocity when R_c^* is 5/8 times the SAW wavelength.

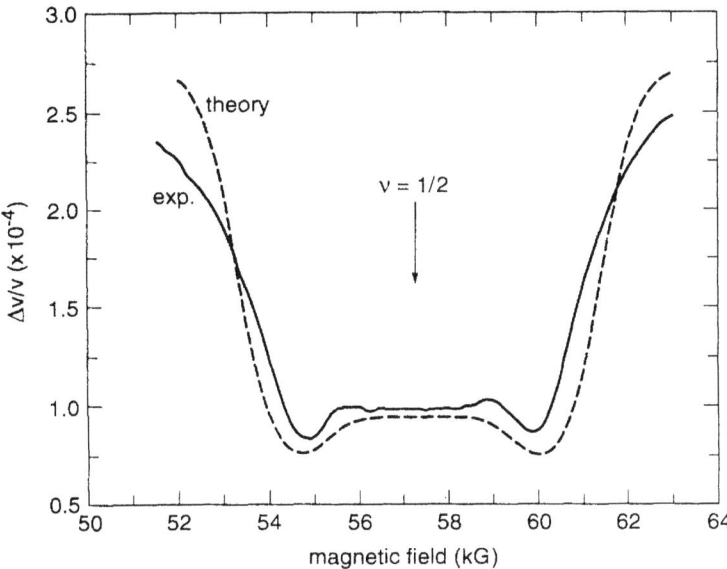

Fig. 10.4. Shift in the sound wave velocity due to its interaction with the compressible state in the vicinity of $\nu = 1/2$. The surface acoustic wave frequency is 8.5 GHz. The theoretical curve from the CFCS calculation of Halperin, Lee and Read [231] has been broadened to account for disorder. Source: R. L. Willett, R. R. Ruel, K. W. West, and L. N. Pfeiffer, *Phys. Rev. Lett.* **71**, 3846 (1993). (Reprinted with permission.)

Such minima are observed, reproduced in Fig. 10.4. The Boltzmann equation calculations of finite wave vector and finite frequency conductivity by Mirlin and Wölfle [447], taking into account scattering by random magnetic fields, also shows good agreement with experiment.

10.1.4 dc transport in unidirectional periodic potentials

Unidirectional periodic modulation in the electrostatic potential is known to give rise to commensurability effects for *electrons* at very small B (Gerhardts, Weiss, and von Klitzing [182]; Winkler, Kotthaus, and Ploog [688]). Commensurability oscillations in the magnetoresistance have also been observed for two-dimensional electrons subjected to a periodic magnetic field modulation [51, 721], which can be produced, e.g., by superconducting stripes on the surface of a heterstructure. Several experiments have investigated the influence of a one-dimensional density modulation, created, e.g., by imposing a one-dimensional grating by electron beam lithography, on the behavior of the CF Fermi sea at $\nu = 1/2$ and 3/2 (Endo *et al.* [141], Smet *et al.* [602, 604]; Willett *et al.* [686]). For composite fermions, the density modulation also induces a modulation of the effective magnetic field. The problem has also been investigated theoretically (Mirlin *et al.* [448]; von Oppen, Stern, and Halperin [662]; Zwerschke and Gerhardts [754]).

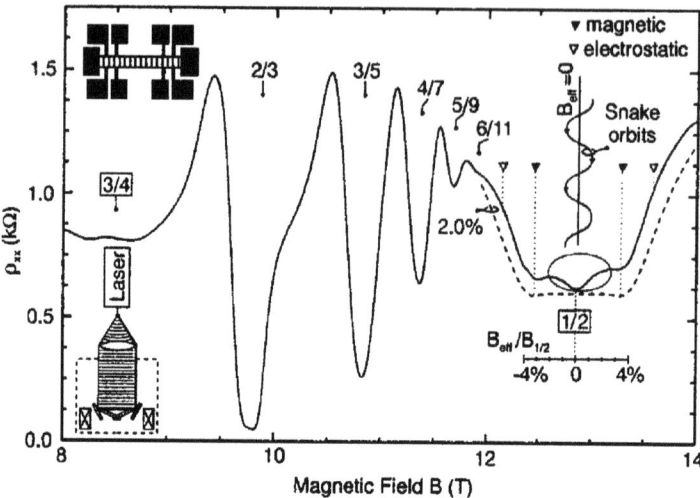

Fig. 10.5. Longitudinal magnetoresistances ρ_{xx} as a function of the external magnetic field in the presence of a periodic modulation of the density imposed by a "holographic illumination." The dashed curve is a fit to data using the theory of Mirlin *et al.* [448]. The arrows show the positions of the primary minima due to magnetic and electrostatic modulations. Source: J. H. Smet, S. Jobst, K. von Klitzing, D. Weiss, W. Wegscheider, and V. Umansky, *Phys. Rev. Lett.* **83**, 2620 (1999). (Reprinted with permission.)

A trace of magnetoresistance from the experiment of Smet *et al.* [604] is shown in Fig. 10.5 for current flowing perpendicular to the weak periodic modulation. Two minima are seen, one on either side of $\nu = 1/2$, which are well explained as the primary commensurability mimina arising from the *magnetic* field modulation. The argument [604] is as follows. The periodic oscillations arise out of commensurability between the cyclotron radius of composite fermions at the Fermi energy, R_c^*, and the period of the lattice, a. The minima in resistance occur at

$$2R_c^* = a\left(j \mp \frac{1}{4}\right),$$
(10.5)

where $j = 1, 2, \ldots$, the minus sign is for periodic electrostatic field, and the plus sign for the periodic magnetic field [183]. The arrows on Fig. 10.5 indicate the positions of the $j = 1$ minimum for both the electrostatic and the magnetic cases. The data shown in this figure as well as on other samples agree with the latter, confirming that charge density oscillations induce oscillations in the effective magnetic field.

The higher order commensurability minima are not seen in Ref. [604], presumably because the mean free path of composite fermions is on the order of the lattice constant. Endo *et al.* [141] study the behavior close to $\nu = 3/2$ under weak unidirectional periodic modulation with much shorter periods (about a factor of three smaller than those in Ref. [604]). As seen in Fig. 10.6, they observe several minima, marked by vertical short lines on the figure, which they identify with commensurability oscillations corresponding to

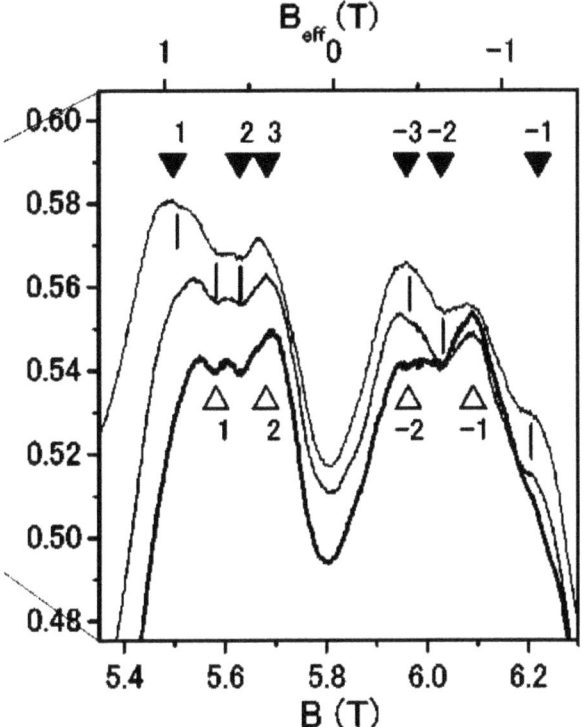

Fig. 10.6. R_{xx} for current perpendicular to a weak unidirectional periodic modulation in the vicinity of $\nu = 3/2$. The triangles show the positions predicted by Eq. (10.5), with positive sign, for a fully polarized CF Fermi sea (solid downward) and a spin-unpolarized CF Fermi sea (open upward). The short vertical lines indicate the positions of the observed minima. Source: A. Endo, M. Kawamura, S. Katsumoto, and Y. Iye, *Phys. Rev. B* **63**, 113310 (2001). (Reprinted with permission.)

$j = 1, 2, 3$ in Eq. (10.5). Most of these agree well with the positions predicted by Eq. (10.5), with positive sign, which are shown by solid triangles for a fully spin-polarized CF Fermi sea (open triangles show the expected positions for a spin-unpolarized CF Fermi sea). A positive magnetoresistance with increasing $|B^*|$ is similar to that observed earlier by Smet *et al.* [602].

10.1.5 Transport through a quantum point contact

A quantum point contact, or a constriction, is a narrow one-dimensional channel, which can be defined on a two-dimensional system by using a single split gate or several split gates. The width of the channel can be changed by varying the potential on the gates. When many transverse subbands are occupied, the conductance of the channel is not quantized. Application of a magnetic field causes edge states to form, which suppresses backscattering

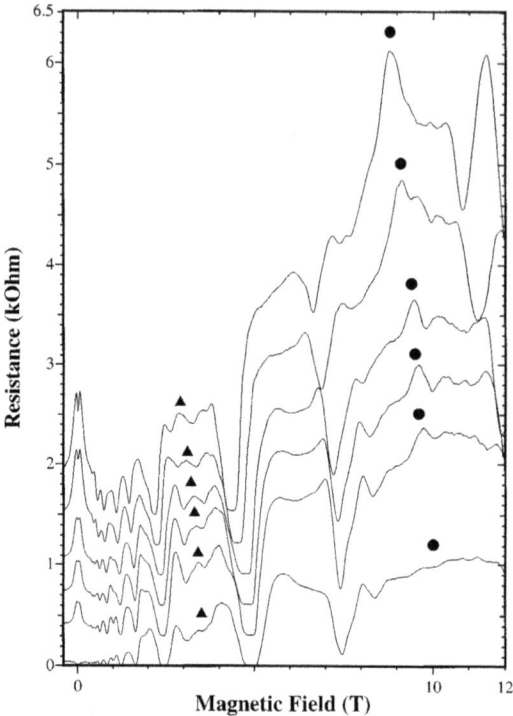

Fig. 10.7. R_{xx} through a quantum constriction at various values of the gate voltage. Circles and triangles indicate filling factors $\nu = 1/2$ and 3/2 in the constriction. A negative magnetoresistance at $B = 0$ indicates suppression of backscattering due to the formation of "skipping orbit" (edge) states. Similar behavior is observed at $\nu = 1/2$ and 3/2. Source: J. E. F. Frost, C.-T. Liang, D. R. Mace, M. Y. Simmons, D. A. Ritchie, and M. Pepper, *Phys. Rev. B* **35**, 9602 (1996). (Reprinted with permission.)

and gives rise to a negative magnetoresistance. Similar negative magnetoresistances are seen by Frost *et al.* [175] and Liang *et al.* [393] for composite fermions. Figure 10.7, taken from Ref. [175], shows a negative magnetoresistance at $\nu = 1/2$ and 3/2, which is an indication of the formation of CF skipping orbit states as $|B^*|$ is turned on by moving away from $B^* = 0$. The approximately symmetric nature of the peak at $\nu = 1/2$ supports the notion that B^* switches sign at $\nu = 1/2$. From the resistance, as well as the pinch-off voltage, Liang *et al.* estimate that the width of the channel is smaller for composite fermions at $\nu = 1/2$ than for electrons at $B = 0$ (see also Ref. [208]). This is understood by noting that variation in the electron density across the channel causes a variation in B^*; $|B^*|$ increases near the edges of channel where the filling factor is smaller than $\nu = 1/2$. The effective channel width for composite fermions, defined by the region where $B^* \approx 0$ (large values of $|B^*|$ bend the composite fermion trajectories and thus inhibit ballistic transport), is therefore narrower than the width of the electronic wave function. Frost *et al.* also investigate transport through a cross junction device and observe a change in the slope of the Hall resistance at $v = 1/2$, which they relate to quenching of Hall resistance at $B = 0$.

10.1.6 Cyclotron resonance

Many of the above experiments measure the semiclassical cyclotron orbits of composite fermions. Kukushkin *et al.* [360, 361] have observed the cyclotron resonance of composite fermions, which is discussed in Section 8.7.3.

10.2 Thermopower

Application of a thermal gradient to an electron system induces a current. When no net current is being drawn from the system, a voltage must be established that produces an electric current that exactly cancels the induced thermal current. The diagonal and off-diagonal components of the thermopower tensor, S_{xx} and S_{xy}, are defined as $E_x = S_{xx} \nabla_x T$ and $E_y = S_{xy} \nabla_x T$, where E_x and E_y are the components of the induced electric field, and $\nabla_x T$ is the temperature gradient, assumed to be along the x direction. The off-diagonal component, which is nonzero in the presence of a magnetic field, is also called the Nernst–Ettinghausen coefficient. Thermopower is the most sensitive electronic transport property of a metal.

The diagonal component is determined by electron diffusion at low temperatures, and by phonon drag at high temperatures. For the low-temperature behavior, for noninteracting electrons, theory [737] predicts that, for temperatures small compared to the cyclotron energy but large compared to disorder, the diffusion component of the diagonal thermopower vanishes at integral fillings, and reaches a maximum at half-filled Landau levels, where it has a universal value of

$$S_{xx}^{d} = \frac{k_B \ln 2}{ev} \approx \frac{60}{v} \; (\mu V/K) \; . \tag{10.6}$$

It is proportional to the entropy per particle, and is independent of the mass. Disorder depresses the value by an amount $k_B T / \Gamma$, where Γ is LL broadening. The behavior in a magnetic field is to be contrasted with the low-temperature behavior at zero magnetic field [726],

$$S_{xx}^{d} = \frac{\pi^2}{3}(2p+1)\frac{k_B^2 T}{eE_F} \; , \tag{10.7}$$

which is obtained from the well-known Mott formula [726, 752]

$$S_{xx}^{d} = \frac{\pi^2}{3}\frac{k_B^2 T}{e}\left(\frac{1}{\sigma(E)}\frac{\partial \sigma}{\partial E}\right)_{E=E_F} \; , \tag{10.8}$$

by expressing the conductivity σ in terms of an energy-dependent relaxation time $\tau(E) \propto E^p$. The parameter p is weakly density dependent, but has a value close to one. Equation (10.7) implies that the diffusion thermopower is proportional to T. The

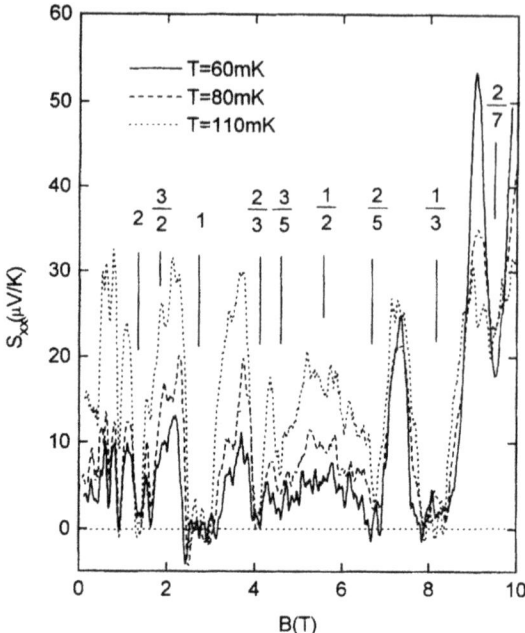

Fig. 10.8. Diagonal component of the thermoelectric power as a function of magnetic field for a hole-doped sample at three different temperatures. Source: X. Ying, V. Bayot, M. B. Santos, and M. Shayegan, *Phys. Rev. B* **50**, 4969 (1994). (Reprinted with permission.)

high-temperature behavior of S_{xx}^{d} is more complicated [149], but the proportionality to the phonon specific heat suggests a T^3 temperature dependence.

Several thermopower measurements have been performed on two-dimensional electron as well as hole systems in the lowest Landau level (Bayot *et al.* [23]; Crump *et al.* [102]; Tieke *et al.* [636]; Ying *et al.* [726]; Zeitler *et al.* [738]), and are well described in terms of composite fermions. Figures 10.8 and 10.9 show the data from the experiment of Ying *et al.* [726]. As expected, the thermopower vanishes at integral fillings, and also at $n/(2n \pm 1)$, at $T = 0$. However, the data at $\nu = 1/2$ and 3/2 are substantially inconsistent with the prediction of Eq. (10.6). The measured S_{xx} is roughly an order of magnitude smaller than the universal value in Eq. (10.6). Also, this equation would predict a ratio of three between the thermopowers at 1/2 and 3/2, whereas the observed ratio is 0.7. Attributing the reduction in the overall amplitude as arising from impurity scattering would require a strongly filling-factor-dependent scattering to explain the ratio $S_{xx}^{\mathrm{d}}(1/2)/S_{xx}^{\mathrm{d}}(3/2) = 0.7$, which is unlikely.

The magnitude of S_{xx}, its temperature dependence, and the value of the ratio $S_{xx}^{\mathrm{d}}(1/2)/S_{xx}^{\mathrm{d}}(3/2)$, are all naturally explained by a model of noninteracting composite fermions at $B^* = 0$. Assuming $p = 1$ in Eq. (10.7), the observed thermopower at $\nu = 1/2$ implies $m^*/m_e = 0.6 \pm 1$, which is consistent with CF mass determined by other means.

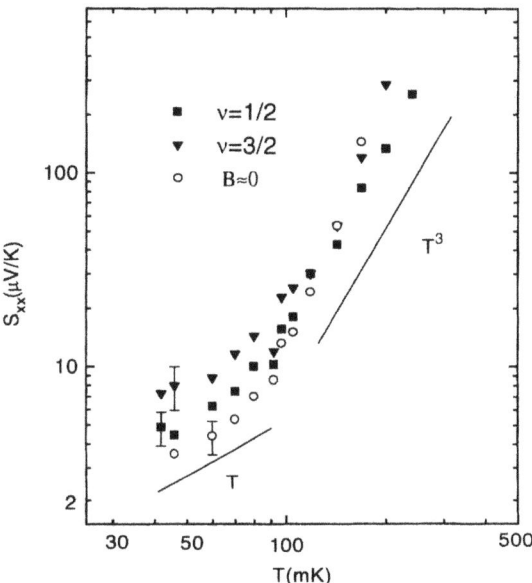

Fig. 10.9. Temperature dependence of S_{xx} at $\nu = 1/2$, 3/2, and at zero magnetic field. Source: X. Ying, V. Bayot, M. B. Santos, and M. Shayegan, *Phys. Rev. B* **50**, 4969 (1994). (Reprinted with permission.)

The ratio is determined by the ratio between CF Fermi energies. The CF Fermi energy $E_F^* = \hbar^2 k_F^2 / 2m^*$ has two parameters that depend on the filling factor: The k_F at $\nu = 1/2$ is determined from the full density, whereas at $\nu = 3/2$ from 1/3 of the density (because the electrons in the full Landau level are inert). The CF mass scales with $\sqrt{B} \sim \sqrt{1/\nu}$. Putting it all together, the CF model predicts the ratio

$$\frac{S_{xx}^{d}(1/2)}{S_{xx}^{d}(3/2)} = \frac{1}{3}\sqrt{3} = 0.58 , \tag{10.9}$$

which is in good agreement with the experimentally observed one. Finally, as seen in Fig. 10.9, the temperature dependence of S_{xx} is consistent with linear, again supporting the description in terms of free fermions at zero effective magnetic field.

Bayot *et al.* [23] extend the thermopower measurements to the FQHE states at $\nu = n/(2n \pm 1)$, and find that the temperature and the magnetic field dependences can be explained by treating the FQHE as the IQHE of composite fermions. The results of their measurements are shown in Fig. 10.10. Bayot *et al.* note: (i) S_{xx} is symmetric with respect to $B^* = 0$. (ii) The minima and maxima occur at integral and half integral values of ν^*. (iii) The peak values of S_{xx} at a fixed magnetic field rise linearly with B^*. (iv) The temperature dependence is fitted well as a sum of the diffusion and phonon terms (the solid line in the

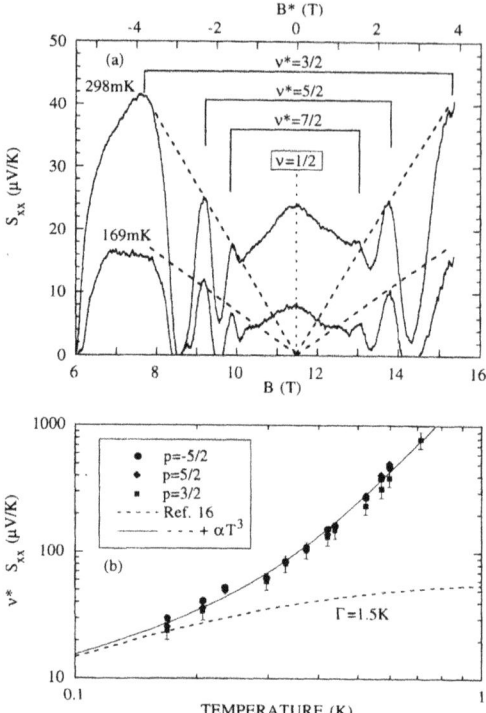

Fig. 10.10. (a) The diagonal thermopower S_{xx} of composite fermions at two temperatures. (b) The plot of v^*S_{xx} at several values of v^*, denoted by p. Ref. 16 here refers to Zawadski and Lassnig [737]. Source: V. Bayot, E. Grivei, H. C. Manoharan, X. Ying, and M. Shayegan, *Phys. Rev. B* **52**, R8621 (1995). (Reprinted with permission.)

lower panel of Fig. 10.10), where the latter is taken as proportional to T^3, and an expression from Ref. [737] is used for the former (the dashed line in the lower panel of Fig. 10.10). The Λ level broadening is a parameter in the fit, which is estimated to be $\Gamma = 1.5$ K, which agrees well with the value deduced from the gaps in transport experiments [428] (see Eq. 8.24).

Zeitler *et al.* [738, 739] and Tieke *et al.* [636] focus on the high-temperature regime, where the thermoelectric power is dominated by phonon drag. The absolute value of S_{xx} in these experiments varies strongly from one sample to another, and also as a function of temperature. However, as shown in Fig. 10.11, S_{xx} is the same at $v = 1/2$ and $3/2$, within experimental error, and S_{xx} at $v = 3/4$ differs by a constant factor. (This result holds for temperatures below which the FQHE features are seen.) The S_{xx} at $v = 1/2$ is related to S at $B = 0$ by a constant factor of five. Furthermore, the same power law behavior $S \sim T^k$ is seen at $B = 0$ and $v = 1/2$, with $k \approx 4$. They interpret all these findings by modeling the system in terms of composite fermions at zero magnetic

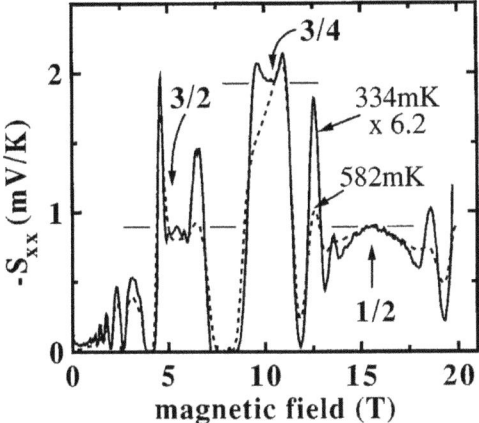

Fig. 10.11. The diagonal component of the thermopower, S_{xx}, at two temperatures, with the data at 334 mK scaled by a factor of 6.2. Source: B. Tieke, U. Zeitler, R. Fletcher, S. A. J. Wiegers, A. K. Geim, J. C. Maan, and M. Henini, *Phys. Rev. Lett.* **76**, 3630 (1996). (Reprinted with permission.)

field. The ratios are interpreted as arising from differences in CF effective masses. That interpretation implies that $m^* = 5m_e$ at $\nu = 1/2$, which is larger than the masses quoted in Section 8.7, but could be consistent with the mass enhancement seen in Shubnikov–de Haas experiments as $\nu = 1/2$ is approached (Section 8.7.2).

Thermal transport of composite fermions at various even denominator fractions has been studied theoretically (Karavolas and Triberis [325–327]; Tsaousidou, Butcher, and Kubakaddi [647]). Another thermodynamic quantity, namely the electronic specific heat, for which the model of noninteracting composite fermions predicts a value of $C = \pi^2 k_B^2 R/3E^*$, is difficult to measure due to the dominance of the lattice contribution [211, 667].

10.3 Spin polarization of the CF Fermi sea

The spin polarization of the CF Fermi sea has been measured in several experiments [131, 173, 174, 357, 437], which feature in Chapter 11. The model of noninteracting composite fermions is found to be remarkably successful in describing the temperature dependence of the spin polarization, as well as the transition from fully to partially spin-polarized states.

10.4 Magnetoresistance at $\nu = 1/2$

For electrons at zero magnetic field, the dominant scattering mechanism for resistance at low temperatures is Coulomb scattering with impurities. A novel feature of composite fermions is that they also experience spatial fluctuations in the effective magnetic field, which cause large-angle scattering and are believed to be the dominant mechanism for resistance [231, 309]. The spatial inhomogeneity in B^* results from density variations due to the random potential created by remote ionized donor impurities. Assuming a random distribution of the impurities, Halperin, Lee, and Read [231] estimate, using the CFCS formulation, that the diagonal resistivity of composite fermions is given by

$$\rho_{xx}^* \approx \frac{\rho_{\text{imp}}}{\rho} \frac{4\pi p^2}{k_{\text{F}}^* de^2} , \qquad (10.10)$$

where ρ_{imp} is the density of the charged impurities in a doping layer at a distance d from the two-dimensional electron layer. For ideal modulation-doped samples, we have $\rho_{\text{imp}} = \rho$. This expression overestimates the measured resistivities by factors of three or more, which might indicate correlations between the ionized donors [231].

Positive magnetoresistance   As mentioned in the beginning of this chapter, the renewed investigations of the nature of the 1/2 state began with the observation of a deep minimum in the resistance at $\nu = 1/2$, implying a positive magnetoresistance with increasing $|B^*|$ [300]. Kalmeyer and Zhang [309] consider the question of whether a true metallic phase can be possible for composite fermions at $\nu = 1/2$. For noninteracting electrons such a phase does not occur, according to the scaling theory of localization [1]. Kalmeyer and Zhang propose that the presence of a random magnetic field suppresses weak localization of composite fermions, thus producing a metal even at zero temperature. (The random magnetic field destroys time reversal symmetry, which is instrumental for weak localization through coherent backscattering.) They argue that, with increasing disorder, the system undergoes a transition into a strongly localized insulating phase. These assertions are supported by numerical simulations on free fermions subject to both a random magnetic field and a random potential disorder. The same simulations also suggest a *positive* magnetoresistance around average $B^* = 0$.

The existence of a metallic state in a random magnetic field is controversial [381, 400, 631, 711]. Khveshchenko has studied the problem within the CFCS approach [333]. Mirlin, Polyakov, and Wölfle [449, 524] and Evers *et al.* [145] investigate the random-magnetic-field induced localization within a semiclassical Drude model, and conclude that the conductivity σ_{xx} drops with increasing average value of $|\bar{B}^*|$; the conductivity is enhanced at $\bar{B}^* = 0$ because of percolating "snake states," which represent transport of composite fermions along regions with $B^* = 0$. Kivelson *et al.* [344] question the validity of these approaches for their neglect of correlation between charge and magnetic flux disorder. They note that even when the regions with positive and negative B^* are equal in area, so the average B^* vanishes, the negative B^* regions correspond to higher CF density, indicating that more composite fermions see negative values of B^* than positive. (To incorporate this effect in a CFCS formulation, one must use the magnetized modified random phase approximation mentioned in Section 5.16.3.) This would produce a negative

Hall conductance for composite fermions at $\nu = 1/2$, for which no evidence has so far been seen. How this contradiction is resolved is not yet known.

Temperature dependence Kang *et al.* [322] measure the temperature dependence of the resistance at $\nu = 1/2$, which they present in terms of a temperature-dependent mobility for composite fermions, the two being related by the expression $\mu^* = 1/(\rho e \rho_{xx})$ ($\rho =$ density) at $B^* = 0$. The CF mobility is approximately two orders of magnitude smaller than the electron mobility at $B = 0$. At low temperatures, the CF mobility is well fitted by the expression

$$\mu^*_{\text{imp+int}} = c_1 + c_2 T^2 + c_3 \ln T , \tag{10.11}$$

where the first two terms on the right hand side arise from impurity scattering for a degenerate CF Fermi sea, and the last from CF–CF interactions. Analogous terms appear for electrons at $B = 0$ [378], except that the last term is negligible for electrons in high-mobility samples but appreciable for composite fermions. The ratio c_1/c_2 is related to the density of states mass at the Fermi energy; Kang *et al.* estimate the mass to be on the order of $m^*/m_e \approx 3$, although it increases from 2 to 6 under illumination. One possible origin for this increase may be enhanced screening from the highly polarizable δ layer containing the dopants, which reduces the strength of the short-range part of the repulsive interaction between electrons. At about 1 K, the measured mobility deviates from the above expression; the difference is interpreted by Kang *et al.* as coming from the scattering of composite fermions by acoustic phonons. The mobility of composite fermions due to CF–acoustic phonon scattering is seen to follow the behavior $\mu^*_{\text{ac}} \sim T^{-3}$ over several orders of magnitude, to be contrasted with the T^{-5} dependence for electrons [528, 622].

Rohkinson, Su, and Goldman [554] measure the CF conductivity at 1/2 and 3/2 in the Corbino geometry. It is well described by the expression

$$\sigma^*_{xx} = \sigma^*_0 + \lambda \frac{e^2}{h} \ln(T/T_0) \tag{10.12}$$

over a range of temperature where the conductivity varies by a factor of 30. Rokhinson *et al.* take the logarithmic correction as evidence of residual interactions between composite fermions, by analogy to a similar correction for two-dimensional electrons in zero magnetic field [378]. In a subsequent study, Rohkinson and Goldman [555] note that the magnetic field dependence of the magnetoresistance cannot be fitted by a simple Drude model, which works well for electrons near $B = 0$, and conclude that the dominant scattering mechanism for composite fermions is different from that for electrons at zero magnetic field. A similar conclusion is reached by Wang *et al.* [664] from the shift in the positions of the fractional quantum Hall plateaus. Liang *et al.* [394, 395] study the effect of disorder, which they control by varying the density. They observe that the positive magnetoresistance at $B^* = 0$ disappears with increasing disorder. Furthermore, in the parameter region where the positive magnetoresistance is absent, the CF conductivity follows the 2D variable-range hopping behavior, $\sigma^*_{xx} \sim \exp[-(T_0/T)^{1/3}]$, which they take as evidence of a metal to insulator transition of composite fermions as a function of increasing disorder.

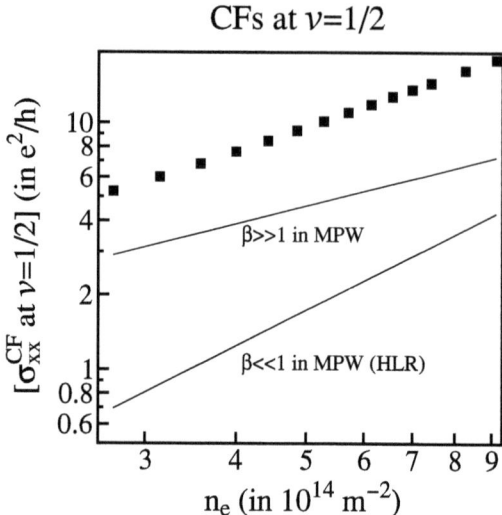

Fig. 10.12. The density dependence of the CF conductivity σ_{xx}^{CF} at $\nu = 1/2$ along with theoretical predictions by Mirlin, Polyakov, and Wölfle [449] (MPW) and Halperin, Lee, and Read [231] (HLR). The parameter n_e is the electron density. The limit $\beta \ll 1$ ($\beta \gg 1$) denotes weak (strong) random magnetic field. The measurements are performed at 0.3 K. Source: C. T. Liang, M. Y. Simmons, D. A. Ritchie, and M. Pepper, *J. Phys. Condens. Matter* **16**, 1095-1101 (2004). (Reprinted with permission.)

Density dependence   Equation (10.10) predicts that the CF resistivity scales with density as $\rho_{xx}^* \sim \rho^{-\alpha}$ with $\alpha = 3/2$, using $k_F^* = \sqrt{4\pi\rho}$. The dependence of the CF resistivity at $\nu = 1/2$ and $3/2$ has been studied as a function of carrier density in high-quality front-gated GaAs heterostructures by Liang *et al.* [396] (also see Cheng *et al.* [73]). They assume that ρ_{imp} remains constant (equal to the density ρ at zero gate voltage) as ρ is varied by gate voltage, and find that ρ_{xx}^* scales approximately inversely with the density, i.e., $\alpha \approx 1.0$, as seen in Fig. 10.12. The exponent lies in the range obtained by the semiclassical Boltzmann equation calculations of Mirlin, Polyakov, and Wölfle [449], who predict $\alpha = 3/2$ for weak spatial randomness in the effective magnetic field B^*, consistent with Eq. (10.10), but $\alpha = 3/4$ for strong spatial randomness.

Scaling with magnetic field: role of density gradient   Apart from the structure due to the IQHE and the FQHE, the longitudinal resistance R_{xx} is seen to rise linearly with B (Hirai *et al.* [252]; Pan *et al.* [498]; Stormer *et al.* [621]). That can be seen either by raising the temperature to a value, say 1.5 K, where most QHE structure is wiped out, or by focusing on the compressible CF Fermi sea states where the diagonal resistance saturates at a nonzero value in the limit $T \to 0$. As seen in Fig. 10.13, the low-temperature resistance of the CF Fermi seas at 1/4, 1/2, 3/4, and 3/2 scales linearly with the magnetic field B. This behavior is inconsistent with Eq. (10.10), which implies, for example, that the resistances at 1/4 and

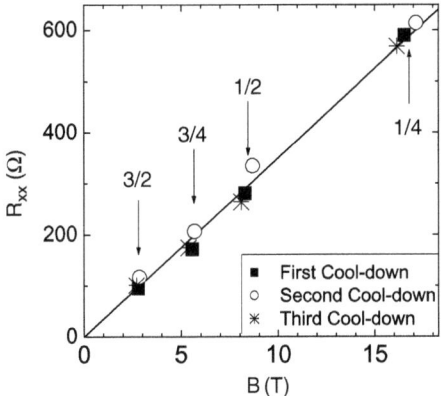

Fig. 10.13. The low-temperature ($T \sim 35$ mK) diagonal resistance of CF Fermi seas at $\nu = 1/4$, 1/2, 3/4, and 3/2 in a sample with mobility 1×10^7 cm^2/V s. Source: W. Pan, H. L. Stormer, D. C. Tsui, L. N. Pfeiffer, K. W. Baldwin, and K. W. West, cond-mat/0601627 (2006). (Reprinted with permission.)

1/2 differ by a factor of 4 rather than 2. Pan *et al.* [498] attribute the linear dependence to the presence of a small unintentional density gradient in the sample, which was discussed in Section 4.7 in the context of the resistivity rule and the observed quantization of R_{xx}. From the resistivity rule in Eq. (4.50), $R_{xy} \sim B$ implies $R_{xx} \sim B$. Such analysis raises questions regarding the relevance of the theoretically predicted local resistivities, e.g., Eq. (10.10), to the measured ones.

10.5 Compressibility

The compressibility K of a 2DES is given by

$$\frac{1}{K} = -A \left(\frac{dP}{dA} \right)_N ,$$

(10.13)

where P is the pressure, A is the area of the 2DES, and N is the total number of particles. By standard manipulations (see, for example, Mahan [416]), it can be related to the chemical potential μ as

$$\frac{1}{K} = \rho^2 \frac{\partial \mu}{\partial \rho} ,$$

(10.14)

where ρ is the areal density, and the chemical potential is related to the total energy E by $\mu = \partial(E/A)/\partial\rho$. The compressibility is also related to the static longitudinal dielectric function $\epsilon(q, \omega = 0)$ as

$$\lim_{q \to 0} \epsilon(q, \omega = 0) = 1 + V(q)\rho^2 K ,$$

(10.15)

where $V(q)$ is the Fourier transform of the Coulomb interaction. The compressibility of a system is a fundamental thermodynamic quantity.

The capacitance between the 2DES and a metal gate depends on the geometry, and also on the compressibility of the 2DES, and therefore can be used to measure the compressibility. Dorozhkin *et al.* [121, 122] measure the capacitance in the vicinity of $\nu = 1/2$ as a function of the carrier density, and find that the difference between the capacitances at $\nu = 1/2$ and zero magnetic field is insensitive to the carrier density. Furthermore, the magnitude of the difference is well explained by assuming free electrons, for which $d\mu/d\rho = \pi\hbar^2/m_b$ at $B = 0$ and $d\mu/d\rho = 0$ at $\nu = 1/2$ (the latter result follows from the LL degeneracy for free electrons at $\nu = 1/2$). The authors take this observation to mean that the contributions to compressibility from interactions are nearly identical for the Fermi sea at $B = 0$ and the CF Fermi sea at $B^* = 0$. This result is not understood. Eisenstein, Pfeiffer, and West [139] use a modified capacitive technique, in which the compressibility is related to the fraction of an ac field applied by the gate that leaks through the layer (measured by a second layer), to measure the compressibility in the lowest Landau level, as well as the chemical potential discontinuity at the fractional filling $\nu = 1/3$.

11

Composite fermions with spin

So far we have considered fully spin-polarized states, as appropriate in the limit of very large magnetic fields. This chapter concerns the role of spin. The lowest Landau level problem including the spin degree of freedom is defined by

$$H\Psi' = E\Psi',$$ (11.1)

$$H = \frac{e^2}{\epsilon} \sum_{j<k} \frac{1}{|\mathbf{r}_j - \mathbf{r}_k|} + g\mu_e B S_z,$$ (11.2)

$$\Psi' = A[\Psi[\{z_j\}] \alpha_1 \ldots \alpha_{N_\uparrow} \beta_{[1]} \ldots \beta_{[N_\downarrow]}].$$ (11.3)

Here α and β are the spin-up and spin-down spinors, A is the antisymmetrization operator, $[j] \equiv N_\uparrow + j$, g is the Landé g factor, $\mu_e = e\hbar/2m_e c$ is the Bohr magneton for the electron (m_e is the electron mass in vacuum), and

$$S_z = \sum_j S_z^j$$ (11.4)

is the z component of the total spin. We use the symbol Ψ' for the full wave function and Ψ for the spatial part, the two related as in Eq. (11.3).

The spin degree of freedom is frozen when the Zeeman splitting is large compared to the interaction energy. Let us see when that is true. The Zeeman splitting, i.e., the energy to flip the spin of an electron, is given by

$$E_Z = 2g\mu_e B S_z^j = g\mu_e B.$$ (11.5)

For an electron in vacuum, $g \approx 2.0$, giving $E_Z = \hbar e B/m_e c$, which is precisely equal to the cyclotron energy in vacuum. The situation, however, is different for electrons caught inside the band structure of a material. For example, in GaAs, the g factor is renormalized to $g \approx -0.4$ (note the sign change) and the cyclotron energy is given by $\hbar\omega_c = \hbar e B/m_b c$

where $m_b = 0.067m_e$ is the band mass of the electron. As a result, the Zeeman splitting is strongly suppressed, with

$$\frac{E_Z}{\hbar \omega_c} \approx \frac{1}{70} \quad \text{(GaAs)}. \tag{11.6}$$

The Zeeman energy is actually rather small compared to the Coulomb energy $e^2/\epsilon\ell$ under typical experimental conditions. For $B = 9\,\text{T}$, we have $(e^2/\epsilon\ell)/E_Z \approx 55$. More appropriate is to compare the Zeeman energy with the effective cyclotron energy of composite fermions (a representative value for which can be taken to be $0.1e^2/\epsilon\ell$, the $\nu = 1/3$ gap), which is also often larger than, or comparable to, the Zeeman splitting. Therefore, it may sometimes be energetically favorable for a nonzero fraction of particles to reverse their spins, provided they can gain more in interaction energy than they lose in Zeeman energy.

Because the Coulomb interaction is spin independent (it only depends on the position of the electrons), the Hamiltonian commutes with both $\boldsymbol{S}^2$ and $\mathcal{S}_z$, and the eigenstates Ψ' satisfy

$$\boldsymbol{S}^2 \Psi' = S(S+1)\Psi', \tag{11.7}$$

$$\mathcal{S}_z \Psi' = S_z \Psi'. \tag{11.8}$$

The last term in Eq. (11.2) can be replaced by $g\mu_e BS_z$; it does not affect the eigenfunctions, and contributes $g\mu_e BS_z$ to the eigenenergies. Thus, the solutions for $E_Z \neq 0$ can be obtained from those for $E_Z = 0$. (The rotation symmetry in the spin space is destroyed by spin–orbit coupling, which is rather small in GaAs and is not considered here.)

We close the introductory section with a comment on the particle–hole symmetry in the lowest Landau level, which refers to the possibility of describing the system in terms of either electrons or holes. For a fully polarized system, interacting electrons at ν are equivalent to interacting holes at $1 - \nu$, and the filling factors in the range $1 > \nu > 1/2$ are related to filling factors $0 < \nu < 1/2$. Including the spin degree of freedom, the particle–hole symmetry relates

$$\nu \Longleftrightarrow 2 - \nu. \tag{11.9}$$

11.1 Controlling the spin experimentally

The states or excitations that involve reversed spins are sensitive to the Zeeman energy. The Zeeman energy can be varied experimentally in several ways.

Tilting the sample A clever trick for changing the Zeeman energy, while leaving everything else unaffected (at least to the lowest order), involves application of a magnetic field paralled to the two-dimensional electron system. This is accomplished conveniently by tilting the magnetic field, while adjusting its magnitude to keep the normal component unchanged. These experiments are referred to as "tilted-field" experiments. The filling factor remains constant, for it depends only on the normal component $B_\perp$, but, being proportional to the total magnetic field B, the Zeeman energy increases. For a system with zero width, the component of the magnetic field parallel to the plane affects only the Zeeman energy. The experimental samples have a nonzero thickness, and the parallel component of the magnetic field also couples, in principle, to the orbital motion of the electron. Because the thickness is typically small, a good first approximation is to neglect the nonzero thickness effects, which we make throughout this chapter (although one must be mindful of this approximation under extreme tilts).

Changing the density The Zeeman energy at a given filling factor can be varied by changing the density (by tuning the voltage on a front or a back gate). Both the Zeeman and the Coulomb energies change in this case.

Varying the Landé *g* factor The g factor in GaAs, -0.44, is different from $g = 2.0$ in vacuum because of corrections arising from band structure and spin–orbit coupling [250, 557]. It can be tuned by quantum confinement (as in a GaAs quantum well) and/or by application of hydrostatic pressure. The g factor of bulk GaAs is estimated to pass through zero at about 17 kbar. For quantum wells, $g = 0$ is obtained at a width-dependent pressure; g is close to its bulk value for 20 nm quantum wells, but passes through zero, under normal pressure, for a quantum well width of ≈ 5.5 nm [606]. Different semiconductor hosts can also have wide variations in the values of the g factor; FQHE in non-GaAs samples is, at the moment, not very well developed, however.

11.2 Violation of Hund's first rule

In spite of the smallness of the Zeeman energy, the states for $\nu < 1$ may be expected to be fully spin polarized for the following reason. Let us consider first the limit of zero Zeeman energy and recall the well-known Hund's first rule, which has been quite successful in predicting the ground state spin in atomic physics. According to this rule, of all states with the same kinetic energy, the ground state is the one with maximum spin polarization; whenever possible, the ground state is fully spin polarized. The fully polarized state typically has the lowest interaction energy, because a Pauli repulsion is already built into the spatial part of the wave function due to antisymmetry. This might suggest that for LL fillings less than one, when the kinetic energy of electrons is suppressed, the electron system should be fully spin polarized even at zero Zeeman energy.

That, however, is not the case. Tilted field experiments already indicated in the 1980s that the FQHE states are sometimes not fully spin polarized. Certain FQHE states were found to disappear with increasing Zeeman energy, suggesting less than full spin polarization. Clark *et al.* [90] observed a difference between the behaviors of $\nu = 4/3$ and $\nu = 5/3$; the former first disappears and then reappears with tilting, while the latter remains largely unaffected. Eisenstein *et al.* [136] observed a sharp change in the measured gap at $\nu = 8/5$

as a function of the tilt, first decreasing and then increasing, which was interpreted as a transition from a spin-unpolarized state into a maximally spin-polarized one. A collapse of the $\nu = 5/2$ FQHE with tilt was also initially interpreted as implying a non-fully spin-polarized state (Eisenstein *et al.* [135]), but is now believed to be a result of the formation of an anisotropic phase (Lilly *et al.* [399]; Pan *et al.* [490]), similar to that observed in higher Landau levels without tilt, and thought to be related to the formation of stripes. However, the observations where the FQHE first disappears and then *re*appears with increasing Zeeman energy make a compelling case for the involvement of spin.

Exact diagonalization studies, with E_Z set equal to zero, also showed convincingly (Xie, Guo, and Zhang [707]; Yoshioka [730]; Zhang and Chakraborty [741]) that the ground state is often partially spin polarized or unpolarized. See, for example, Fig. 11.1. In fact, the ground state is almost never fully spin polarized for $\nu < 2$. Hund's maximum spin rule is badly violated.

The experiments and numerical studies mentioned above were only the tip of the iceberg, and elegant experimental work in the 1990s, to be described below, has revealed a much more complete picture of the role of spin in the FQHE. We see below that many facts are straightforwardly explained, both qualitatively and quantitatively, by the CF theory.

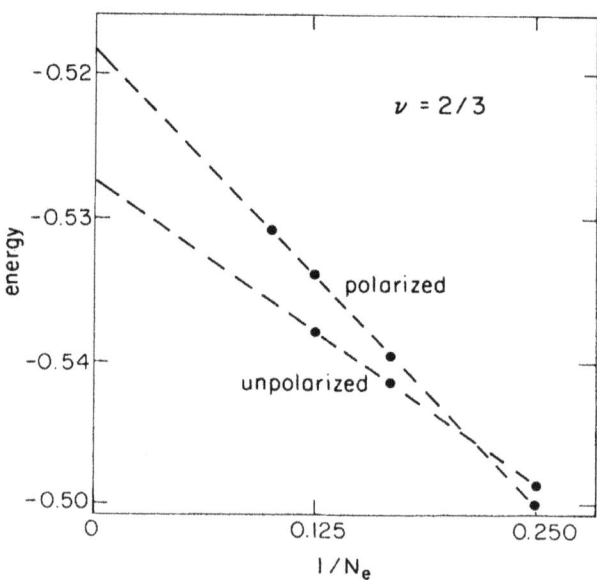

Fig. 11.1. The ground state energy per particle for the fully spin-polarized and the spin-singlet states at $\nu = 2/3$ as a function of $1/N_e$, N_e being the number of electrons. The energies, quoted in units of $e^2/\epsilon\ell$, are obtained from exact diagonalization on the spherical geometry. The spin-singlet state has lower energy in the limit $N_e \to \infty$. Source: X. C. Xie, Y. Guo, and F. C. Zhang, *Phys. Rev. B* **40**, 3487 (1989). (Reprinted with permission.)

11.3 Mean-field model of composite fermions with a spin

Let us begin by recalling the situation for noninteracting electrons. Gaps appear at integral filling factors, $\nu = n$, given by

$$n = n_\uparrow + n_\downarrow, \tag{11.10}$$

where $n_\uparrow$ is the number of occupied spin-up Landau bands and $n_\downarrow$ is the number of occupied spin-down Landau bands. This state is denoted by

$$\Phi_{n_\uparrow, n_\downarrow} = \Phi_{n_\uparrow} \Phi_{n_\downarrow}. \tag{11.11}$$

We are not writing explicitly the full wave function including the spinor coordinates; that is done carefully in a later section. Which $(n_\uparrow, n_\downarrow)$ state becomes the ground state depends on the Zeeman energy, and transitions from one ground state to another can be affected by varying the Zeeman energy relative to the cyclotron energy.[1]

Assuming that composite fermions are formed even at low Zeeman splittings, and can be taken as nearly independent particles with an effective mass (both these assumptions will need to be tested against experiments and exact results), their physics is the same as that described in the previous paragraph, with "electrons" replaced by "composite fermions," "Landau levels" by "Λ levels," "filling factor" by "effective filling factor," and "cyclotron energy" by "effective cyclotron energy." Figure 11.2 illustrates the physics schematically for the integral filling $\nu^* = 4$ of composite fermions. The IQHE of composite fermions

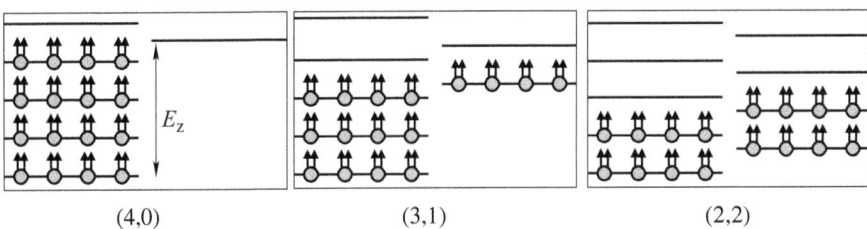

(4,0)   (3,1)   (2,2)

Fig. 11.2. Schematic view of the evolution of the FQHE state at $\nu = 4/9$ or $4/7$, which maps into $\nu^* = 4$ filled Λ levels, as a function of the Zeeman energy, E_Z. The horizontal lines are the Λ levels, with the spin-up Λ levels on the left in each panel and the spin-down Λ levels on the right. The energy difference between the lowest Λ levels on the left and right in each panel is equal to the Zeeman energy. The left-most panel shows the situation at a very high Zeeman energy, when only the spin-up Λ levels are occupied, producing a fully spin-polarized state. The middle panel shows an intermediate Zeeman energy, with three spin-up and one spin-down Λ levels occupied, corresponding to a partially spin-polarized state. The last panel shows the the spin-singlet state obtained in the low Zeeman energy limit, wherein two spin-up and two spin-down Λ levels are occupied. The states are denoted by $(n_\uparrow, n_\downarrow) = (4,0)$, $(3,1)$, and $(2,2)$.

---

[1] Although this would require an extreme tilting in GaAs, for which the Zeeman energy is tiny compared to the cyclotron energy.

occurs at $\nu^* = n = n_\uparrow + n_\downarrow$, which corresponds, as before, to electron fillings

$$\nu = \frac{n}{2pn \pm 1} = \frac{n_\uparrow + n_\downarrow}{2p(n_\uparrow + n_\downarrow) \pm 1}. \tag{11.12}$$

The spin polarization of the state is given by

$$\gamma_e = \frac{n_\uparrow - n_\downarrow}{n_\uparrow + n_\downarrow}, \tag{11.13}$$

and the wave function for the FQHE state is

$$\Psi_{\frac{n}{2pn+1}} = \mathcal{P}_{LLL} \Phi_{n_\uparrow, n_\downarrow} \Phi_1^{2p}, \tag{11.14}$$

or

$$\Psi_{\frac{n}{2pn-1}} = \mathcal{P}_{LLL} \Phi^*_{n_\uparrow, n_\downarrow} \Phi_1^{2p}. \tag{11.15}$$

Particle–hole symmetry in the lowest Landau level implies FQHE at

$$\nu = 2 - \frac{n}{2pn \pm 1}. \tag{11.16}$$

Which of these states will be the ground state for a given set of parameters? The quantitative answer from the CF theory would require a careful calculation with microscopic wave functions, which we come to later; here, we wish to see what qualitative understanding can be achieved by working with the mean-field model of nearly free composite fermions.

Let us first consider a very small but nonzero E_Z. (The case of $E_Z = 0$ is considered in Section 11.5.) When n is an even integer, the minimum energy state (the relevant energy being the effective cyclotron energy) is the one in which

$$n_\uparrow = n_\downarrow = \frac{n}{2}, \qquad n = \text{even}, \tag{11.17}$$

which is a spin-singlet state. For odd n we have

$$n_\downarrow = \frac{(n-1)}{2}, \qquad n_\uparrow = \frac{(n+1)}{2}, \qquad n = \text{odd}, \tag{11.18}$$

where the ground state is partially polarized, with spin polarization given by Eq. (11.13). Thus, the general rule at very small E_Z is: For filling factors with even numerators, the ground state is a spin singlet, whereas for filling factors with odd numerators, the ground state has a nonzero spin polarization. At the Laughlin fraction $\nu = 1/(2p+1)$, the ground state is fully spin polarized even at a very small E_Z.

What happens as the Zeeman energy is increased? Now we must consider other values of $n_\uparrow$ and $n_\downarrow$. For example, at $\nu = 4/9$, which corresponds to $n = 4$ of composite fermions, there are three possibilities: $(n_\uparrow, n_\downarrow) = (2, 2), (3, 1)$, and $(4, 0)$. The unpolarized $(2, 2)$ and the fully polarized $(4, 0)$ have been considered above, but for an intermediate range of Zeeman

Table 11.1. *FQHE states with spin*

ν	n	$(n_\uparrow, n_\downarrow)$	polarization	wave function
$\frac{1}{3}$	1	$(1,0)$	1	$\mathcal{P}_{\mathrm{LLL}} \Phi_{1,0} \Phi_1^2$
$\frac{2}{5}$	2	$(1,1)$	0	$\mathcal{P}_{\mathrm{LLL}} \Phi_{1,1} \Phi_1^2$
$\frac{2}{5}$		$(2,0)$	1	$\mathcal{P}_{\mathrm{LLL}} \Phi_{2,0} \Phi_1^2$
$\frac{3}{7}$	3	$(2,1)$	$\frac{1}{3}$	$\mathcal{P}_{\mathrm{LLL}} \Phi_{2,1} \Phi_1^2$
$\frac{3}{7}$		$(3,0)$	1	$\mathcal{P}_{\mathrm{LLL}} \Phi_{3,0} \Phi_1^2$
$\frac{4}{9}$	4	$(2,2)$	0	$\mathcal{P}_{\mathrm{LLL}} \Phi_{2,2} \Phi_1^2$
$\frac{4}{9}$		$(3,1)$	$\frac{1}{2}$	$\mathcal{P}_{\mathrm{LLL}} \Phi_{3,1} \Phi_1^2$
$\frac{4}{9}$		$(4,0)$	1	$\mathcal{P}_{\mathrm{LLL}} \Phi_{4,0} \Phi_1^2$
$\frac{2}{3}$	2	$(1,1)$	0	$\mathcal{P}_{\mathrm{LLL}} \Phi_{1,1}^* \Phi_1^2$
$\frac{2}{3}$		$(2,0)$	1	$\mathcal{P}_{\mathrm{LLL}} \Phi_{2,0}^* \Phi_1^2$
$\frac{3}{5}$	3	$(2,1)$	$\frac{1}{3}$	$\mathcal{P}_{\mathrm{LLL}} \Phi_{2,1}^* \Phi_1^2$
$\frac{3}{5}$		$(3,0)$	1	$\mathcal{P}_{\mathrm{LLL}} \Phi_{3,0}^* \Phi_1^2$

Notes: Various states at fractions of the form $\nu = n/(2n \pm 1)$.
Spin polarization is defined as $\gamma_e = (n_\uparrow - n_\downarrow)/(n_\uparrow + n_\downarrow)$.

energies, the state $(3,1)$ also becomes relevant (Fig. 11.2). In general, upon increasing E_Z, the ground state spin changes discontinuously when the Λ levels of up and down spins cross one another. After going through several different polarizations, depending on the filling factor under consideration, the fully polarized state becomes the ground state, as shown schematically in Fig. 11.2.

To summarize the predictions from the mean-field CF theory: The inclusion of spin at low Zeeman energies does not introduce new fractions. (The residual interactions between composite fermions, neglected here, may generate new fractions, but these are expected to be weak.) In general, however, many FQHE states can occur at each of the old fractions. The CF theory predicts, for any given fraction, the number of distinct states, their energy ordering, and their spin polarizations. Partially spin-polarized states are obtained naturally in the CF theory. We see in Section 11.5 that Hund's first rule makes a different prediction for the ground state spin when applied to composite fermions rather than electrons, and is found to work quite well; composite-fermionization thus takes precedence over Hund's rule. "The fractional table" (Table 7.1) only depicts the $(n, 0)$ states and must be modified to include the spin physics. The new states can be incorporated by adding a third axis representing the Zeeman energy; the thickness of the table will increase from the left to right, with unit thickness at far left, because only one spin polarization is possible at $\nu = 1/(2p + 1)$.

At the mean-field level, we can equate the cyclotron energies to $\hbar e B^*/m^* c$ to express the values of E_Z at the transitions in terms of a single mass parameter. Such a model is seen to be quite reasonable below.

A consideration of composite fermions in a negative B^* is necessary for a complete picture of the spin physics in the FQHE. As discussed in Chapter 5, the fully polarized FQHE states in the filling factor range $1 > \nu > 1/2$ can be viewed in two ways, either as particle–hole symmetric partners of $\nu = n/(2n+1)$, or in terms of composite fermions in negative B^*. In contrast, for non-fully polarized states, particle–hole symmetry relates states in the range $1/2 > \nu > 0$ to those in the range $2 > \nu > 3/2$. The only way of understanding the partially polarized states at $\nu = n/(2n-1)$ and their particle–hole partners $2 - n/(2n-1)$ is in terms of composite fermions in negative B^*. An example is the spin-singlet state at $\nu = 2/3$, which is $\nu^* = 2$ of composite fermions in negative B^* ($\Phi_{2/3} = \mathcal{P}_{\mathrm{LLL}} \Phi_1^2 \Phi_{1,1}^*$).

11.4 Microscopic theory

We earlier showed that the composite-fermionization of a wave function Φ with a definite orbital angular momentum (i.e., multiplication by Φ_1^{2p} followed by LLL projection) leaves the angular momentum invariant. When the spin degree of freedom is relevant, we construct wave functions for composite fermions in identical manner:

$$\Psi' = \mathcal{P}_{\mathrm{LLL}} \Phi_1^{2p} \Phi', \tag{11.19}$$

where in the spherical geometry

$$\Phi_1 = \prod_{j<k} (u_i v_j - v_i u_j), \tag{11.20}$$

and in the disk geometry

$$\Phi_1 = \prod_{j<k} (z_j - z_k). \tag{11.21}$$

Φ' is the full wave function of electrons at the effective magnetic field, including both the spatial coordinates and spins; the spatial part will be denoted by Φ. Φ_1 is the fully antisymmetric wave function of one filled Landau level of "spinless" electrons. The factor ϕ_1^{2p} attaches $2p$ vortices to each electron to convert it into a composite fermion. Each composite fermion sees $2p$ vortices on every other composite fermion, independent of spin. It may, in principle, be possible to write wave functions in which pairs of electrons with different spins see different vortices than pairs of electrons with the same spin, but, as seen below, the above wave functions provide a quantitative account of experimental observations.

In the spherical geometry, we continue to define the up and down directions for spin relative to the z-axis, and not relative to the direction of the radial magnetic field. In other words, the Zeeman energy term in the Hamiltonian will be taken as $g\mu_e B S_z$ rather than $g\mu_e \mathbf{B} \cdot \mathbf{S}$ (the two being unequal in the spherical geometry). The problem is mathematically well defined, and gives the same thermodynamic limit as the planar geometry. A term like $g\mu_e \mathbf{B} \cdot \mathbf{S}$ will induce spin–orbit coupling in the spherical geometry, which is an unnecessary complication.

The CF construction conserves the spin quantum numbers in going from Φ' to Ψ'. Let us assume that Φ' is an eigenstate of the spin operators $\mathbf{S}^2$ and S_z:

$$\mathbf{S}^2 \Phi' = S(S+1)\Phi', \tag{11.22}$$

$$S_z \Phi' = S_z \Phi'. \tag{11.23}$$

Then we have

$$\begin{aligned}
\mathbf{S}^2 \Psi' &= \mathbf{S}^2 \mathcal{P}_{\mathrm{LLL}} \Phi_1^{2p} \Phi' \\
&= \mathcal{P}_{\mathrm{LLL}} \Phi_1^{2p} \mathbf{S}^2 \Phi' \\
&= \mathcal{P}_{\mathrm{LLL}} \Phi_1^{2p} S(S+1)\Phi' \\
&= S(S+1)\Psi'.
\end{aligned} \tag{11.24}$$

We have used that Φ_1^2 commutes with $\mathbf{S}^2$, because Φ_1^2 is independent of spin, and the projection operator also commutes with spin, as is clear from its form in Eq. (5.55), which involves only the angular momentum operators. Similarly, we have

$$S_z \Psi' = S_z \Psi'. \tag{11.25}$$

11.4.1 Wave functions with spin

In this subsection we discuss some general properties of wave functions for a collection of many identical particles with spin. Of interest are the conditions satisfied by an eigenstate of S_z and $\mathbf{S}^2$. The spin-up and spin-down spinors α and β satisfy

$$\langle \alpha_j | \alpha_k \rangle = \delta_{jk}, \qquad \langle \beta_j | \beta_k \rangle = \delta_{jk}, \qquad \langle \alpha_j | \beta_k \rangle = 0. \tag{11.26}$$

A standard representation of α_j and β_j is

$$\alpha_j = \begin{pmatrix} 1 \\ 0 \end{pmatrix}_j, \qquad \beta_j = \begin{pmatrix} 0 \\ 1 \end{pmatrix}_j. \tag{11.27}$$

The up and down directions are defined relative to a given axis, which is conveniently chosen to be the z-axis (also the direction of the physical magnetic field, whenever such a field is present).

Two particles The most familiar case is that of two particles, for which the wave function can always be expressed as a product of a spin part and a spatial part. The eigenstates of $\mathcal{S}_z$ and $\mathcal{S}^2$ are given by

$$\Psi'_{0,0}(\mathbf{r}_1,\mathbf{r}_2) = \chi_S(\mathbf{r}_1,\mathbf{r}_2)\frac{1}{\sqrt{2}}[\alpha_1\beta_2 - \beta_1\alpha_2], \tag{11.28}$$

$$\Psi'_{1,0}(\mathbf{r}_1,\mathbf{r}_2) = \chi_A(\mathbf{r}_1,\mathbf{r}_2)\frac{1}{\sqrt{2}}[\alpha_1\beta_2 + \beta_1\alpha_2], \tag{11.29}$$

$$\Psi'_{1,1}(\mathbf{r}_1,\mathbf{r}_2) = \chi_A(\mathbf{r}_1,\mathbf{r}_2)\alpha_1\alpha_2, \tag{11.30}$$

$$\Psi'_{1,-1}(\mathbf{r}_1,\mathbf{r}_2) = \chi_A(\mathbf{r}_1,\mathbf{r}_2)\beta_1\beta_2. \tag{11.31}$$

Here, the subscript of Ψ' displays the spin quantum numbers S, S_z. When the spin part is antisymmetric then the spatial part is symmetric, and vice versa. The former case corresponds to a spin singlet ($S = 0$), whereas the latter to a spin triplet ($S = 1$).

Many particles For $N \geq 3$ the wave function, in general, may not be written in a product form. Let us consider the situation in which $N_\uparrow$ electrons have up spin, and $N_\downarrow = N - N_\uparrow$ electrons have down spin. To construct a basis function for a given set of N orbitals, we place the first $N_\uparrow$ electrons in the single particle orbitals $\eta_1(\mathbf{r}_1)\alpha_1, \ldots, \eta_{N_\uparrow}(\mathbf{r}_{N_\uparrow})\alpha_{N_\uparrow}$, the remaining $N_\downarrow$ electrons in $\zeta_{[1]}(\mathbf{r}_{[1]})\beta_{[1]}, \ldots, \zeta_{[N_\downarrow]}(\mathbf{r}_{[N_\downarrow]})\beta_{[N_\downarrow]}$, and antisymmetrize with respect to an exchange of spins *and* the coordinates of any two particles. (We use the convenient notation $[j] = N_\uparrow + j$.) This produces the Slater determinant

$$\Phi' = \frac{1}{\sqrt{N!}} \begin{vmatrix} \eta_1(\mathbf{r}_1)\alpha_1 & \eta_1(\mathbf{r}_2)\alpha_2 & . & . & \eta_1(\mathbf{r}_N)\alpha_N \\ \eta_2(\mathbf{r}_1)\alpha_1 & \eta_2(\mathbf{r}_2)\alpha_2 & . & . & \eta_2(\mathbf{r}_N)\alpha_N \\ . & . & . & & \\ . & . & . & & \\ . & . & . & & \\ \zeta_{[N_\downarrow]}(\mathbf{r}_1)\beta_1 & \zeta_{[N_\downarrow]}(\mathbf{r}_2)\beta_2 & & & \zeta_{[N_\downarrow]}(\mathbf{r}_N)\beta_N \end{vmatrix}$$

$$= \sqrt{\frac{N_\uparrow!N_\downarrow!}{\sqrt{N!}}}\mathcal{A}[\Phi(\{\mathbf{r}_j\})\alpha_1 \cdots \alpha_{N_\uparrow}\beta_{[1]} \cdots \beta_{[N_\downarrow]}], \tag{11.32}$$

where

$$
\Phi = \frac{1}{\sqrt{N_\uparrow!}}
\begin{vmatrix}
\eta_1(\mathbf{r}_1) & \eta_1(\mathbf{r}_2) & \cdot & \cdot & \cdot \\
\eta_2(\mathbf{r}_1) & \eta_2(\mathbf{r}_2) & \cdot & \cdot & \cdot \\
\cdot & & \cdot & & \\
\cdot & & & \cdot & \\
\cdot & & & & \cdot
\end{vmatrix}
$$

$$
\times \frac{1}{\sqrt{N_\downarrow!}}
\begin{vmatrix}
\zeta_{[1]}(\mathbf{r}_{[1]}) & \zeta_{[1]}(\mathbf{r}_{[2]}) & \cdot & \cdot & \cdot \\
\zeta_{[2]}(\mathbf{r}_{[1]}) & \zeta_{[2]}(\mathbf{r}_{[2]}) & \cdot & \cdot & \cdot \\
\cdot & & \cdot & & \\
\cdot & & & \cdot & \\
\cdot & & & & \cdot
\end{vmatrix} .
\tag{11.33}
$$

A antisymmetrizes the wave function with respect to an exchange of both the spin and the spatial coordinates of two particles. The second step in Eq. (11.32) follows by considering the coefficient of a given spin configuration.[2] Obviously, all η's, or all ζ's, must be distinct, but some of the η's may be the same as some of the ζ's.

The general wave function for "spinful" electrons is a linear superposition of such Slater determinant basis states, given by

$$
\Phi' = A[\Phi[\{z_j\}] \, \alpha_1 \cdots \alpha_{N_\uparrow} \beta_{[1]} \cdots \beta_{[N_\downarrow]}],
\tag{11.35}
$$

where Φ, in general, has a complicated form. While any wave function of this type is a valid wave function insofar as the Pauli principle is concerned, our interest is in wave functions that are eigenstates of $\mathcal{S}^2$ and $\mathcal{S}_z$. The antisymmetrization produces terms in which different particles will have up and down spins, but the number $N_\downarrow$ or $N_\uparrow$ is the same for each term, indicating that Φ' is already an eigenstate of S_z. We now derive what properties $\Phi[\{z_j\}]$ must satisfy in order for Φ' to be an eigenstate of $\mathcal{S}^2$.

Without any loss of generality, the spatial part $\Phi[\{z_j\}]$ is taken to be antisymmetric with respect to the exchange of coordinates of two like spin particles. It can always be brought into this form, because any part symmetric with respect to such an interchange is eliminated by the antisymmetrization operator. To see this, if Φ is not antisymmetric with respect to the exchange $z_1 \leftrightarrow z_2$, we write it as a sum of a symmetric and an antisymmetric parts:

$$
\Phi = \frac{1}{2}[e + (1, 2)]\Phi + \frac{1}{2}[e - (1, 2)]\Phi,
\tag{11.36}
$$

---

[2] Alternatively, we can use the Laplace expansion of a determinant. Let us take the first m rows of an $n \times n$ matrix A. From these, we choose m columns to form a submatrix, B_j, and denote the complementary $(n - m) \times (n - m)$ submatrix C_j. The Laplace expansion of the determinant is given by

$$
\det A = \sum_j \epsilon_\mathrm{P} \det B_j \det C_j,
\tag{11.34}
$$

where ϵ_P is the sign of the permutation P bringing the m columns of B_j followed by the $n - m$ columns of C_j into the original order.

where (i, j) permutes the coordinate labels z_i and z_j, and e is the identity operator. The symmetric part is annihilated upon the overall antisymmetrization of the wave function. Proceeding in this manner, the desired form for Φ results.

11.4.2 Fock's cyclic condition

The state with $S_z = S$ is called the "highest weight" state. Given an eigenstate with $S_z = S$, a multiplet of eigenstates with $S_z < S$ can be constructed by successive applications of the S_z lowering operator. These have identical interaction energy, because the spin commutes with the Coulomb term in the Hamiltonian; the energy splitting of the states in a given spin multiplet is an integral multiple of the Zeeman splitting.

The requirement that Φ' possess definite symmetry properties under rotation in the spin space (i.e., have good spin quantum numbers) restricts the form of the spatial part Φ. An acceptable Φ must satisfy, for the highest weight vector, what is known as Fock's cyclic condition, which we now derive. We assume in the following that $N_\uparrow \geq N_\downarrow$.

Theorem 11.1 Φ' *has spin* $S = S_z$ *if and only if* $\Phi(\{r_j\})$ *is annihilated by an attempt to antisymmetrize a spin-down electron* [l] *with respect to the spin-up electrons, that is*

$$\left[e - \sum_{k=1}^{N_\uparrow} (k, [l]) \right] \Phi[\{z_j\}] = 0, \tag{11.37}$$

where the identity operator e *makes no permutations and* $(k, [l])$ *is a label exchange operator, exchanging* r_k *and* $r_{[l]}$. *This condition is known as Fock's cyclic condition* [234].

Proof We begin by showing that the choice of [l] is arbitrary, i.e., if the above equation is true for a given [l], then it is true for all [m].

$$0 = ([l], [m]) \left[e - \sum_{k=1}^{} (k, [l]) \right] \Phi[\{z_j\}]$$

$$= ([l], [m]) \left[e - \sum_{k=1}^{} (k, [l]) \right] ([l], [m])([l], [m]) \Phi[\{z_j\}]$$

$$= -([l], [m]) \left[e - \sum_{k=1}^{} (k, [l]) \right] ([l], [m]) \Phi[\{z_j\}]$$

$$= - \left[e - \sum_{k=1}^{} (k, [m]) \right] \Phi[\{z_j\}] \tag{11.38}$$

for arbitrary [m].

Let us first assume that Φ' is an eigenstate of $\boldsymbol{S}^2$ and $\mathcal{S}_z$ with eigenvalues $S(S+1)$ and S. The spin raising operator must annihilate this state. The total spin is

$$\boldsymbol{S} = \sum_j \boldsymbol{S}_j \tag{11.39}$$

and the spin raising and lowering operators are defined as

$$\mathcal{S}^+ = \mathcal{S}^x + i\mathcal{S}^y = \sum_j (\mathcal{S}_j^x + i\mathcal{S}_j^y) = \sum_j \mathcal{S}_j^+, \tag{11.40}$$

$$\mathcal{S}^- = \mathcal{S}^x - i\mathcal{S}^y = \sum_j (\mathcal{S}_j^x - i\mathcal{S}_j^y) = \sum_j \mathcal{S}_j^-, \tag{11.41}$$

$$\mathcal{S}_j^- \alpha_j = \beta_j, \qquad \mathcal{S}_j^- \beta_j = 0, \tag{11.42}$$

$$\mathcal{S}_j^+ \alpha_j = 0, \qquad \mathcal{S}_j^+ \beta_j = \alpha_j. \tag{11.43}$$

We have

$$0 = \mathcal{S}^+ A[\Phi \alpha_1 \dots \alpha_{N_\uparrow} \beta_{[1]} \dots \beta_{[N_\downarrow]}]$$

$$= \sum_{l=1}^{N_\downarrow} A[\Phi \alpha_1 \dots \alpha_{N_\uparrow} \alpha_{[l]} \beta_{[1]} \dots \beta_{[l-1]} \beta_{[l+1]} \dots \beta_{[N_\downarrow]}]$$

$$= \sum_{l=1}^{N_\downarrow} A[\Phi_{[l]} \alpha_1 \dots \alpha_{N_\uparrow} \alpha_{[l]} \beta_{[1]} \dots \beta_{[l-1]} \beta_{[l+1]} \dots \beta_{[N_\downarrow]}] \tag{11.44}$$

where

$$\Phi_{[l]} = \left[e - \sum_{k=1}^{N_\uparrow} (k, [l]) \right] \Phi[\{z_j\}]. \tag{11.45}$$

The last step follows because, as shown earlier, the antisymmetrization forces Φ to be antisymmetric under the exchange of the coordinates of electrons of like spin. Now, $\Phi_{[l]}$ is antisymmetric under the exchange of any two coordinates from the set $\{r_1, \dots, r_{N_\uparrow}, r_{[l]}\}$, or any two coordinates from the set $\{r_{[1]}, \dots, r_{[l-1]}, r_{[l+1]}, \dots, r_{[N_\downarrow]}\}$. Therefore, the coefficient of a given spin configuration, say $\alpha_1 \dots \alpha_{N_\uparrow} \alpha_{[l]} \beta_{[1]} \dots \beta_{[l-1]} \beta_{[l+1]} \dots \beta_{[N_\downarrow]}$, is $\Phi_{[l]}$, which must vanish to ensure that the spin raising operator annihilates the state. This produces Eq. (11.37).

Alternatively, let us assume that Φ satisfies the Fock cyclic condition, and show that the state Φ' is an eigenstate of $\mathbf{S}^2$ with eigenvalue $S_z(S_z + 1)$ [194]. For this purpose we use the explicit form

$$\mathbf{S}^2 = \mathcal{S}_z^2 + \frac{1}{2}(\mathcal{S}^+ \mathcal{S}^- + \mathcal{S}^- \mathcal{S}^+). \tag{11.46}$$

The last term can be written as

$$\frac{1}{2}\sideset{}{'}\sum_{k,l}(\mathcal{S}_k^+ \mathcal{S}_l^- + \mathcal{S}_k^- \mathcal{S}_l^+) + \frac{1}{2}\sum_{k}(\mathcal{S}_k^+ \mathcal{S}_k^- + \mathcal{S}_k^- \mathcal{S}_k^+)$$

$$= \frac{1}{2}\sideset{}{'}\sum_{k,l}(\mathcal{S}_k^+ \mathcal{S}_l^- + \mathcal{S}_k^- \mathcal{S}_l^+) + \sum_{k}(\mathcal{S}_k^2 - \mathcal{S}_{kz}^2),$$

where the prime denotes the condition $l \neq k$. For particles with spin 1/2, $\mathcal{S}_k^2 - \mathcal{S}_{kz}^2 = \frac{1}{2}(\frac{1}{2} + 1) - (\frac{1}{2})^2 = \frac{1}{2}$, which gives

$$\mathbf{S}^2 = \mathcal{S}_z^2 + \frac{N}{2} + \frac{1}{2}\sideset{}{'}\sum_{k,l}(\mathcal{S}_k^+ \mathcal{S}_l^- + \mathcal{S}_k^- \mathcal{S}_l^+). \tag{11.47}$$

When $\mathbf{S}^2$ acts on Φ', the action of the first two terms on the right hand side of the preceding equation is straightforward. In the sum, the terms which attempt to raise the spin of a spin-up electron or lower the spin of a spin-down electron do not contribute. The remaining terms are

$$\sum_{k=1}^{N_\uparrow}\sum_{l=1}^{N_\downarrow} \mathcal{S}_{[l]}^+ \mathcal{S}_l^-. \tag{11.48}$$

$\mathcal{S}_{[l]}^+ \mathcal{S}_k^-$ changes the spins of the kth and $[l]$th particles in $\alpha_1 \dots \alpha_{N_\uparrow} \beta_{[1]} \dots \beta_{[N_\downarrow]}$. However, we may now exchange the spin *and* the coordinates of the two particles inside the antisymmetrization, which gives an additional negative sign. Therefore, we get

$$\mathbf{S}^2 A[\Phi[\{z_j\}] \, \alpha_1 \dots \alpha_{N_\uparrow} \beta_{[1]} \dots \beta_{[N_\downarrow]}] = A[\chi[\{z_j\}] \, \alpha_1 \dots \alpha_{N_\uparrow} \beta_{[1]} \dots \beta_{[N_\downarrow]}]. \tag{11.49}$$

Here

$$\chi[\{z_j\}] = \left[\left(\frac{N_\uparrow - N_\downarrow}{2} \right)^2 + \frac{N}{2} \right] \Phi[\{z_j\}] - \sum_{k=1}^{N_\uparrow} \sum_{l=1}^{N_\downarrow} (k, [l]) \Phi[\{z_j\}]$$

$$= \left[\left(\frac{N_\uparrow - N_\downarrow}{2} \right)^2 + \frac{N}{2} \right] \Phi[\{z_j\}] - \sum_{l=1}^{N_\downarrow} \Phi[\{z_j\}]$$

$$= \left[\left(\frac{N_\uparrow - N_\downarrow}{2} \right)^2 + \frac{N}{2} \right] \Phi[\{z_j\}] - N_\downarrow \Phi[\{z_j\}]$$

$$= \left(\frac{N_\uparrow - N_\downarrow}{2} \right) \left(\frac{N_\uparrow - N_\downarrow}{2} + 1 \right) \Phi[\{z_j\}], \tag{11.50}$$

where we have used the cyclic condition. It follows that Φ' is an eigenstate of $\mathcal{S}^2$ with eigenvalue $S(S+1)$ with

$$S = \frac{N_\uparrow - N_\downarrow}{2} = S_z. \tag{11.51}$$

$\square$

11.4.3 Examples

To summarize the result from the previous subsection, a state Φ' in Eq. (11.35) has a good symmetry under rotation in the spin space if:

- The spatial part Φ is antisymmetric to an exchange of two coordinates belonging to particles with like spins.
- Φ is annihilated if we further attempt to antisymmetrize a spin-down electron with respct to the spin-up electrons (the cyclic symmetry). Here we have assumed $N_\uparrow \geq N_\downarrow$.

For strongly correlated systems, construction of wave functions satisfying these conditions is not always obvious. The CF theory will tell us how to do that for the incompressible FQHE states and their excitations. Most relevant for our purpose are product states of the type

$$\Phi = \Phi_\uparrow \Phi_\downarrow, \tag{11.52}$$

where $\Phi_\uparrow$ and $\Phi_\downarrow$ are antisymmetric, either single Slater determinants or linear combinations of Slater determinants. In general, whenever all single particle orbitals participating in $\Phi_\downarrow$ are *fully* occupied in $\Phi_\uparrow$, Φ is annihilated by any attempt to antisymmetrize a spin-down electron with respect to the spin-up electrons, because the spin-up orbital in question is already occupied. Φ' is therefore an eigenstate of the spin with $S = S_z$. Such states appear more often than one might first imagine. Some examples are:

(i) States in which all particles have spin up. In this case, Φ is fully antisymmetric, and the wave function has the simple factorized form

$$\Phi' = \Phi[\{z_j\}]\,\alpha_1 \ldots \alpha_N. \tag{11.53}$$

It is annihilated by the spin raising operator. This is the wave function that we have been considering prior to this chapter.

(ii) States at $\nu = 1 + \nu_\downarrow$ in which the spin-up electrons fill all single particle orbitals in the lowest Landau level, whereas the spin-down electrons only partially occupy the lowest Landau level:

$$\Phi = \Phi_{1\uparrow}\Phi_{\nu_\downarrow}. \tag{11.54}$$

Any choice for $\Phi_{\nu_\downarrow}$ produces a definite spin, provided it is restricted to the lowest Landau level.

(iii) Incompressible states of the kind

$$\Phi = \Phi_{n_\uparrow,n_\downarrow} \equiv \Phi_{n_\uparrow}\Phi_{n_\downarrow}, \tag{11.55}$$

which have $n_\uparrow$ spin-up Landau levels fully occupied and $n_\downarrow$ spin-down Landau levels fully occupied.

(iv) Particle–hole excitations of the form

$$\Phi = \Phi_{n_\uparrow}^{\mathrm{p-h}}\Phi_{n_\downarrow}. \tag{11.56}$$

11.4.4 Microscopic wave functions for FQHE

Composite-fermionization of Φ produces

$$\Psi' = \mathcal{P}_{\mathrm{LLL}}\Phi_1^2 A[\Phi[\{z_j\}]\,\alpha_1 \ldots \alpha_{N_\uparrow}\beta_{[1]}\ldots\beta_{[N_\downarrow]}]$$
$$= A[\Psi[\{z_j\}]\,\alpha_1 \ldots \alpha_{N_\uparrow}\beta_{[1]}\ldots\beta_{[N_\downarrow]}], \tag{11.57}$$

with

$$\Psi = \mathcal{P}_{\mathrm{LLL}}\Phi_1^2\Phi, \tag{11.58}$$

because both Φ_1^2 and the projection operator, being completely symmetric with respect to particle exchange (see Eq. (5.55) for the projection operator), commute with antisymmetrization. The spatial part of the wave function for a state with nontrivial spin structure is thus constructed in the same manner as before. In particular, the wave functions for the incompressible ground states at the fractions $\nu = \frac{n}{2pn+1}$ are given by

$$\Psi_{\frac{n}{2pn+1}} = \mathcal{P}_{\mathrm{LLL}}\Phi_1^2\Phi_{n_\uparrow,n_\downarrow}, \tag{11.59}$$

where $n = n_\uparrow + n_\downarrow$.

Projection into the lowest Landau level is handled as before. For method II (Section 5.14.6), we get

$$\Psi = \mathcal{P}_{LLL} \begin{vmatrix} \eta_1(\mathbf{r}_1) & \eta_1(\mathbf{r}_2) & \cdot & \cdot & \cdot \\ \eta_2(\mathbf{r}_1) & \eta_2(\mathbf{r}_2) & \cdot & \cdot & \cdot \\ \cdot & & \cdot & & \\ & \cdot & & \cdot & \\ \cdot & \cdot & & & \cdot \end{vmatrix}$$

$$\times \begin{vmatrix} \zeta_{[1]}(\mathbf{r}_{[1]}) & \zeta_{[1]}(\mathbf{r}_{[2]}) & \cdot & \cdot & \cdot \\ \zeta_{[2]}(\mathbf{r}_{[1]}) & \zeta_{[2]}(\mathbf{r}_{[2]}) & \cdot & \cdot & \cdot \\ \cdot & & \cdot & & \\ & \cdot & & \cdot & \\ \cdot & \cdot & & & \cdot \end{vmatrix} \prod_{j<k} (z_j - z_k)^{2p}$$

$$= \begin{vmatrix} \mathcal{P}_{LLL}\eta_1(\mathbf{r}_1)\mathcal{J}_1^p & \mathcal{P}_{LLL}\eta_1(\mathbf{r}_2)\mathcal{J}_2^p & \cdot & \cdot & \cdot \\ \mathcal{P}_{LLL}\eta_2(\mathbf{r}_1)\mathcal{J}_1^p & \mathcal{P}_{LLL}\eta_2(\mathbf{r}_2)\mathcal{J}_2^p & \cdot & \cdot & \cdot \\ \cdot & & \cdot & & \\ & \cdot & & \cdot & \\ \cdot & \cdot & & & \cdot \end{vmatrix}$$

$$\times \begin{vmatrix} \mathcal{P}_{LLL}\zeta_{[1]}(\mathbf{r}_{[1]})\mathcal{J}_{[1]}^p & \mathcal{P}_{LLL}\zeta_{[1]}(\mathbf{r}_{[2]})\mathcal{J}_{[2]}^p & \cdot & \cdot & \cdot \\ \mathcal{P}_{LLL}\zeta_{[2]}(\mathbf{r}_{[1]})\mathcal{J}_{[1]}^p & \mathcal{P}_{LLL}\zeta_{[2]}(\mathbf{r}_{[2]})\mathcal{J}_{[2]}^p & \cdot & \cdot & \cdot \\ \cdot & & \cdot & & \\ & \cdot & & \cdot & \\ \cdot & \cdot & & & \cdot \end{vmatrix}. \quad (11.60)$$

The Coulomb energy is given by the expectation value

$$E = \frac{\langle \Psi' | \sum_{i<j} V(r_{ij}) | \Psi' \rangle}{\langle \Psi' | \Psi' \rangle}$$

$$= \frac{\langle \Psi | \sum_{i<j} V(r_{ij}) | \Psi \rangle}{\langle \Psi | \Psi \rangle}. \quad (11.61)$$

To see the last step, we write Ψ' explicitly in its antisymmetrized form, which contains $\binom{N}{N_\uparrow}$ terms with different sets of spin-up and spin-down electrons. Because the interaction is spin independent, the spin part can be commuted through it, and annihilates the matrix element for cross terms. All the diagonal terms make the same contribution in the numerator as well as in the denominator. Canceling the factor $\binom{N}{N_\uparrow}$ produces the second equality above. The energy and other spin-independent quantities of physical interest only require the spatial part of the wave function.

11.5 Comparisons with exact results: resurrecting Hund's first rule

The CF theory has been tested for situations involving spin for small systems, where exact results can be obtained numerically. Beginning with the limit $E_Z = 0$ is convenient, and also provides maximum contrast from the fully spin-polarized states we have been considering in earlier chapters. As usual, the comparisons proceed by first asking if the CF theory correctly predicts the quantum numbers of the ground state. We then proceed to test its quantitative validity.

FQHE states with several spin polarizations were obtained in Section 11.3. For $\nu = n/(2pn \pm 1)$, with n even, which maps into $\nu^* = n$ of composite fermions, a unique ground state is obtained even when $E_Z = 0$. Let us reconsider the case of $\nu = n/(2pn \pm 1)$ with n odd. The ground state has $(n-1)/2$ Λ levels fully occupied, with both spins, contributing a filling $n - 1$. The remaining composite fermions reside in the $[(n+1)/2]$th Λ level. For $E_Z = 0$, each composite fermion can have either spin, and the ground state has a large degeneracy.

Further progress can be made by applying Hund's maximum spin rule to composite fermions (Wu and Jain [278, 704]; Onoda *et al.* [486]). Hund's rule is a consequence of the residual interactions between composite fermions, and thus incorporates the interaction between composite fermions at the simplest level. For $\nu = n/(2pn \pm 1)$ with n odd, Hund's rule tells us that composite fermions in the topmost Λ level arrange to have the maximum spin, which can be accomplished by placing them all in spin-up states. (Of course, their spins could have been chosen to point in any arbitrary direction.) That fixes the state uniquely at these filling factors.

To take a concrete example, let us consider the "electron atom" with $(N, 2Q) = (6, 9)$. Hund's maximum spin rule applied to electrons predicts a fully spin-polarized state with $S = 3$, because a sufficient number of single particle orbitals are available. Exact diagonalization tells us, however, that the spin of the ground state is $S = 1$. Let us now analyze the problem in terms of composite fermions and see how Hund's rule is rescued. From $Q^* = Q - N + 1$, the system maps into a "CF atom" $(N, 2Q^*) = (6, -1)$. The sign of Q^* only determines the direction of the magnetic field, which is of no relevance to the issue of ground state quantum numbers. At $2|Q^*| = 1$ the lowest level degeneracy is $2(2|Q^*| + 1) = 4$, which takes care of four of the composite fermions. We put the remaining two in the next level, and the application of Hund's rule gives $S = 1$. In general, Hund's rule is found to work well when applied to composite fermions (Exercise 11.4). Figure 11.3 shows a comparison between the exact ground state spin and the CF prediction.

Table 11.2 compares the CF theory with exact diagonalization results for several systems for which the CF theory predicts filled shell ground states (i.e. $\nu^* = n$). The L and S quantum numbers are predicted correctly by the CF theory, and the projected wave functions of Eq. (11.59) are accurate at the same level as the fully polarized ones. The LLL state in the study of Table 11.2 is obtained by a "hard-core projection," defined as $\Psi_{n/(2n-1)} = \Phi_1 \mathcal{P}_{\text{LLL}} \Phi_1 \Phi^*_{n_\uparrow, n_\downarrow}$. Because of the explicit factor Φ_1, this wave function guarantees that no two electrons can coincide, independent of their spin. In other words, it has vanishing

Table 11.2. *Testing CF theory for partially spin polarized FQHE states*

ν	N	Q	S	L	overlap	E_{CF}	E_{exact}
	4	5/2	0	0	0.999 83	$-0.600\,94$	$-0.601\,02$
2/3	6	4	0	0	0.998 96	$-0.574\,20$	$-0.574\,38$
	8	11/2	0	0	0.998 05	$-0.561\,58$	$-0.561\,82$
3/5	5	4	3/2	0	1	$-0.534\,94$	$-0.534\,94$
	8	13/2	2	0	0.991 75	$-0.522\,76$	$-0.523\,04$

Notes: CF predictions obtained from Eq. (11.59).
Zeeman energy set to zero.
S is total spin.

Source: Wu, Dev, and Jain [703].

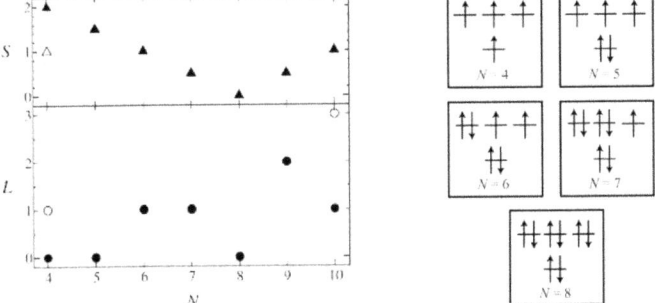

Fig. 11.3. Left panel: The spin and the orbital angular momentum quantum numbers (solid triangles and solid circles, respectively) of the exact ground states of N particles at $Q = N - 1$, which corresponds to $Q^* = Q - N + 1 = 0$ (i.e., $B^* = 0$). The open symbols show the prediction from the CF theory; they are visible only when the prediction disagrees with the exact result. Right panel: Schematic occupation of the CF Λ level orbitals. Recall that for $Q^* = 0$ the lowest Λ level is an s-shell, the second Λ level is a p-shell, and so on. From M. Onoda, T. Mizusaki, T. Otsuka, and H. Aoki, *Physica B: Condensed Matter* **298**, 173 (2001). (Reprinted with permission.)

energy for a delta function interaction

$$V_{\text{HC}} = V_0 \sum_{j<k} \delta^{(2)}(r_j - r_k). \tag{11.62}$$

(For fully polarized electrons this interaction is equivalent to no interaction at all, because the Pauli principle guarantees that the spatial part of the wave function vanishes at coincident points. For partially polarized states, the wave function in general does not vanish when two electrons with unlike spin have the same coordinates.)

The reason why Hund's rule fails for electrons is as follows: We consider first electrons in an atomic shell in zero magnetic field, and assume a strong delta function repulsive interaction. (Coupling to other shells is neglected.) The fully polarized state has zero interaction energy, because the spatial part of the wave function is fully antisymmetric, vanishing when two electrons coincide. This is a simple way of understanding why the repulsive interaction leads to maximum spin. Returning to electrons in the lowest Landau level, for $\nu < 1$, we can write the wave function as

$$\Psi' = \Phi_1 \Psi'_B, \qquad (11.63)$$

where Ψ'_B is a wave function for spinful bosons at effective flux

$$Q^B = Q - \frac{N-1}{2}, \qquad (11.64)$$

and is symmetric under exchange of two particles. All wave functions Ψ' of this type have zero energy for the delta function interaction of Eq. (11.62), because the factor Φ_1 explicitly builds in the hard-core condition. The spin and angular momentum quantum numbers of Ψ are the same as those of Ψ_B, which can take a large range of values. As a result, wave functions with maximum spin are not picked out by the hard-core constraint. The hard-core condition does not tell us what the ground state spin is, with one exception discussed in the next paragraph.

For the special case of $\nu = 1$, we have $\Psi'_B = 1$. Here the hard-core ground state is unique, which is completely antisymmetric in spatial coordinates, and therefore fully spin-polarized. The hard-core condition thus gives a fully spin polarized state at $\nu = 1$.

One may wonder about the applicability of Hund's *second* rule, which states that the ground state angular momentum has the maximum value consistent with the first rule. This rule also fails completely when applied to electrons. It works much better for composite fermions (Onoda *et al.* [486]; Rezayi and Read [549]; Fig. 11.3), but is not as reliable as the first rule (as is also the case in atomic physics). We note that the L quantum number is predicted correctly when the first rule produces a filled shell, in which case we have $L = 0$.

11.6 Phase diagram of the FQHE with spin

In a noteworthy experiment, Du and collaborators [126] determine the phase diagram of the FQHE in the filling factor range $2 > \nu > 1$ as a function of the Zeeman energy. The reason for focusing on this filling factor range, as opposed to $1 > \nu > 0$, is that it allows one to begin with small Zeeman energies without tilting, thus enabling access to a larger range of parameters (the Zeeman energy can only be increased by tilting the sample). The $2 > \nu > 1$ region is conveniently viewed in terms of holes at a filling of $2 - \nu$ in the lowest Landau level, with holes capturing vortices to make composite fermions. For example, the $\nu = 8/5$ FQHE state is the $2/5$ FQHE of holes, which, in turn, is two filled Λ levels of composite fermions built from the holes of the lowest Landau level. In an extensive study

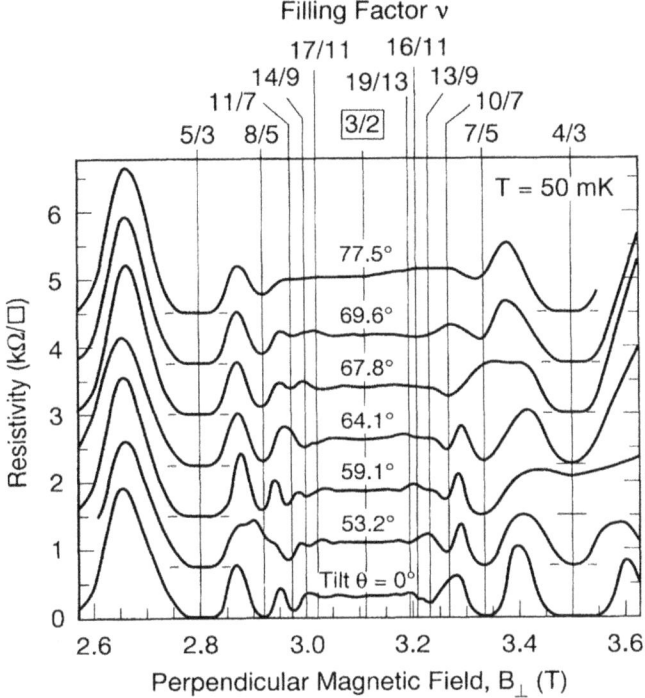

Fig. 11.4. Angular dependence of the magnetoresistance at filling factors near $\nu = 3/2$. $B_{tot} = B_\perp / \cos\theta$ is the total magnetic field at angle θ. A magnetoresistance minimum signifies an incompressible FQHE state. Source: R. R. Du, A. S. Yeh, H. L. Stormer, D. C. Tsui, L. N. Pfeiffer, and K. W. West, *Phys. Rev. Lett.* **75**, 3926 (1995). (Reprinted with permission.)

of the behavior of many FQHE states at $\nu = 2 - n/(2n \pm 1)$ as a function of tilt, Du *et al.* discover that the longitudinal resistance oscillates as a function of the Zeeman energy, indicating transitions between FQHE states (Fig. 11.4). For example, the FQHE state at 4/3 disappears at a tilt of 59.1° but reappears at higher tilts. The reader may check for several fractions (4/3, 5/3, 7/5, 8/5) that the number of transitions is consistent with the prediction of the CF theory. (Yeh *et al.* [723] studied the spin physics for [4]CFs in the filling factor region $1/3 > \nu \geq 1/5$, analyzed by Yoshioka [732]; we confine our discussion below to [2]CFs, which have been studied more extensively.)

The spin-singlet states of composite fermions at $\nu = 2/3$ and 2/5 have also been demonstrated by reducing the g factor by application of hydrostatic pressure (Holmes *et al.* [257,258]). Nicholas *et al.* [477] study the evolution of ρ_{xx} as a function of hydrostatic pressure up to 13.4 kbar, where g falls to a level of approximately 0.1. With decreasing g, odd numerator states show a reduction in strength, with 5/3 and 7/5 completely disappearing, whereas even numerator states become stronger, and new states such as 4/5 and 6/5 appear.

These qualitative observations are consistent with the expectations of the CF theory. A quantitative analysis of their results indicates a factor of two enhancement for the CF g^* factor at $v = 5/3$, which they attribute to exchange interactions.

Kukushkin *et al.* [357] deduce the spin polarization of the FQHE state as a function of the Zeeman energy from the photoluminescence signal produced by the recombination of electrons with photoexcited holes bound to acceptor atoms. The measured polarization as a function of B, reproduced in Fig. 11.5, demonstrates preference for certain values of polarization, which are consistent with those predicted by the model of nearly free composite fermions (Table 11.1). This experiment, as well as the one by Du *et al.*, allows a determination of the critical Zeeman energies where transitions between differently polarized states take place.

The thermodynamic limits for the Coulomb energies of variously polarized FQHE states at $v = n/(2n + 1)$ have been computed (Park and Jain [500, 506]) using wave functions of Eq. (11.59) for spinful composite fermions, projected into the lowest Landau level by method II (Section 5.14.6). The results shown in Fig. 11.6 confirm, microscopically, the energy ordering predicted by the nearly free composite fermion model, namely that the energy increases with the degree of spin polarization. Theoretical values of the Zeeman energy where the transitions take place can be readily obtained from the Coulomb energies of Fig. 11.6. Equating the Coulomb energy difference between two successive CF states at $v = n/(2n + 1)$, $\delta e^2/\epsilon\ell$, to the Zeeman energy difference, E_Z/n, gives the Zeeman energy at the transition point to be $E_Z/(e^2/\epsilon\ell) = n\delta$. Figure 11.7 shows the theoretical values for $E_Z/(e^2/\epsilon\ell)$ along with experimental values from Du *et al.* [126] and Kukushkin *et al.* [357]. (Particle–hole symmetry has been used for the data of Du *et al.* [126].) The differences between the critical Zeeman energies from the two experiments can be taken as a measure of the sensitivity of the critical Zeeman energies to various sample-specific parameters. The agreement between theory and experiment is satisfactory in view of the following observations: (i) The theory neglects nonzero thickness, LL mixing, and disorder effects. (ii) The phase boundary is determined from energy differences that are rather small, roughly two orders of magnitude smaller than the ground state energies.

Figure 11.5 also reveals certain weak plateaus, for example, at $\gamma_e = 1/2$ for $v = 2/3$ or $2/5$, which are not predicted by the model of nearly free composite fermions. Freytag *et al.* [173] measure the spin polarization at $v = 2/3$ by an NMR technique, and find a phase transition from the fully polarized state into a partially polarized state with $\gamma_e \sim 3/4$. These results are not fully understood, but stability at these values of spin polarization must be caused by the residual interaction between composite fermions. Murthy [465] proposes to explain the intermediate polarizations in terms of a combined charge and spin density wave order of composite fermions. Mariani *et al.* [430], on the other hand, consider two degenerate CF Fermi seas with opposite spin polarizations (as would be appropriate for two coincident half-filled Landau levels) and argue, using the CFCS approach, that gauge fluctuations mediate an attractive interaction between composite fermions, which induces a spin-singlet pairing of composite fermions, opening a gap to produce FQHE.

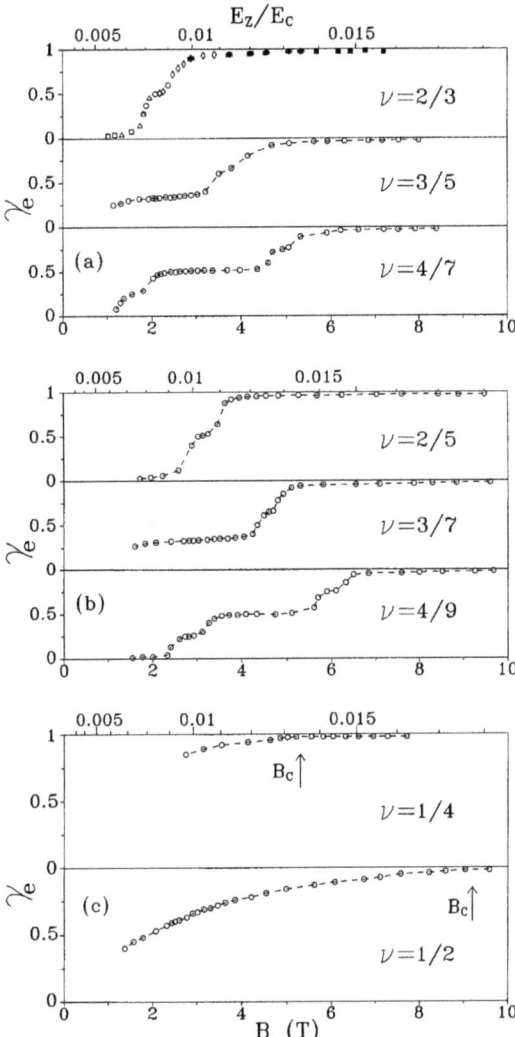

Fig. 11.5. Magnetic field dependence of the electron spin polarization γ_e, measured for various FQHE states by a photoluminescence technique [357]. γ_e is given by $(\rho_\uparrow - \rho_\downarrow)/(\rho_\uparrow + \rho_\downarrow)$ where $\rho_\uparrow$ $(\rho_\downarrow)$ is the spin-up (spin-down) electron density. (a) γ_e of the FQHE states at $\nu = n/(2n-1)$, (b) γ_e of the FQHE states at $\nu = n/(2n+1)$, and (c) γ_e of the CF Fermi seas at $\nu = 1/2$ and $1/4$. Different symbols for $\nu = 2/3$ are from different samples. Source: I. V. Kukushkin, K. von Klitzing, and K. Eberl, *Phys. Rev. Lett.* **82**, 3665 (1999). (Reprinted with permission.)

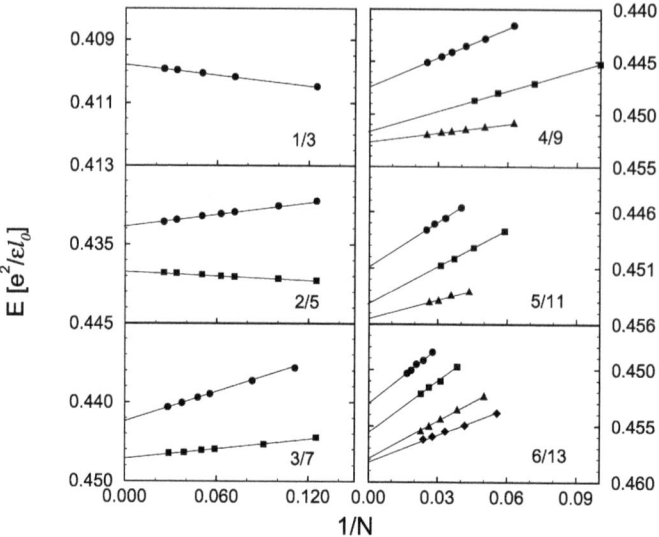

Fig. 11.6. Thermodynamic extrapolations for the energies per particle for various FQHE states of composite fermions carrying two vortices. The spin polarization decreases from top to bottom in each panel. In other words, the top straight line refers to $(n, 0)$, the one below that to $(n - 1, 1)$, and so on. Spherical geometry is used. Source: K. Park and J. K. Jain, *Solid State Commun.* **119**, 291 (2001). (Reprinted with permission.)

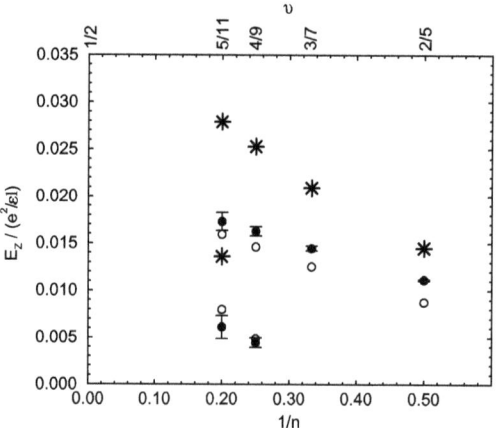

Fig. 11.7. The critical Zeeman energies for certain magnetic transitions at $v = n/(2n + 1)$. The circles are from theory (Park and Jain [500]), the stars are from the transport experiments of Du *et al.* [126], and the filled squares are from the optical experiments of Kukushkin *et al.* [357]. The x-axis is proportional to B^*, and the filling factor is shown on the top axis. From K. Park and J. K. Jain, *Phys. Rev. Lett.* **80**, 4237 (1998). (Reprinted with permission.)

11.7 Polarization mass

Du *et al.* [126, 626] (Fig. 11.8), and many subsequent experiments, successfully interpret the experimental results in terms of a nearly free CF model, with the CF mass and the CF g factor treated as parameters to be determined empirically. The renormalization of the CF g factor, called g^*, is a measure of the strength of the residual interactions between composite fermions. The quantity g^* can be expected to be robust, because there *is* a bare g^*, equal to the electron g factor (-0.44 for GaAs). We stress that the quantitative predictions of the CF theory can be compared to experiments without any need for defining a mass or a g factor, as done in the previous section. While it would certainly be intuitively satisfying if

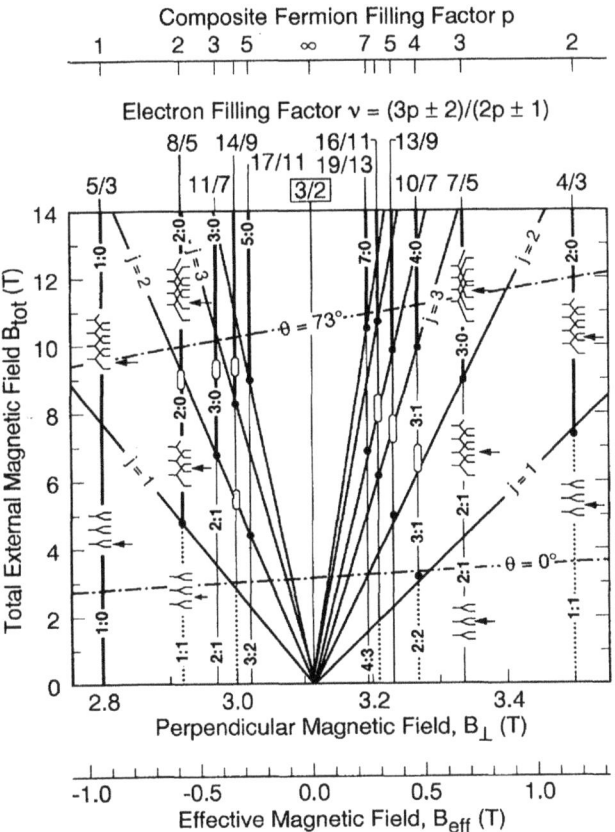

Fig. 11.8. The solid dots indicate the positions where transitions between differently polarized FQHE states take place. The lines are fits from the free CF theory, with the g^* factor and the effective mass m^* treated as empirical parameters. The symbol B_{eff} is used for the effective magnetic field, B^*. Source: R. R. Du, A. S. Yeh, H. L. Stormer, D. C. Tsui, L. N. Pfeiffer, and K.W. West *Phys. Rev. Lett.* **75**, 3926 (1995) (Reprinted with permission.)

a nearly free fermion model were to provide a reasonably good semi-quantitative account of the phenomenology, the validity of such a model is not a necessary consequence of the CF theory.

There is no reason why the effective mass that determines the spin polarization ought to be the same as the mass that governs transport gaps, because the two measurements probe different properties of the system. The former has to do with the energy difference between uniform ground states, while the latter is related to the energy required to create a CF particle–hole pair. The former, called the "polarization mass," is deduced theoretically as follows (Park and Jain [500]): For a non-interacting fermion system, a completely polarized Fermi sea is obtained for $E_Z \geq E_F$, where E_F is the Fermi energy of a *fully polarized* Fermi sea. The theoretical intercept in Fig. 11.9 gives the CF Fermi energy $E_F^* = 0.022 V_C$. Noting that $k_F = 1/\ell$ at $\nu = 1/2$ (for a fully polarized Fermi sea), we can write

$$E_F^* \equiv \frac{\hbar^2 k_F^2}{2m_p^*} = \frac{m_e}{m_p^*} \frac{1}{2} \hbar \Omega_C, \tag{11.65}$$

($\hbar \Omega_C$ is the cyclotron energy of an electron in vacuum) which gives

$$\frac{m_p^*}{m_e} = 23 \frac{\hbar \Omega_c}{V_C} = 0.60\sqrt{B[\text{T}]} \quad \text{(theory)}, \tag{11.66}$$

where we have used the relations in Section 8.7. To test if this effective mass is consistent with the phase boundary away from $\nu = 1/2$, the empty squares in Fig. 11.9 show the transition positions that this value of m_p^* leads to in a free-CF model. The qualitative, and

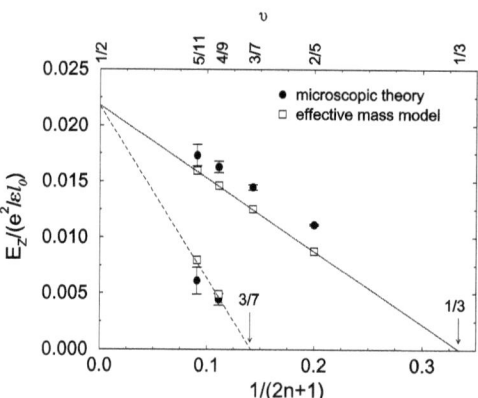

Fig. 11.9. Critical Zeeman energy E_Z (filled circles) at which the spin polarization of the FQHE state is predicted to change. The empty squares are obtained in a model of nearly free composite fermions, with the mass fixed from the y-intercept, as discussed in the text. The microscopic theory and the nearly independent CF model are in good agreement. Note that critical E_Z lines from the nearly independent CF model, dotted and dashed, have an x-intercept at $\nu = 1/3$ and 3/7. Source: K. Park and J. K. Jain, *Solid State Commun.* **119**, 291 (2001). (Reprinted with permission.)

even semi-quantitative, agreement with the microscopic results provides an a-posteriori justification of the effective mass model. It should be noted, however, that m_p^* has significant ν dependence. For example, at $\nu = 2/5$, the energy difference between the fully polarized and the spin-singlet states of $0.0055e^2/\epsilon\ell$ produces $m_p^*/m_e \approx 0.4\sqrt{B[T]}$. A striking implication of the calculations is that the polarization mass of composite fermions is different from, and much larger than, the activation mass of Eq. (8.23).

The polarization mass of composite fermions at $\nu = 1/2$ has been determined by several methods. Kukushkin *et al.* [357] estimate its value to be

$$\frac{m_p^*}{m_e} = 0.75\sqrt{B[T]} \quad \text{(Kukushkin)} \tag{11.67}$$

from the magnetic field where the CF Fermi sea at $\nu = 1/2$ becomes fully spin polarized (Fig. 11.5, bottom panel), as they vary the density with the help of a back gate. Melinte *et al.* [437] study the temperature dependence of spin polarization of the 1/2 CF Fermi sea, with the spin polarization determined from an NMR measurement of the Knight shift of ^{71}Ga nuclei. For zero tilt, the experimental results are very well described by a free CF model (Fig. 11.10) with an effective mass

$$\frac{m_p^*}{m_e} = 0.65\sqrt{B[T]} \quad \text{(Melinte)}. \tag{11.68}$$

Using the same effective mass, the agreement is somewhat worse for nonzero tilts, suggesting a tilt dependence for the polarization mass.

Freytag *et al.* [174] also measure temperature dependence of the spin polarization of the CF Fermi sea at $\nu = 1/2$ by NMR techniques (as in Ref. [437]) and compare their results to theoretical predictions of Park and Jain [500] and Shankar [583]. They obtain measurements of m_p^*, g^*, as well as the CF Fermi energy E_F^*. Full polarization is reached at $E_Z/(e^2/\epsilon\ell) = 0.022$, which agrees well with the theoretical prediction in Fig. 11.7 and implies a Fermi energy of composite fermions of $E_F^* \approx 3$ K. The thermal depolarization of the CF Fermi sea is also in satisfactory agreement with the expectation from the model of free composite fermions, although with a mass that depends on the tilt angle. The agreement worsens above $\sim$2 K; this temperature is comparable to the CF Fermi energy, and therefore presumably gives a temperature scale beyond which composite fermions begin to dissociate. Freytag *et al.* also study spin polarization slightly away from $\nu = 1/2$ and note that the behavior is symmetric around $\nu = 1/2$ for the fully polarized CF Fermi sea (obtained at large tilt angles), but asymmetric features appear at smaller tilt angles. That indicates a lack of full particle–hole symmetry for the partially polarized CF Fermi sea, which is inconsistent with the model of free fermions and indicates relevance of residual interactions between composite fermions.

Dementyev *et al.* [113] analyze the temperature dependence of the spin polarization at several tilt angles. They find a partially spin-polarized CF Fermi sea at zero tilt (at $B = 5.52$ T), consistent with Kukushkin *et al.* [357]. They successfully fit the temperature dependence in a two parameter model, wherein residual interactions between composite

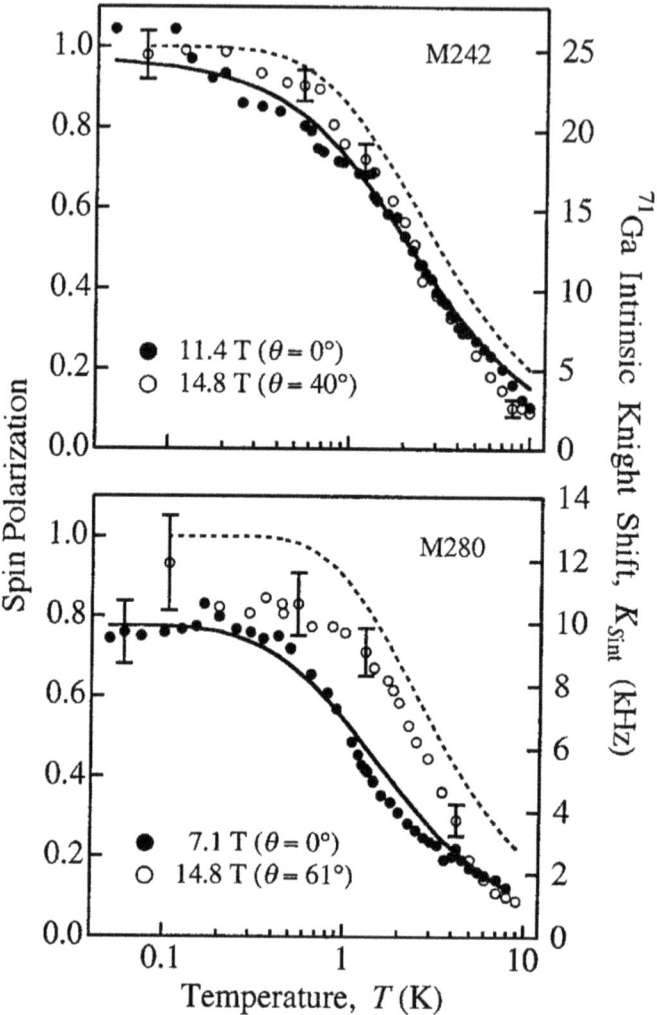

Fig. 11.10. Temperature dependence of the spin polarization of the 1/2 CF Fermi sea of two different samples at zero tilt (filled circles) and a nonzero tilt (empty circles). Solid line is the prediction from a noninteracting fermion model with $m_p^*/m_e = 2.2 \pm 0.2$ for the upper panel and $m_p^*/m_e \approx 1.7 \pm 0.2$ for the lower panel, which provide the best fit to the zero tilt data. Source: S. Melinte, N. Freytag, M. Horvatić, C. Berthier, L. P. Lévy, V. Bayot, and M. Shayegan, *Phys. Rev. Lett.* **84**, 354 (2000). (Reprinted with permission.)

fermions are included through a Stoner correction; this correction includes the exchange interaction through an effective Zeeman energy, leading to an enhancement of the spin susceptibility. The Stoner term (J) and the CF mass ($m_p^{'*}$) are obtained from the best fits. (This mass is not the same mass as m_p^*; the latter also incorporates, approximately, exchange corrections.) The experimental value for the mass at zero tilt is given by

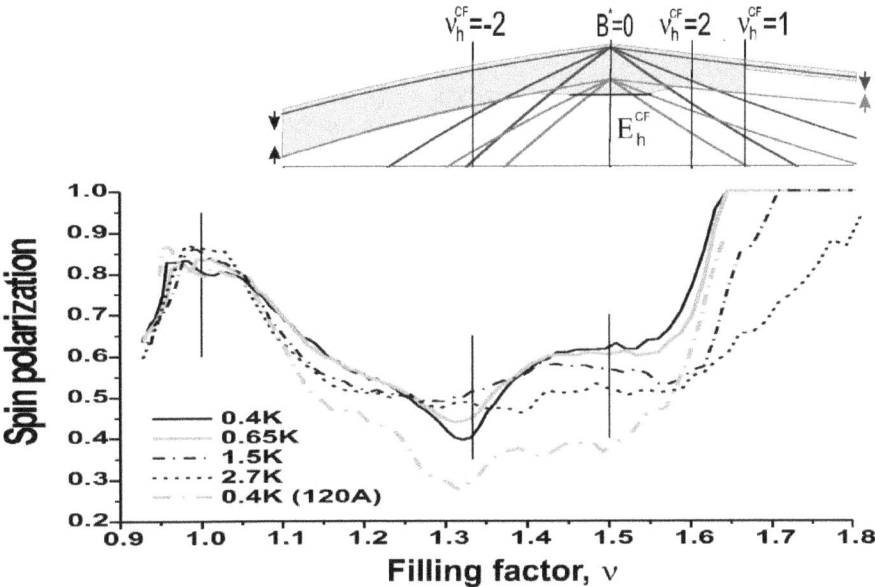

Fig. 11.11. Spin polarization of holes as a function of the filling factor for several temperatures. The inset shows the schematic picture for the Λ level structure of composite fermions (made of holes) in the filling factor region $1 < v < 2$. Source: R. Chughtai, V. Zhitomirsky, R. J. Nicholas, and M. Henini, *Phys. Rev. B* **65**, 161305(R) (2002). (Reprinted with permission.)

$m_p^{\prime*}/m_e = 0.34\sqrt{B[\text{T}]}$. Dementyev *et al.* [113] argue that the fit is not as good with $J = 0$, and note that different parameters are needed for different tilt angles.

Chughtai *et al.* [77] investigate the spin polarization at $v = 3/2$, which is equivalent to holes at $v_h = 2 - v = 1/2$. They determine directly the spin polarization of the holes by measuring, with interband optical reflectivity, the hole population in the $0\uparrow$ and $0\downarrow$ Landau bands. The results, shown in Fig. 11.11, are consistent with the CF theory. There is a partially polarized flat region around $v_h = 1/2$, and the system becomes fully polarized for $v_h < 1/3$, as expected for composite fermions in the lowest Λ level. The minimum at $v_h = 2/3$ corresponds to a spin-singlet state. From the spin polarization of 60% at $v_h = 1/2$, taking the g factor to be -0.44, Chughtai *et al.* obtain the polarization mass of $\sim 1.9m_e$. With the hole density of 1.47×10^{11} cm^{-2} (one-third of the electron density), this is equivalent to a polarization mass of

$$\frac{m_p^*}{m_e} = 0.55\sqrt{B[\text{T}]} \qquad \text{(Chughtai)}. \qquad (11.69)$$

Dujovne *et al.* [131] study transition of the CF Fermi sea from fully spin polarized to partially spin polarized, by measuring, through inelastic Raman scattering, the energy of the lowest spin-flip excitation at $v = 1/2$. As seen in Fig. 11.12, the energy of the lowest spin-reversed excitation approaches either a nonzero value or zero in the limit $v \to 1/2$,

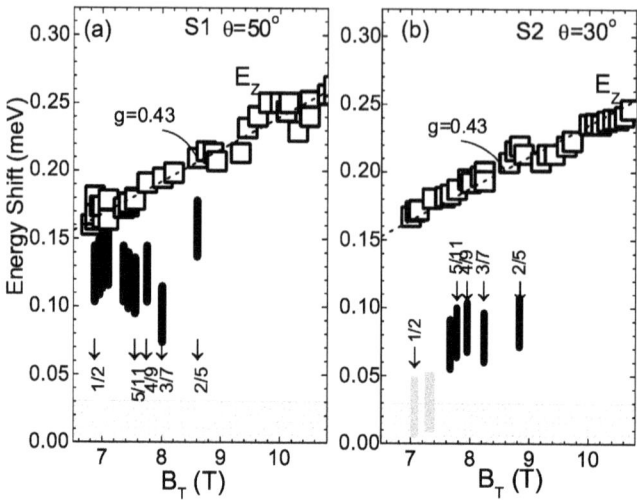

Fig. 11.12. Energies of low-lying spin-flip excitations as a function of the filling factor (shown on the figure) at two different tilt angles in two samples. E_Z is the Zeeman splitting. The shaded region at the bottom depicts the energy region inaccessible in the experiment. Source: I. Dujovne, A. Pinczuk, M. Kang, B. S. Dennis, L. N. Pfeiffer, and K. W. West, *Phys. Rev. Lett.* **95**, 056808 (2005). (Reprinted with permission.)

depending on experimental conditions. The former indicates a fully polarized CF Fermi sea, whereas the latter a partially polarized one. They estimate from such data that the transition occurs at $E_Z/V_C = 0.0165$, which is consistent with the values found in Refs. [113, 174, 357, 359, 437] by independent methods, as well as with the theoretical prediction [500].

Du *et al.* [126] analyze their transport experiments in a free composite fermion model, with the CF mass and the CF g factor as free parameters, and find, as seen in Fig. 11.8, a good fit with $g^*m^*/2m_e = 0.132 + 0.025B^*[\mathrm{T}]$. Taking $g^* = 0.44$, $B^* = 0$, and $B = 3.1\,\mathrm{T}$ (magnetic field at $\nu = 3/2$), this gives

$$\frac{m_p^*}{m_e} = 0.34\sqrt{B[\mathrm{T}]} \quad (\mathrm{Du}) \tag{11.70}$$

for the CF Fermi sea at $\nu = 3/2$. The value is of the same order as those mentioned in the preceding paragraphs, but somewhat lower. A combination of several factors contributes to the difference: LL mixing is more significant at $\nu = 3/2$ than at $\nu = 1/2$; the experiments are performed under tilted magnetic fields; perhaps most importantly, the mass is not measured at 3/2 directly but extrapolated from the nearby FQHE states (notably, the mass compares favorably to the theoretical polarization mass of $0.4\sqrt{B[\mathrm{T}]}$ at $\nu = 2/5$). The polarization mass exhibits no enhancement (cf. the Shubnikov–de Haas mass in Section 8.7.2) as the CF Fermi sea is approached.

As a final note in this section, the distinction between the polarization and the activation masses of composite fermions helps resolve, to an extent, an earlier puzzle regarding the spin polarization of the CF Fermi sea at $\nu = 1/2$ (Park and Jain [500]). The fraction of reversed spins is given by $0.5(1 - E_Z/E_F^*) = 0.5 - (m^*/m_e)(E_Z/\hbar\Omega_c)$, where $E_Z/\hbar\Omega_c \approx 1/5$ for parameters of GaAs. Using the activation mass for m^*, the CF sea is predicted to be close to being *un*polarized for typical experimental parameters; even for a field as high as $B = 9\,\text{T}$, a simple calculation shows that approximately 45% electrons have reversed spin. That contradicts the experimental measurements of the Fermi wave vector, which indicate a close to fully polarized CF sea [208, 320, 599, 683]. (Experimental evidence is that even the CF sea at $\nu = 3/2$ at $B \sim 3\,\text{T}$ is nearly fully spin polarized [321].) The apparent discrepancy between theory and experiments is resolved [500] if one properly uses the polarization mass for determining the spin polarization of the CF Fermi sea. A completely polarized sea would be obtained for $B > 17\,\text{T}$, and 14% (29%) of spins are reversed at $B = 9\,\text{T}$ ($3\,\text{T}$). We stress that these numbers are to be taken only as crude estimates, given the neglect of several effects in the theory. Experimentally, the 1/2 CF Fermi sea appears to be fully polarized typically beyond $\sim 10\,\text{T}$ for GaAs systems.

11.8 Spin-reversed excitations of incompressible states

In the 1980s, exact diagonalization studies at $\nu = 1/3$ (Chakraborty, Pietiläinen, and Zhang [56]; Rezayi [544]) already indicated that, at low Zeeman splittings, the lowest energy excitation involves spin reversal. The spin-flip excitations are identified experimentally by ascertaining the dependence of their energy on the magnetic field: the energy of a spin-reversed excitation contains a term proportional to B in addition to the usual $\sqrt{B_\perp}$ dependent term arising from interactions. Several experiments (Clark *et al.* [90]; Eisenstein *et al.* [136]; Engel *et al.* [143]) identified spin-reversed excitations in tilted-field experiments.

A coherent understanding of the spin-reversed excitations is achieved through the CF theory, which clarifies that these are simply CF excitons, or particle–hole pairs of composite fermions, in which the spin of the excited composite fermion is reversed. The Λ level index of the excited composite fermion may increase, remain the same, or decrease. That, combined with multiple ground states with different spin polarizations, produces a remarkably rich variety of excitations. These have been studied by inelastic light scattering as well as transport.

11.8.1 Neutral spin-flip CF-excitons

The spin-wave mode exists, on general grounds, for all states with nonzero spin polarization; it is the Goldstone mode associated with the spontaneous breaking of the rotation symmetry in the spin space (whereby the total spin picks a direction). According to the Larmor theorem, the energy of the spin-wave mode is equal to the bare Zeeman splitting in the small wave vector limit. The spin-wave excitation at $\nu = 1/3$ was observed experimentally by Pinczuk *et al.* [515] (Fig. 11.13). From an analysis of the temperature dependences of the spin-wave

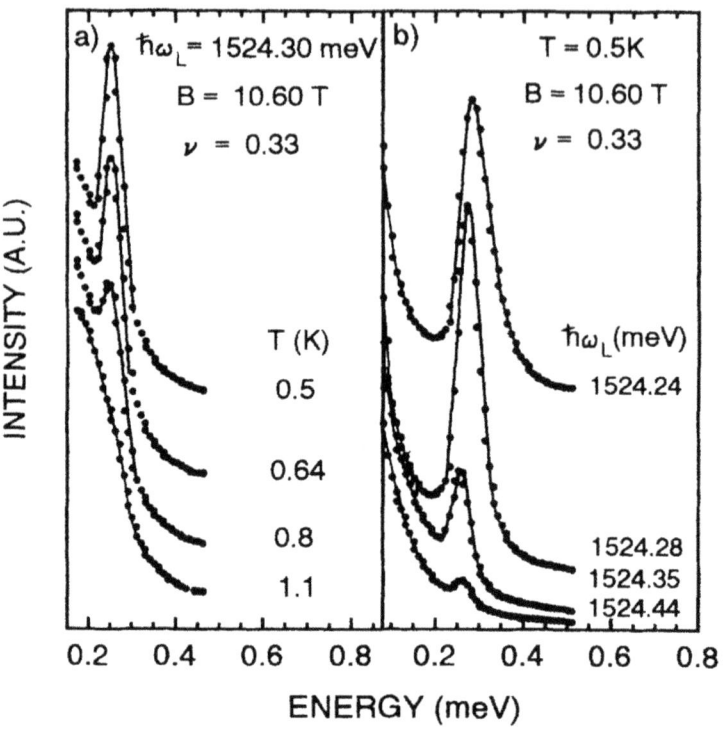

Fig. 11.13. The sharp peak is identified as the long-wavelength spin-wave excitation in resonant light scattering spectra at $\nu = 1/3$. Panels (a) and (b) show the dependence on temperature and the incident photon energy, respectively. The energy of the mode (0.26 meV) is equal to the Zeeman energy $E_Z = g\mu_B B$ with $g = -0.43$. Source: A. Pinczuk, B. S. Dennis, L. N. Pfeiffer, and K. W. West, *Phys. Rev. Lett.* **70**, 3983 (1993). (Reprinted with permission.)

as well as exciton modes, Davies *et al.* [109] conclude that the collapse of the correlations in the 1/3 FQHE state occurs due to thermal excitations of the long-wavelength spin waves. Recently, substantial progress has been made in extending the experiments to other fractions, e.g., $\nu = 2/5$ and 3/7, where a richer structure emerges. Kang *et al.* [323] report observation of spin-reversed modes other than the spin wave at these fractions. At $\nu = 2/5$, they observe a mode with energy approximately equal to $2E_Z$; the near absence of $\sqrt{B_\perp}$ dependence of the energy indicates that it is not modified by the Coulomb interaction. Another striking observation is of a mode at $\nu = 3/7$ that has an energy *smaller* than E_Z, roughly $0.4E_Z$.

At $\nu = 1/3$, which maps into $\nu^* = 1$ of composite fermions, the lowest energy spin-reversed mode is the one in which the composite fermion flips its spin while remaining within the lowest Λ level. Nakajima and Aoki [471] show, by comparison with exact diagonalization results [544], that the CF theory provides an excellent description of the spin-wave mode at $\nu = 1/3$. For general fillings, the possibility of several kinds of modes

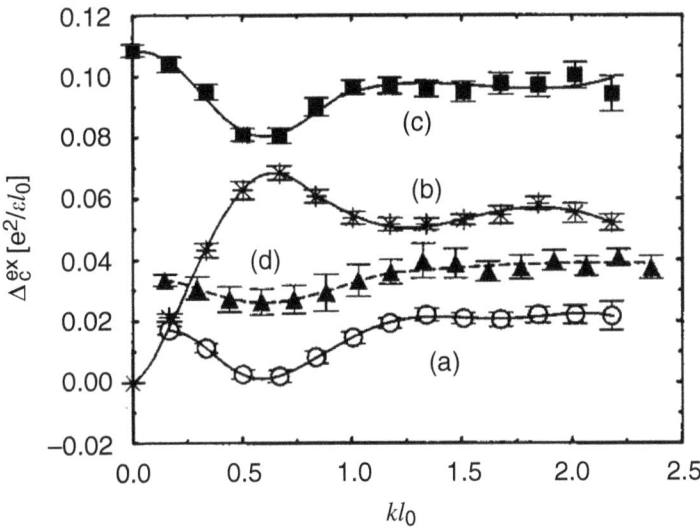

Fig. 11.14. The curves labeled (a), (b), and (c) show the dispersions of three spin-reversed modes for $N = 30$ particles for the Coulomb potential $V(r) = e^2/\epsilon r$. The ground state is assumed to be fully polarized. The error bars indicate the estimated statistical error in Monte Carlo. Δ_c^{ex} denotes the Coulomb energy of the exciton measured relative to the ground state; the Zeeman energy $E_Z = |g|\mu_B B$ must be added to it to obtain the full energy of the spin-reversed excitation. The curve (d) shows the energy of the spin reversed excitation $0\downarrow \to 1\uparrow$ of the *unpolarized* 2/5 state for $N = 38$. (The Zeeman splitting must be *subtracted* from it.) Source: S. S. Mandal and J. K. Jain, *Phys. Rev. B* **63**, 201310(R) (2001). (Reprinted with permission.)

is obvious in the CF theory. Consider, for example, the fully polarized state at $\nu = 2/5$, wherein two Λ levels, labeled 0 and 1, are fully occupied. Neglecting excitations involving the third Λ level, the three possible low-energy excitations are: (i) $1\uparrow \to 0\downarrow$; (ii) $1\uparrow \to 1\downarrow$; and (iii) $0\uparrow \to 0\downarrow$. The last two conserve the Λ level index, whereas the composite fermion lowers its Λ level index in the first. The energy dispersions of these spin-reversed excitations have been evaluated numerically (Mandal and Jain [422]). The three Coulomb eigenvalues obtained in this way are shown in Fig. 11.14, labeled (a), (b), and (c). For the fully spin polarized state at $\nu = 3/7$, the Coulomb Hamiltonian is diagonalized in the subspace defined by six modes: $2\uparrow \to 0\downarrow$; $2\uparrow \to 1\downarrow$; $1\uparrow \to 0\downarrow$; $2\uparrow \to 2\downarrow$; $1\uparrow \to 1\downarrow$; and $0\uparrow \to 0\downarrow$, to obtain the excitation spectrum [422].

The familiar spin-wave mode is recovered at small wave vectors, as expected. These calculations also demonstrate the existence of a low-energy "spin-flip roton." For zero thickness, the energies of the spin-flip rotons at $\nu = 2/5$ (for the $1\uparrow \to 0\downarrow$ mode) and $\nu = 3/7$ ($2\uparrow \to 0\downarrow$) are estimated to be $-0.0024(18)e^2/\epsilon l_0$ and $-0.0091(36)e^2/\epsilon l_0$, respectively. At $\nu = 2/5$, the interaction energy of the spin-flip roton is close to zero, implying that its total energy is close to E_Z. At $\nu = 3/7$, the interaction energy of the roton is negative, which, for typical parameters, leads to a total energy of approximately $0.5E_Z$. The spin-flip roton is identified [422] with the low-energy excitation observed by Kang *et al.* at this filling factor [323], both because no other such low-energy mode is known, and because the calculated energy of the spin-flip roton is in reasonable agreement with the observed energy. Many qualitative features of the results, including the spin rotons, are obtained in a theoretical study of magnetoexcitations by Murthy [464].

Dujovne *et al.* [129, 130] study spin-reversed excitations in the filling factor range $2/5 \geq \nu \geq 1/3$ by resonant light scattering, as shown in Figs. 11.15 and 11.16. In addition to the spin-wave excitation at E_Z, they observe two modes below E_Z, labeled SF^- and SF^+, which both involve spin flip. These modes are absent at $\nu = 1/3$. The lower-energy mode SF^- is strongest close to $\nu = 2/5$, and is interpreted as the spin-flip roton in the dispersion of the collective mode $1\uparrow \rightarrow 0\downarrow$ of composite fermions. Its energy agrees with the calculations of Ref. [422]. SF^+ is strongest close to $\nu = 1/3$, and is interpreted as the spin-reversed $1\uparrow \rightarrow 0\downarrow$ excitation out of the sparsely populated second Λ level. The two modes coexist in an intermediate filling factor. These modes provide evidence for the Λ level structure

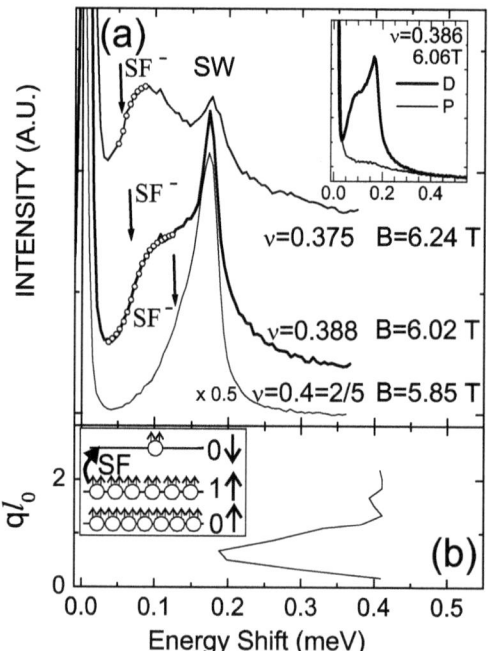

Fig. 11.15. (a) Inelastic light scattering spectra at filling factors at or slightly below $\nu = 2/5$. The spin-wave mode is marked by SW. The spin-flip mode labeled SF^- is interpreted as the deep spin-flip roton in the dispersion of the $1\uparrow \rightarrow 0\downarrow$ excitation. (b) Theoretical dispersion of the lowest energy spin-flip excitation, taken from Ref. [422]. The inset shows the spin-split Λ level structure. Source: I. Dujovne, A. Pinczuk, M. Kang, B. S. Dennis, L. N. Pfeiffer, and K. W. West, *Phys. Rev. Lett.* **90**, 036803 (2003). (Reprinted with permission.)

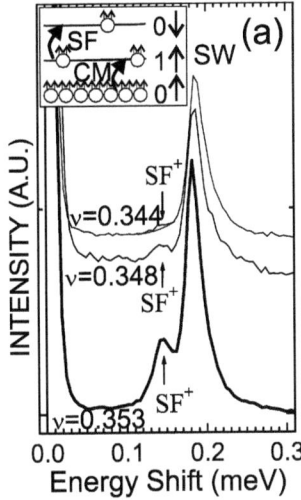 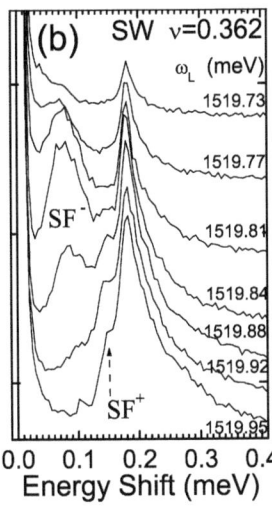

Fig. 11.16. Inelastic light scattering spectra at several filling factors slightly greater than $\nu = 1/3$. The spin-flip mode is marked SF$^+$. The inset shows the spin-split Λ level structure. Source: I. Dujovne, A. Pinczuk, M. Kang, B. S. Dennis, L. N. Pfeiffer, and K. W. West, *Phys. Rev. Lett.* **90**, 036803 (2003). (Reprinted with permission.)

of composite fermions. Dujovne *et al.* [129] regard an abrupt change in the behavior at an intermediate filling between 1/3 and 2/5 as a manifestation of the residual interactions between composite fermions in the second Λ level, which causes an as-yet-unexplained qualitative change in the character of the state.

Gallais *et al.* [177, 178] study the line widths and the scattering intensities of the spin-flip mode as a function of the filling factor near either side of $\nu = 1/3$. Denoting the CF filling factor by $\nu^* = 1 + \bar{\nu}^*$, they find sharp Raman peaks with a constant broadening close to $\nu^* = 1$, but a substantial increase in broadening for $\bar{\nu}^* > 0.2$. The scattering intensities vary linearly with $\bar{\nu}^*$ close to $\nu^* = 1$, as expected in a mean-field model of noninteracting composite fermions, but deviate from such behavior for $\bar{\nu}^* > 0.2$. These results demonstrate enhanced interactions between composite fermions for $\bar{\nu}^* > 0.2$. From the line width, the energy scale of the inter-CF interaction is determined to be approximately 0.1 meV at $\bar{\nu}^* \approx 0.5$. They also do not find a significant loss of spin polarization in this region, indicating that the state remains fully spin polarized. In addition, a discontinuous change in the energy of the spin-flip mode at $\nu = 1/3$ is observed, consistent with different energy scales associated with ^{2}CFs and ^{4}CFs.

11.8.2 Charged spin-reversed excitations

The Arrhenius behavior of the resistance yields the gap to creating a far separated pair of CF–quasiparticle and CF–quasihole. Many studies have measured spin-dependent gaps

[90, 127, 136, 143, 576]. These measurements can be interpreted in terms of a renormalized g factor for composite fermions (g^*).

Du *et al.* [127] measure transport gaps at several filling factors near $\nu = 3/2$ as a function of the total magnetic field B_{tot} (Fig. 11.17), changed by tilting the sample. The g factor of composite fermions is determined directly from the slope of the linear dependence of the gap on B_{total} using the relation $\partial \Delta / \partial B_{total} = g^* \mu_B$. They obtain $g^* \approx 0.6$ at $\nu = 5/3$ and $g^* \approx 0.75$ for $\nu = 4/3$. From the temperature dependence of the spin polarization of incompressible states, Kukushkin *et al.* [358] ascertain the Zeeman splitting for composite fermions and find a factor of two enhancement for g^* for composite fermions relative to electrons. Schulze-Wischeler *et al.* [576] vary the carrier density by infrared

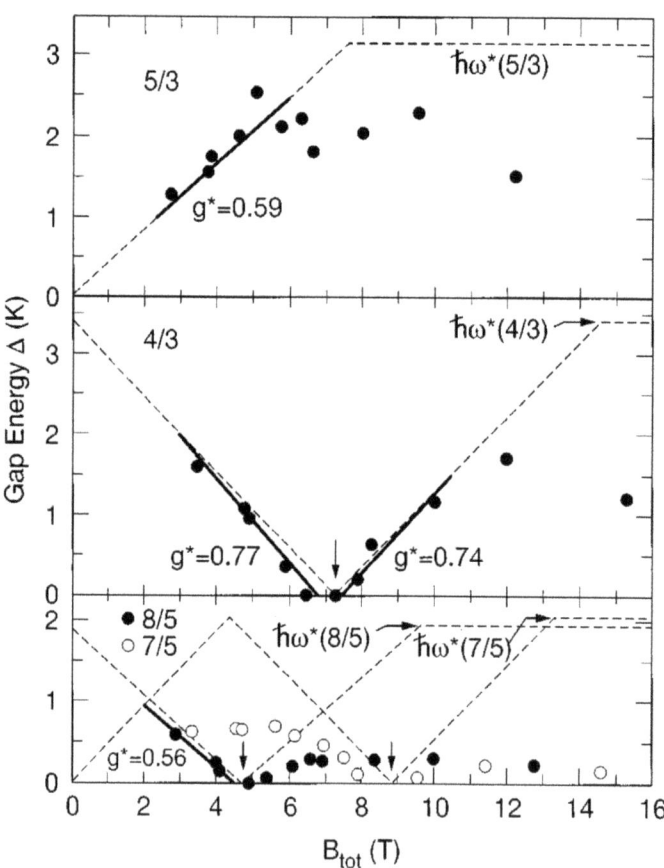

Fig. 11.17. The energy gaps at $\nu = 5/3, 4/3, 8/5$, and $7/5$ as a function of the total magnetic field B_{tot}. The g factor for composite fermions, g^*, determined from the slope, is shown on the plots. Source: R. R. Du, A. S. Yeh, H. L. Stormer, D. C. Tsui, L. N. Pfeiffer, and K. W. West, *Phys. Rev. B* **55**, R7351 (1997). (Reprinted with permission.)

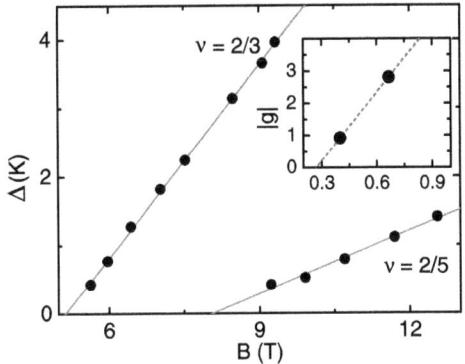

Fig. 11.18. Activation gaps at 2/3 and 2/5 as a function of the magnetic field. The linear dependence allows a determination of the CF g factor. Source: F. Schulze-Wischeler, E. Mariani, F. Hohls, and R. J. Haug *Phys. Rev. Lett.* **92**, 156401 (2004). (Reprinted with permission.)

light illumination. The gap varies linearly with the magnetic field (at a fixed filling factor), as seen in Fig. 11.18, which gives $g^* \approx 0.92$ and $g^* \approx 2.8$ at $\nu = 2/5$ and $2/3$, respectively; this implies a heavier renormalization of g^* by interactions than seen in other experiments, and also a stronger filling factor dependence. The origin of the differences between various experimental values is not understood at present. It could possibly be due to different experimental methods (tilting vs. density variation), the filling factor dependences of parameters, and sample specific features. Yet another complication is that the g^* relevant for the spin-flip excitation gaps (or the temperature dependence of the spin polarization) is not necessarily the same as the g^* governing the transition points, in the same way as the activation mass differs from the polarization mass.

The temperature dependence of the spin polarization of the incompressible states has also been measured by several groups [331, 359, 437]. Either thermal depolarization or polarization is seen depending on whether the zero-temperature ground state is spin polarized or unpolarized. For the IQHE states, the spin-wave excitations, with minimum energy equal to the Zeeman gap, dominate the low-temperature behavior; particle–hole pair excitations have too high an energy to be relevant. The situation can be different in the FQHE. Let us consider, for example, $\nu = 1/3$. The contribution to depolarization from the Zeeman energy excitation is proportional to $\exp(-E_Z/k_B T)$, whereas that from the CF-quasiparticle–quasihole excitation is proportional to $\exp(-\Delta^S/2k_B T)$; the extra factor of 1/2 in the latter is due to greater entropy of the CF-quasiparticle–quasihole excitation [414]. (Both excitations are parts of the spin-flip CF-exciton branch; they correspond to small or the large wave vector limits.) The quantity Δ^S, the energy to create a far separated pair of CF-quasiparticle (with reversed spin) and quasihole, is a combination of the Zeeman energy and the Coulomb energy. When the latter is less than twice the Zeeman energy (i.e., $\Delta^S < 2E_Z$), which is often the case for experimental parameters (see the theoretical discussion below), the low-temperature behavior is governed by the

CF-quasiparticle–quasihole excitations rather than the spin-wave excitations [414]. This provides a method for determining the energy gap Δ^S from spin polarization measurements. This energy is often larger than the gap to creating a spin-conserving pair of charged quasiparticle and quasihole, and therefore not accessible in the transport experiments. The spin polarization measurements, on the other hand, are only sensitive to spin-flip excitations.

The measurements of Khandelwal *et al.* [331] and Melinte *et al.* [437] at $\nu = 1/3$ confirm the $\exp(-\Delta^S/2k_BT)$ dependence, with Δ^S slightly less than $2E_Z$. Kukushkin *et al.* [359] find (Fig. 11.19) that at low temperatures, the temperature dependence of the electron spin polarization of the incompressible states at 1/3 and 2/3 exhibits an activated behavior, given by

$$\gamma_e = 1 - 2\exp\left(-\frac{\Delta^S}{2k_BT}\right) \tag{11.71}$$

for fully polarized states, and $\gamma_e = \exp(-\Delta^S/k_BT)$ for spin-unpolarized states. The measured Δ^S is considerably larger than the activation gap from transport experiments, as well as the Zeeman splitting, consistent with its interpretation as a far separated CF-quasiparticle–quasihole pair. Mariani *et al.* [431] perform a careful theoretical analysis of the temperature dependence of the spin polarization measured by Kukushkin *et al.* [357–359] and find agreement to within 10% with the CF model over a range of B and T, with the polarization mass as a fitting parameter, and a filling-factor-dependent value for $g^*m_p^*$.

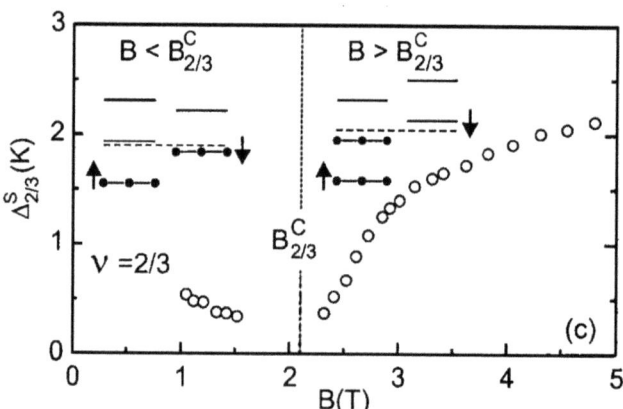

Fig. 11.19. The gap governing the Arrhenius behavior of the spin polarization at $\nu = 2/3$. The qualitative change in the behavior at the critical magnetic field $B_{2/3}^c = 2.1$ T is consistent with a transition from a spin-unpolarized state to a fully spin-polarized state. The Λ level diagrams for composite fermions are shown in the insets. Source: I. V. Kukushkin, J. H. Smet, K. von Klitzing, and K. Eberl, *Phys. Rev. Lett.* **85**, 3688 (2000). (Reprinted with permission.)

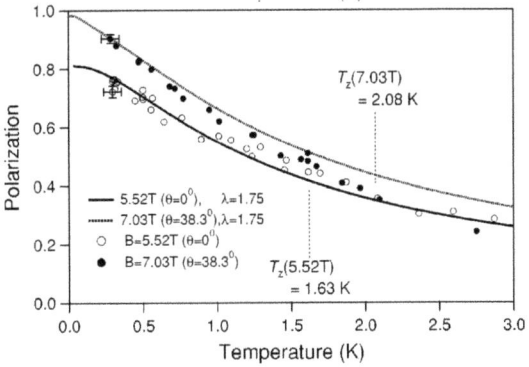

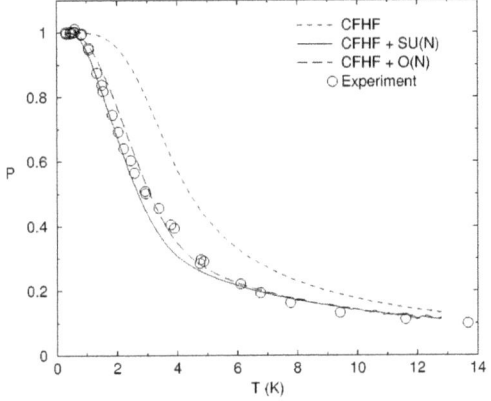

Fig. 11.20. The temperature dependence of the spin polarization (P) at $\nu = 1/2$ (upper panel) and $\nu = 1/3$ (lower panel). The circles are experimental results from Dementyev *et al.* [113] (upper panel, which also shows polarization at a nonzero tilt) and Khandelwal *et al.* [331] (lower panel). The curves are obtained from the Murthy–Shankar theory. The upper panel shows the result of the Hartree–Fock calculation for the CF Fermi sea. In the lower panel, the curve labeled CFHF is from the CF Hartree–Fock approximation, and the other two curves include the effect of CF spin waves in two large N expansions. The nonzero thickness modification of the interaction has been incorporated through a parameter λ, although the results are largely insensitive to it. Source: G. Murthy and R. Shankar, *Rev. Mod. Phys.* **75**, 1101 (2003) for the upper panel (which is an improvement on the earlier calculations from R. Shankar, *Phys. Rev. Lett.* **84**, 3946 (2000)) and G. Murthy, *J. Phys. Condens. Matter* **12**, 10543 (2000) for the lower panel. (Reprinted with permission.)

The Murthy–Shankar theory has been used to obtain a quantitative description of the magnetic phase transitions of the CF states, including a calculation of the temperature dependence of their spin polarization [466, 468, 582, 583]. Figure 11.20 shows comparisons between theory and experiment at $\nu = 1/2$ and $1/3$. The Hartree–Fock approximation works well for the CF Fermi sea, but overestimates the spin polarization of the gapped states. Murthy [466] finds that a sophisticated self-consistent treatment of interacting

Table 11.3. *Energy gaps for spin-reversed excitations*

ν	$\Delta_C^{\uparrow\downarrow} = \Delta^{\uparrow\downarrow} - \Delta_Z \left[\frac{e^2}{\epsilon\ell}\right]$	$\Delta^{\uparrow\uparrow} \left[\frac{e^2}{\epsilon\ell}\right]$	$B_c[T]$	$B_{tr}[T]$
1/3	0.0740(24)	0.106(3)	28	—
2/5	0.0235(55)	0.058(5)	33	3.5
3/7	0.0022(94)	0.047(4)	56	5.8
4/9	−0.0402(135)	0.035(6)	157	7.4

Notes: Magnetic field in tesla, energy in units of $e^2/\epsilon\ell$.

Results are for $V(r) = e^2/\epsilon r$.

Effects of nonzero thickness, Landau level mixing, and disorder are not considered.

Crossover magnetic field B_c is given for parameters appropriate for GaAs; spin-reversed excitation has the lowest energy for fields below B_c.

B_{tr} is magnetic field at which a transition into a partially polarized ground state takes place.

Spin-reversed gap at $\nu = 1/3$ is consistent with that of Rezayi [562] and Chakraborty, Pietilainen, and Zhang [57].

Gaps $\Delta_C^{\uparrow\uparrow}$ are taken from Jain and Kamilla [292] and B_{tr} from Park and Jain [514].

Source: Mandal and Jain [436].

spin waves produces good agreement for the latter (over a surprisingly wide range of temperatures, including temperatures above the typical activation gap at $\nu = 1/3$), indicating the importance of spin waves.

The values of spin-reversed gaps have been calculated theoretically for fully spin-polarized FQHE states at $\nu = n/(2n + 1)$, wherein the $0\uparrow, 1\uparrow, \ldots, (n - 1)\uparrow$ Λ levels are occupied. The lowest-energy spin-flip excitation corresponds to the $(n - 1)\uparrow \rightarrow 0\downarrow$ transition of a composite fermion. The interaction components of the spin reversed transport gaps (for the excitation consisting of a far separated pair of a CF-quasiparticle and a CF-quasihole) are given in Table 11.3 for several filling factors, along with $\Delta^{\uparrow\uparrow}$, the gap to creating the spin conserving excitation $(n - 1) \uparrow \rightarrow n \uparrow$. A comparison of $\Delta^{\uparrow\downarrow}$ and $\Delta^{\uparrow\uparrow}$ indicates that the lowest energy gap may involve spin flip up to fairly large magnetic fields, the theoretical values of which are denoted by B_c (Table 11.3). The state is fully polarized only for magnetic fields higher than B_{tr}, the theoretical estimates for which are also given in Table 11.3. The theoretical numbers in the table are only crude estimates. For one things, they deal with an "ideal" model; for example, the nonzero thickness reduction of the interelectron interaction lowers the interaction energy of all states, hence also $\Delta_C^{\uparrow\downarrow}$, and thereby disfavors spin-reversed excitations. Also, the crossover magnetic field B_c is a function of the difference between two energy gaps, which are themselves differences between two eigenenergies, making its theoretical estimate rather unreliable.

The Chern–Simons theory has also been applied to non-fully spin polarized FQHE states (Hansson *et al.* [237, 238]; Lopez and Fradkin [403]; Mandal and Ravishankar [420, 421]).

11.9 Summary

Comparisons with computer experiments on small systems indicate (Section 11.5) that the CF theory is quantitatively as accurate for non-fully polarized states as it is for fully polarized ones. Here is a summary of how the CF theory correlates with experimental findings.

- The CF theory provides a unified qualitative account of a vast number of experimental facts involving the spin degree of freedom. The only exceptions are the weak "half-polarized" states, not predicted by the model of *noninteracting* composite fermions.
- The data from each individual experiment can be modeled, phenomenologically, in terms of free composite fermions, with the polarization mass and g factor treated as fitting parameters.
- Many experiments deduce the polarization mass from the spin polarization of the CF Fermi sea. The mass values are in good agreement with the theoretical prediction given in Eq. (11.66). The quantitative agreement is better than that for the activation mass. That is not surprising. The polarization mass is obtained from the measurement of a thermodynamic quantity, and can, therefore, be expected to be more robust. That, we believe, is the reason behind consistency between different measurements of m_p^* from the spin polarization of the CF Fermi sea. The activation mass, on the other hand, involves localized charged excitations, the energies of which are more sensitive to disorder.
- The experimental polarization mass (at zero tilt) is a factor of four to five larger than the experimental activation mass $m_a^*/m_e = (0.15-0.20)\sqrt{B[\mathrm{T}]}$ reported by Du *et al.* [123].
- Measurements of m_p^* and g^* from the polarization of the FQHE states (as opposed to the CF Fermi sea), and from the temperature dependences of the spin-flip gaps, produce a larger range of values, which also depend on the filling factor. These are all of similar magnitude, however.
- The variance between different experiments can be expected from the following facts: (i) The critical Zeeman energies, where transitions between differently spin-polarized states occur, are governed by very small energy differences between the competing states, which, in turn, depend sensitively on sample specific quantities. (ii) The spin-flip gaps depend on these factors as well as on disorder. There is no reason why the corrections to the spin-flip gaps due to disorder should not be as substantial as those for the gaps of fully polarized states (Chapter 8).

Interestingly, the spin physics of composite fermions is richer than that of electrons in the IQHE. While phases with various spin polarizations are, in principle, possible for the latter as well, only the smallest spin polarization is seen in experiments, because the LL spacing is much larger than the Zeeman energy. The richness of the phase diagram of the FQHE is a consequence of the fortunate coincidence that, for typical experimental parameters, the Zeeman energy is comparable to the Λ level spacing (the effective cyclotron energy of composite fermions).

11.10 Skyrmions

This section is concerned with the nature of charged excitations at $\nu = 1$ for $E_Z = 0$, and their relevance to experiments where E_Z is very small but nonzero. For large Zeeman energies, the ground state at $\nu = 1$ is fully polarized, and its excitations can be understood perfectly well in a single particle language. (The wave functions of the excitations are approximately the same as those in a free fermion theory, although the energies can depend significantly on the interaction.) As explained earlier, at $\nu = 1$ the ground state is fully polarized even at $E_Z = 0$ due to exchange effects. However, here the massless spin waves renormalize the charged excitations. Lee and Kane [379] noted the relevance of a topological "skyrmion" excitation in this limit, which has a nontrivial spin texture equivalent to spin wrapping around the unit sphere. In earlier exact diagonalization studies, Rezayi [544,547] had found the unexpected result that, for $E_Z = 0$, the removal of a single electron from the $\nu = 1$ state, or the addition of a single electron to it, causes a macroscopic change in the spin, and the state becomes a spin singlet. Sondhi *et al.* [607] identify this spin-singlet state with the "hole skyrmion" or the "particle skyrmion".[3] The name skyrmion is motivated by field theoretical descriptions (Chern–Simons field theory and the nonlinear sigma model) of Refs. [379] and [607], which are beyond the scope of this book, and not necessary for understanding many essential properties of the skyrmion. The microscopic treatment below follows MacDonald, Fertig, and Brey [413] and Moon *et al.* [453].

11.10.1 Low-energy spectrum

The interaction in Eq. (11.62) is often useful to work with, for which the hard-core states of Eq. (11.63) have zero energy and all other states have nonzero energy. The hard-core states will plausibly have lower Coulomb energies than other states. We first identify the quantum numbers of the hard-core states for the hole skyrmion, and then obtain the energy ordering of the states for the Coulomb interaction.

Theorem 11.2 Let us take $N = 2M$ to be an even number for simplicity. Then, for $Q = N/2 = M$, the hard-core states occur at $(L, S) = (0, 0), (1, 1), \ldots, (M, M)$, with a single multiplet at each value.

Proof The hard-core states have the form given in Eq. (11.63), where the bosonic wave function Ψ'_B is to be constructed at the effective flux

$$Q^B = \frac{1}{2}. \tag{11.72}$$

The problem thus reduces to determining the possible quantum numbers of $2M$ bosons in four single-particle orbitals:

$$(l_z, s_z) = \left(\frac{1}{2}, \frac{1}{2}\right), \left(-\frac{1}{2}, \frac{1}{2}\right), \left(\frac{1}{2}, -\frac{1}{2}\right), \left(-\frac{1}{2}, -\frac{1}{2}\right). \tag{11.73}$$

---

[3] These are also referred to as skyrmion and antiskyrmion, respectively.

Let us denote the number of bosons in these orbitals by $N_{\uparrow\uparrow}$, $N_{\downarrow\uparrow}$, $N_{\uparrow\downarrow}$, and $N_{\downarrow\downarrow}$. The state

$$N_{\uparrow\uparrow} = N, \ N_{\downarrow\uparrow} = N_{\uparrow\downarrow} = N_{\downarrow\downarrow} = 0 \tag{11.74}$$

has $S_z = M = L_z$. In general, we have

$$N_{\uparrow\downarrow} + N_{\downarrow\downarrow} = M - S_z. \tag{11.75}$$

The right hand side is the z component of the spin measured relative to the fully polarized $N_{\uparrow\uparrow} = N$ state, and the left hand side tells us that each spin flip changes the z component of the total spin by one unit. Similarly, we have

$$N_{\downarrow\uparrow} + N_{\downarrow\downarrow} = M - L_z. \tag{11.76}$$

Let us now count the total number of distinct configurations that produce a given (L_z, S_z). The only independent parameter is $N_{\downarrow\downarrow}$; the other occupations can be determined uniquely in terms of $N_{\downarrow\downarrow}$ with the help of the above equations. Furthermore, because the occupation numbers $N_{\downarrow\uparrow}$ and $N_{\downarrow\downarrow}$ cannot be negative, it follows that

$$N_{\downarrow\downarrow} = 0, 1, \ldots, \min(M - L_z, M - S_z). \tag{11.77}$$

The total number of configurations at (L_z, S_z) is therefore $1 + \min(M - L_z, M - S_z)$.

We begin with $L_z = S_z = M$. There is a single configuration here. This implies a single multiplet with $(L, S) = (M, M)$, which has one state at all values of (L_z, S_z) with $L_z = -M, -M + 1, \ldots, M - 1, M$ and $S_z = -M, -M + 1, \ldots, M - 1, M$. Now let us ask if a multiplet occurs with $(L, S) = (M - 1, M)$. For this purpose, we count the number of configurations that give $(L_z, S_z) = (M - 1, M)$; if the number of configurations is one, then there is no multiplet with $(L, S) = (M - 1, M)$, because the configuration belongs to the $(L, S) = (M, M)$ multiplet. For k configurations there are $k - 1$ multiplets with $(L, S) = (M - 1, M)$. It can be seen that there is only one configuration that gives $(L_z, S_z) = (M - 1, M)$, because $1 + \min(M - L_z, M - S_z) = 1$. Similarly there is no multiplet with $(L, S) = (M, M - 1)$. Next we come to $(L, S) = (M - 1, M - 1)$. There are two configurations that give $(L_z, S_z) = (M - 1, M - 1)$, because of the relation $1 + \min(M - L_z, M - S_z) = 2$. That implies a single multiplet with $(L, S) = (M - 1, M - 1)$. In general, if we start with $L_z = S_z$, then decreasing either L_z or S_z by unity does not change the number of configurations, but if both are decreased by unity, then the number of configurations increases by one unit. This proves that there exist multiplets at $(L, S) = (0, 0), (1, 1), \ldots, (M, M)$.

To confirm that no states are missing, we note that the total number of ways N identical bosons can be arranged in four states is

$$\binom{N + 3}{3}. \tag{11.78}$$

On the other hand, the degeneracy of an (L, S) multiplet is $(2L + 1)(2S + 1)$, which gives

$$\sum_{L=0}^{M}(2L + 1)^2.$$ (11.79)

The two are equal.   □

For the hard-core interaction of Eq. (11.62), all skyrmion states are degenerate at zero energy, by construction. The Coulomb interaction lifts the degeneracy. The energy ordering of these states can be obtained using the CF theory as follows [278]. The mapping into composite fermions takes us into the low-energy Hilbert space. Let us consider the system with a hole, that is with $Q = N/2$. We have already seen that the low-energy states are $L = S$ multiplets, with $(L, S) = (0, 0), \ldots, (N/2, N/2)$. The CF transformation maps it into $Q^* = Q - N + 1 = -(N - 2)/2$. The sign of Q^* is of no relevance, because it amounts to changing the direction of the magnetic field, which leaves the eigenspectrum unchanged while the eigenfunctions become complex conjugated. We throw away the sign, and make a particle–hole transformation in the lowest Landau level, which relates N to $N' = 2(2|Q^*| + 1) - N = N - 2$. This gives a system of $N - 2$ particles at monopole strength $(N - 1)/2$, which again has a single hole at $\nu = 1$. However, now the low-energy states are $(L, S) = (0, 0), \ldots, (\frac{N-2}{2}, \frac{N-2}{2})$. This tells us that the $(L, S) = (N/2, N/2)$ multiplet of the original system had the highest energy. An iteration of this procedure tells us that the energy of the skyrmion multiplets increases with S (or L) quantum number. This ordering, also seen in exact diagonalization studies (Fig. 11.21 [710]), is crucial for the relevance of skyrmions in actual experiments.

At sufficiently large Zeeman energies, the ordinary hole, (M, M), is the lowest-energy state; electrons are fully spin polarized in this state. The above considerations tell us that, for $E_Z = 0$, the spin-singlet state $(L, S) = (0, 0)$ has the lowest energy. At nonzero Zeeman energies, the spin of the lowest-energy state is determined by a competition between the Coulomb and the Zeeman energies. As the Zeeman energy is increased, level crossings take place between various skyrmion states, and the ground state shifts progressively to higher S. We consider $E_Z = 0$ for most of this section; the Zeeman contribution to the energy, obviously relevant for experiment, is considered in Section 11.10.4.

What about the particle skyrmion, obtained by *adding* an electron to $\nu = 1$? Here, we can formulate the problem in terms of $N - 2 \equiv 2M$ holes; assuming a hard-core interaction between holes gives the same multiplets as above.

11.10.2 Trial wave function

We focus on the hole skyrmion for simplicity, and construct an approximate but very good wave function for the multiplet

$$(S, L) = \left(\frac{N}{2} - K, \frac{N}{2} - K\right),$$ (11.80)

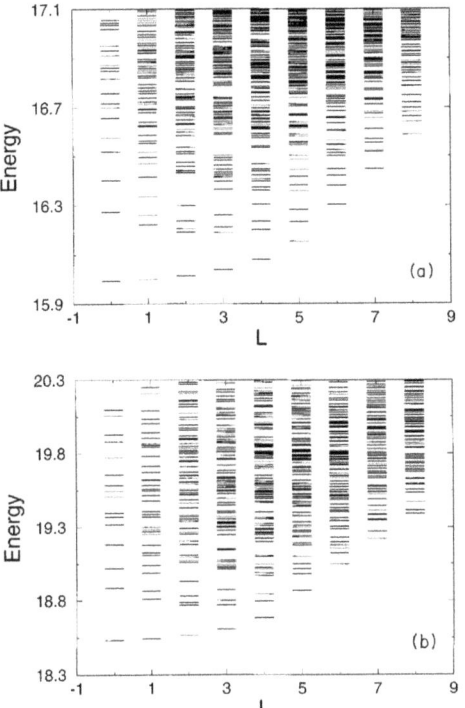

Fig. 11.21. Exact spectra for $N = 10$ electrons with $E_Z = 0$, for (a) $Q = 5$ (hole), and (b) $Q = 4$ (particle). Only states with $S \leq 5$ are shown. The lowest branch has $L = S$. Source: X. C. Xie and S. He, *Phys. Rev. B* **53**, 1046 (1996). (Reprinted with permission.)

for the situation when K is finite (although possibly large compared to one) and $N \to \infty$. (This subsection closely follows Refs. [413] and [453].) The strategy will be to map into a boson problem according to Eq. (11.63), and construct a state with $L_z = S_z = \frac{N}{2} - K$ that is annihilated by the angular momentum and spin raising operators. That gives us the highest weight state for the $(L, S) = (\frac{N}{2} - K, \frac{N}{2} - K)$ multiplet, from which other states of the multiplet can be obtained, if needed, with the help of the lowering operators. The electronic state can be obtained by multiplication by Φ_1, which conserves the spin and angular momentum quantum numbers. Let us denote the basis states in the boson problem by $|N_{\uparrow\uparrow}, N_{\downarrow\downarrow}, N_{\uparrow\downarrow}, N_{\downarrow\uparrow}\rangle$. From our earlier discussion, the basis states with $L_z = S_z = \frac{N}{2} - K$ have the form $|N - 2K + k, k, K - k, K - k\rangle$, where $2K - N \leq k \leq K$. (The total occupation is N.) The spin and angular momentum raising operators are given by

$$S^+ = b^\dagger_{\uparrow\uparrow} b_{\uparrow\downarrow} + b^\dagger_{\downarrow\uparrow} b_{\downarrow\downarrow} \tag{11.81}$$

$$L^+ = b^\dagger_{\uparrow\uparrow} b_{\downarrow\uparrow} + b^\dagger_{\uparrow\downarrow} b_{\downarrow\downarrow}. \tag{11.82}$$

Let us apply the spin raising operator

$$S^+|N - 2K + k, k, K - k, K - k\rangle$$

$$= \sqrt{(N - 2K + k + 1)(K - k)}|N - 2K + k + 1, k, K - k - 1, K - k\rangle$$

$$+ \sqrt{k(K - k + 1)}|N - 2K + k, k - 1, K - k, K - k + 1\rangle$$

$$\approx \sqrt{(N - 2K + k + 1)(K - k)}|N - 2K + k + 1, k, K - k - 1, K - k\rangle,$$

$$(11.83)$$

where we have used the familiar relations $b^\dagger|n\rangle = \sqrt{n+1}|n+1\rangle$ and $b|n\rangle = \sqrt{n}|n-1\rangle$ for boson creation and annihilation operators, and the second term has been dropped in the final expression because it is of order $O(N^{-1/2})$ relative to the first one. It then follows that the spin raising operator annihilates the state to leading order in $\sqrt{N}$ if we set either $k = K$ or $k = 2K - N$. The same is true of the angular momentum raising operator. The choice $k = 2K - N$ is not consistent with our assumption that K is finite, so we choose $k = K$. This produces the state $|N - K, K, 0, 0\rangle$, which has no bosons in states with $l_z \neq s_z$; such a coupling between the spin and the angular momentum for the skyrmion was manifest earlier through the equality of the L and S quantum numbers of the multiplets. The wave function

$$\Psi_K^{\text{h-sk}} = \Phi_1|N - K, K, 0, 0\rangle \qquad (11.84)$$

is thus a good representation of the hole skyrmion with $(L, S) = (\frac{N}{2} - K, \frac{N}{2} - K)$ in the limit $N \to \infty$ but with $K = \text{finite}$. We use it as a trial wave function for arbitrary K.

In the spherical geometry, we have $Y_{1/2,1/2,1/2} \sim u$ and $Y_{1/2,1/2,-1/2} \sim v$. The unnormalized wave function for the hole skyrmion, therefore, is

$$\Psi_K^{\text{h-sk}} = \Phi_1 \sum_{\{i_1,\dots,i_K\}} \left[v_{i_1} \dots v_{i_K} u_{j_1} \dots u_{j_{N-K}} \beta_{i_1} \dots \beta_{i_K} \alpha_{j_1} \dots \alpha_{j_{N-K}} \right], \qquad (11.85)$$

where the sum is over distinct particle indices $\{i_1, \dots, i_K\}$, and the j's are the remaining particles.

In analogy to the theory of superconductivity, where working with the grand canonical ensemble is convenient, we consider the wave function

$$\Psi^{\text{h-sk}}(\lambda) = \sum_{K=0}^{\infty} \lambda^K \Psi_K^{\text{h-sk}}$$

$$= \Phi_1 \prod_{j=1}^{N} (u_j \alpha_j + \lambda v_j \beta_j)$$

$$= \Phi_1 \prod_{j=1}^{N} \begin{pmatrix} u_j \\ \lambda v_j \end{pmatrix}. \qquad (11.86)$$

For large λ the sum over K is dominated by terms in a narrow range of K, the value of which depends on λ.

For $\lambda = 0$ this wave function reduces to the fully polarized state

$$\Psi^{h-sk}(\lambda = 0) = \prod_{j=1}^{N}(u_j\alpha_j)\Phi_1$$

$$= \left(\prod_j u_j\right)\Phi_1(\alpha_1\ldots\alpha_N). \tag{11.87}$$

Because Φ_1 is a Slater determinant of states $u^{2Q'}$, $u^{2Q'-1}v, \ldots, uv^{2Q'-1}$, $v^{2Q'}$ with $2Q = N - 1$, the above state is a Slater determinant of u^{2Q}, $u^{2Q-1}v, \ldots, uv^{2Q-1}$ with $2Q = N$. It is a filled LL state except for a hole in the state $v^{2Q} \sim [\sin(\theta/2)e^{-i\phi}]^{2Q}$, which is localized at the south pole.

In the disk geometry, the wave functions analogous to Eqs. (11.85) and (11.86) are given by (Exercise 11.8)

$$\Psi_K^{h-sk} = \Phi_1 \sum_{i_1,\ldots,i_K} \left[z_{j_1}\ldots z_{j_{N-K}}\beta_{i_1}\ldots\beta_{i_K}\alpha_{j_1}\ldots\alpha_{j_{N-K}}\right] \tag{11.88}$$

$$\Psi^{h-sk}(\lambda) = \Phi_1 \prod_j (z_j\alpha_j + \lambda\beta_j) = \Phi_1 \prod_{j=1}^{N}\binom{z_j}{\lambda}, \tag{11.89}$$

where

$$\Phi_1 = \prod_{j<k}(z_j - z_k)\exp\left[-\frac{1}{4}\sum_l |z_l|^2\right]. \tag{11.90}$$

In the limit of $\lambda = 0$, $\Psi^{h-sk}(\lambda)$ of Eq. (11.89) reduces to the familiar hole. To gain a feel for the meaning of λ, we employ a mapping into the one-component plasma (reader: the remainder of this subsection borrows results and terminology from Section 12.4):

$$|\Psi^{h-sk}|^2 = \exp[-\beta E\{z_j\}], \tag{11.91}$$

where $\beta = 1$, and

$$E\{z_j\} = -2\sum_{j<k}\ln|z_j - z_k| - \sum_j \ln(|z_j|^2 + \lambda^2) + \frac{1}{2}\sum_j |z_j|^2. \tag{11.92}$$

The first term on the right hand side gives the interaction between "particles" of charge -1, and the last term gives the interaction between the "particles" of charge -1 with a uniform background of charge density $\bar{\rho} = (2\pi \ell^2)^{-1}$. The second term is the energy of "particles" in an external potential

$$V(r) = \ln(r^2 + \lambda^2). \tag{11.93}$$

From $\nabla^2 V(r) = -4\pi\rho(r)$, this potential is produced by an impurity charge density

$$\rho_{\text{imp}}(r) = -\frac{\lambda^2}{\pi} \frac{1}{(r^2 + \lambda^2)^2}. \tag{11.94}$$

Because of the overall charge neutrality of the plasma, we expect that the particles screen the impurity potential to produce an overall excess charge density $\delta\rho(r) = -\rho_{\text{imp}}(r)$, which gives the charge density of "particles"

$$\delta\rho(r) = \frac{\lambda^2}{\pi} \frac{1}{(r^2 + \lambda^2)^2}, \tag{11.95}$$

which is also the charge density of electrons, in units of the electron charge. The parameter λ thus represents the size of the skyrmion. Exercise 11.10 proves that the net charge deficiency is precisely equal to e, which is a consistency check on the validity of the wave function $\Psi^{\text{h-sk}}$. The net charge is independent of the size of the skyrmion. (Equation 11.95 is not correct, in general, because the screening of the impurity charge is locally perfect only if $\lambda/\ell \gg 1$, but the total charge deficiency in the plasma near the impurity should be equal to the total impurity charge for arbitrary λ/ℓ.)

11.10.3 Skyrmion on sphere

The $(L, S) = (0, 0)$ state has $S_z = L_z = 0$, which therefore is represented by the wave function in Eq. (11.86) with $\lambda = 1$, for which $\langle S_z \rangle = \langle L_z \rangle = 0$:

$$\Psi^{\text{h-sk}}(\lambda = 1) = \Phi_1 \prod_{j=1}^{N} \binom{u_j}{v_j}. \tag{11.96}$$

This wave function has a pictorial interpretation, which is most easily seen by switching the direction of the magnetic field to point radially inward, which amounts to a complex conjugation of the wave function. Because the factor Φ_1 does not alter the spin of the function it multiplies, we look at

$$\prod_{j=1}^{N} \binom{u_j^*}{v_j^*} = \prod_{j=1}^{N} \binom{\cos(\theta_j/2)e^{-i\phi_j/2}}{\sin(\theta_j/2)e^{i\phi_j/2}}. \tag{11.97}$$

This is precisely the spinor representation, in which the spin points radially outward (Exercise 11.10):

$$\langle \boldsymbol{S}(\boldsymbol{\Omega}) \rangle = \frac{1}{2}\boldsymbol{\Omega}. \tag{11.98}$$

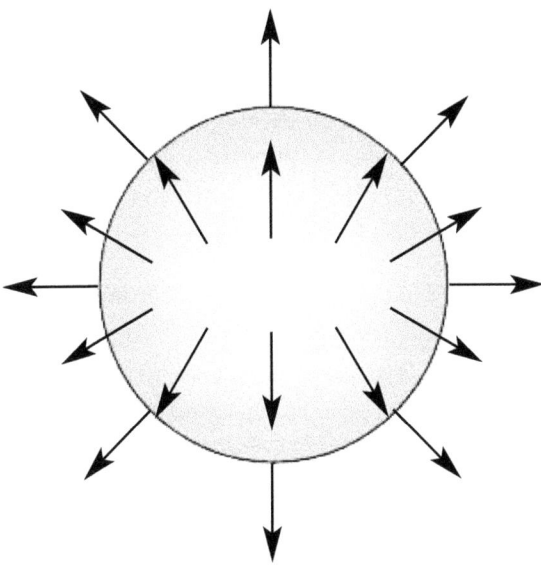

Fig. 11.22. Skyrmion on a sphere. The spin direction depends on the spatial location, pointing radially outward.

The $S = 0$ skyrmion thus resembles a "hedgehog" in the spherical geometry (Fig. 11.22).

When an electron is dragged around on the sphere, its spin rotates adiabatically to keep its direction aligned with $\mathbf{\Omega}$, to minimize the exchange energy due to ferromagnetic coupling to nearby electrons. Its phase therefore gets two contributions; one is the Aharonov–Bohm phase due to the magnetic field emanating radially from the magnetic monopole of strength $Q = -(N - 1)/2$ at the center (remember we have temporarily reversed the magnetic field direction), and the other is the Berry phase associated with the rotation of its spin. Section 3.10.7 showed that the Berry phase associated with the rotation of spin $S = 1/2$ is equivalent to the Aharonov–Bohm phase of an electron moving under the influence of a magnetic monopole of strength $Q = 1/2$. The total phase associated with the electron motion is then as though it sensed an effective magnetic monopole of strength $Q_{\text{eff}} = Q + 1/2$. The effective degeneracy of the lowest "Landau level" therefore is $2|Q_{\text{eff}}| + 1 = N - 1$, which gives a fully occupied Landau level.

Why is the skyrmion state preferred for $E_Z = 0$?[4] Let us consider a particle skyrmion in the lowest Landau level, obtained by adding one electron to the fully spin-polarized $\nu = 1$ state. In the maximally polarized state, the additional particle is placed in a spin-reversed orbital. It has wrong spin relative to the nearby electrons, which costs exchange energy. Furthermore, it causes a localized charge disturbance, thereby increasing the direct

---

[4] The skyrmion structure in the spherical geometry should not be confused with what a $\mathbf{B} \cdot \mathbf{S}$ type Zeeman coupling term would produce. We have set the Zeeman energy to zero. Even when nonzero, the Zeeman splitting is proportional to S_z in our model, so the hedgehog spin texture has a high energy compared to the fully polarized state.

Coulomb interaction energy. Nature takes care of both problems by creating the skyrmion texture, which is stable (filled LL type) with uniform density, and for which nearby electrons have their spins almost aligned everywhere. Thus, when a spin-reversed electron is added, nearby electrons flip their spins to lower the exchange energy. At $E_Z = 0$ half of the electrons in the system flip their spins to produce a spin-singlet state.

11.10.4 Finite Zeeman energy: baby skyrmions

Let us consider the hole skyrmion $(N, 2Q) = (N, N)$, with N even. When $E_Z = 0$, the $(L, S) = (0, 0)$ state has the lowest energy. For a nonzero Zeeman energy, no matter how small, this state is irrelevant, because it has an extremely high, $O(N)$ energy. For large Zeeman energies, the $(L, S) = (N/2, N/2)$ state wins; it has no wrong spins and an $O(1)$ interaction energy. At a general nonzero Zeeman energy, therefore, only a finite number of spins may be flipped for the lowest-energy state. The Zeeman energy thus rapidly shrinks the infinitely large skyrmion of $E_Z = 0$ to a finite size excitation with only a few reversed spins. Even though the resulting excitation is far from the fully fledged skyrmion of $E_Z = 0$, it is customary to continue to call it a (baby) skyrmion, provided its spin content is greater than that of the large E_Z excitation.

The qualitative difference between "$E_Z = 0$" and "E_Z very small" is reminiscent of the Nagaoka theorem [469] for the Hubbard model, with the on-site repulsion characterized by U and the hopping by t. The theorem states, for a square lattice, that when $U = \infty$, the ground state of the system with exactly one hole has the maximum spin (i.e., is a ferromagnet). For nonzero U but $U/t \gg 1$, the ground state at half filling (no hole) is an antiferromagnet. This might suggest that, in the limit of $U/t \to \infty$, the introduction of a single hole causes a transition from an antiferromagnet to a ferromagnet. "Large U" and "infinite U" behave differently, however, both at half filling and one hole away from half filling. For infinite U, there is no antiferromagnetism at half filling. For large but finite U, the addition of a hole only produces a finite ferromagnetic domain inside an antiferromagnet.

A determination of the size of the skyrmion, or the number of spin flips per additional flux quantum, requires the knowledge of skyrmion energy as a function of S. The Hartree–Fock method of Fertig *et al.* [155] proceeds as follows: The many particle state in Eq. (11.89) is a Slater determinant of single particle orbitals

$$\chi_m = (z^{m+1}\alpha + \lambda z^m \beta) \exp\left[-\frac{1}{4}|z|^2\right], \tag{11.99}$$

which are linear superpositions of the LLL orbitals $\eta_{0,m+1}\alpha$ and $\eta_{0,m}\beta$. A more general Hartree–Fock wave function is constructed as

$$\Psi^{\text{h-sk}} = \prod_{m=0}^{\infty} (u_m b_m^\dagger + v_m a_{m+1}^\dagger)|0\rangle, \tag{11.100}$$

where $b_m^\dagger$ creates an electron in the spin-down state and $a_m^\dagger$ in the spin-up state. The variational parameters are determined in Ref. [155], as a function of the Zeeman energy, by minimizing the total energy of this state. Alternatively, the energy of the skyrmion wave function can be computed by Monte Carlo or exact diagonalization [310]. The outcome of these studies is that, for typical experimental situations, the elementary charge-one excitation involves $\approx$2–7 spin flips, demonstrating that the exchange Coulomb interaction does indeed affect the spin content of the excitation.

11.10.5 Observations of skyrmions

Using the optically pumped NMR technique, Barrett *et al.* [21] measure the Knight shift, which is proportional to the spin polarization of the electron system. These measurements show that as the filling factor moves away from $\nu = 1$, the spin polarization drops faster than what would be expected in a picture of noninteracting electrons. The latter would predict, for example, that for $\nu < 1$ the system remains fully spin polarized (for any nonzero Zeeman energy). Figure 11.23 shows that each additional flux quantum involves a reversal of approximately 3.6 spins, which is in accordance with the theoretical estimates by Fertig *et al.* [155] for the experimental parameters.

These results are consistent with the spin polarization measurements of Aifer, Goldberg, and Broido [4] by magnetoabsorption spectroscopy, which distinguishes the occupancy of the spin-up and spin-down states (Fig. 11.24). Very close to $\nu = 1$, however, the spin polarization is found to be relatively flat in Fig. 11.24.

The spin content of the quasiparticles also appears in the Zeeman energy dependence of the activation gap, which involves an excitation of a particle–hole pair. Figure 11.25 shows the activation gap measured by Schmeller *et al.* [572] as a function of the tilt (keeping the perpendicular component of the magnetic field fixed) at $\nu = 1$. A model of noninteracting electrons would predict a tilt-dependent part of the gap proportional to E_Z, but the experiment is consistent with a dependence of $7.0E_Z$. That corresponds to approximately 3.5 spin flips each for the particle and the hole.

Maude *et al.* [433] produce a very small g factor by application of hydrostatic pressure. From the Zeeman energy dependence of the activation gap, they see evidence for skyrmions with as many as 30 reversed spins. They also confirm that the spin activation gap persists all the way to zero Zeeman splittings, as expected.

The experiment of Aifer *et al.* [4], as well other recent experiments (Khandelwal *et al.* [332]; see also Fig. 11.11), show a flat region or a tilted plateau in polarization near filling factor one at low temperatures, which appears to be inconsistent with the picture of skyrmions being the relevant excitations. It has been suggested [332] that this is a consequence of skyrmion localization by disorder.

11.10.6 Composite fermion skyrmion

Because $\nu = 1/(2p + 1)$ is analogous to $\nu^* = 1$ of composite fermions, one might expect a skyrmion of composite fermions here [310]. Indeed, exact diagonalization studies have

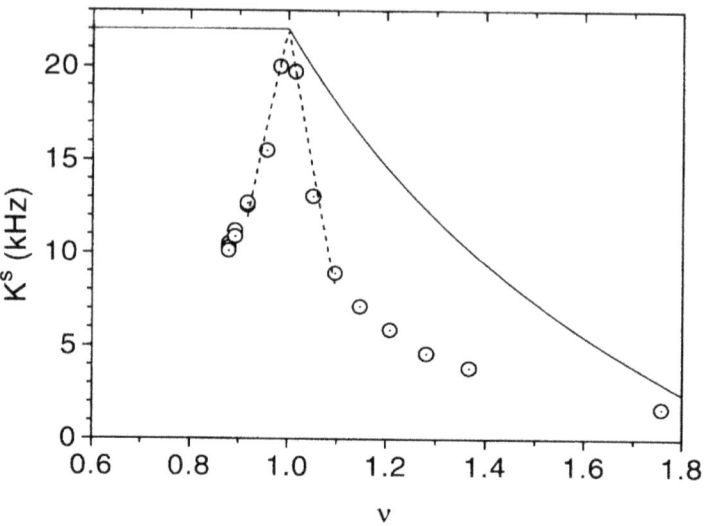

Fig. 11.23. Filling factor dependence of the Knight shift near $\nu = 1$. The solid line is the prediction from the noninteracting electron model, whereas the dashed line is predicted with the assumption that 3.6 spins are reversed for each additional flux quantum. Source: S. E. Barrett, G. Dabbagh, L. N. Pfeiffer, K. W. West, and R. Tycko, *Phys. Rev. Lett.* **74**, 5112 (1995). (Reprinted with permission.)

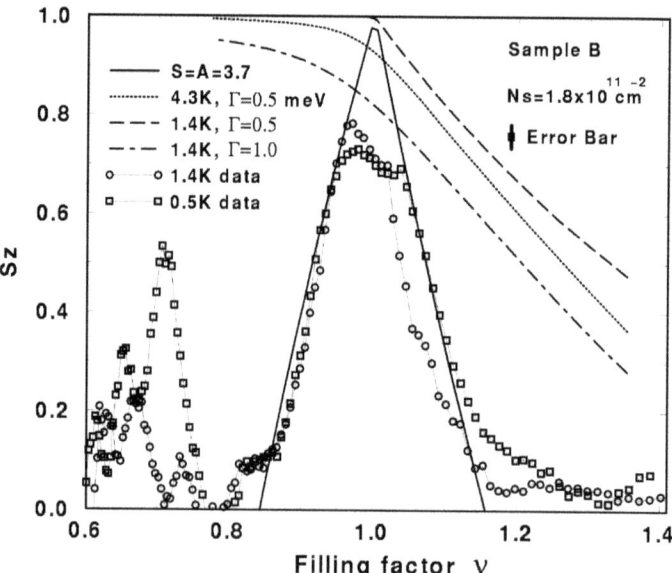

Fig. 11.24. Filling factor dependence of the spin polarization deduced from interband optical spectroscopy. The solid line is the theoretical prediction assuming 3.7 spin flips per particle, whereas the dashed, dotted, and dash-dotted lines are the expected behavior without skyrmions, for several different parameters. Source: E. H. Aifer, B. B. Goldberg, and D. A. Broido, *Phys. Rev. Lett.* **76**, 680 (1996). (Reprinted with permission.)

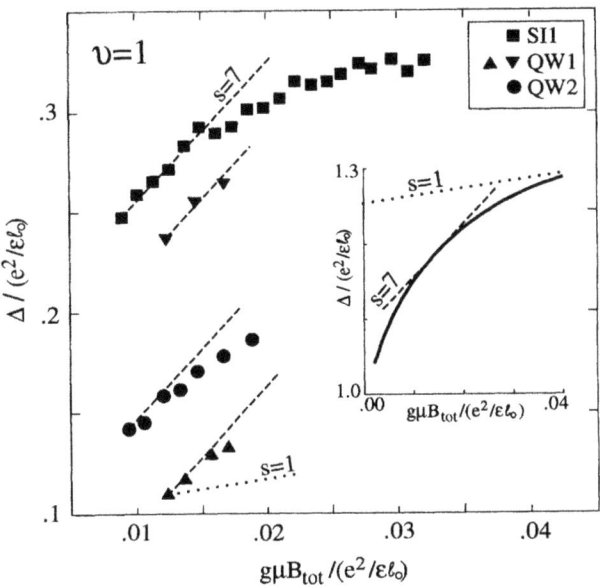

Fig. 11.25. Zeeman energy dependence of the activation gap at $\nu = 1$ for several samples. The dashed line corresponds to seven spin flips, whereas the dotted line to a single spin flip. The inset shows the theoretical prediction of Ref. [155]. No skyrmion is found at $\nu = 3$ and $\nu = 5$. Source: A. Schmeller, J. P. Eisenstein, L. N. Pfeiffer, and K. W. West, *Phys. Rev. Lett.* **75**, 4290 (1995). (Reprinted with permission.)

shown that the low-energy part of the $E_Z = 0$ spectrum at one flux quantum away from 1/3 (where we expect the ^{2}CF skyrmion) has qualitative resemblance to that at one flux quantum away from $\nu = 1$, with a low-energy band of multiplets with $L = S$ (see, for example, Figs. 3, 4, and 5 of Wójs and Quinn [696]). For a more quantitative test, one constructs the wave function

$$\Psi^{\text{CF-sk}} = \Phi_1^{2p} \Psi^{\text{el-sk}}. \qquad (11.101)$$

For small systems one may take $\Psi^{\text{el-sk}}$ as the exact wave function of the electron skyrmion at $\nu = 1$, whereas for large N one may use wave functions of the type in Eq. (11.85). For six particles, the exact Coulomb energies of the $\nu = 1/3$ quasihole and quasiparticle skyrmions at $(L, S) = (0, 0)$ are -0.4367 and -0.4620 in the conventional units of $e^2/\epsilon\ell$; the ^{2}CF skyrmion wave function predicts -0.4362 and -0.4618, respectively [310]. Table 11.4 gives results for the entire CF-quasihole skyrmion branch. The agreement confirms the formation of skyrmions in computer experiments. The conditions for the observation of CF skyrmions, however, are more stringent than those for the electron skyrmions at $\nu = 1$, because the interaction between composite fermions at $\nu = 1/(2p+1)$, which is responsible for stabilizing the CF skyrmion, is much weaker than the interaction between electrons at

Table 11.4. *Testing CF-quasihole skyrmion*

(L,S)	Exact Coulomb eigenstate	Hard-core trial state
(0,0)	−0.4367	−0.4362
(1,1)	−0.4364	−0.4359
(2,2)	−0.4357	−0.4354
(3,3)	−0.4337	−0.4333

Notes: Comparison of exact Coulomb energy of low-energy states at $\nu = 1/3$ for $N = 6$ at $E_Z = 0$ with the CF-quasihole skyrmion state, Eq. (11.101), obtained by multiplying hard-core states of hole skyrmion at $\nu = 1$ by Φ_1^2.
Energies per particle in units of $e^2/\epsilon\ell$.

Source: Kamilla, Wu, and Jain [311].

$\nu = 1$. Theoretical estimates based on trial wave functions of the type in Eq. (11.85) indicate [310] that, for $g = 0.44$, the ^{2}CF-quasihole skyrmion at $\nu = 1/3$ occurs only at magnetic fields below ≈ 1.6 T. Wójs and Quinn [696] determine the critical Zeeman splittings for ^{2}CF-quasihole and quasiparticle skyrmions, from extrapolations of exact results, to be $0.0050e^2/\epsilon\ell$ and $0.0093e^2/\epsilon\ell$, respectively, which translate into critical magnetic fields of $\sim$1 T for GaAs parameters. The asymmetry between the quasiparticle and quasihole ^{2}CF skyrmions can be understood from the observation that vortices are attached only to electrons, not to holes.

The CF skyrmions would have been academic but for the possibility that the g factor can be reduced by application of pressure. At very small Zeeman energies, Leadley *et al.* [377] observe excitations consisting of three reversed spins at $\nu = 1/3$, which they interpret as baby ^{2}CF skyrmions. In contrast, no evidence is seen for a skyrmion at $\nu = 2/5$. Optically pumped NMR studies show evidence of a very small spin associated with the excitations of $\nu = 1/3$ ($\sim$0.1 spin flips per quasihole) even at magnetic fields as high as 12 T (Khandelwal *et al.* [331]); the origin of this subtle effect is not yet understood.

Doretto *et al.* [120] have evaluated the dispersion of the spin exciton at $\nu = 1/3$ and 1/5 using the Murthy–Shankar theory. Neglecting excitations across Λ levels, they suggest that CF-quasihole and CF-quasiparticle skyrmions form a bound state.

Exercises

11.1 Confirm that Φ' is normalized in both forms of Eq. (11.32).

11.2 Demonstrate the Fock cyclic condition explicitly for $N = 4$ particles at $\nu = 2$, that is, for

$$\Phi_{1,1} = \prod_{j<k=1}^{2} (z_j - z_k) \prod_{r<s=1}^{2} (w_r - w_s),$$

where z_j and w_r denote the coordinates of spin-up and spin-down electrons, respectively.

11.3 For the fully polarized state with $S = S_z = N/2$, the spin part and the spatial part factorize, the former being fully symmetric and the latter fully antisymmetric. States with $S = N/2$ and $0 \leq S_z < N/2$ can be obtained by repeated applications of the spin lowering operator. Show that these also have a factorized form, and determine the spatial part for arbitrary S_z.

11.4 Determine the spin quantum numbers for the following systems on sphere by applying Hund's rule (i) to electrons, and (ii) to composite fermions: $(N, Q) = (6, 4)$ [0]; $(N, Q) = (6, 4.5)$ [1]; $(N, Q) = (6, 5)$ [0]; $(N, Q) = (6, 5.5)$ [1]; $(N, Q) = (6, 6)$ [1]; $(N, Q) = (8, 5)$ [1]; $(N, Q) = (8, 5.5)$ [0]; $(N, Q) = (8, 6)$ [1]; $(N, Q) = (8, 6.5)$ [2]. Compare the answers with the exact ground state spin for $E_Z = 0$ given in square brackets for each system (Refs. [278, 704]).

11.5 What spin and angular momentum quantum numbers does the CF theory predict for $(N, Q) = (9, 8)$ and $(N, Q) = (10, 9)$? Compare with the exact ground state quantum numbers in Fig. 11.3.

11.6 Show that the relation between the monopole strength Q and the particle number N for the incompressible FQHE state at $\nu = n/(2pn \pm 1)$, in which $n_\uparrow$ spin-up Λ levels and $n_\downarrow$ spin-down Λ levels are occupied, is given by

$$Q = N \left(p \pm \frac{1}{2n} \right) - \left(p \pm \frac{n_\uparrow^2 + n_\downarrow^2}{2n} \right). \tag{E11.1}$$

11.7 Consider $N = 4$ bosons at $Q^B = 1/2$. Make a table with columns corresponding to L_z and the rows to S_z. ($L_z, S_z = -2, -1, 0, 1, 2$.) Construct explicit configurations and fill in the table the total number of states for each (L_z, S_z). Deduce from it the quantum numbers of various multiplets.

11.8 Consider a hole skyrmion in the disk geometry. The relevant orbital quantum number now is m, the eigenvalue of L_z. For $\nu = 1$, the orbitals with $m = 0, 1, \ldots, N - 1$ are occupied. For a single hole, the orbitals $m = 0, 1, \ldots, N$ are available. Show that, through $\Psi' = \prod_{j<k}(z_j - z_k)\Psi'_B$, the fermionic system maps into a bosonic system in which N bosons can occupy four states, $0\uparrow, 0\downarrow, 1\uparrow, 1\downarrow$, where the number refers to the value of m and the arrow to spin. Following the reasoning leading to Eqs. (11.85) and (11.86), derive Eqs. (11.88) and (11.89).

11.9 Show that the expectation value of spin $\mathcal{S} = \frac{1}{2}\sigma$ (σ are the Pauli matrices) in the state

$$\begin{pmatrix} \cos(\theta/2)e^{-i\phi/2} \\ \sin(\theta/2)e^{i\phi/2} \end{pmatrix} \tag{E11.2}$$

is $\mathbf{\Omega}/2$.

11.10 Determine the total charge deficiency for a hole skyrmion represented by the wave function given in Eq. (11.89) by making use of Eq. (11.95).

12

Non-composite fermion approaches

Analogies play a central role in physics. The CF theory was motivated by the analogy between the fractional and the integral quantum Hall effects. Other analogies were proposed prior to the CF theory. As discussed in Section 5.5, the Laughlin wave function was motivated by the Jastrow wave functions used earlier for ^{4}He superfluids. A "composite boson" approach modeled the physics of the Laughlin wave function after Bose–Einstein condensation. A "hierarchy" approach proposed understanding the FQHE at non-Laughlin fractions by using the Laughlin wave function as the basic building block. Both the hierarchy and composite boson approaches take the Laughlin wave function as their starting point. These ideas, as seen below, are distinct from the CF theory.

This chapter also presents certain other non-CF works. The simple Jastrow form of the Laughlin wave function allows for certain technical simplifications; in particular, a mapping into a classical plasma enables alternative, but $1/m$-specific, derivations of certain properties that were obtained in earlier chapters by other means. In addition, we describe two quantitative approaches for excitations without using composite fermions: Laughlin's wave functions for the quasiparticle and quasihole at $\nu = 1/m$, and Girvin, MacDonald, and Platzman's "single-mode approximation" for neutral excitations.

We also briefly outline the hydrodynamic theory of Conti and Vignale, and Tokatly, which treats the correlated liquid state in the lowest Landau level as a continuous elastic medium, and formulates its collective dynamics in terms of the displacement field. Quantization of this field yields bosonic excitation modes, analogous to the phonons of a crystal. This approach has been applied to the bulk and the edge of incompressible and compressible states.

12.1 Hierarchy scenario

An appealing hierarchical construction was advanced by Haldane [221] and Halperin [230] in an attempt to deal with the non-Laughlin odd-denominator fractions. The basic idea is to build Laughlin-like "daughter" states of the fractionally charged, anyonic quasiparticles of a given "parent" state, and thus construct new states iteratively starting from the Laughlin fractions. All odd-denominator fractions are thus produced. The fractions at any level of the hierarchy are expressed as continued fractions; for example, the fractions at the

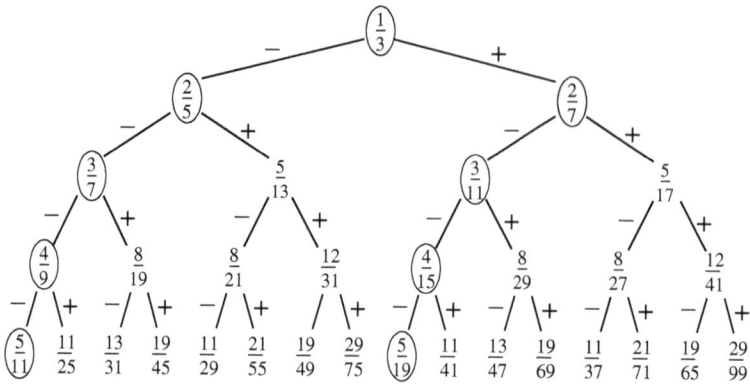

Fig. 12.1. Hierarchical family tree stemming from 1/3. The observed fractions are circumscribed.

fifth level of the hierarchy are given by

$$f = \cfrac{1}{m \pm \cfrac{1}{p_2 \pm \cfrac{1}{p_3 \pm \cfrac{1}{p_4 \pm \cfrac{1}{p_5}}}}}, \qquad (12.1)$$

where p_j is an even integer. Taking the simplest values $m = 3$ and $p_j = 2$, the hierarchy generates the fractions shown in Fig. 12.1, which have a family tree structure with two children at each generation. The 2^{n-1} fractions at the nth level correspond to different choices of signs in the continued fraction. The convention used in the figure is that the levels of the hierarchy evolve downward, with the positive sign in Eq. (12.1) taking us south-east and the negative sign taking us south-west.

Even though the innovative hierarchy idea appeared natural at first, compatible with the Landau philosophy of working with the deviations from a known ground state, inconsistencies with experiment became evident as more facts were gathered. Essentially, such an approach is expected to be valid *close* to the starting point (the Laughlin state), but an explanation of experimental phenomena in FQHE requires us to go far from it. Let us summarize the problems. (Because it has not been possible to make reliable quantitative predictions in this approach, the comparisons with experiment are of a qualitative nature.)

(a) A fundamental aspect of the hierarchy approach is a parent–daughter relation between fractions. Because each state has two equally plausible daughters (one constructed from quasiparticles and the other from quasiholes), this manifests through a family tree structure for fractions. The appearance of *sequences* of fractions in actual experiments reveals a different underlying structure. (b) The order of stability of fractions predicted by the hierarchy is incompatible with experiment. One would expect all fractions on a given level to be roughly equally stable. However, while some fractions at very deep levels have been observed, a large number of fractions at earlier levels have not. For example, the tenth generation of the $1/3$ family tree consists of 2^9 fractions (assuming only two daughters at each generation), but only one of them ($10/21$) has been observed. (c) Even if one confines oneself to the observed fractions, the gaps decrease much more slowly than implied by the general considerations of the hierarchy scenario, because a daughter, being a *fractional* quantum Hall state of the quasiparticles, is expected to be much weaker than the parent. (d) For the same reason, the very observation of a large number of fractions is inconsistent with the hierarchy idea.

The CF theory is not a hierarchy. No parent–daughter relationship exists between the fractions of a given sequence $v = n/(2pn \pm 1)$, just as none does between the integers of the IQHE. Composite fermions allow us to understand $3/7$ perfectly well without making any reference to $2/5$ (or to any other FQHE state); if one insisted on ascribing a relation to them, the two would be siblings rather than daughter and parent. In the CF theory, all fractions are explained on an equal footing, each fraction stands on its own, and $1/3$ is not the mother of all fractions.[1]

From a theoretical perspective, the hierarchy scenario runs into a conceptual impediment that cannot be overcome within this approach. To illustrate, let us begin with $v = 1/3$. As the magnetic field is varied, quasiparticles are produced, one for each additional flux quantum. Due to their nonzero size, however, they are not ideal anyons. They have a well-defined braiding statistics only when the overlap between them is negligible, which, as seen in Chapter 9 (see Figs. 9.5 and 9.6), requires them to be ≥ 10 magnetic lengths apart. At shorter distances, significant corrections to the braiding statistics appear. That presents a difficulty for hierarchy. Simple counting shows that to reach $v = 2/5$, which is supposed to be the first "daughter" state, it would be necessary to create half as many quasiparticles as the total number of electrons. At such high densities, the quasiparticles are so strongly overlapping (the interelectron separation is ~ 3.5 magnetic lengths at $v = 1/3$) that the model of quasiparticles with well-defined fractional charge and fractional braiding statistics becomes questionable. An explanation of $10/21$, for which experimental evidence exists (see, for example, Ref. [494]), would require that a macroscopically large number of quasiparticles of $1/3$ make a daughter state at $2/5$; a macroscopically large number of quasiparticles of $2/5$ produce a daughter at $3/7$; and the process be repeated seven more times. If the very first link is broken, the subsequent generations remain unborn.

The point is that the statistics of an emergent entity is not absolute (unlike the Fermi statistics of the electron) but has a limited range of validity. As an illustration, let us consider the more familiar example of a collection of another set of composite particles: ^{4}He atoms. When they are far from one another, they are well described as bosons, and indeed exhibit Bose–Einstein condensation. Let us now imagine applying pressure to bring ^{4}He atoms

---

[1] Some fractions, such as $4/11$, which are understood as *fractional* QHE of composite fermions, are sometimes thought of as the second level of a CF hierarchy. That hierarchy, however, is distinct from the Haldane–Halperin hierarchy being discussed here. The latter also produces these fractions by appropriate choices of parameters in Eq. (12.1).

closer to one another. Eventually, under extreme pressure, when they begin to overlap strongly, the system goes into a "plasma" phase, wherein electrons are no longer bound to their parent nuclei (as in a white dwarf). A theory in terms of bosonic ^{4}He variables is no longer valid, and indeed no Bose–Einstein condensation would occur. A description of the plasma phase is impossible without knowing the constituents of the ^{4}He boson, and must be formulated in terms of electrons and ^{4}He nuclei.

The anyon language is similarly inapplicable for high densities of the FQHE quasiparticles.[2] The CF theory identifies the more fundamental entities – composite fermions – which provide a valid description at low energies in the entire relevant filling factor range. (For sufficiently high energy states, the composite fermion description also becomes inadequate and one must revert to electrons.)

Haldane's version of the hierarchy [221, 226] assigns, nominally, bosonic braiding statistics to the quasiparticles and quasiholes, but otherwise produces the same structure as the Halperin hierarchy. As underscored by Haldane, the hierarchy construction is valid only under the assumption that the interaction between the quasiparticles is sufficiently strongly repulsive at short distances, i.e., is dominated by the short-range pseudopotentials. That is not borne out by explicit calculation (Section 6.7), which shows that the interaction is only weakly repulsive at short distances, and sometimes even attractive.

The reader may encounter in the literature statements to the effect that the CF and the hierarchy approaches are "equivalent" [538]. That is incorrect, for reasons explained above. The claims of equivalence are based on the concurrence of the two approaches in one narrow aspect, namely in the values of fractional local charge and fractional braiding statistics for quasiparticles. However, not much significance ought to be attached to this fact, given that a determination of these values does not require an understanding of the physical origin of the FQHE; they can be derived from general principles (Su [629]) simply by assuming incompressibility at a fractional filling factor. They are a consequence, not a cause, of incompressibility, and, therefore, it is not surprising that different theories produce the same values for them. The CF and hierarchy approaches attribute different physics to the origin of the FQHE, and have contrasting testable consequences. (Also see the last paragraph of the next section.) In particular, the hierarchy approach does not contain composite fermions and the vast body of physics that follows from them.

12.2 Composite boson approach

Girvin and MacDonald [193] interpreted Laughlin's wave function as a Bose condensate, a notion that was further developed by Zhang, Hansson, and Kivelson [743] and Read [537], and later came to be known as the "composite boson" approach. The idea, roughly, is as follows: For the Laughlin wave function, it appears natural to take the bound state of an electron and an *odd* number (m) of vortices as the fundamental object, which is called a

---

[2] Its validity is not entirely obvious even in the dilute limit. For ^{4}He atoms, the hard-core repulsion protects their bosonic statistics at low densities. The interaction between the FQHE quasiparticles, on the other hand, is rather weak, and sometimes even attractive (Fig. (6.6)). It is thus ineffective in suppressing overlapping configurations even for dilute densities.

composite boson. Crudely, this amounts to writing the Laughlin wave function as

$$\Psi_{1/m} = \prod_{j<k} (z_j - z_k)^m \Psi_B(\{z_i\}) , \qquad (12.2)$$

where $\Psi_B(\{z_i\}) = \exp[-\sum_j |z_j|^2/4]$. Since $\Psi_B(\{z_i\})$ is symmetric under exchange and everywhere positive, it is tempting to view $\Psi_{1/m}$ as a Bose condensate of composite bosons.

This is to be contrasted with the CF interpretation as one filled Λ level of composite fermions, as expressed in Eq. (5.37), and reproduced here for convenience:

$$\Psi_{1/m} = \prod_{j<k} (z_j - z_k)^{2p} \Phi_1 , \qquad (12.3)$$

with $m = 2p + 1$. From the CF perspective, it is no more than a coincidence that the Pauli correlations in Φ_1 appear through the binding of precisely one vortex to each electron (which is not the case if Φ_1 is replaced by Φ_n), which combines with the $2p$ vortices of composite fermions to produce a total of $2p + 1$ vortices bound to each electron.

The analogies of Bose–Einstein condensation and the integral quantum Hall effect are equally plausible for the Laughlin ground state wave function at $\nu = 1/m$. A consideration of the physics beyond the $\nu = 1/m$ ground state clarifies that the two views are distinct. The bosonic approach does not extend to fractions other than $\nu = 1/m$.

This offers a curious example of how different approaches may appear "natural" depending on one's perspective. The composite boson interpretation seems most obvious if one takes the Laughlin wave function as the point of departure. Even someone with an overactive imagination would have no reason to write it as in Eq. (5.37) and interpret it as one filled Λ level of composite fermions. On the other hand, if one takes the analogy between the FQHE and the IQHE as the guiding principle, then the composite fermion physics and Eq. (5.37) appear natural.

The question often arises why electrons capture an even, rather than an odd, number of vortices. What bound states are formed is often a complicated issue even in few-body systems, the resolution of which requires detailed microscopic calculations and a careful comparison with experiments. Some insight into why composite fermions are preferred can be gained from the observation that the Pauli repulsion between fermions helps produce desirable correlations.[3]

Some may find it disappointing that the FQHE does not lend itself to an explanation in the familiar language of Bose–Einstein condensation. On the other hand, it can be argued that the lack of BEC and order parameter makes the physics of this quantum liquid more unusual and interesting.

---

[3] Bosons with strong hard-core repulsive interaction in one dimension emulate fermions for the same reason. The ground state wave function is given by $\Psi^B = |\Psi^F|$, where Ψ^F is the wave function for the Fermi sea (Girardeau [187]). Another example is that of interacting *bosons* in the lowest Landau level, which can be produced by rotating a bosonic trap; they are found, theoretically, to capture a single vortex and behave like composite fermions [69, 96, 98, 427, 541–543, 656, 679].

To summarize: It has not been possible to translate the hierarchy or the composite boson ideas into reliable microscopic theories that are amenable to quantitative tests. Additionally, a large and growing body of experimental facts is incompatible with these ideas. Prominent among these are: the FQHE at non-Laughlin fractions; similarity between the FQHE and the IQHE; appearance of sequences of fractions; filling factor dependence of gaps; existence of composite fermions; the compressible CF Fermi sea ($\nu = 1/2$); the effective magnetic field; the paired CF state ($\nu = 5/2$); non-fully spin-polarized FQHE; and a plethora of excitations and other phenomena.

12.3 Response to Laughlin's critique

Laughlin expressed reservations about the CF theory in his 1998 Nobel Lecture (Laughlin [372]), which have caused some confusion and, therefore, deserve clarification.[4] To quote:

Fractional quantum Hall quasiparticles are the elementary excitations of a distinct state of matter that cannot be deformed into noninteracting electrons without crossing a phase boundary. That means that they are different from electrons in the only sensible way we have of defining different, and in particular are not adiabatic images of electrons the way quasiparticle excitations of metals and band insulators are. Some composite fermion enthusiasts claim otherwise – that these particles are nothing more than screened electrons (Jain, 1989) – but this is incorrect. The alleged screening process always runs afoul of a phase boundary at some point, in the process doing some great violence to the ground state and low-lying excitations. I emphasize these things because there is a regrettable tendency in solid-state physics to equate an understanding of nature with an ability to model, an attitude that sometimes leads to overlooking or misinterpreting the higher organizing principle actually responsible for an effect. In the case of the integral or fractional quantum Hall effects, the essential thing is the accuracy of quantization. No amount of modeling done on any computer, existing or contemplated, will ever explain this accuracy by itself. Only a thermodynamic principle can do this. The idea that the quasiparticle is only a screened electron is unfortunately incompatible with the key principle at work in these experiments. If carefully analyzed it leads to the false conclusion that the Hall conductance is integrally quantized.

We take the two principal objections raised in this paragraph one by one. Composite fermions have indeed been described as "screened electrons." Screening is perturbative in many familiar applications, but it has been stressed (see Section 9.4 and Ref. [201]) that the screening of an electron into a composite fermion, which occurs through the binding of exactly an even number of vortices to electrons, is quantized and nonperturbative. A vortex cannot be attached adiabatically – either all or none of it is bound to an electron. Composite fermions are thus topologically distinct from electrons and cannot be obtained from electrons in any perturbative treatment. This is also the main point of the topological Chern–Simons approaches. Therefore, the objection following from an assumed adiabatic connectivity between composite fermions and electrons is not valid. (The last statement

---

[4] This section owes a great deal to many discussions with Fred Goldhaber.

in Laughlin's critique cannot be addressed in the absence of explicit details of the stated analysis.)

Let us next turn to the question of "modeling" versus "understanding." Laughlin rightly notes that modeling cannot replace understanding. That would be a serious criticism, for example, against presenting the exact numerical solutions of the Schrödinger equation as an "explanation" of the FQHE. That is not the case with the composite fermion theory. In fact, the question, "What thermodynamic principle explains the vast body of experimental facts pertaining to the FQHE and other related phenomena?", has only one answer: "The formation of composite fermions." Composite fermions provide an understanding of the origin of incompressibility at fractional fillings and a multitude of other facts without resorting to any computer modeling. The usefulness of computer modeling lies in that it yields a detailed and undeniable confirmation of the CF principle. Being exact, computer studies are especially powerful in the FQHE, which is why Laughlin tested and confirmed his own wave function in numerical calculations on small systems [369]. The statement about computer calculations being incapable of explaining the accuracy of the Hall quantization is valid, but again not an objection to the CF theory. The CF principle produces incompressibility at certain special fractional fillings; such incompressibility, as first pointed out by Laughlin, leads to a precisely quantized Hall resistance. The exactness of Hall quantization is not related to the quantitative accuracy of the wave functions.

Laughlin does acknowledge in the same article the explanation of the half-filled Landau level state as a Fermi sea of composite fermions. To quote: "And of course there is the discovery of the strange Fermi surface at half-filling and its explanation in terms of composite fermions by Bert Halperin, Patrick Lee, and Nick Read (1993) that is now defining the intellectual frontier in this field."

Laughlin also stresses elsewhere in the article the appearance of an induced gauge interaction between the "quasiparticle defects" of an incompressible FQHE state, revealed by their fractional braiding statistics. He suggests the possibility that the gauge interaction of the standard model of particle physics might similarly be an emergent phenomenon in a more fundamental theory that does not postulate this gauge interaction at the outset. This is an interesting idea. But the braiding statistics of the quasiparticles defects, which is feeble and as yet unobserved,[5] is itself a consequence of a much more robust emergent gauge interaction between composite fermions (Section 9.8.2). The latter causes a substantial effective reduction of the external magnetic field, which has been verified through its numerous experimental consequences. The "quarks" of the FQHE are not the quasiparticle defects of an incompressible state of composite fermions but composite fermions themselves.[6]

---

[5] As discussed in Section 9.8.3, the effect of the statistical gauge interaction between distant quasiparticles is negligible compared with the influence of the strong magnetic field perpendicular to the Hall plane. We might envision making the statistical interaction strong by creating a high density of quasiparticles, but then the whole idea of fractional braiding statistics falls apart.

[6] Yang and Mills originally applied the concept of non-Abelian gauge interaction to the "wrong" particles, the nucleons, rather than quarks (which were discovered only later).

12.4 Two-dimensional one-component plasma (2DOCP)

Due to its simple Jastrow form, the Laughlin wave function is amenable to a mapping
into the statistical mechanics of a two-dimensional one-component classical plasma. The
probability of finding particles in a configuration $\{z_j\}$ is proportional to

$$|\Psi(\{z_j\})|^2 = \exp\left[2m \sum_{k<j} \ln|z_k - z_j| - \frac{1}{2\ell^2} \sum_k |z_k|^2 \right] . \tag{12.4}$$

This quantity can be interpreted as the Boltzmann probability factor for a classical gas
of charged "particles" in a uniform neutralizing background in two dimensions, namely a
two-dimensional one-component plasma (2DOCP). These "particles" live in a strictly two-
dimensional space, and are distinct from electrons, which are three-dimensional particles
confined to move in two dimensions. To distinguish the two, we refer to the former as
"particles" (with quotation marks). The number of "particles" is the same as the number of
electrons, along with an exact correspondence between their positions. But the "particles"
do not have the same charge and interaction as the electrons.

To see the mapping between the two problems, we recall some facts from 2D classical
electrodynamics. Gauss's law in two dimensions is

$$\nabla^2 V(\boldsymbol{r}) = -4\pi\rho(\boldsymbol{r}) . \tag{12.5}$$

The potential of a point charge of unit strength at the origin satisfies

$$\nabla^2 V(\boldsymbol{r}) = -4\pi\delta^{(2)}(\boldsymbol{r}) . \tag{12.6}$$

The identity

$$\nabla^2 \ln(r) = 2\pi\delta^{(2)}(\boldsymbol{r}) , \tag{12.7}$$

is analogous to $\nabla^2(1/r) = -4\pi\delta^{(3)}(\boldsymbol{r})$ of three dimensions, and gives the form of the
Coulomb potential in two dimensions as

$$V(\boldsymbol{r}) = -2\ln r . \tag{12.8}$$

Proof We prove Eq. (12.7). We have

$$\nabla^2 \ln r = \frac{1}{r}\frac{\partial}{\partial r}\left(r\frac{\partial}{\partial r}\ln r \right) = 0 \tag{12.9}$$

for $r \neq 0$. (For $r = 0$, the quantity inside the brackets is ill defined.) Now consider a closed loop C around the origin enclosing an area σ. Then

$$\int_{\sigma} d\boldsymbol{a} \, \nabla^2 \ln r = \int_{\sigma} d\boldsymbol{a} \, \nabla \cdot (\nabla \ln r) \tag{12.10}$$

$$= \oint_C \nabla \ln r \times d\boldsymbol{l} \tag{12.11}$$

$$= \oint_C \frac{\hat{r}}{r} \times (dr \, \hat{r} + r \, d\theta \, \hat{\theta}) \tag{12.12}$$

$$= 2\pi \hat{z} \, . \tag{12.13}$$

Here, as usual, the area element $d\boldsymbol{a}$ points perpendicular to the plane (in the z direction), and we have used Stokes's theorem:[7]

$$\int_{\sigma} d\boldsymbol{a} \, \nabla \cdot \boldsymbol{A} = \oint_C \boldsymbol{A} \times d\boldsymbol{l} \, . \tag{12.15}$$

□

For a uniform charge density, we have

$$\nabla^2 V(r) = -4\pi \bar{\rho} \, , \tag{12.16}$$

the solution for which is

$$V(r) = -\pi \bar{\rho} r^2 \, , \tag{12.17}$$

because $\nabla^2 r^2 = (\partial_x^2 + \partial_y^2)(x^2 + y^2) = 4$.

The electrostatic energy of particles of charge e_j interacting with a uniform background charge density $\bar{\rho}$ is given by

$$E(\{\boldsymbol{r}_j\}) = -2 \sum_{j<k} e_j e_k \ln |\boldsymbol{r}_j - \boldsymbol{r}_k| - \pi \bar{\rho} \sum_i e_i |\boldsymbol{r}_i|^2 \, . \tag{12.18}$$

For "particles" of charge

$$e_j = -1 \tag{12.19}$$

in a uniform background charge density

$$\bar{\rho} = \frac{1}{2\pi m \ell^2} \, , \tag{12.20}$$

---

[7] This is the two-dimensional version of the divergence theorem. It can be derived from the usual form of Stokes's theorem

$$\int_{\sigma} (\nabla \times \boldsymbol{A}) \cdot d\boldsymbol{a} = \oint_C \boldsymbol{A} \cdot d\boldsymbol{l} \tag{12.14}$$

by substituting $\boldsymbol{A} \to \boldsymbol{A} \times \boldsymbol{B}$, where $\boldsymbol{B}$ is a constant vector in the 2D plane.

we have

$$E(\{r_j\}) = -2 \sum_{j<k} \ln |r_j - r_k| + \frac{1}{2m\ell^2} \sum_i |r_i|^2 \ . \tag{12.21}$$

We can therefore write

$$|\Psi_{1/m}(\{z_j\})|^2 = \exp[-\beta E(\{z_j\})] \ , \tag{12.22}$$

with

$$\beta = m \ . \tag{12.23}$$

Various diagonal correlation functions for $\Psi_{1/m}$ are the same as the corresponding correlation functions of the 2DOCP. For example, the pair correlation function is given by

$$g(r_1, r_2) = \frac{N(N-1)}{\rho^2} \int d^2 r_3 \dots d^2 r_N |\Psi_{1/m}(r_1, \dots, r_N)|^2$$

$$= \frac{N(N-1)}{\rho^2} \int d^2 r_3 \dots d^2 r_N \exp[-\beta E(r_1, \dots, r_N)] \ . \tag{12.24}$$

What does the analogy to the 2DOCP tell us?

- It is intuitively obvious that the classical plasma is overall charge neutral. The wave function Ψ, therefore, describes a uniform density system.
- The charge neutrality implies that the number density of the "particles" is equal to $\bar\rho$. Because the "particles" and electrons have the same number density, the density of electrons is also $\bar\rho = 1/(2\pi m\ell^2)$. The filling factor of the state described by $\Psi_{1/m}$ is

$$\nu = \frac{\rho\phi_0}{B} = 2\pi \ell^2 \rho = \frac{1}{m} \ . \tag{12.25}$$

- The overall charge neutrality, by itself, does not imply that the state is a liquid; that would require a calculation of the pair correlation function. One can, however, make certain statements from the intuitive insight gained by mapping into the 2DOCP. Because the exponent m is inversely proportional to the temperature, we expect the "particles" to form a crystal for large m, which would imply a crystal state also for electrons. As m is reduced (i.e., the temperature is increased), the crystal eventually melts. From Monte Carlo simulations of the 2DOCP [48] this transition is known to occur at $m \approx 72$. The Laughlin wave function with $m > 72$ thus describes a crystal. This crystal has no relevance to the actual crystal state at low fillings (more on that in Chapter 15), because the Laughlin wave function loses its validity well before the filling factor reaches such small values.

12.5 Charged excitations at $\nu = 1/m$

We introduced in Section 5.5 Laughlin's wave function for the $\nu = 1/m$ ground state. He also wrote trial wave functions for the quasihole and quasiparticle excitations of this state.

12.5.1 Laughlin's trial wave functions

Laughlin showed that an excited state can be obtained by inserting a point flux tube into an incompressible state and adiabatically increasing the flux through it from zero to one flux quantum (Section 9.3.4). This thought experiment produces, in principle, the exact excitation, and suggests a construction for approximate trial wave functions for it. A single particle state evolves according to

$$\eta_l \rightarrow \eta_{l\pm 1} , \tag{12.26}$$

where

$$\eta_l = (2\pi 2^l l!)^{-1/2} z^l e^{-\frac{1}{4}|z|^2} , \tag{12.27}$$

and the sign depends on the direction of the inserted flux, which has been taken to be at the origin. An approximate trial wave function can be constructed by first expressing the ground state wave function in terms of basis functions:

$$\Psi_{1/m} = \sum_{\{l_j\}} C_{\{l_j\}} \left(\prod_j \eta_{l_j}(z_j) \right) \tag{12.28}$$

and then making the above replacement for each particle. The resulting wave function is still too complicated. Laughlin further simplified the problem by not worrying about the prefactors but only the exponents, to write

$$\Psi_{1/m}^{L-qh} = e^{-\frac{1}{4}\sum_l |z_l|^2} \left(\prod_i z_i \right) \prod_{j<k} (z_j - z_k)^m \tag{12.29}$$

for the quasihole, and

$$\Psi_{1/m}^{L-qp} = e^{-\frac{1}{4}\sum_l |z_l|^2} \left(\prod_i \frac{\partial}{\partial z_i} \right) \prod_{j<k} (z_j - z_k)^m \tag{12.30}$$

for the quasiparticle.[8] These describe a quasihole and a quasiparticle at the origin.

As discussed in Section 9.3.5, the local charge of the quasiparticle or the quasihole can be calculated from its wave functions, which, in general, requires evaluation of multi-dimensional integrals. The plasma analogy provides a simple method for obtaining the local charge of the quasihole at $\nu = 1/m$. For a quasihole at r_0, with $\Psi_{1/m}^{L-qh} = \prod_j (z_j - z_0) \Psi_{1/m}$, we have

$$|\Psi_{1/m}^{L-qh}|^2 = e^{-\beta E} , \tag{12.31}$$

where $\beta = m$, and

$$E(\{r_j\}) = -2 \sum_{j<k} \ln |r_j - r_k| - \frac{2}{m} \sum_j \ln |r_j - r_0| + \frac{1}{2m\ell^2} \sum_i |r_i|^2 . \tag{12.32}$$

---

[8] This excitation was originally called a "quasielectron" [369]. Because it is fundamentally distinct from an electron, we prefer the name "quasiparticle."

This is the energy of "particles" of charge -1 in a uniform background, as before, but with an additional external charge $-1/m$ at r_0. The "particles" are repelled by this external charge, and because the plasma screens completely, precisely $1/m$ of a "particle" is missing from the vicinity of r_0. In the electron system, precisely $1/m$ of an electron is missing from the neighborhood of r_0, giving the fractional charge $e^* = e/m$. The local charge of Laughlin's *quasiparticle* ($\Psi_{1/m}^{L-qp}$) at $\nu = 1/m$ cannot be obtained by this method, but Monte Carlo studies [345] have demonstrated that it has the correct charge $e^* = -e/3$.

The plasma analogy does not apply to the general FQHE states, or even to improved wave functions at $\nu = 1/m$. (Recall that the Laughlin wave function, although very good, is not exact.) Wave function independent derivations at the fractions $\nu = n/(2pn \pm 1)$ are given in Section 5.10 for uniform density, and in Section 9.3 for the local charges of the quasiparticle and quasihole.

12.5.2 Comparison with the CF theory

The CF-quasiparticles and CF-quasiholes for the general FQHE state are described in Section 5.9.4 (Eqs. 5.60 and 5.61). We consider here the special cases of CF-quasiparticle and CF-quasihole at $\nu = 1/m$, shown pictorially in Fig. 12.2, and ask how their wave functions compare with Laughlin's.

A single quasihole The wave function for a CF-quasihole is related to the wave function of the state at $\nu = 1$ with one electron removed. If we remove the electron at the origin (i.e., from the zero angular momentum orbital), the wave function is given by

$$\Phi_1^{hole} = \begin{vmatrix} z_1 & z_2 & z_3 & \cdot & \cdot \\ z_1^2 & z_2^2 & z_3^2 & \cdot & \cdot \\ \cdot & \cdot & \cdot & \cdot & \cdot \\ \cdot & \cdot & \cdot & \cdot & \cdot \end{vmatrix} \exp\left[-\frac{1}{4} \sum_i |z_i|^2 \right] \tag{12.33}$$

$$= \left(\prod_j z_j \right) \prod_{j<k} (z_j - z_k) \exp\left[-\frac{1}{4} \sum_i |z_i|^2 \right]. \tag{12.34}$$

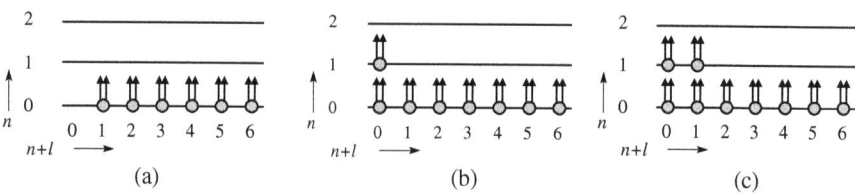

Fig. 12.2. (a) A single CF-quasihole at the origin. (b) A single CF-quasiparticle at the origin. (c) Two CF-quasiparticles at the origin. The disk geometry is assumed, and the filling factor is $\nu = 1/3$. The angular momentum is denoted by l, and the Λ level index by n.

The wave function for the CF-quasihole at the origin is, then,

$$\Psi^{CF-qh}_{\frac{1}{2p+1}} = \left(\prod_j z_j\right) \Psi_{\frac{1}{2p+1}}, \tag{12.35}$$

which matches Laughlin's ansatz in Eq. (12.29), but has a physical interpretation as a missing composite fermion. The wave function is obtained more elegantly than the steps leading to Eq. (12.29).

A single quasiparticle A CF-quasiparticle is a single composite fermion in an otherwise empty Λ level, shown in Fig. 12.2(b). The wave function for the state containing an additional electron in the second Landau level (at the origin) is

$$\Phi^{particle}_1 = \begin{vmatrix} z_1^* & z_2^* & z_3^* & \cdot & \cdot \\ 1 & 1 & 1 & \cdot & \cdot \\ z_1 & z_2 & z_3 & \cdot & \cdot \\ \cdot & \cdot & \cdot & \cdot & \cdot \\ \cdot & \cdot & \cdot & \cdot & \cdot \end{vmatrix} \exp\left[-\frac{1}{4}\sum_i |z_i|^2\right]$$

$$= \sum_{i=1}^N (-1)^{i+1} z_i^* \prod_{j<k}' (z_j - z_k) \exp\left[-\frac{1}{4}\sum_i |z_i|^2\right]$$

$$= \left[\sum_{i=1}^N \frac{z_i^*}{\prod_j'(z_i - z_j)}\right] \Phi_1, \tag{12.36}$$

where the prime denotes the condition $j, k \neq i$. To obtain the wave function for the CF-quasiparticle, we composite-fermionize $\Phi^{particle}_1$ by first multiplying it by $\prod_{j<k}(z_j - z_k)^{2p}$ and then projecting the product into the lowest Landau level using the methods in Section 5.14. This yields

$$\Psi^{CF-qp}_{\frac{1}{2p+1}} = \exp\left[-\frac{1}{4}\sum_i |z_i|^2\right] \left[\sum_{i=1}^N \frac{2\frac{\partial}{\partial z_i}}{\prod_j'(z_j - z_i)}\right] \prod_{l<m}(z_l - z_m)^{2p+1}$$

$$= \sum_{i=1}^N \frac{2(2p+1)\sum_k'(z_i - z_k)^{-1}}{\prod_j'(z_j - z_i)} \Psi_{\frac{1}{2p+1}}. \tag{12.37}$$

(The presence of $(z_j - z_k)$ factors in the denominator is not a problem because they are canceled by similar factors in the numerator.) The CF-quasiparticle wave function is different from Laughlin's. The comparisons in Table 12.1 and in Fig. 12.3 demonstrate this wave function to be better. Similar conclusions are reached by other studies (Girlich and M. Hellmund [188]; Kasner and Apel [328]; Melik-Alaverdian and Bonesteel [436]).

Several other qualitative facts, including the asymmetry between the quasiparticle and the quasihole, find an intuitive explanation in CF theory. These are discussed in Section 5.9.4 in a more general context.

Table 12.1. *Testing two trial wave functions for the $v = 1/3$ quasiparticle*

N	D	CF	Laughlin
3	3	1	1
4	11	0.9969	0.9987
5	46	0.9930	0.9967
6	217	0.9941	0.9885
7	1069	0.9828	0.9651
8	5529	0.9671	0.9365

Notes: Exact wave function for quasiparticle excitation obtained from diagonalization of Coulomb Hamiltonian. Its overlap with trial wave functions Eq. (12.30) (Laughlin) and Eq. (12.37) (CF).

Disk geometry with symmetric gauge is used. D is dimension of Fock space in the lowest Landau level.

Source: Dev and Jain [116]

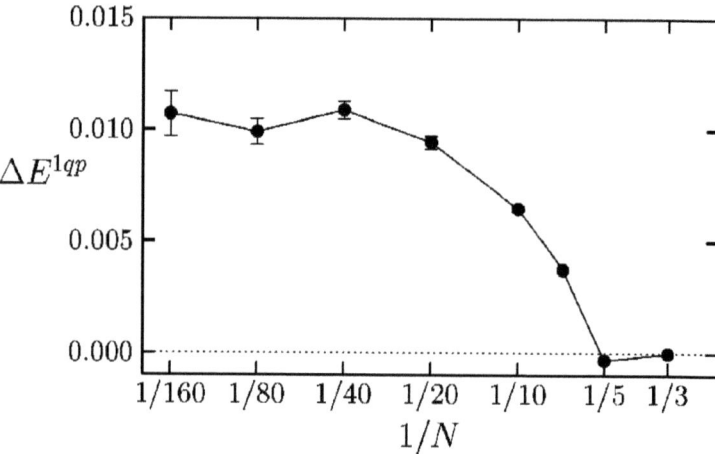

Fig. 12.3. $\Delta E^{1qp} = E_{\mathrm{L}}^{1qp} - E_{\mathrm{CF}}^{1qp}$ is the difference between the energies of two trial wave functions for the quasiparticle at $v = 1/3$ given in Eqs. (12.30) and (12.37), for up to $N = 160$ electrons. E_{L}^{1qp} is the energy of Laughlin's quasiparticle wave function, and E_{CF}^{1qp} is the energy of the CF-quasiparticle. The former is approximately $0.011e^2/\epsilon\ell$ higher in the thermodynamic limit, which is about 15% of the energy of the quasiparticle ($\sim 0.07e^2/\epsilon\ell$). Source: G.-S. Jeon and J. K. Jain, *Phys. Rev. B* **68**, 165346 (2003). (Reprinted with permission.)

Two quasiparticles A generalization of Laughlin's construction to two quasiparticles suggests the wave function (see, for example, Kjønsberg and Myrheim [345])

$$\Psi_L^{2\text{-L-qp}} = e^{-\sum_j |z_j|^2/4} \prod_l \left(2\frac{\partial}{\partial z_l}\right)\left(2\frac{\partial}{\partial z_l}\right)\prod_{j<k}(z_j - z_k)^3 . \tag{12.38}$$

In contrast, two CF-quasiparticles at the origin, shown in Fig. 12.2(c), are described by the wave function

$$\Psi^{2\text{-CF-qp}} = \mathcal{P}\prod_{j<k}(z_j - z_k)^2 \begin{vmatrix} z_1^* & z_2^* & \cdots \\ z_1^* z_1 & z_2^* z_2 & \cdots \\ 1 & 1 & \cdots \\ z_1 & z_2 & \cdots \\ \vdots & \vdots & \cdots \\ z_1^{N-3} & z_2^{N-3} & \cdots \end{vmatrix}$$

$$\times \exp\left[-\frac{1}{4}\sum_j |z_j|^2\right]. \tag{12.39}$$

The explicit LLL projected form can be written down as before, but is not shown here. The collection of two CF-quasiparticles has lower energy than that of two Lauglin-quasiparticles. The energy difference is estimated [292] to be $\approx 0.16e^2/\epsilon\ell$, which is roughly equal to twice the energy required to create a single CF-quasiparticle. A qualitative difference appears between the two approaches at the level of two quasiparticles. The two wave functions considered above have different total angular momenta, $L = 3N^2 - 7N + 4$ (CF) and $L = 3N^2 - 7N$ (Laughlin). (The largest occupied single electron orbital has the same angular momentum for the two states, though.) Construction of a two quasiparticles state with an angular momentum $L = 3N^2 - 7N + 4$ is not obvious within Laughlin's approach.

Kjønsberg and Myrheim [345] calculate the braiding statistics of the 1/3 quasiparticles using the Laughlin wave function and note (Fig. 12.4) that it does not produce a well-defined value in the limit when the quasiparticles are far separated, indicating that this wave function does not capture the long-distance behavior of the actual quasiparticle sufficiently accurately for this purpose. CF quasiparticles, on the other hand, possess a well-defined braiding statistics (Kjønsberg and Leinaas [346]; Jeon, Graham, and Jain [293, 294]).

12.6 Neutral excitations: Girvin–MacDonald–Platzman theory

The neutral excitations of FQHE states are understood as CF-excitons (Fig. 5.4), which have been confirmed to be extremely accurate (Fig. 6.1). This section presents another model for the neutral excitations.

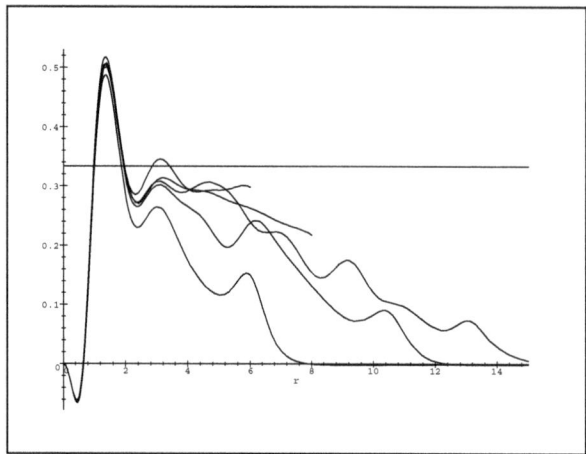

Fig. 12.4. The braiding statistics parameter (shown on the y-axis) for the quasiparticle at $v = 1/3$, using Laughlin's trial wave function (Eq. 12.30), determined from a Monte Carlo evaluation of the Berry phase. The x-axis label is r, which is related to the distance d between the quasiparticles as $d/\ell = 2\sqrt{2}\, r$. The lowest curve is for 20 electrons, the next for 50, and the third curve from the bottom is for 75 electrons. The results for 100 and 200 electrons terminate at $r = 8$ and $r = 6$ due to numerical problems, and the height of the horizontal line is 1/3. Source: H. Kjønsberg and J. Myrheim, *Int. J. Mod. Phys. A* **14**, 537 (1999). (Reprinted with permission.)

A trial wave function for a phonon-like excitation of a Bose superfluid can be constructed as (Bijl [31]; Feynman [160])

$$\phi_k = \frac{1}{\sqrt{N}} \rho_k \phi_{\text{ground}} , \tag{12.40}$$

where ϕ_{ground} is the ground state wave function, and

$$\rho(r) = \sum_{j=1}^{N} e^{-ik \cdot r_j} \tag{12.41}$$

is the density operator. Clearly, ϕ_k describes a collective "density wave" excitation, analogous to the "phonon" of a crystal. Because ϕ_k is closely related to the ground state wave function, it is likely to have favorable correlations and low energy. Its orthogonalilty to ϕ_{ground},

$$\langle \phi_{\text{ground}} | \phi_k \rangle = \frac{1}{\sqrt{N}} \int d^2 r \, e^{-ik \cdot r} \langle \phi_{\text{ground}} | \rho(r) | \phi_{\text{ground}} \rangle = 0 , \tag{12.42}$$

follows, for $k \neq 0$, because $\langle \phi_{\text{ground}} | \rho(r) | \phi_{\text{ground}} \rangle = $ constant for a spatially uniform ground state. The energy of this wave function can be related to the static structure factor, which, in turn, can be obtained from neutron scattering experiments. A peak in the structure factor produces a minimum in the dispersion, which is called the roton minimum. Such a minimum is observed experimentally, but the energy at the minimum is a factor of two lower than that predicted by the above wave function. A quantitative understanding of the dispersion of the collective excitation has been one of the triumphs of the theory of superfluidity, which has required inclusion of backflow corrections (Feynman and Cohen [162]) and detailed variational and Green's function Monte Carlo studies (see, for example, Manousakis and Pandharipande [429]).

The above prescription allows construction of a trial wave function for an excited state from the knowledge of the ground state wave function. It may seem that we are getting excited states for free, but that is not the case. We are making the assumption that the low-energy neutral excitation is a density wave, which may seem natural, but must be verified.

Girvin, MacDonald, and Platzman (GMP) [191, 192] extend this idea to the neutral excitation of the FQHE, and we describe below their theory. We begin by writing the wave function

$$\phi_k = \frac{1}{\sqrt{N}} \rho_k \Psi(r_1, \ldots, r_N) \tag{12.43}$$

where Ψ is an ansatz for the FQHE ground state wave function. Because the number of electrons remains unchanged, this describes a neutral excitation. A shortcoming of this wave function is that it is not restricted to the lowest Landau level, which makes it inappropriate in very high magnetic fields. Since we are constructing a *trial* wave function, we project it into the lowest Landau level to obtain the GMP wave function:

$$\bar{\phi}_k = \mathcal{P}_{\text{LLL}} \phi_k$$

$$= \frac{1}{\sqrt{N}} \bar{\rho}_k \Psi(r_1, \ldots, r_N) , \tag{12.44}$$

with

$$\bar{\rho}(r) = \mathcal{P}_{\text{LLL}} \sum_{j=1}^{N} e^{-ik \cdot r_j} \mathcal{P}_{\text{LLL}}$$

$$= \mathcal{P}_{\text{LLL}} \sum_{j=1}^{N} \exp\left[-\frac{i}{2} k \bar{z}_j\right] \exp\left[-\frac{i}{2} \bar{k} z_j\right] \mathcal{P}_{\text{LLL}}$$

$$= \sum_{j=1}^{N} \exp\left[-ik \frac{\partial}{\partial z_j}\right] \exp\left[-\frac{i}{2} \bar{k} z_j\right] , \tag{12.45}$$

where $z = x - iy$, $k = k_x - ik_y$, and it is understood that the derivatives do not act on the Gaussian factor. This is known as the "single-mode approximation" (SMA), and is equivalent to assuming that the oscillator strength is exhausted by a single excitation.

All we need to do now is evaluate the energy of the GMP wave function. That can be done exactly in numerical studies on finite systems. Another method is based on the result, proved below, that, as in in the case of helium superfluid, the energy of the GMP wave function can be expressed entirely as a function of the static structure factor of the ground state. The latter, in turn, can be computed with the help of either a candidate wave function or some other approximate technique.

It is convenient to work with $\bar{V}$, the projected Coulomb energy. The projection is most easily carried out in the Fourier space, where

$$V = \frac{1}{2} \int \frac{d^2q}{(2\pi)^2} V(q) \sum_{i \neq j} e^{iq \cdot (r_i - r_j)}$$

$$= \frac{1}{2} \int \frac{d^2q}{(2\pi)^2} V(q) \left(\rho_q^\dagger \rho_q - N \right) . \tag{12.46}$$

A result from Exercise 12.7 (Eq. E12.18) gives the projected interaction:

$$\bar{V} = \frac{1}{2} \int \frac{d^2q}{(2\pi)^2} V(q) \left(\bar{\rho}_q^\dagger \bar{\rho}_q - N e^{-q\bar{q}/2} \right) . \tag{12.47}$$

The energy expectation value of the GMP wave function, measured from the ground state energy, is given by

$$\Delta_k = \frac{\langle \bar{\phi}_k | \bar{V} - E_0 | \bar{\phi}_k \rangle}{\langle \bar{\phi}_k | \bar{\phi}_k \rangle}$$

$$= \frac{\bar{f}(k)}{\bar{S}(k)} , \tag{12.48}$$

where

$$E_0 = \frac{\langle \Psi | \bar{V} | \Psi \rangle}{\langle \Psi | \Psi \rangle}, \tag{12.49}$$

$$\bar{S}(k) = \frac{1}{N} \langle \Psi | \bar{\rho}_k^\dagger \bar{\rho}_k | \Psi \rangle, \tag{12.50}$$

and

$$\bar{f}(k) = \frac{1}{N} \langle \Psi | \bar{\rho}_k^\dagger [\bar{V}, \bar{\rho}_k] | \Psi \rangle . \tag{12.51}$$

$\bar{S}(k)$ is the projected structure factor and $\bar{f}(k)$ is the projected oscillator strength.

We first consider $\bar{S}(k)$. The quantity that is most readily calculated from a given wave function is the structure factor

$$S(k) = \frac{1}{N} \langle \Psi | \rho_k^\dagger \rho_k | \Psi \rangle = \frac{1}{N} \langle \Psi | \mathcal{P}_{LLL} (\rho_k^\dagger \rho_k) | \Psi \rangle , \tag{12.52}$$

where the last equation follows because Ψ is assumed to be strictly in the lowest Landau level. $\bar{S}(k)$ can be obtained from $S(k)$ using the relation

$$\bar{S}(k) = S(k) - (1 - e^{-k\bar{k}/2}) , \tag{12.53}$$

which follows from Eq. (E12.18) derived in Exercise 12.7.

Using $\bar{\rho}_q^\dagger = \bar{\rho}_{-q}$, the expression for the projected oscillator strength can be rewritten as

$$\bar{f}(k) = \frac{1}{2N} \langle \Psi | [\bar{\rho}_k^\dagger, [\bar{V}, \bar{\rho}_k]] | \Psi \rangle . \tag{12.54}$$

The double commutator can be evaluated with the help of Eqs. (12.47) and (E12.17) to give

$$\bar{f}(k) = \frac{1}{2} \sum_q V(q) \left(e^{\frac{q^* k}{2}} - e^{\frac{qk^*}{2}} \right) \tag{12.55}$$

$$\times \left[\bar{S}(q) e^{-\frac{k^2}{2}} \left(e^{-\frac{k^* q}{2}} - e^{-\frac{kq^*}{2}} \right) + \bar{S}(k+q) \left(e^{\frac{k^* q}{2}} - e^{\frac{kq^*}{2}} \right) \right] .$$

Thus, the energy of the GMP excitation, Eq. (12.48), can be obtained from the knowledge of the static structure factor.

Let us ask what the SMA gives for $\nu = 1$. This state has no excitations within the lowest Landau level, as is indicated by the vanishing of $\bar{S}_k$. We therefore switch back to the "unprojected" SMA, for which

$$\Delta_k = \frac{f(k)}{S(k)} , \tag{12.56}$$

with

$$S(k) = \frac{1}{N} \langle \Psi | \rho_k^\dagger \rho_k | \Psi \rangle , \tag{12.57}$$

and

$$f(k) = \frac{1}{N} \langle \Psi | \rho_k^\dagger [H, \rho_k] | \Psi \rangle , \tag{12.58}$$

where $H = K+V$ is the sum of kinetic and interaction energies. Since the density commutes with V (without projection), the commutator reduces to $[K, \rho_k]$, which can be evaluated (Exercise 12.9) to give

$$f(k) = \frac{\hbar^2 k^2}{2m_b} . \tag{12.59}$$

The gap is then given by

$$\Delta_k = \frac{\hbar^2 k^2}{2m_b S(k)} = \frac{\hbar^2 k^2}{2m_b [1 - e^{k^2 \ell^2 / 2}]} . \tag{12.60}$$

The $k\ell \to 0$ limit,

$$\Delta_k = \frac{\hbar^2}{m_b \ell^2} = \hbar \omega_c , \tag{12.61}$$

is consistent with Kohn's theorem, which states that, in the presence of a magnetic field, given the exact ground state of the interacting system (Ψ), the state $\rho_k \Psi$ is an exact eigenstate, in the limit $k \to 0$, with eigenenergy $\hbar \omega_c$.

12.6.1 SMA on a sphere

To construct the GMP wave function in the spherical geometry, we need an appropriate form for the density operator. In the planar geometry, we defined

$$\rho(\mathbf{r}) = \sum_{j=1}^{N} \delta^{(2)}(\mathbf{r} - \mathbf{r}_j) = \sum_{j=1}^{N} \int \frac{d^2q}{(2\pi)^2} e^{i\mathbf{q}\cdot(\mathbf{r}-\mathbf{r}_j)} \equiv \int \frac{d^2q}{(2\pi)^2} e^{i\mathbf{q}\cdot\mathbf{r}} \rho_{\mathbf{q}} . \qquad (12.62)$$

In the spherical geometry, we write

$$\rho(\mathbf{\Omega}) = \sum_{j=1}^{N} \delta^{(2)}(\mathbf{\Omega} - \mathbf{\Omega}_j)$$

$$= \sum_{j=1}^{N} \sum_{L=0}^{\infty} \sum_{M=-L}^{L} Y_{L,M}^*(\mathbf{\Omega}) Y_{L,M}(\mathbf{\Omega}_j)$$

$$\equiv \sum_{L=0}^{\infty} \sum_{M=-L}^{L} Y_{L,M}^*(\mathbf{\Omega}) \rho_{LM} , \qquad (12.63)$$

where

$$\rho_{LM} = \sum_{j=1}^{N} Y_{L,M}(\mathbf{\Omega}_j) . \qquad (12.64)$$

The GMP wave function for the excitation is then given by

$$\phi_{LM} = \frac{1}{\sqrt{N}} \mathcal{P}_{LLL} \rho_{LM} \Psi(\mathbf{\Omega}_1, \ldots, \mathbf{\Omega}_N) , \qquad (12.65)$$

where $\mathcal{P}_{LLL}$ projects the state into the lowest Landau level.

Recall that the angular momentum operator for $Q = 0$ is

$$\mathbf{L} = \mathbf{\Lambda} + Q\mathbf{\Omega} = \mathbf{\Lambda} . \qquad (12.66)$$

Because the spherical harmonics transform as vectors under rotations generated by $\mathbf{\Lambda}$, and commute with $\mathbf{\Omega}$, we have

$$[L_z, \rho_{LM}] = M \rho_{LM} \qquad (12.67)$$

and

$$[L_\pm, \rho_{LM}] = \sqrt{L(L+1) - M(M \pm 1)} \, \rho_{LM \pm 1} . \qquad (12.68)$$

It further follows that

$$\mathbf{L}^2 \rho_{LM} \Psi = L(L+1) \rho_{LM} \Psi \qquad (12.69)$$

and

$$L_z \rho_{LM} \Psi = M \rho_{LM} \Psi , \qquad (12.70)$$

Table 12.2. *Roton gaps from the CF and the Girvin–MacDonald–Platzman theories*

ν	CF	GMP
1/3	0.063(3)	0.078
1/5	0.0095(6)	0.017
1/7	0.0009(5)	0.0063

Notes: Energy of the primary roton in units of $e^2/\epsilon\ell$. CF energies obtained by extrapolation of finite system results.

Same ground state wave functions used in both studies.

Source: Jain and Kamilla [281]; Girvin, MacDonald, and Platzman [191, 192].

where we have used that Ψ has $L = 0$. The GMP wave function thus has a definite symmetry under rotation.

To compare the energy Δ_L for systems with different sizes, or with Δ_q of the planar geometry, one converts L into q with the help of the relation [226] $L = qR$, where R is the radius of the sphere. It can be verified that, with this identification, the energy of the (unprojected) GMP excitation on the sphere at $\nu = 1$ becomes identical to Eq. (12.60).

12.6.2 Testing the SMA

The SMA needs a ground state wave function. In exact diagonalization studies, we can use the exact ground state wave function to construct the GMP wave function, and evaluate its energy exactly. Figure 12.5 shows the dispersion of the GMP excitation (solid dots) for $\nu = 1/3$, 2/5 and 3/7, evaluated by and He, Simon and Halperin [247] and Platzman He [521]. The spherical geometry is used in these studies.

Calculation of the GMP dispersion for larger systems requires the static structure factor (or the pair correlation function), which is obtained from a trial wave function for the ground state. Initially, Girvin, MacDonald, and Platzman applied SMA to $\nu = 1/(2p+1)$ because of the availability of the Laughlin wave function. Dispersions of the GMP mode at other fractions of the type $\nu = n/(2n+1)$ have also been calculated (Park and Jain [505]; Scarola, Park and Jain [567]), using the ground state wave functions of Eq. (5.32). Table 12.2 gives the minimum energy required to create a neutral excitation, evaluated both from the GMP and the CF-exciton models. These studies show that the GMP theory works well at low and intermediate wave vectors, especially at the Laughlin fractions, where it successfully predicts a roton minimum.

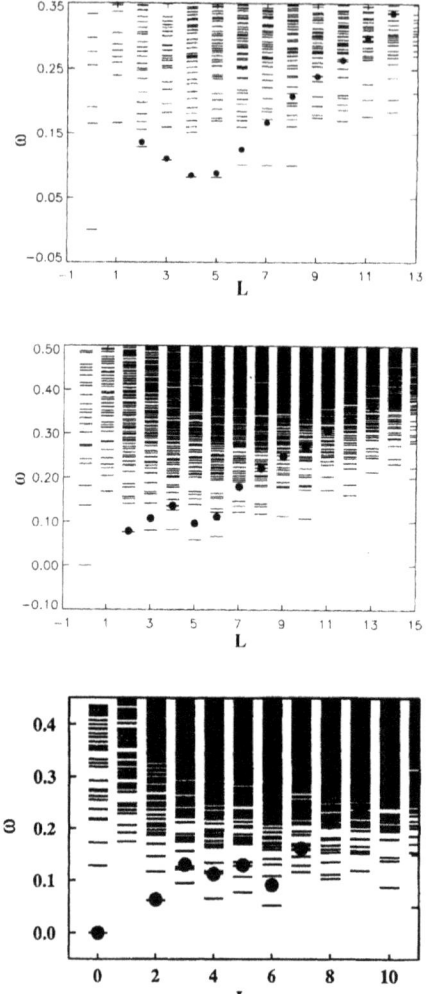

Fig. 12.5. Energy dispersion of the GMP excitation (dots) at $\nu = 1/3$ for 9 particles, at $\nu = 2/5$ for 10 particles, and at $\nu = 3/7$ for 12 particles. The dashes show the exact Coulomb eigenenergies. Sources: P. M. Platzman and S. He, *Phys. Rev. B* **49**, 13674 (1994); S. He, S. H. Simon, and B. I. Halperin, *Phys. Rev. B* **50**, 1823 (1994). (Reprinted with permission.)

12.7 Conti–Vignale–Tokatly continuum-elasticity theory

Classical liquids and solids also exhibit a collective response to external perturbations, which can be described, in the long-wavelength limit, by treating them as a continuous elastic medium. In the well-developed classical elasticity theory, the dynamics of the system is described in terms of the displacement field, $\boldsymbol{u}(\boldsymbol{r}, t)$, the deviation of a volume element from its equilibrium position. Its equations of motion involve the bulk modulus and the shear modulus, which are to be fixed phenomenologically by comparing to either experiments or microscopic calculations.

Conti and Vignale [95, 196, 657] take the view that the LLL electron liquid can also be treated analogously, but with frequency and wave vector dependent visco-elastic constants. The current density and the density can be expressed in terms of u (for small perturbations) as

$$j(r,t) = \rho_0 \partial_t u(r,t) \tag{12.71}$$

$$\delta\rho = -\rho_0 \nabla \cdot u , \tag{12.72}$$

where $v = \partial_t u$ is the velocity field and ρ_0 is the unperturbed density, which is uniform in space and constant in time. Equation (12.72) follows from the linearized continuity equation

$$\partial_t \rho + \nabla \cdot j \simeq \partial_t \delta\rho + \rho_0 \nabla \cdot v = 0 . \tag{12.73}$$

From an analysis of the equations of motion for $u(r,t)$ from classical elasticity theory in the limit of high magnetic field, Conti and Vignale conclude that the bulk modulus B must diverge as

$$B \propto \frac{1}{q^4} \tag{12.74}$$

at small wave vectors to produce a nonzero gap for the exciton in the long-wavelength limit. Furthermore, from a calculation of the static structure factor, they relate the shear modulus S (which they assume to be independent of frequency) to Δ, the $q = 0$ gap of the exciton:

$$S = \frac{1-\nu}{4\nu} \rho_0 \Delta . \tag{12.75}$$

By taking S to be the shear modulus of the Wigner crystal, $S = 0.097\,75\rho_0\sqrt{\nu}e^2/\epsilon\ell$ [35], this predicts $\Delta = 0.11, 0.044,$ and $0.025e^2/\epsilon\ell$ at $\nu = 1/3, 1/5,$ and $1/7$, which are in decent agreement with the exact diagonalization results.

Conti and Vignale [95] write a Lagrangian for the displacement field (the extremization of which produces the equations of motion) and quantize the displacement field in the standard manner to obtain a Hamiltonian of independent bosons with wave vector dependent energy:

$$H = \sum_q \hbar\omega_q \left(b_q^\dagger b_q + \frac{1}{2} \right) . \tag{12.76}$$

The boson is identified with the intra-LL collective mode (or the inter-Λ level CF exciton). The FQHE liquid is thus dynamically equivalent to a set of noninteracting bosons, which are analogous to the phonons of a Wigner crystal.

Tokatly [638, 639] has assigned different q, ω dependences to the elastic moduli. He notes that Kohn's theorem, which requires stress to vanish for rigid displacements, forbids divergences in the $q \to 0$ limit. The f-sum rule, on the other hand, requires finiteness of the bulk and shear moduli in the $\omega \to \infty$ limit. He has shown that a consistent picture can be obtained by taking a constant bulk modulus K, and a shear modulus of the form

$$S(\omega, q \to 0) \propto \frac{\omega^2}{\omega^2 - \Delta^2} \tag{12.77}$$

with a divergence at $\omega \to \Delta$. It vanishes in the $\omega \to 0$ limit, as expected for a liquid. With these choices the f-sum rule and Kohn's theorem can be satisfied, and the collective mode dispersion has a roton minimum.

The Conti–Vignale and the Tokatly theories can be shown [640] to be related, in the long-wavelength limit and for frequencies close to Δ, by a canonical transformation that preserves the form of the equations of motion but transfers the divergence from the shear modulus to the bulk modulus or vice versa. They predict identical relation (Eq. 12.75) between the high-frequency shear modulus and the gap. Tokatly has also derived the continuum elasticity theory starting from the CFCS theory, by linearizing the equation of motion for the Wigner function (which is the same as the semiclassical Boltzmann equation).

12.8 Search for a model interaction

Model interactions have been constructed for which some simple FQHE wave functions are the exact ground states. We discuss four cases.

12.8.1 $\nu = 1/3$

Haldane [221] constructs a model interaction for which the Laughlin wave function for the $\nu = 1/m$ state is the exact nondegenerate ground state. The model is based on the observation that, at $\nu = 1/m$, Laughlin's wave function is the unique wave function that contains no pairs with relative angular momentum less than m. To see this, we only need to note that, apart from the Gaussian factor, the wave function of a pair with relative angular momentum m is given by $(z_i - z_j)^m$. How does this help? Let us consider the model

$$V_l = 0, \qquad l \geq m, \tag{12.78}$$

where V_l are the Haldane pseudopotentials. The actual values of the nonzero V_l's are not relevant, except that they are taken to be positive, as would be the case for a repulsive interaction. For this model, the energy of pairs with relative angular momenta $l \geq m$ is identically zero. In particular, the energy of the Laughlin wave function vanishes. The Laughlin wave function is thus an eigenstate for this model. It is also a ground state, since negative energies are not possible.

We still need to prove uniqueness. The general wave function excluding pairs of angular momentum $l < m$ has the form

$$F_S[\{z_j\}] \prod_{i<k} (z_i - z_k)^m \exp\left[-\sum_l \frac{|z_l|^2}{4\ell^2}\right], \tag{12.79}$$

where $F_S[\{z_j\}]$ is an arbitrary symmetric polynomial. At $\nu = 1/m$ we must have $F = 1$; otherwise, F supplies additional powers of z_j, thereby decreasing the filling factor. This proves that, at $\nu = 1/m$, Laughlin's wave function is the unique ground state for the model Hamiltonian.

Trugman and Kivelson [646] construct a real-space interaction for which Laughlin's $\nu = 1/3$ wave function is the ground state

$$V_{TK}(\boldsymbol{r}) = \alpha \nabla^2 \delta^{(2)}(\boldsymbol{r}) . \tag{12.80}$$

The seemingly strange Laplacian of a delta function can be a defined through a limiting procedure, but there is no need to do that. The interaction is perfectly well defined in terms of matrix elements. The Haldane pseudopotentials for this interaction are given by

$$\begin{aligned} V_l &= \langle l | V_{TK}(\boldsymbol{r}) | l \rangle \\ &= \int d^2r |\psi_l(\boldsymbol{r})|^2 \alpha \nabla^2 \delta^{(2)}(\boldsymbol{r}) \\ &= \int d^2r \, \alpha \delta^{(2)}(\boldsymbol{r}) \frac{\partial}{\partial r} r \frac{\partial}{\partial r} \frac{|\psi_l(\boldsymbol{r})|^2}{r} , \end{aligned} \tag{12.81}$$

where $\boldsymbol{r}$ is the relative coordinate, and $\psi_l(\boldsymbol{r})$ is the wave function for two electrons in relative angular momentum l state. The matrix element vanishes provided $|\psi_l(r)|^2 \propto r^{2+\epsilon}$ as $r \to 0$, with $\epsilon > 0$. Because $\psi_l(r) \sim r^l$, we have $V_l = 0$ for $l \geq 2$. The rest of the proof follows as before. (We recall that the even pseudopotentials are not relevant for fully spin-polarized electrons.)

The above discussion also implies that for $\nu > 1/3$, *all* pairs cannot possibly have a relative angular momentum of three or greater. Therefore, the argument cannot be generalized to filling factors $\nu = n/(2n+1)$ with $n > 1$, even though, as demonstrated by numerical calculations (Gros and MacDonald [217]), the hard-core model exhibits FQHE at these filling factors.

12.8.2 $\nu = 1/2$

Greiter, Wen, and Wilczek [213] write a Hamiltonian involving a three-body interaction for which the Moore–Read Pfaffian wave function (Eq. 7.11) is the exact ground state. To see this, let us consider (charged) *bosons* in the lowest Landau level. Any bosonic wave function that vanishes when any two bosons coincide is a zero-energy eigenstate of the interaction $H = V \sum_{i<j} \delta^{(2)}(\boldsymbol{r}_i - \boldsymbol{r}_j)$. The lowest degree polynomial with that property (in addition to being symmetric under exchange) is the $\nu = 1/2$ wave function $\prod_{j<k}(z_j - z_k)^2$ (suppressing the Gaussian factor for simplicity), because the lowest degree polynomial that vanishes upon particle coincidence is $\prod_{j<k}(z_j - z_k)$, and the minimum one must do to symmetrize it is to supply an additional factor of $\prod_{j<k}(z_j - z_k)$, the lowest degree polynomial that is completely antisymmetric. Many zero energy eigenstates of H of the form

$F[\{z_i\}] \prod_{j<k}(z_j - z_k)^2$, $F[\{z_i\}]$ symmetric under exchange, exist for $\nu < 1/2$, and none for $\nu > 1/2$. Greiter, Wen, and Wilczek consider a less restrictive interaction:

$$H_{\mathrm{Pf}} = V \sum_{i<j<k} \delta^{(2)}\left(r_i - r_j\right) \delta^{(2)}\left(r_i - r_k\right) , \qquad (12.82)$$

which imposes a penalty only when *three* bosons coincide. The $\nu = 1$ bosonic Pfaffian wave function

$$\Psi_1^{\mathrm{Pf,boson}} = \mathrm{Pf}\left(\frac{1}{z_i - z_j}\right) \prod_{i<j}(z_i - z_j) \exp\left[-\frac{1}{4}\sum_k |z_k|^2\right] \qquad (12.83)$$

has the property that it vanishes when three bosons coincide (although not when two do), because the Pfaffian factor only removes pairwise zeroes from $\prod_{j<k}(z_j - z_k)$. The wave function is, thus, a zero-energy eigenstate of H_{Pf}. (This property remains valid for the multi-quasihole wave functions discussed in Section 9.9.) Further thought will convince the reader that this is the lowest degree polynomial that vanishes upon three-boson coincidences, and thus the unique bosonic ground state of H_{Pf} at $\nu = 1$. $\Psi_1^{\mathrm{Pf,boson}}$ can thus be obtained by exact diagonalization. The fermionic wave function $\Psi_{1/2}^{\mathrm{Pf}}$ is the ground state for a three-body interaction that involves appropriate derivatives of the delta function interaction.

The three-body interaction of Eq. (12.82) has a natural generalization to the spherical geometry (Read and Rezayi [539]). The $\nu = 1$ bosonic and the the $\nu = 1/2$ fermionic Pfaffian states correspond to $2Q = N - 2$ and $2Q = 2N - 3$, respectively. For bosons, the closest approach of three particles corresponds to the maximum total angular momentum for the triplet, which is $L_{\max} = 3Q$ (Q being the orbital angular momentum for each boson). An elimination of such configurations is equivalent to an avoidance of triplet coincidences. The Hamiltonian in the spherical geometry can thus be equivalently written as

$$H = V \sum_{i<j<k} P_{ijk}(L_{\max}) , \qquad (12.84)$$

where $P_{ijk}(L_{\max})$ is the projection operator onto a triplet of orbital angular momentum $L_{\max}$. This Hamiltonian imposes a penalty for triplet angular momentum of $L_{\max} = 3Q$, and obtains the Pfaffian as the exact zero-energy ground state. For fermions, the closest approach of three particles corresponds to the maximum total angular momentum $L_{\max} = 3Q - 3$; an elimination of such configurations produces the spherical version of the Moore–Read wave function at $\nu = 1/2$.

12.8.3 $\nu = 2/5$

A model can be constructed for which the unprojected wave function $\Phi_2 \Phi_1^2$ for the $\nu = 2/5$ ground state is exact [275]. The model truncates the Hilbert space to the lowest two Landau levels ($n = 0$ and $n = 1$), takes them to be degenerate, and considers the Trugman–Kivelson interaction, the zero-energy eigenstates of which vanish with a third-order zero when two particles approach one another. The most general antisymmetric wave function of this type is $\Phi_1^2 \Phi_{\nu^*}$, where Φ_{ν^*} is an antisymmetric function with at most one power of $\bar{z}_j$ (aside from the Gaussian factor), i.e., Φ_{ν^*} is restricted to the lowest two Landau levels. For $\nu^* < 2$ there are many choices for such a wave function; for $\nu^* = 2$ there is a unique choice (Φ_2); and for $\nu^* > 2$ no such wave function exists. $\Phi_1^2 \Phi_2$ is thus the unique ground state for this model at $\nu = 2/5$. Rezayi and MacDonald [548] find, for a six-particle system, that the ground state evolves smoothly as the Landau level spacing is varied from zero to infinity, indicating the absence of a level-crossing transition for this system during the process of adiabatic projection into the lowest Landau level.

12.8.4 Spin-singlet $\nu = 1/2$

The Haldane–Rezayi wave function [224, 225] of Eq. (E7.6) avoids relative angular momentum $m = 1$ for all pairs – for like-spin electrons due to the Fermi statistics, and for unlike-spin electrons due to the Jastrow factor. It does allow, however, pairs with relative angular momentum $m = 0$. It is an exact zero-energy eigenstate for the interaction $V_j = V_1 \delta_{j1}$, which is called the "hollow-core" interaction model, because it does not penalize two electrons at their closest approach (because $V_0 = 0$).

Summary A model that can be solved exactly to produce the qualitative phenomenology of the FQHE does not exist. We have sometimes succeeded, as in the popular television quiz show "Jeopardy," in formulating a question to which an already-known simple wave function is the answer. This approach has limited applicability, however. To summarize what we have so far: (a) One model gives only one fraction. (b) Only certain simple wave functions are obtained.[9] (c) No model has been solved for all eigenstates and eigenenergies even at a single filling factor; even the first excited state eludes exact solution. The most important use of such models has been in numerically generating certain trial wave functions by exact diagonalization. Fortunately, a wave function derives its legitimacy not from being the exact solution for a model interaction, but from being an accurate representation of the actual (Coulomb) solution.[10]

Exercises

12.1 A superficial "derivation" of the Laughlin wave function is as follows. Perform a Chern–Simons transformation in which m flux quanta are attached to each electron, and show that the electron wave function is related to the wave function in the transformed problem, Ψ_{CS}, by

$$\Psi = \prod_{j<k} \left(\frac{z_j - z_k}{|z_j - z_k|} \right)^m \Psi_{CS} . \tag{E12.1}$$

---

[9] In principle, we can always concoct a model for which a given wave function is the ground state. For example, any wave function $|\Psi\rangle$ is the exact, nondegenerate ground state of the Hamiltonian $H = -|\Psi\rangle\langle\Psi|$. This information, however, is of no use.
[10] "I would rather be approximately right than exactly wrong." A Wall Street broker.

Demonstrate that the assumptions: (i) Ψ_{CS} has no vortices (i.e., is everywhere non-negative), and (ii) Ψ is in the lowest Landau level, uniquely fix the form of Ψ_{CS} and yield the Laughlin wave function for Ψ. (Note that the form of the interaction did not play any role. A "real" derivation would tell us, for the Coulomb interaction, why Ψ is not the ground state at large m, and what are the corrections to it for small m.)

12.2 Consider a Laughlin-like wave function for bosons

$$\Psi_B = \prod_{j<k} |z_j - z_k|^m \exp\left[-\frac{1}{4}\sum_l |z_l|^2\right]. \tag{E12.2}$$

This wave function has the same charge density as $\Psi_{1/m}$, therefore describes a state of charged bosons at $\nu = 1/m$. Consider the off-diagonal element of the one-particle reduced density matrix (Appendix H)

$$\rho_1(z, z') = N\frac{\int d^2z_2, \ldots, d^2z_N \Psi_B^*(z, z_2, \ldots, z_N)\Psi_B(z', z_2, \ldots, z_N)}{\int d^2z_1, \ldots, d^2z_N \Psi_B^*(z_1, z_2, \cdots, z_N)\Psi_B(z_1, z_2, \ldots, z_N)}. \tag{E12.3}$$

The plasma analogy enables a determination of its long-distance behavior. Show that $e^{(m/2)\ln|z-z'|}\rho(z, z')$ is proportional to the partition function for the classical problem which has charge -1 particles at z_j and charge $-1/2$ impurities at fixed positions z and z'. Because of the complete screening of the impurities by the plasma, this function ought to be independent of z and z' in the limit $|z - z'| \to \infty$, implying that in this limit we have $\rho(z, z') \sim |z - z'|^{-m/2}$, i.e., an algebraic off-diagonal long-range order (ODLRO). For the fermionic wave function $\Psi_{1/m}$, the additional phases destroy the algebraic ODLRO, producing a Gaussian fall off for $\rho_1(z, z')$. (Source: Girvin and MacDonald [193].)

12.3 This exercise concerns the result that Jastrow wave functions can exhibit Bose–Einstein condensation, which is crucial if they are to be meaningful for a study of BEC. Consider the following wave function for bosons (Bijl [31]; Jastrow [289]):

$$\Psi_N = e^{-\frac{\beta}{2}\sum_{i<j} u(r_{ij})}, \tag{E12.4}$$

where $u(r) = \infty$ for $r < a$ (which implies vanishing probability of two bosons approaching closer than distance a, as appropriate for an infinite repulsive interaction at short distances), and has a power law behavior for $r > a$. The parameter β depends on the strength of the interaction. Define the normalization factor

$$Q_N = \int d\mathbf{r}_1 \cdots d\mathbf{r}_N\, e^{-\beta\sum_{j<k} u(r_{jk})}, \tag{E12.5}$$

which can be interpreted, apart from a factor of $N!$, as the classical partition function of a system of particles with an interaction $2u(r)$, with β interpreted as the inverse temperature. The number of bosons in the zero momentum state is given by (Appendix H)

$$n_0 = V \lim_{|r-r'|\to\infty} \langle \hat{\psi}^\dagger(r)\hat{\psi}(r')\rangle$$

$$= V\frac{1}{V^2} \int' dr\, dr'\, \langle \hat{\psi}^\dagger(r)\hat{\psi}(r')\rangle$$

$$= \frac{N}{V}\frac{\mathcal{Q}_{N+1}}{\mathcal{Q}_N}, \tag{E12.6}$$

where the prime on the integral sign denotes the condition $|r - r'| > a$ and

$$\mathcal{Q}_{N+1} \equiv \int' dr\, dr'\, dr_2 \ldots dr_N\, \Psi(r,r_2,\ldots r_N)\Psi(r',r_2,\ldots r_N) . \tag{E12.7}$$

(i) Show that

$$\mathcal{Q}_{N+1} \geq \mathcal{Q}_{N+1}\, e^{-\phi-\Delta}, \tag{E12.8}$$

where ϕ and Δ are positive, finite numbers defined by

$$\sum_{j=2}^N u(|r - r_j|) \geq -\phi \tag{E12.9}$$

$$\min[u(r)] = -\Delta . \tag{E12.10}$$

These conditions are not satisfied by all $u(r)$. Equation (E12.10) requires that $u(r)$ be bounded from below, and Eq. (E12.9) imposes a constraint on the long-range form of $u(r)$, derived below.

(ii) Next, show that

$$\frac{(N + 1)\mathcal{Q}_N}{\mathcal{Q}_{N+1}} = e^{\beta\mu} = z , \tag{E12.11}$$

where μ is the chemical potential of the equivalent classical problem, and z is the activity.

(iii) Derive the inequality

$$\frac{n_0}{N} \geq \frac{N}{V} z^{-1} e^{-2\phi-\Delta} . \tag{E12.12}$$

If the chemical potential does not diverge, which is the case for many reasonable interactions $u(r)$, and ϕ and Δ are finite numbers, then the system exhibits BEC.

(iv) Obtain the condition that ϕ is finite provided $|u(r)| \leq A/r^{d+\epsilon}$ in d dimensions.[11] (Source: Reatto [540].)

12.4 Show that the two methods of LLL projection (Section 5.14) produce the same wave function (apart from an overall constant) for a single CF-quasiparticle at $\nu = 1/(2p + 1)$.

---

[11] An unexpected corollary is what is known as "supersolidity" (Chester [74]). For many choices of $u(r)$ which satisfy the conditions for ODLRO, the equivalent classical system is known to describe a crystal at sufficiently large β, or low temperatures (which melts at a certain critical value of β as β is decreased). That demonstrates, in principle, the possibility of BEC in a crystal, i.e., of the coexistence of diagonal and off-diagonal long-range orders.

12.5 Consider the spin-singlet state

$$\Psi_{2/5} = \Phi_{1,1}\Phi_1^2 , \tag{E12.13}$$

which is a generalization of the Laughlin wave function to a two-component system (Halperin [229]). Using the plasma analogy:
(i) Show that $\Psi_{2/5}$ has uniform density.
(ii) Obtain the charge of the vortex

$$\prod_{j=1}^{N/2}(z_j - \eta)\Psi_{2/5} . \tag{E12.14}$$

(iii) Show that $\Psi_{2/5}$ is the unique zero-energy ground state in the lowest Landau level (assuming zero Zeeman energy) for the hard-core interaction

$$V_0 = V_1 > 0, \qquad V_2 = V_3 = \cdots = 0 . \tag{E12.15}$$

12.6 The "hole" wave function at $\nu = 1$, in which one electron is missing from the $m = 0$ orbital (Fig. 12.2), is given by Eq. (12.35) with $p = 0$.
(i) Show that, for a finite N, this wave function is not unique at the corresponding total angular momentum.
(ii) Now model this state in terms of composite fermions carrying two vortices, and show that a minimization of the CF kinetic energy produces a unique wave function. Write this wave function (without explicit projection).

   This wave function has been shown to be a much better representation of the exact state than the "hole" of Fig. 12.2 (Jeon *et al.* [298]). The wave function of Oaknin *et al.* [483], reproduced in Eq. (E3.33), is also excellent.

12.7 Derive the following properties of the projected density operator:

$$\bar{\rho}_q^\dagger = \bar{\rho}_{-q}, \tag{E12.16}$$

$$[\bar{\rho}_k, \bar{\rho}_q] = (e^{\bar{k}q/2} - e^{k\bar{q}/2})\bar{\rho}_{k+q}, \tag{E12.17}$$

$$\mathcal{P}_{\mathrm{LLL}}\rho_q^\dagger \rho_q = \bar{\rho}_q^\dagger \bar{\rho}_q + (1 - e^{-q\bar{q}/2}). \tag{E12.18}$$

Hint: Use the identity $e^A e^B = e^B e^A e^{[A,B]}$ (Appendix B) to reorder terms.

12.8 Using the symmetry under $k \to -k$, show that the oscillator strength can be written as a double commutator:

$$f(k) = \frac{1}{2N}\langle\Psi|[\rho_k^\dagger, [H, \rho_k]]|\Psi\rangle . \tag{E12.19}$$

Since the density commutes with V (without projection), the commutator reduces to $[K, \rho_k]$, where $K = \sum_j \pi_j^2/2m_{\mathrm{b}}$, and $\pi = p + eA/c$. Evaluate the commutators to derive Eq. (12.59).

12.9 Kohn's theorem is derived in this exercise. Consider

$$H = \frac{1}{m_b} \sum_j \pi_j^2 + V, \qquad (\text{E}12.20)$$

where $\pi = p + eA/c$, and define $\Pi = \sum_j \pi_j$, the kinetic momentum of the whole system. Now calculate the commutator

$$[H, \Pi_\pm] = \pm \hbar \omega_c \Pi_\pm , \qquad (\text{E}12.21)$$

where

$$\Pi_\pm = \Pi_x \pm i \Pi_y . \qquad (\text{E}12.22)$$

This, incidentally, is the operator form for the Lorentz equation for the whole system:

$$\frac{d\Pi}{dt} = \frac{i}{\hbar}[H, \Pi] = -\frac{e}{m_b c} \Pi \times B . \qquad (\text{E}12.23)$$

Show that $\Pi_+ \Psi$ is an exact eigenstate with energy $\hbar \omega_c$ above the ground state. Further, using the symmetric gauge, show that Π_+ is proportional to the center of mass ladder operator $A^\dagger$ of Eq. (3.247). (Source: Kohn [351].)

12.10 Use the explicit form for the spherical harmonics, L_z and $L_\pm$ (Eqs. 3.131 and 3.130) to confirm the commutation relations in Eqs. (12.67) and (12.68) for a single particle with $L = 1$.

13

Bilayer FQHE

In this chapter we consider what happens when two FQHE planes are brought in close proximity, as shown in Fig. 13.1. Interest in this system arises because it can be experimentally realized, because it provides a handle that allows us to study the interplay between the inter and intra-layer correlations, and because *inter*layer correlations bring about new structures that have no analog in a single layer. For simplicity, we assume that the tunneling between the planes is negligibly small, and (except in Section 13.5) that the Zeeman energy is sufficiently large that the electron spin is completely frozen. The validity of these assumptions ought to be carefully examined for each experiment in question. Also, we only consider the situation with equal densities in the two layers.

The layer index is also called "pseudospin," labeled either 1, 2 or ↑, ↓. The wave function is analogous to that for the real spin

$$\Psi'_\nu = A[\Psi_\nu[\{z_j\}, \{w_r\}]\alpha_1 \ldots \alpha_{N_1}\beta_1 \ldots \beta_{N_2}] , \tag{13.1}$$

where A is the antisymmetrizer, particles labeled $j = 1, \ldots, N_1$ are in the left layer (pseudospin up) and those with $r = 1, \ldots, N_2$ are in the right layer (pseudospin down), and α and β are pseudospinors. The coordinates are defined as $z_j = x_j - iy_j$ and $w_r = x_r - iy_r$. In the absence of tunneling, the number of particles in each layer is fixed, which, in the pseudospin language, means that $S_z = (N_1 - N_2)/2$ is fixed. The spatial part $\Psi[\{z_j\}, \{w_r\}]$ is chosen to be antisymmetric with respect to an exchange of coordinates of two particles particles within the same layer. Following the standard practice in the literature, we define the filling factor for the bilayer as the sum of the filling factors for the individual layers:

$$\nu = \nu_1 + \nu_2 . \tag{13.2}$$

Care must be exercised in using the pseudospin language. The problem may seem completely analogous to that of spinful electrons in a single layer, but it is not. The reason is that the effective two-dimensional interelectron interaction is explicitly pseudospin dependent: For electrons in the same layer, it is

$$V_{\uparrow\uparrow}(r_{12}) = V_{\downarrow\downarrow}(r_{12}) = \frac{e^2}{\epsilon r_{12}} , \tag{13.3}$$

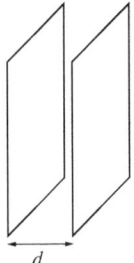

Fig. 13.1. A bilayer system.

whereas two electrons in different layers have the interaction energy

$$V_{\uparrow\downarrow}(r_{12}) = V_{\uparrow\downarrow}(r_{12}) = \frac{e^2}{\epsilon\sqrt{r_{12}^2 + d^2}} . \tag{13.4}$$

Here, r_{12} is the projection of the distance in the plane, and d is the interlayer separation. The interaction can be written as

$$V_{\sigma_1\sigma_2}(r_{12}) = \frac{V(r_{\uparrow\uparrow}) + V(r_{\uparrow\downarrow})}{2} + \frac{V(r_{\uparrow\uparrow}) - V(r_{\uparrow\downarrow})}{2}\boldsymbol{\sigma}_1 \cdot \boldsymbol{\sigma}_2 , \tag{13.5}$$

where $\boldsymbol{\sigma}$ is the Pauli spin matrix. This form makes it apparent that the Hamiltonian $\sum_{j<k} V(r_{jk})$ does not commute with the total pseudospin $(\sum_j \boldsymbol{\sigma}_j)^2$ for $d/\ell \neq 0$, and, therefore, the energy eigenstates do not have a well-defined total pseudospin quantum number. The z component of the pseudospin, proportional to the difference in the numbers of particles in the two layers, is a good quantum number, however.

13.1 Bilayer composite fermion states

13.1.1 Two limits

In the trivial limit $d/\ell \to \infty$, particles in different layers are uncorrelated. The bilayer state is given simply by the product of two independent single-layer composite fermion states.

The use of pseudospin language is most useful in the other limit, $d/\ell = 0$. (Please, understand that this theoretical limit does not give a single layer, but two coincident planes, still with no tunneling.) In this case, the full SU(2) symmetry is restored, because of the layer independence of the interaction. The problem of "spinless" electrons in a bilayer system with $d/\ell = 0$ becomes formally identical to that of spinful electrons in a single layer with the Zeeman energy set equal to zero, the physics of which is well understood, as explained in Section 11.3. To summarize, at filling factors $\nu = n/(2pn + 1)$, the ground state is a spin singlet for even n and partially spin polarized at odd n, with the wave function given by

$$\Psi_{\frac{n}{2pn+1}} = \Phi_1^{2p}\Phi_{n_\uparrow,n_\downarrow} , \tag{13.6}$$

where $n_\uparrow$ and $n_\downarrow$ are given by $n/2$ for even n and $(n \pm 1)/2$ for odd n. Here $\Phi_{n_\uparrow, n_\downarrow}$ is the product of two Slater determinant wave functions at fillings $n_\uparrow$ and $n_\downarrow$.

These results carry over to the bilayer at $d/\ell = 0$, with the real spin replaced by the pseudospin. For the pseudospin singlet state the densities in the two layers must be equal, because it necessarily has $S_z = 0$. The state with $S \neq 0$ can have many S_z values, however, and hence also many choices for N_1 and N_2. The special case with $S \neq 0$ and $S_z = 0$ (i.e., the total pseudospin lies in the plane of the bilayer) has equal densities in two layers. We stress that the FQHE at $d/\ell = 0$ originates because of interlayer correlations, incorporated through the factor Φ_1^2 that does not discriminate on the basis on the layer index. Without such correlations, the state in each layer at $\nu_1 = \nu_2 = n/[2(2pn + 1)]$ would be a compressible state.

13.1.2 Intermediate separations

We now proceed to construct an interpolation scheme for arbitrary values of d/ℓ. Let us consider a FQHE state at $d/\ell = 0$, say, a pseudospin-singlet incompressible state at $\nu = n/(2pn + 1)$. Now, in a thought experiment, let us increase d/ℓ to a nonzero, but very small, value. Since the FQHE state has a gap to charged excitations, we might expect that the gap survives for sufficiently small d/ℓ. That strongly indicates that interlayer correlations can cause new structure for $d/\ell \neq 0$.

On the other hand, for very large separations, two compressible states result. Thus, when we reduce the layer separation from $d/\ell = \infty$ to $d/\ell = 0$, at some small interlayer separation each electron in one layer develops $2p$ zeroes with respect to each electron in the other layer. A natural conjecture is that the interlayer zeroes appear one by one, rather than all at once. This leads to a consideration of the following class of wave functions (Scarola and Jain [569]):

$$\Psi_{(\nu_1'^{-1}\nu_2'^{-1}m)} = \prod_{r,j}(z_j - w_r)^m \Psi_{\nu_1'}[\{z_k\}]\Psi_{\nu_2'}[\{w_s\}] \,, \tag{13.7}$$

where the fully antisymmetric wave function $\Psi_{\nu'}$ is the single layer wave function at filling factor ν', and m is the number of interlayer zeroes. (The filling factor of the product state is derived below; for $m \neq 0$, ν_j' is *not* equal to the filling factor ν_j of the jth layer. Also, the notation is slightly different from that used in Ref. [569].) The state described by $\Psi_{(\nu_1'^{-1}\nu_2'^{-1}m)}$ will be denoted by $(\nu_1'^{-1}\nu_2'^{-1}m)$. This state is incompressible when both $\Psi_{\nu_1'}$ and $\Psi_{\nu_2'}$ are incompressible, and compressible if either or both of $\Psi_{\nu_1'}$ and $\Psi_{\nu_2'}$ are compressible. These wave functions are expected to have low energies because they build intralayer correlations through the CF theory, and interlayer correlations through an additional Jastrow factor. The interlayer exponent m can be either an even or an odd integer; the overall antisymmetry is taken care of by the explicit antisymmetrization in Eq. (13.1).

We ought to bear in mind that z_j and w_r denote the (x, y) coordinates *within* their respective 2D sheets. The condition $z_j = w_r$ does not imply two spatially coincident particles, but rather two particles directly across one another, and the factor $(z_j - w_r)$ ensures that z_j avoids the point directly across w_r in its own layer.

Halperin [229] had earlier generalized the Laughlin wave function to a two-component system:

$$\Psi_{(m'm''m)}[\{z_i\}, \{w_r\}] = \prod_{j<k=1}^{N_1} (z_j - z_k)^{m'} \prod_{r<s=1}^{N_2} (w_r - w_s)^{m''} \prod_{j,r} (z_j - w_r)^m$$

$$\times \exp\left[-\frac{1}{4} \left(\sum_{j=1}^{N_1} |z_j|^2 + \sum_{r=1}^{N_2} |w_r|^2 \right) \right], \tag{13.8}$$

where m' and m'' are odd integers. The wave functions of Eq. (13.7) reduce to the Halperin wave functions when $v'_1 = 1/m'$ and $v'_2 = 1/m''$, but can deal with a wider class of bilayer states, both incompressible and compressible. While all Halperin $(m'm''m)$ wave functions are, in principle, allowed for $d/\ell \neq 0$, only those satisfying the Fock condition are valid candidates for $d/\ell = 0$.

The (mmm) wave function, m odd, is nothing but Laughlin's single layer $v = 1/m$ wave function, with half of the coordinates interpreted as belonging to one layer and the rest to the other layer. In the pseudospin language, half of the electrons have spin up and half spin down, i.e., $S_z = 0$. However, since the spatial part is fully antisymmetric with respect to an exchange of any two coordinates, independent of its spin, the spin part of the wave function must be fully symmetric. The wave function thus describes a fully pseudospin-polarized state with $S = N/2$ and $S_z = 0$. Taking all particles to be in the same layer (w's replaced by z's) would produce $S = N/2$ and $S_z = N/2$. Different values of S_z can be obtained by choosing different proportions of z's and w's, but the full antisymmetry of the spatial part guarantees $S = N/2$.

We determine the filling factor of the state $(v'_1^{-1} v'_2^{-1} m)$ in the symmetric gauge in the planar geometry, leaving a derivation of the same result in the spherical geometry to Exercise 13.2. In the symmetric gauge, the wave function describes a uniform state of electrons in two disks, one in each layer. The sizes of the disks are determined by the largest powers of z_j and w_r, given by (apart from unimportant order-one corrections) $N_1/v'_1 + mN_2$ and $N_2/v'_2 + mN_1$, respectively. To ensure that electrons in the two layers occupy the same area, N_1 and N_2 must be related by

$$N_1 v'_1^{-1} + mN_2 = N_2 v'_2^{-1} + mN_1 . \tag{13.9}$$

The filling factors of the individual layers are given by

$$v_1^{-1} = v'_1^{-1} + m\frac{N_2}{N_1}, \tag{13.10}$$

$$v_2^{-1} = v'_2^{-1} + m\frac{N_1}{N_2}, \tag{13.11}$$

and the total filling factor is $\nu = \nu_1 + \nu_2$.

We specialize below to the situation when the densities in the two layers are equal, i.e., $N_1 = N_2$. From the preceding considerations, this implies $\nu'_1 = \nu'_2 \equiv \nu'$, $\nu_1 = \nu_2 = \nu/2$, and the total filling factor is given by

$$\nu = \nu_1 + \nu_2 = \frac{2\nu'}{1 + m\nu'} . \tag{13.12}$$

Since we are interested in enumerating the states at a given total filling factor ν, we write

$$\nu' = \frac{\nu}{2 - m\nu} . \tag{13.13}$$

Thus, for a given ν, the possible states are

$$\left(\frac{2 - m\nu}{\nu} \ \frac{2 - m\nu}{\nu} \ m \right) . \tag{13.14}$$

For a single layer, the prominent FQHE states $\Psi_{\nu'}$ occur at

$$\nu' = \frac{n'}{2pn' \pm 1} , \tag{13.15}$$

where n' is a positive or a negative integer. These produce the bilayer states

$$\left(\frac{2pn' \pm 1}{n'} \ \frac{2pn' \pm 1}{n'} \ m \right) \tag{13.16}$$

at the prominent bilayer fractions

$$\nu = \frac{2n'}{(2p + m)n' \pm 1} . \tag{13.17}$$

The corresponding wave functions are given in Eq. (13.7).

Let us take some examples of the possible states as we decrease the interlayer separation from $d/\ell = \infty$ to $d/\ell = 0$:

$$\nu = 1 : \quad (220) \rightarrow (111) \tag{13.18}$$
$$\nu = \tfrac{1}{2} : \quad (440) \rightarrow (331) \rightarrow (222) \tag{13.19}$$
$$\nu = \tfrac{1}{3} : \quad (660) \rightarrow (551) \rightarrow (442) \rightarrow (333) \tag{13.20}$$

$$\nu = \tfrac{2}{5}: \quad (550) \to (441) \to (332) \tag{13.21}$$

$$\nu = \tfrac{2}{3}: \quad (330) \to (221) \to (\bar{1}\bar{1}2) \tag{13.22}$$

$$\nu = \tfrac{3}{5}: \quad \begin{pmatrix} 10 & 10 \\ 3 & 3 \end{pmatrix} 0 \to \begin{pmatrix} 7 & 7 \\ 3 & 3 \end{pmatrix} 1 \tag{13.23}$$

$$\nu = \tfrac{4}{7}: \quad \begin{pmatrix} 7 & 7 \\ 2 & 2 \end{pmatrix} 0 \to \begin{pmatrix} 5 & 5 \\ 2 & 2 \end{pmatrix} 1 \to \begin{pmatrix} 3 & 3 \\ 2 & 2 \end{pmatrix} 2 \tag{13.24}$$

Comments: (i) The $(\bar{1}\bar{1}2)$ state at $\nu = 2/3$ is shorthand for

$$\Psi_{2/3} = \mathcal{P}_{\mathrm{LLL}} \prod_{j,r} (z_j - w_r)^2 [\Phi_1^2 \Phi_1^*]_1 [\Phi_1^2 \Phi_1^*]_2 . \tag{13.25}$$

In other words, "$\bar{1}$" represents the $\nu = 1$ wave function $\Phi_1^2 \Phi_1^*$ rather than Φ_1; the latter would also give the desired filling factor, but then the wave function would not be valid for $d/\ell = 0$, because it does not satisfy the Fock condition.

(ii) For $\nu = 3/5$, the state at $d/\ell = 0$ is somewhat complicated and not shown in Eq. (13.23). In Chapter 11 we saw for the single layer problem with $E_Z = 0$ that the ground state at this filling is partially spin polarized. We encountered in that chapter the wave function with $S = S_Z$, but that is not suitable for the symmetric bilayer problem with equal number of electrons in each layer. One can, in principle, use the spin lowering operator to construct a state with $S_Z = 0$, after which the S_z index can be interchanged with the pseudospin (or layer) index.

(iii) The states in which interlayer correlations are stronger than intralayer correlations are unphysical. Such states are automatically eliminated if we do not continue beyond the $d/\ell = 0$ state that we know from Chapter 11. For example, for $\nu = 1/3$, we stop at (333) and do not consider (224).

(iv) The states described above, of course, do not exhaust possible bilayer states (even in the absence of real spin), but merely represent the most straightforward extension of the CF theory to bilayers. While the two limiting cases are well tested, the intermediate bilayer CF states must be verified.

13.1.3 Numerical investigations

Many numerical studies have investigated bilayer states with equal numbers of particles in each layer. Chakraborty and Pietiläinen [57] study an eight particle system (four in each layer) at $\nu = 1/2$ and find an incompressible state for $d/\ell \approx 2.0$. Yoshioka, MacDonald, and Girvin [731] compare the exact wave functions with the Halperin ($m'm''m$) wave functions at several fillings ($\nu = 1, 1/2, 1/3, 3/5$) and find transitions between them; the (331) wave function has a high overlap with the exact state at $d/\ell \approx 1.5$ for six particles. He *et al.* [243] also include the effect of interlayer tunneling for bilayer states at $\nu = 1, 2/3$, and $1/2$ and find, with increasing d/ℓ, a destruction of the $\nu = 1$ state and a transition between two

FQHE states at $\nu = 2/3$; their optimal parameters for $\nu = 1/2$ FQHE are consistent with those in Ref. [731].

Numerical studies based on the CF theory have also been performed. The phase diagrams as a function of d/ℓ has been determined from the thermodynamic limits for the energies of various wave functions in Eq. (13.7) as a function of d/ℓ at $\nu = 1$, 1/2, 1/3 and 2/5 (Scarola and Jain [569]). The phase diagrams consist of alternating incompressible and compressible states at these fractions (that is often, but not always, the case; e.g. at $\nu = 4/7$ the transition is from an incompressible state to another incompressible state), and interlayer correlations are unimportant when d/ℓ is larger than a few (typically 3 to 4) magnetic lengths. For example, (331) is stable for $1.7 < d/\ell < 3.0$.

Nakajima and Aoki [472, 473] investigate the collective mode excitation at $\nu = 1/m$, m odd, in terms of pseudospin waves of composite fermions, by analogy to their earlier work [471] on spin waves in a single layer at the same fractions. They study a system of five electrons at a layer separation of $d/\ell = 1.0$ in the presence of nonzero interlayer tunneling and conclude that the exact results are well described in terms of a composite fermion system with reduced interactions but unchanged interlayer tunneling.

Shibata and Yoshioka [588] apply the density matrix renormalization group method (which combines a real-space renormalization group with exact diagonalization) to the $\nu = 1$ bilayer system containing up to 24 electrons to obtain the ground state wave function and the excitation gap. They see evidence for a smooth crossover (rather than a first-order phase transition) from independent CF Fermi seas at large separations to an interlayer coherent state at $d/\ell \approx 1.6$.

Several bilayer states and their topological properties have also been studied using the CFCS theory (Ho [255]; Lopez and Fradkin [403]; Rajaraman [534]).

13.2 1/2 FQHE

At $\nu = 1/2$, the two limiting states are compressible. At $d/\ell = 0$, (222) is a pseudospin unpolarized CF Fermi sea, and at $d/\ell = \infty$ the state (440) has a $\nu = 1/4$ CF Fermi sea in each layer. An incompressible state, described by the Halperin (331) wave function, appears possible, however, at an intermediate separation.

FQHE has been observed by Suen *et al.* [630] and Eisenstein *et al.* [137] at $\nu = 1/2$ in bilayer systems, as seen in Figs. 13.2 and 13.3. The former experiment studies a wide quantum well (68 nm wide), which has been shown, by a self-consistent Hartree–Fock calculation, to effectively act as a bilayer system. A 1/2 plateau is seen in the approximate parameter range $2.6 < d/\ell < 8$ with an activation gap on the order of 200 mK. The numerical calculations described in the previous section are not appropriate for this experiment. The experiment of Eisenstein and collaborators employs a double quantum well geometry with a thin but high barrier to suppress interlayer tunneling, and finds that the 1/2 FQHE disappears beyond $d/\ell \geq 3.0$, consistent with the theoretical estimates. Theories also predict a lack of 1/2 FQHE for $d/\ell < 1.7$, a regime not investigated in this experiment.

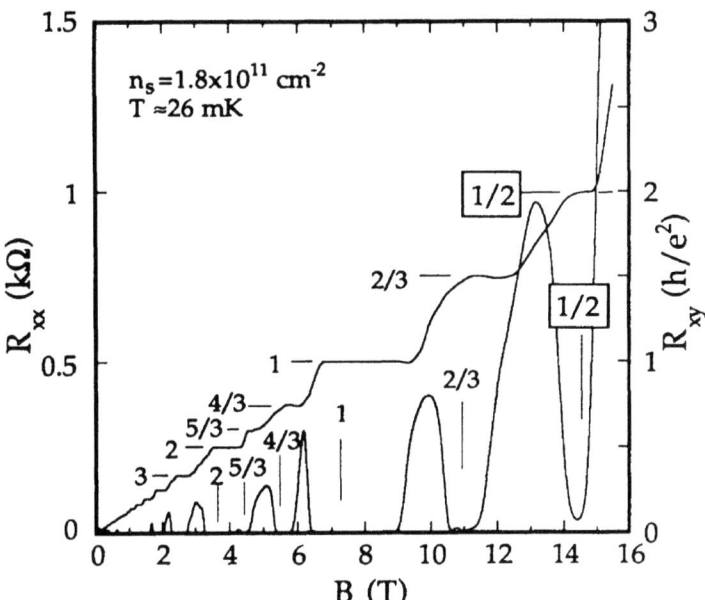

Fig. 13.2. Observation of the 1/2 FQHE in a bilayer system. Source: Y. W. Suen, L. W. Engel, M. B. Santos, M. Shayegan, and D. C. Tsui, *Phys. Rev. Lett.* **68**, 1379 (1992). (Reprinted with permission.)

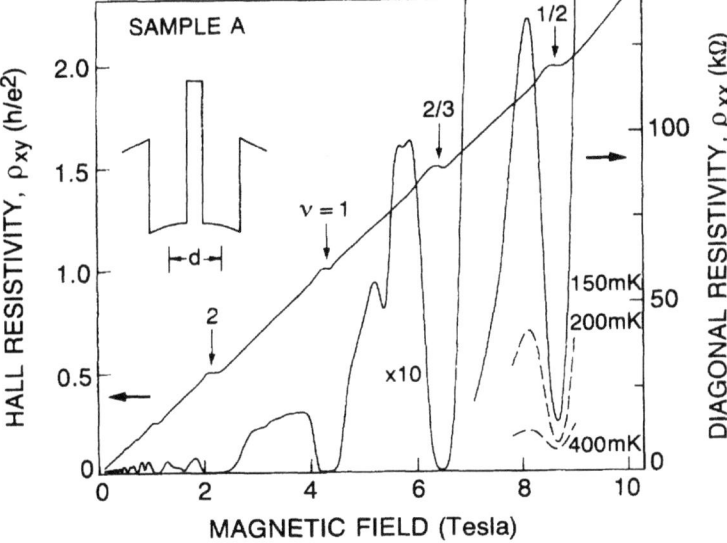

Fig. 13.3. Observation of the 1/2 FQHE in a bilayer system. Source: J. P. Eisenstein, G. S. Boebinger, L. N. Pfeiffer, K. W. West, and S. He, *Phys. Rev. Lett.* **68**, 1383 (1992). (Reprinted with permission.)

The calculations of He *et al.* [245], performed for the actual experimental parameters and sample geometries, show quantitative agreement with both experiments.

We recall from Section 7.4.2 that the FQHE at 1/2 is also possible in a single layer for certain interactions (for example, those described by the second LL Coulomb pseudopotentials), with the Moore–Read Pfaffian wave function providing a representation for its ground state. Greiter, Wen, and Wilczek [214] note a close connection between the bilayer Halperin (331) wave function and the single layer Moore–Read wave function: the latter is obtained upon fully antisymmetrizing the spatial part of the former. The correspondence is established as follows:

$$
A\left[\prod_{i<j}\left[(z_i - z_j)^3(w_i - w_j)^3\right]\prod_{i,j}(z_i - w_j)\right]
$$

$$
= \Phi_1^2\, A\left[\frac{\prod_{i<j}(z_i - z_j)(w_i - w_j)}{\prod_{i,j}(z_i - w_j)}\right]
$$

$$
= (-1)^{n(n-1)/2}\,\Phi_1^2\, A\,\det\frac{1}{z_i - w_j}
$$

$$
= (-1)^{n(n-1)/2}\,\Phi_1^2\,\sum_{P}\epsilon_P A\prod_j\frac{1}{z_j - w_{Pj}}
$$

$$
= (-1)^{n(n-1)/2}\,n!\,\Phi_1^2\, A\prod_j\frac{1}{z_j - w_j}
$$

$$
= (-1)^{n(n-1)/2}\,n!\,\Phi_1^2\,\mathrm{Pf}\frac{1}{Z_j - Z_k}\,. \tag{13.26}
$$

Here, z_j and w_j are coordinates for $n = N/2$ particles in the two layers and Z_j labels *all* electrons. The symbol A denotes antisymmetrization with respect to *all* coordinates. The fully antisymmetric function Φ_1 is a defined as

$$
\Phi_1 = \prod_{i<j}\left[(z_i - z_j)(w_i - w_j)\right]\prod_{i,j}(z_i - w_j) = \prod_{i<j}(Z_i - Z_k)\,, \tag{13.27}
$$

and an identity due to Cauchy [410],

$$
\det\frac{1}{z_i - w_j} = (-1)^{n(n-1)/2}\frac{\prod_{i<j}(z_i - z_j)(w_i - w_j)}{\prod_{i,j}(z_i - w_j)}\,, \tag{13.28}
$$

has been used. The symbol P refers to all permutations, and $\epsilon_P = +1$ ($\epsilon_P = -1$) for even (odd) permutations. In the fifth line of Eq. (13.26), we have noted that all permutations yield the same contribution, and, in the last step, used the definition of the Pfaffian. The Gaussian factors are suppressed for simplicity.

13.3 v = 1: interlayer phase coherence

To test if a given wave function describes a Bose–Einstein condensate, we determine if it exhibits off-diagonal long-range order (ODLRO). No ordinary ODLRO appears in the FQHE (Section 8.1). Certain bilayer FQHE states may, however, possess a subtle kind of ODLRO, which is the topic of this section. Specifically, we discuss the possibility of an *excitonic* ODLRO in the Halperin (*mmm*) wave functions at $v = 1/m$, and the new physics that such an ODLRO entails. A more detailed account of this topic, along with many relevant references, can be found in the review article by Girvin and MacDonald [194]. The general definition of ODLRO and its connection to Bose–Einstein condensation are given in Appendix H.

13.3.1 ODLRO in Halperin's (mmm) wave function

We demonstrate the presence of the following ODLRO:

$$\lim_{|r-r'|\to\infty} \langle \hat{\psi}_{ex}^{\dagger}(r')\hat{\psi}_{ex}(r) \rangle \neq 0 , \tag{13.29}$$

where

$$\hat{\psi}_{ex}(r) \equiv \hat{\psi}_2^{\dagger}(r)\hat{\psi}_1(r) . \tag{13.30}$$

Here, subscripts 1 and 2 refer to the two layers. The operator $\hat{\psi}_{ex}(r)$ destroys an electron at r in layer 1 and creates an electron at r in layer 2; in other words, it destroys (or, depending on convention, creates) an exciton at r. The above ODLRO can thus be interpreted as Bose–Einstein condensation of excitons. Equation (13.29) is established as follows:

$$\langle \hat{\psi}_1^{\dagger}(r')\hat{\psi}_2(r')\hat{\psi}_2^{\dagger}(r)\hat{\psi}_1(r) \rangle$$

$$= -\langle \hat{\psi}_1^{\dagger}(r')\hat{\psi}_2^{\dagger}(r)\hat{\psi}_2(r')\hat{\psi}_1(r) \rangle$$

$$= -N(N-1) \int \prod_{j=2}^{N}(d^2x_j \, d^2y_j)$$

$$\times \Psi^*(r',x_2\ldots x_N; r,y_2\ldots y_N)\Psi(r,x_2\ldots x_N; r',y_2\ldots y_N)$$

$$= N(N-1) \int \prod_{j=2}^{N}(d^2x_j \, d^2y_j)|\Psi(r,x_2\ldots x_N; r',y_2\ldots y_N)|^2$$

$$= (\rho_{1/2})^2 g(r-r')$$

$$\to (\rho_{1/2})^2 . \tag{13.31}$$

Here, we have denoted the coordinates of electrons in two layers by x_j and y_k, and made use of Eq. (F.30) and the full antisymmetry of the Halperin wave function. The last equality follows from the expression Eq. (8.8) for the pair correlation function. The quantity $\rho_{1/2} = 1/4\pi\ell^2$ is the density in a *single* layer (reader: why?). Exploiting the antisymmetry property, the off-diagonal matrix element can thus be expressed as a familiar diagonal correlation function. By definition, the pair correlation function $g(r) \to 1$ in the limit $r \to \infty$ for any liquid, demonstrating that the right hand side is nonzero in this limit (the last step).

13.3.2 The Fertig bilayer wave function at $v = 1$

Fertig [154] introduced the following second-quantized wave function for the bilayer $v = 1$ state:

$$\Psi_{\text{Fertig}} = \prod_l \frac{1}{\sqrt{2}} (c_{1l}^{\dagger} + c_{2l}^{\dagger})|0\rangle \tag{13.32}$$

$$= \prod_l \frac{1}{\sqrt{2}} (1 + c_{2l}^{\dagger} c_{1l})|0'\rangle , \tag{13.33}$$

where $|0\rangle$ is the vacuum state with no particles, and

$$|0'\rangle \equiv \prod_l c_{1l}^{\dagger}|0\rangle \tag{13.34}$$

has layer 1 fully occupied and layer 2 fully unoccupied. The operator c_{1l} (c_{2l}) destroys an electron in layer 1 (2) in the angular momentum l orbital, where $l = 0, 1, \ldots$ The form in Eq. (13.32) is reminiscent of the BCS wave function for a superconductor.

That the Fertig wave function describes a state at $v = 1$ is apparent from the observation that, in Eq. (13.32), one and only one electron occupies each angular momentum orbital (although it can be in either layer). Is the Fertig wave function a representation of the (111) state? The two are of course not *identical*, because the particle numbers in the individual layers are conserved in (111) but not in the Fertig wave function. It is possible, however, that they are equivalent in the same sense as canonical and grand canonical ensembles are. Should that be the case, the Fertig wave function would lead to much simplification. That the number of particles in each individual layer is not a sharp quantum number in Eq. (13.32) has no physical significance. It is simply a matter of technical convenience, and does not imply, or require, any tunneling between layers. One must be careful, however, only to calculate quantities that conserve the number of particles in each layer independently.

The correspondence between the Fertig and Halperin wave functions is not immediately obvious. We note that both have the same ODLRO:

$$\langle \Psi_{\text{Fertig}}|\hat{\psi}_2^{\dagger}(r)\hat{\psi}_1(r)|\Psi_{\text{Fertig}}\rangle = \sum_{l,l'} \eta_l^*(z)\eta_{l'}(z)\langle \Psi_{\text{Fertig}}|c_{2l}^{\dagger}c_{1l'}|\Psi_{\text{Fertig}}\rangle$$

$$= \sum_l \frac{1}{2}|\eta_l(z)|^2 \langle 0|(c_{1l} + c_{2l})c_{2l}^{\dagger}c_{1l}(c_{1l}^{\dagger} + c_{2l}^{\dagger})|0\rangle$$

$$= \frac{1}{4\pi \ell^2} = \rho_{1/2} , \tag{13.35}$$

where we have used $\eta_l(z) = (2\pi 2^l l!)^{-1/2} z^l \exp(-|z|^2/4)$ (with $\ell = 1$). This quantity is analogous to the BCS order parameter, and consistent with the ODLRO in the (111) state. Jeon and Ye [297] have shown that a projection of the Fertig wave function onto the $N_1 = N_2$ sector produces precisely the Halperin (111) wave function.

The SU(2) pseudospin rotation symmetry makes the bilayer system at $d/\ell = 0$ equivalent to the Heisenberg problem, in which spins can rotate freely in three dimensions. The ODLRO in the Halperin (111) wave function, that represents the ground state of a $v = 1$ bilayer system for $d/\ell = 0$, is equivalent to pseudospin ferromagnetism. Because of the continuous nature of the symmetry, however, the ODLRO is destroyed at any nonzero temperature, as stipulated by the Hohenberg–Mermin–Wagner theorem.

Finite interlayer separation causes quantum fluctuations that alter details of the above wave function, and also change the symmetry of the problem to U(1) (or the XY model, in which the spins rotate in the $x-y$ plane only). The symmetry is made explicit by writing for the ground state

$$\Psi_{\text{Fertig}} = \prod_l \frac{1}{\sqrt{2}} (c_{1l}^\dagger + e^{i\phi} c_{2l}^\dagger) |0\rangle \tag{13.36}$$

and noting that the energy is independent of ϕ. A spatial modulation in the phase costs energy, and the low-energy physics is described by an effective action

$$H = \frac{\rho_s}{2} \int d^2 r |\nabla \phi|^2 , \tag{13.37}$$

which is familiar from the Kosterlitz–Thouless theory of phase transitions in two-dimensional models in the XY universality class. Although *true* ODLRO is still destroyed at any nonzero temperature, superfluidity and Josephson-like effects are possible (Ezawa and Iwazaki [147, 148]; Wen and Zee [674]). Fertig shows [154], within a Hartree–Fock formalism, that the pseudospin wave dispersion changes from $\omega_k \sim k^2$ at $d/\ell = 0$ to $\omega_k \sim k$ at $d/\ell \neq 0$, consistent with the presence of a Goldstone mode related to the broken U(1) symmetry. Analogous "valley waves" were considered earlier by Rasolt, Halperin, and Vanderbilt [536] for the quantum Hall effect in multivalley semiconductors.

13.3.3 Experimental case for excitonic superfluidity

To begin with, establishing the (111) state itself is important: it must be distinguished, experimentally, from the ordinary, uncorrelated $v = 1$ state that would result from the integral quantum Hall effect in the symmetric subband of the bilayer system in the presence of nonzero interlayer tunneling. Murphy *et al.* [460] study the transition from an incompressible state to a compressible state at $v = 1$ as a function of d/ℓ. This dimensionless parameter can be varied either by making many bilayers with different d, or, more conveniently, by tuning the electron density in a given bilayer, which alters the value of ℓ at $v = 1$. Figure 13.4 demonstrates that the $v = 1$ quantum Hall effect is destroyed by tuning d/ℓ. From the phase diagram so determined, shown in the inset of Figure 13.4, Murphy *et al.*

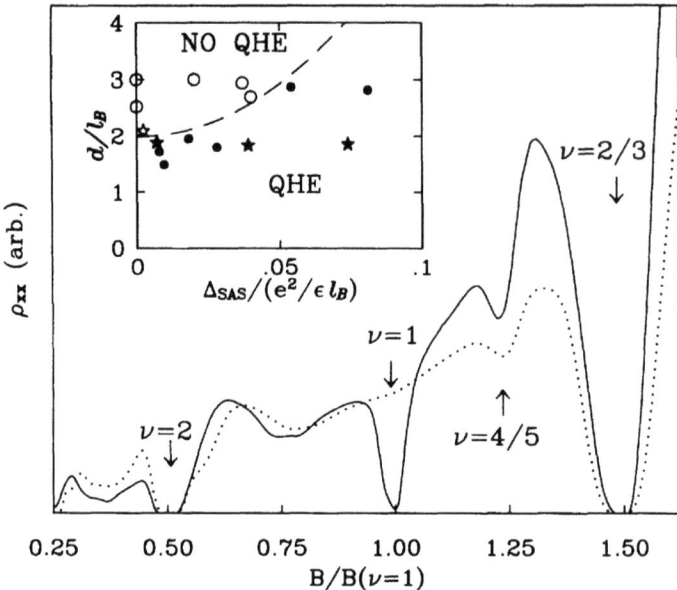

Fig. 13.4. Longitudinal resistivity ρ_{xx} for two bilayer samples with different values of d/ℓ. The dashed line in the inset shows the estimated boundary for the $\nu = 1$ QHE state in the parameter space defined by the interlayer separation d and the symmetric–antisymmetric gap Δ_{SAS}. The magnetic field has been scaled by its value at $\nu = 1$ to facilitate comparison. Source: S. Q. Murphy, J. P. Eisenstein, G. S. Boebinger, L. N. Pfeiffer, and K. W. West, *Phys. Rev. Lett.* **72**, 728-731 (1994). (Reprinted with permission.)

conclude that the (111) state occurs for $d/\ell < 2.0$ even in the limit of vanishing interlayer tunneling. (The x-axis on the inset of the figure shows Δ_{SAS}, the gap between the symmetric and antisymmetric subbands of the bilayer system, which is proportional to the interlayer tunneling. Δ_{SAS} is calculated self-consistently in a Hartree approximation.)

Spielman *et al.* [609] investigate interlayer transport in a bilayer system at $\nu = 1$ with negligible interlayer tunneling, which requires contacting each layer separately. At large values of d/ℓ, they observe the Coulomb gap discussed in Section 8.9, which strongly suppresses the tunnel conductance at small biases. However, the behavior changes qualitatively when d/ℓ falls below ~1.8. As seen in Fig. 13.5, a sharp peak appears at zero bias, in addition to the Coulomb blockade peaks at nonzero voltage biases. The parameter regime for the new peak is roughly consistent with that for the interlayer coherent (111) state at $\nu = 1$, suggesting a correlation between the two. The enhancement of interlayer tunneling in the (111) state can be understood qualitatively by noting that the presence of a hole directly across each electron in this state facilitates tunneling. Spielman *et al.* argue that the sharpness of the peak indicates a collective mechanism for tunneling, possibly related to the Goldstone mode (pseudospin wave) of the broken symmetry state. It is noted that the

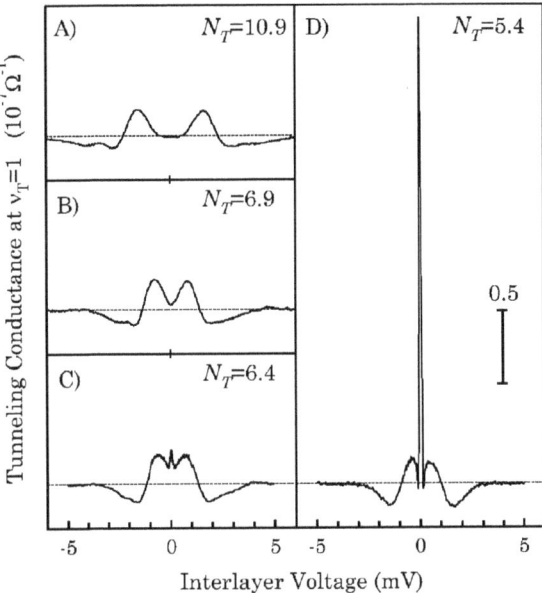

Fig. 13.5. Tunneling conductance dI/dV as a function of the interlayer voltage at total filling of $\nu = 1$ for several values of d/ℓ. The value of d/ℓ is changed by varying ℓ at $\nu = 1$, by using systems with different densities. (The density is given by $N_T \times 10^{10}$ cm^{-2}; N_T for each trace is shown on the figure.) Source: I. B. Spielman, J. P. Eisenstein, L. N. Pfeiffer, and K. W. West, *Phys. Rev. Lett.* **84**, 5808-5811 (2000). (Reprinted with permission.)

interlayer tunneling must be small, to ensure the (111) state, but nonzero, to allow electron transport between the two layers. Interpretation of the experiment as a Josephson effect is complicated by the fact that the tunneling does not occur between *two* weakly coupled (111) states, but rather between two parts of a single (111) state.

An experimental observation of the superfluid phase must deal with the charge neutral nature of the condensing objects (excitons), which do not carry a net current. Using an ingenious experimental geometry, Kellogg *et al.* [330] and Tutuc *et al.* [652] study transport of the neutral excitons in a "counterflow" configuration, in which currents in the two layers are equal in magnitude but flow in opposite directions. The experiment by Kellogg *et al.* is performed on GaAs electron bilayers separated by $d/\ell = 1.58$, and the one by Tutuc *et al.* on GaAs hole bilayers with $d/\ell = 1.33$. As seen in Fig. 13.6, the parallel and counterflow configurations show qualitatively different behaviors at $\nu = 1$; the Hall resistance is quantized for the former but has a minimum for the latter. From the temperature dependence, both the Hall and the longitudinal resistances show approximately activated behavior, presumably vanishing in the $T \to 0$ limit. The counterflow conductivity $\sigma_{xx}^{\text{counterflow}}$, which is obtained from the longitudinal and the Hall resistances by matrix inversion, increases rapidly with decreasing temperature. These results are taken to support the formation of excitonic superfluidity at $T = 0$.

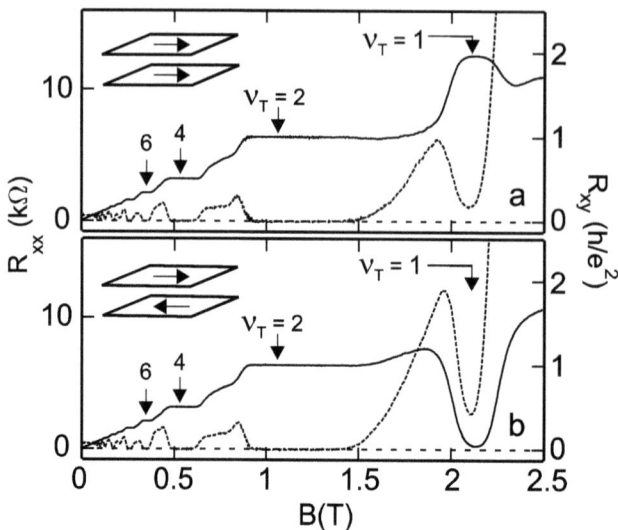

Fig. 13.6. Hall and longitudinal resistances in a bilayer system at $T = 50$ mK. The traces in the upper panel are for currents flowing in the same direction in both layers; the lower panel depicts measurements in the counterflow configuration (with the two layers contacted individually). Solid curves show the Hall resistance and dotted ones the longitudinal resistance. The center-to-center separation between the layers is $d/\ell = 1.58$. Source: M. Kellogg, J. P. Eisenstein, L. N. Pfeiffer, and K. W. West, *Phys. Rev. Lett.* **93**, 036801 (2004). (Reprinted with permission.)

While these experiments strongly support superfluid-like behavior, they also demonstrate the absence of *perfect* excitonic superfluidity at nonzero temperatures in these bilayer systems. In spite of a sharp resonance, the height of the interlayer tunneling peak remains finite in the limit of vanishing temperature. Theory predicts a finite temperature Kosterlitz–Thouless transition [453, 674, 715], below which the *Ohmic* resistance in the counterflow channel should vanish, resulting in a non-linear I–V; no evidence of non-Ohmic behavior has been seen so far. Several theories speculate that the experimentally observed behavior is consistent with a disordered excitonic superfluid [19, 157, 158, 585, 614, 665].

13.4 Composite fermion drag

A current flowing in one layer induces a current in the other layer as a result of interlayer momentum relaxation by Coulomb interaction (when the interlayer separation is not too large) or phonon exchange (for large separations). If no current is being drawn through the other layer, a voltage is eventually established that produces a current to cancel the induced current. The induced voltage in the "passive" layer divided by the current in the "active" layer is called the drag resistance, or the trans-resistance. For simple Fermi liquids, drag vanishes at $T = 0$, because of the lack of phase space allowing inelastic scatterings.

Using the CFCS framework, Sakhi [564] and Ussishkin and Stern [653] predict that for the CF Fermi sea ($\nu = 1/2$ in each layer) the Coulomb drag resistance is proportional to $T^{4/3}$, as opposed to T^2 for ordinary Fermi liquids. The drag of composite fermions has been investigated by Lilly *et al.* [397] in the weakly coupled region outside the (111) part of the phase diagram in Fig. 13.4. They do not see clear evidence of $T^{4/3}$ dependence, and more surprisingly, discover that the drag extrapolates to a nonzero value in the $T \to 0$ limit. That might suggest that the actual bilayer state in this region might not be weakly coupled. The observation might be related to a CFCS calculation by Bonesteel, McDonald, and Nayak [34] that suggests that the CS gauge field mediated interaction between composite fermions is attractive and induces an interlayer pairing at any layer separation. Such CF pairing in bilayer systems has been considered by other groups (Foster, Bonsteel, and Simon [171]; Lozovik and Ovchinnikov [407]; Morinari [459]; Ussishkin and Stern [654]; Zhou and Kim [747]). The situation is further complicated by the experimental observation (Spielman *et al.* [610]) of the sensitivity of the phase boundary between the (111) and the compressible states to the nuclear spin polarization, indicating that the weakly coupled compressible state may not be fully spin polarized for the experimental parameters (see Section 13.5).

Jörger *et al.* [308] investigate composite fermion drag in several bilayers with different layer separations. Their results are in good agreement with the calculations of Ussishkin and Stern [653]. The theoretical expressions are not very sensitive to the CF mass, and produce equally satisfactory fits for m^*/m_e in a range of $0.084\sqrt{B[\text{T}]}$ to $0.25\sqrt{B[\text{T}]}$.

Zelakiewicz *et al.* [736] perform *phonon* drag measurements on composite fermions, which probe the response of the CF Fermi sea under large wave vector scattering. For this purpose, they consider two layers at a large separation (500 nm), where interlayer Coulomb scattering is negligible and phonons provide the dominant mechanism for interlayer drag. This is supported by the observation that the temperature dependence of the drag resistance scales with a much higher exponent ($\sim T^{3.7}$) than that for the Coulomb drag. They find that the drag in this limit is qualitatively similar to that observed for electrons at zero magnetic field, but deviates from the behavior expected from a completely noninteracting model of composite fermions. For example, the peak in the plot of ρ_D/T^2 vs. T (ρ_D is the drag resistance) is shifted relative to the position predicted from the noninteracting model, and also does not have the expected dependence on the density.

13.5 Spinful composite fermions in bilayers

Most theoretical studies of composite fermions in bilayers have neglected, for simplicity, the spin degree of freedom. The spin is surely frozen at sufficiently high Zeeman energies, but whether that is true for typical experimental parameters is not obvious. This neglect of spin has been questioned by the experiment of Spielman *et al.* [610], mentioned above.

Kumuda *et al.* [354] study the bilayer 2/3 state and find that the noninteracting CF model provides a qualitative account of the phase diagram when both the real and the pseudospins are active. Their system contains two layers separated by 23 nm, with density in each layer controllable individually. At $\nu = 2/3$, which maps into $\nu^* = 2$ of composite fermions, the eight relevant energy levels are: $(0/1, b/a, \uparrow/\downarrow)$, where 0 and 1 denote the Λ level index, b and a label the pseudospin index ("bonding" and "antibonding" states of the bilayer), and ↑ and ↓ are the spin quantum numbers. The relevant energies are the Zeeman splitting (proportional to the total magnetic field), the CF cyclotron energy ($\sim\sqrt{B_\perp}$), and the pseudo-Zeeman splitting (equal to the bonding–antibonding gap Δ_{BAB}). The last one is the same as the symmetric–antisymmetric gap Δ_{SAS} for equal densities (balanced case), but can be varied by creating a density difference between the layers, or by application of a magnetic field parallel to the layers. By changing the total density, the density difference, and the parallel component of the magnetic field, Kumuda et al. [354] identify several phases of composite fermions, including SP-PU, SP-PP, and SU-PP (SP, spin-polarized; PU, pseudospin-unpolarized; PP, pseudospin-polarized; SU, spin-unpolarized). Their activation energy measurements are shown in Fig. 13.7. While the noninteracting model of composite fermions provides a qualitative account of the observations,

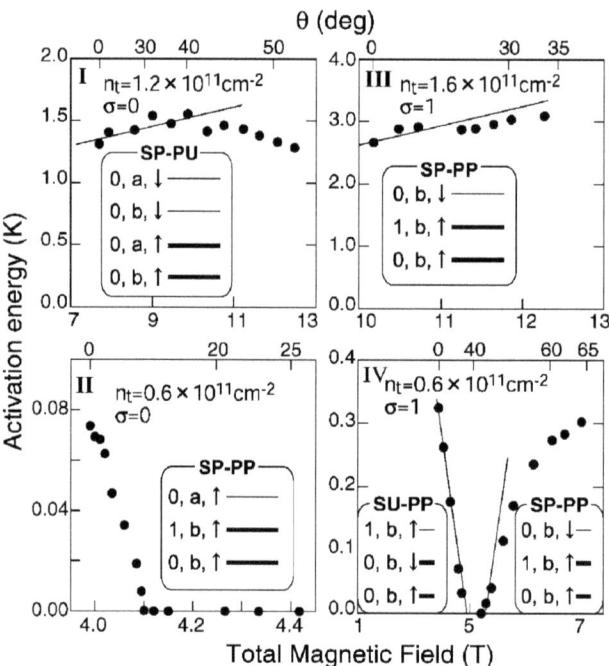

Fig. 13.7. Activation gap as a function of B_{total} (varied by tilting the sample), in several regions of the phase diagram, for a bilayer system at $\nu = 2/3$. The parameter $\sigma = (n_1 - n_2)/n_t$ is a measure of the density imbalance, with n_1 and n_2 being the densities in the two layers and $n_t = n_1 + n_2$ the total density. The straight lines have the dependence $\pm g\mu_B B_{tot}$. The various phases are identified by the dependence of the gap on B_{tot}; their Λ level diagrams are displayed in the insets, with the occupied Λ levels shown as thick lines. The dependence of the gaps in panels I, III and IV is consistent with the identifications shown in the insets; the collapse of the gap in panel II indicates that it is associated with Δ_{BAB}, which decreases rapidly under the application of a parallel magnetic field. Source: N. Kumada, D. Terasawa, Y. Shimoda, H. Azuhata, A. Sawada, Z. F. Ezawa, K. Muraki, T. Saku, and Y. Hirayama, Phys. Rev. Lett. **89**, 116802 (2002). (Reprinted with permission.)

Kumuda *et al.* take a renormalization of the Δ_{BAB} as a signature of the residual interactions between composite fermions.

Exercises

13.1 List the possible bilayer states at $\nu = 3/7$. What is the state for $d/\ell = 0$? What can you say about the relative strengths of various bilayer incompressible states shown in Eqs. (13.18)–(13.24), given our knowledge of the order of stability of the single layer FQHE states?

13.2 In the spherical geometry, a bilayer is modeled as two coincident spherical surfaces, but with the interaction between two electrons depending on whether they are on the same or different spheres. The bilayer wave function for the state

$$\left(\frac{2pn+1}{n} \frac{2pn+1}{n} m\right) \tag{E13.1}$$

is given by

$$\Psi = \prod_{j,r}(u_jv_r - v_ju_r)^m [\Psi_{\frac{n}{2pn+1}}]_1[\Psi_{\frac{n}{2pn+1}}]_2 . \tag{E13.2}$$

(The subscripts 1 and j refer to one layer and 2 and r to the other.) Using that $\prod_{j,r}(u_jv_r - v_ju_r)$ corresponds to monopole strength $Q_J = N/2$, derive the monopole strength Q for Ψ and show that the filling factor is given by

$$\nu = \lim_{N\to\infty} \frac{N}{2Q} = \frac{2n}{(2p+m)n+1} . \tag{E13.3}$$

Further, add one electron to each layer and, following Section 9.3.1, show that the local charge of a single CF-quasiparticle is

$$e^* = \frac{e}{(2p+m)n+1} . \tag{E13.4}$$

(Source: Scarola and Jain [569])

13.3 Derive the charge of a quasihole of the $(m'm''m)$ state using the plasma analogy of Section 12.4. (Source: Girvin and MacDonald [194])

13.4 Consider a system of charged bosons in a single layer, confined to the lowest Landau level. (The problem is not as hypothetical as one might initially think: neutral bosons in rapidly rotating optical traps behave similarly to charged bosons in a magnetic field [96,427,541,656].) A natural ansatz for the wave functions of incompressible states is

$$\Psi = \Phi_n\Phi_1^{2p-1} , \tag{E13.5}$$

which describe the physics that bosons capture an odd number $(2p - 1)$ of vortices each to transmute into composite fermions.

(i) Derive the filling factors of these wave functions.

(ii) Follow Section 8.1 to prove an absence of ODLRO in these states. (The lack of ODLRO in the FQHE is thus not a consequence of the particle statistics, but of the LLL physics.)

(iii) For a delta function repulsive interaction, show that Φ_1^2, the Laughlin state for bosons, is the exact ground state.

14

Edge physics

Let us consider fully spin-polarized electrons in one dimension. Because of the Pauli principle, each electron is confined inside a box marked by its two neighbors on either side. As an electron moves, it collides with the neighboring electrons and, through a two-way domino effect, the dynamics becomes collective. (This is to be contrasted with higher dimensions where electrons can budge sideways to make way for a moving electron.) Electron-like quasiparticles are no longer well defined in one dimension, resulting in a breakdown of Landau's Fermi liquid theory. An understanding of the effect of interactions requires a nonperturbative treatment. Interacting liquids in one dimension are called Tomonaga–Luttinger (TL) liquids [408, 643]. We see in this chapter that the FQHE edge constitutes a realization of a Tomonaga–Luttinger liquid. (Certain organic or blue bronze type conductors have stacks of weakly coupled 1D chains, which behave as independent 1D conductors at high temperatures, when the coupling is irrelevant, but become three dimensional at low temperatures.)

14.1 QHE edge = 1D system

We have learned that excitations cost a nonzero energy in a pure QHE system. That is indeed true for the spherical samples of computer experiments, but not quite true for a sample in the laboratory. Excitations with arbitrarily low energies *are* available at the boundary of a QHE system. The dynamics of these excitations is equivalent to that of a one-dimensional system. The analogy to a 1D system is not immediately obvious in real space (because electrons in the QHE system move on a 2D surface), but can be established rigorously in the Fourier space, as shown in Fig. 14.1. Let us consider a QHE system that is confined by a slowly varying potential $V(y)$ in the y direction but is translationally invariant along the x direction. The energy of the LLL single particle state labeled by the wave vector k_x is then given by $(1/2)\hbar\omega_c + V(k_x\ell^2)$, because its spatial location is related to k_x by $y \approx k_x\ell^2$ (Fig. 3.1) with an appropriate choice of the origin. Taking the Fermi energy as shown in Figure 14.1 and linearizing the dispersion at the Fermi energy produces the so-called Luttinger model shown by dashed lines, with the energy rising linearly with $|k_x|$.[1] This ought to be a satisfactory

---

[1] In the Tomonaga model, the dashed lines stop at the point where they meet at $y = 0$. Extending them all the way down gives the Luttinger model. The latter has twice as many levels at each k_x as the original model, but the "error" does not matter as it occurs deep below the Fermi energy.

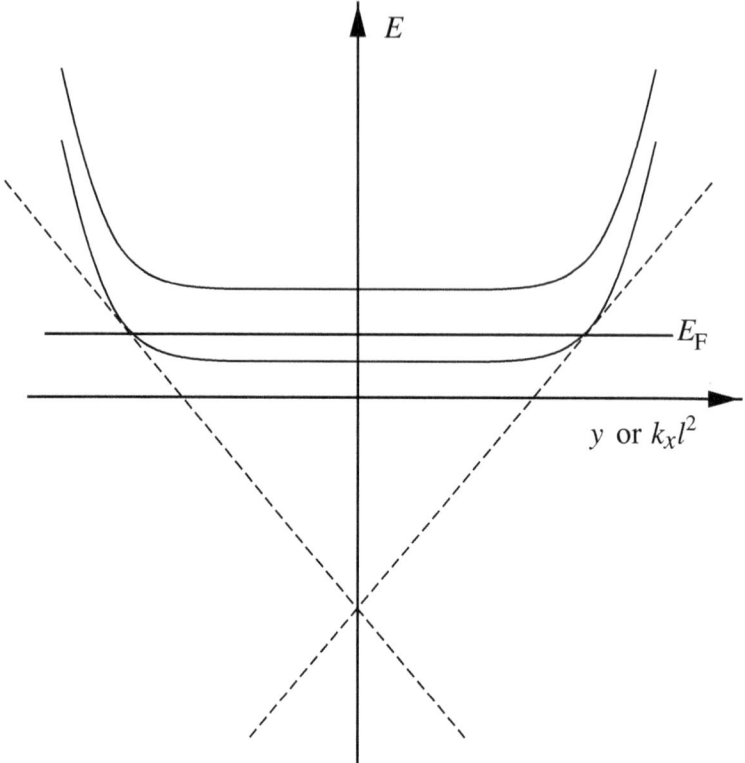

Fig. 14.1. Mapping FQHE edge states into the Luttinger model.

model at sufficiently low temperatures and energies, when the relevant excitations at the Fermi energy remain within the linear regime. Electrons move in different directions on the two dashed lines, often labeled as R or L (right or left moving electrons).

Two simplifications occur in the QHE problem. First, we can take electrons to be fully spin polarized. Second, only one dashed line needs to be considered for situations that do not involve interaction or tunneling between the opposite edges, as appropriate when the edges are spatially far separated. (In ordinary 1D liquids in zero magnetic field, the L and R electrons inhabit the same location in space, so the interaction between them may not be neglected.) This is called a "chiral" TL liquid of effectively spinless electrons (where the qualifier chiral simply indicates that all electrons move in the same direction).

14.2 Green's function at the IQHE edge

The time ordered Green's function is defined as [196, 416] (Appendix G)

$$G(\mathbf{r}, t; \mathbf{r}', 0) = -i\langle 0|T\hat{\psi}(\mathbf{r}, t)\hat{\psi}^\dagger(\mathbf{r}', 0)|0\rangle, \tag{14.1}$$

where T is the time ordering operator, $\hat{\psi}$ and $\hat{\psi}^\dagger$ at $t = 0$ are given by

$$\hat{\psi}(r) = \sum_{n,m} \eta_{nm}(r) c_{nm},$$ (14.2)

$$\hat{\psi}^\dagger(r) = \sum_{n,m} \eta^*_{nm}(r) c^\dagger_{nm},$$ (14.3)

the sum is over a complete set of single particle orbitals, and the ground state $|0\rangle$ is assumed to be normalized to unity. We mostly work with the equal-time Green's function, from which the full Green's function can be obtained straightforwardly (as shown later). Consider the $t \to 0^-$ limit of the time ordered Green's function:

$$\lim_{t \to 0^-} G(r, t; r', 0) = i\langle 0| \hat{\psi}^\dagger(r') \hat{\psi}(r)|0\rangle \equiv G^<(r, r') = i\rho_1(r', r).$$ (14.4)

(We use the same symbol for both the equal-time and unequal-time Green's functions; they are differentiated by the number of parameters in the argument.) The quantity $\rho_1(r', r)$ is the off-diagonal matrix element of the one-particle density matrix, defined as

$$\rho_1(r', r) = \frac{N \int d^2r_2 \ldots d^2r_N \, \Psi^*(r', r_2, \ldots, r_N) \Psi(r, r_2, \ldots, r_N)}{\int d^2r_1 \ldots d^2r_N \, \Psi^*(r_1, r_2, \ldots, r_N) \Psi(r_1, r_2, \ldots, r_N)}.$$ (14.5)

Section 8.1 showed that for a uniform density, isotropic FQHE state, when r' and r lie deep in the bulk, we have

$$\rho_1(r', r) = \frac{\nu}{2\pi} \exp\left[-\frac{|z - z'|^2}{4} + \frac{z\bar{z}' - z'\bar{z}}{4} \right].$$ (14.6)

This implies that the Green's function has a Gaussian decay in the bulk of a QHE system, as expected for an incompressible system.

Let us now ask what happens if r and r' are located at the edge of a QHE system, taking, for illustration, noninteracting electrons at $\nu = 1$, for which the ground state has all orbitals with $m = 0, 1, \ldots, N - 1$ occupied and the orbitals with $m \geq N$ empty. We choose the values

$$z = Re^{-i\theta}, \qquad z' = Re^{-i\theta'},$$ (14.7)

where

$$R = \sqrt{2N}$$ (14.8)

is the radius of the first unoccupied orbital. We then have

$$\rho_1^{\text{edge}}(r',r) = \langle 0|\hat{\psi}^\dagger(r')\hat{\psi}(r)|0\rangle$$

$$= \sum_{m,m'=0}^{\infty} \eta_{m'}^*(r')\eta_m(r)\langle 0|c_{m'}^\dagger c_m|0\rangle$$

$$= \sum_{m=0}^{N-1} \frac{\bar{z}'^m z^m}{2\pi 2^m m!} e^{-\frac{1}{2}R^2}$$

$$= \frac{e^{iN(\theta'-\theta)}}{2\pi} \sum_{m=1}^{N} \frac{1}{(N-m)!} \left(\frac{R^2}{2}\right)^{N-m} e^{-\frac{1}{2}R^2} e^{im(\theta-\theta')}. \tag{14.9}$$

For $\theta' \neq \theta$, the dominant contribution to the sum comes from small m; the rapid fluctuations in the phase $m(\theta' - \theta) \bmod 2\pi$ cause a destructive interference for $m \gg 1$. With the help of the Stirling formula for large n,

$$\ln n! = \left(n + \frac{1}{2}\right) \ln n - n + \ln\sqrt{2\pi} + O(n^{-1}), \tag{14.10}$$

and the relation $N = R^2/2$, we obtain

$$(N-m)! = \sqrt{\pi}R \left(\frac{R^2}{2}\right)^{N-m} e^{-\frac{1}{2}R^2}[1 + O(N^{-1})]. \tag{14.11}$$

Dropping the $O(N^{-1})$ terms in the square brackets, substituting in Eq. (14.9), and extending the upper limit in the sum in Eq. (14.9) to ∞ (which causes vanishing correction for large N), we get

$$\rho_1^{\text{edge}}(r',r) \sim (e^{i(\theta'-\theta)} - 1)^{-1}. \tag{14.12}$$

(For $\theta = \theta'$, the above equation does not produce the correct answer, namely the density, but that is not a problem as this equation was derived under the assumption $\theta \neq \theta'$.) To see how it depends on distance, we note for $r \neq r'$

$$|\rho_1^{\text{edge}}(r',r)| \sim \frac{1}{|\sin([\theta'-\theta]/2)|}. \tag{14.13}$$

As expected for our circular droplet, the result is invariant under the replacement $\theta \to 2\pi - \theta$. The arc distance between r and r' is $|x - x'| = R|\theta' - \theta|$, and the relevant limit with no curvature effects is $|x - x'|/R = |\theta' - \theta| \ll 1$. Here, we have

$$|\rho_1^{\text{edge}}(x',x)| \sim \frac{1}{|x - x'|}, \tag{14.14}$$

where x denotes the one-dimensional coordinate along the edge. The point of the calculation is that, in contrast to the behavior in the bulk, the Green's function decays only as a power law with distance along the edge. Such a power law behavior is characteristic of a gapless system, for example a Fermi liquid. In fact, the inverse distance behavior is characteristic of a noninteracting Fermi liquid in one dimension.

So far we have considered a noninteracting system, which is a Landau Fermi liquid and for which the Green's function can be obtained exactly. The system turns into a TL liquid for interacting electrons. The long-distance behavior retains its power law character when interactions are turned on, but the exponent becomes different from unity.

A convenient way to model TL liquids is through the trick of bosonization, introduced in the next section. Application of this technique to the FQHE edge suggests, at first, a uniqueness, which leads to sharp, experimentally testable predictions.

14.3 Bosonization in one dimension

In one dimension, the problem of fermions can be mapped into a problem of bosons (or vice versa), with every operator in the fermionic problem possessing an image in the bosonic problem. This should not be taken to mean that particle statistics is unimportant in one dimension. The exchange symmetry of the wave function still makes all the difference in the physical properties. The difference arises because a given operator has completely different physical meanings in the two problems.

While the mapping from fermions into bosons is exact, its real usefulness stems from the fact that fermions with a certain class of interactions map into noninteracting bosons, enabling an exact solution of the problem. That already brings out a qualitative difference from higher dimensions, where the interacting electron system (often) maps into a weakly interacting system of *fermionic* quasiparticles. In one dimension, the true particles (objects that are weakly interacting) are certain bosonic objects (which are seen below to be particle–hole density waves). The qualitative distinction between the true particles and electrons is a sign of nonperturbative physics. This section illustrates certain basic features of bosonization, following closely the excellent review by von Delft and Schoeller [658]. We also borrow from Stone [615, 616].

We consider a chiral system of spinless fermions, that is, with all fermions moving in the same direction. This might appear to be an artificial model, but is precisely what is needed for the QHE edge problem. Also, bosonization for the more general problem, which may have many species of fermions moving in both directions, can be accomplished by bosonizing each component separately. The mapping into a system of bosons rests on the fact that this system has the same Fock space as bosons in one dimension. For a proof of this statement, consideration of noninteracting particles is sufficient, because interaction does not alter the Fock space.

For a system of length L in one dimension ($-L/2 < x < L/2$), periodic boundary conditions restrict the wave vector to discrete values:

$$k = \frac{2\pi}{L} m_k \equiv \bar{k} m_k,$$ (14.15)

where m_k is an integer and

$$\bar{k} = \frac{2\pi}{L} \tag{14.16}$$

is the unit wave vector. We assume a linear dispersion

$$\epsilon_k = v_F k, \tag{14.17}$$

where v_F is the Fermi velocity, which is set to unity ($v_F = 1$) below. Further, the Fermi wave vector is taken at $k_F = 0$, so for the ground state of noninteracting fermions, all states with $k \leq 0$ are occupied and all states with $k > 0$ are unoccupied:

$$n_k = 1, \quad k \leq 0, \tag{14.18}$$

$$n_k = 0, \quad k > 0, \tag{14.19}$$

where n_k is the occupation factor of the orbital labeled by k. This defines the "vacuum" state, denoted by $|V\rangle$.

The wave vector k extends all the way to $-\infty$. To avoid infinities, we work with normal ordered quantities. Normal ordering, denoted by $: :$, is defined as

$$: \hat{O} := \hat{O} - \langle V|\hat{O}|V\rangle. \tag{14.20}$$

In other words, various observables are measured relative to their vacuum expectation value. We define, for example,

$$N = \sum_{k=-\infty}^{\infty} : c_k^\dagger c_k := \sum_{k=-\infty}^{\infty} (c_k^\dagger c_k - \langle V|c_k^\dagger c_k|V\rangle) \tag{14.21}$$

and

$$H = \sum_k k : c_k^\dagger c_k := \sum_k k(c_k^\dagger c_k - \langle V|c_k^\dagger c_k|V\rangle). \tag{14.22}$$

We sometimes encounter differences of sums that are individually divergent; in such situations, care must be taken to properly normal order each term. Relabeling of the wave vector (through a change of variables) in a sum can be done only for normal ordered quantities. Consider, for example, the difference between the number of particles in two states: the vacuum state $|V\rangle$ and the state $|V'\rangle$ in which all $k \leq \bar{k}$ are occupied. The latter, obtained by adding a fermion to the $k = \bar{k}$ state, has precisely one more particle than the vacuum state. Let us see how a mindless change of variables in non-normal ordered sums can lead to a wrong answer:

$$\Delta N = \sum_{k=-\infty}^{\infty} \langle V'|c_k^\dagger c_k|V'\rangle - \langle V|c_k^\dagger c_k|V\rangle$$

$$= \sum_{n=-\infty}^{1} n - \sum_{m=-\infty}^{0} m$$

$$= \sum_{n'=-\infty}^{0} n' - \sum_{m=-\infty}^{0} m$$

$$= 0.$$

The correct result is obtained by normal ordering all quantities at the first step.

A complete basis in the Fock space is constructed by specifying the occupied wave vectors. For example, the vacuum state is denoted by $\{0, -\bar{k}, -2\bar{k}, \ldots\}$. We define a basis state (with the same number of particles as in the vacuum):

$$\{q_j\} \equiv \{q_0, -\bar{k} + q_1, -2\bar{k} + q_2, \ldots, -n\bar{k} + q_n, \ldots\}. \tag{14.23}$$

The wave vectors are taken to be ordered:

$$-n\bar{k} + q_n > -(n+1)\bar{k} + q_{n+1}, \tag{14.24}$$

which implies

$$q_0 \geq q_1 \geq q_2, \ldots, \tag{14.25}$$

$$q_j \geq 0 \qquad \text{for all } j. \tag{14.26}$$

The basis defined by ordered sets of q_j's is complete; all antisymmetric wave functions can be written as linear combinations of these states. The energy of $\{q_j\}$ (measured relative to the vacuum state) is given by

$$E\{q_j\} = \sum_j q_j = \sum_{q=0}^{\infty} n_q q, \tag{14.27}$$

where n_q is the number of times the value q appears in the infinite set $\{q_j\}$. For a finite energy excitation, only a finite number of n_q's are nonzero.

We next define a one-dimensional, chiral, bosonic system, with linear energy dispersion,

$$\epsilon_q = q, \tag{14.28}$$

where we restrict $q > 0$. The number of bosons is not fixed. The bosonic basis states in the Fock space are taken as $\{n_q\}$, where the integer n_q is the occupation number of the state at energy q. The total energy is given by

$$E = \sum_{q=0}^{\infty} n_q q. \tag{14.29}$$

A one-to-one correspondence between the bosonic and the fermionic systems

$$\{q_j\} \Leftrightarrow \{n_q\} \tag{14.30}$$

follows because a basis state of the fermionic system is uniquely labeled by $\{q_j\}$, which has a unique $\{n_q\}$ associated with it; conversely, a bosonic state is completely specified by $\{n_q\}$, which has a unique ordered set $\{q_j\}$ associated with it.

The equivalence can, alternatively, be established by noting that the partition functions for the two problems are identical. We use the result that the number of distinct partitions of an integer M, denoted $P(M)$, can be generated as

$$\frac{1}{\prod_{n>0}(1 - x^n)} = \sum_m P(m)x^m. \tag{14.31}$$

The number of times a given power of x, say x^m, appears in the product (left hand side)

$$(1 + x + x^2 + \cdots)(1 + x^2 + x^4 + \cdots)(1 + x^3 + x^6 + \cdots)\ldots \tag{14.32}$$

is the partition of m. The partition function for electrons at energy

$$E = M\bar{k} \tag{14.33}$$

is given by

$$\mathcal{Z}_F = \sum_E P(E)e^{-\beta E}$$

$$= \sum_M P(M)e^{-\beta \bar{k}M}$$

$$= \prod_{n>0}(1 - e^{-\beta \bar{k}n})^{-1}$$

$$= \mathcal{Z}_B. \tag{14.34}$$

A correspondence between the Fock spaces implies, in principle, a mapping between the operators. Labeling by $|i\rangle$ the basis states in the fermion space, and by $|i)$ the corresponding basis states in the boson space, an operator in the boson problem corresponding to any given operator O_F in the fermion space is constructed as

$$O_B \equiv |i\rangle\langle i|O_F|j\rangle\langle j|, \tag{14.35}$$

which satisfies

$$(i|O_B|j) = \langle i|O_F|j\rangle. \tag{14.36}$$

The rest of this section constructs explicit representations for certain useful operators. More details and references to the literature can be found in von Delft and Schoeller [658].

14.3.1 Bosons in terms of fermions

We first construct the boson creation and annihilation operators in terms of the fermion creation and annihilation operators. For this purpose, we define

$$\rho_q^- = \sum_{k=-\infty}^{\infty} c_{k-q}^\dagger c_k, \qquad q > 0 \tag{14.37}$$

and

$$\rho_q^+ = \sum_{k=-\infty}^{\infty} c_{k+q}^\dagger c_k = (\rho_q^-)^\dagger, \qquad q > 0. \tag{14.38}$$

Both ρ_q^- and ρ_q^+ are automatically normal ordered because, for $q > 0$, the vacuum expectation of $c_{k+q}^\dagger c_k$ or $c_{k-q}^\dagger c_k$ vanishes. Also, they do not change the number of particles:

$$[N, \rho_q^-] = 0 = [N, \rho_q^+]. \tag{14.39}$$

The anticommutation relation $\{c_k, c_{k'}^\dagger\} = \delta_{kk'}$ implies that

$$[\rho_q^-, \rho_{q'}^-] = 0 = [\rho_q^+, \rho_{q'}^+]. \tag{14.40}$$

For the commutator $[\rho_q^-, \rho_{q'}^+]$ we have

$$[\rho_q^-, \rho_{q'}^+] = \sum_{k=-\infty}^{\infty} (c_{k+q-q'}^\dagger c_k - c_{k+q}^\dagger c_{k+q'}). \tag{14.41}$$

For $q \neq q'$ the right hand side is normal ordered, and a change of variables $k \rightarrow k + q'$ produces zero. For $q = q'$ we have

$$[\rho_q^-, \rho_q^+] = \sum_{k=-\infty}^{\infty} (c_k^\dagger c_k - c_{k+q}^\dagger c_{k+q})$$

$$= \sum_{k=-\infty}^{\infty} (: c_k^\dagger c_k : + \langle V | c_k^\dagger c_k | V \rangle - : c_{k+q}^\dagger c_{k+q} : - \langle V | c_{k+q}^\dagger c_{k+q} | V \rangle)$$

$$= \sum_{k=-\infty}^{\infty} (\langle V | c_k^\dagger c_k | V \rangle - \langle V | c_{k+q}^\dagger c_{k+q} | V \rangle) \tag{14.42}$$

$$= \frac{Lq}{2\pi}$$

$$\equiv m_q. \tag{14.43}$$

The boson creation and annihilation operators are defined as

$$b_q^\dagger = \frac{i}{\sqrt{m_q}} \sum_{k=-\infty}^{\infty} c_{k+q}^\dagger c_k, \qquad q > 0 \tag{14.44}$$

$$b_q = -\frac{i}{\sqrt{m_q}} \sum_{k=-\infty}^{\infty} c_{k-q}^\dagger c_k, \qquad q > 0. \tag{14.45}$$

They satisfy

$$[b_q, b_{q'}] = 0, \tag{14.46}$$

$$[b_q^\dagger, b_{q'}^\dagger] = 0, \tag{14.47}$$

$$[b_q, b_{q'}^\dagger] = \delta_{q,q'}. \tag{14.48}$$

Also, b_q or $b_q^\dagger$ does not change the number of particles:

$$[N, b_q] = 0 \tag{14.49}$$

That completes the construction of boson creation and annihilation operators in terms of fermionic operators.

14.3.2 *Fermion operators in terms of bosons*

To translate from fermions into bosons, we need to determine the image of electron operators in the bosonic space. That proceeds, most conveniently, through the definition of the field operators:

$$\hat{\phi}^+(x) = -\sum_{q>0} \frac{1}{\sqrt{m_q}} e^{iqx} \, b_q^\dagger \, e^{-aq/2}, \tag{14.50}$$

$$\hat{\phi}^-(x) = -\sum_{q>0} \frac{1}{\sqrt{m_q}} e^{-iqx} \, b_q \, e^{-aq/2}, \tag{14.51}$$

$$\hat{\phi}(x) = \hat{\phi}^+(x) + \hat{\phi}^-(x) = -\sum_{q>0} \frac{1}{\sqrt{m_q}} (e^{iqx} \, b_q^\dagger + e^{-iqx} \, b_q) e^{-aq/2}. \tag{14.52}$$

The fields $\hat{\phi}^+(x)$ and $\hat{\phi}^-(x)$ are Hermitian conjugates, so $\hat{\phi}(x)$ is Hermitian. The cutoff factor $e^{-aq/2}$ has been introduced to regularize the integrals, but a is set to $a = 0^+$ in the final results. The fields satisfy the following commutators:

$$[\hat{\phi}^+(x), \hat{\phi}^+(x')] = 0 = [\hat{\phi}^-(x), \hat{\phi}^-(x')], \tag{14.53}$$

$$[\hat{\phi}^-(x), \hat{\phi}^+(x')] = \sum_{q>0} \frac{1}{m_q} e^{-q[i(x-x')+a]}$$

$$= -\ln[1 - \exp(-i(2\pi/L)(x - x' - ia))]$$
$$= -\ln[i(2\pi/L)(x - x' - ia)], \tag{14.54}$$

where the limit $|x - x'|/L \to 0$ has been taken in the last step. Using this result, we have

$$[\hat{\phi}(x), \hat{\phi}(x')] = -\ln\left(\frac{y - ia}{-y - ia}\right), \tag{14.55}$$

where $y = x - x'$. For $y > 0$, we have

$$y - ia = |y|e^{-i0}, \qquad -y - ia = |y|e^{i(\pi+0)}, \tag{14.56}$$

which gives

$$\ln\left(\frac{y - ia}{-y - ia}\right) = -i\pi. \tag{14.57}$$

For $y < 0$, on the other hand, we have

$$y - ia = |y|e^{i(\pi+0)}, \qquad -y - ia = |y|e^{-i0}, \tag{14.58}$$

which gives

$$\ln\left(\frac{y - ia}{-y - ia}\right) = i\pi. \tag{14.59}$$

These relations yield

$$[\hat{\phi}(x), \hat{\phi}(x')] = -i\pi\,\epsilon(x - x'), \tag{14.60}$$

$$[\hat{\phi}(x), \partial_{x'}\hat{\phi}(x')] = 2\pi\,i\delta(x - x'), \tag{14.61}$$

with the definition

$$\epsilon(x - x') \equiv \text{sign}(x - x') = \theta(x - x') - \theta(x' - x), \qquad \epsilon(0) \equiv 0. \tag{14.62}$$

The field operator for electrons is defined as

$$\hat{\psi}(x) = \sqrt{\frac{2\pi}{L}} \sum_{k=-\infty}^{\infty} e^{-ikx} c_k, \tag{14.63}$$

with inverse relation

$$c_k = \int_{-L/2}^{L/2} \frac{dx}{\sqrt{2\pi L}} e^{ikx}\hat{\psi}(x). \tag{14.64}$$

The anticommutators follow:

$$\{\hat{\psi}(x), \hat{\psi}^\dagger(x')\} = 2\pi\,\delta(x - x'), \tag{14.65}$$

$$\{\hat{\psi}(x), \hat{\psi}(x')\} = 0 = \{\hat{\psi}^\dagger(x), \hat{\psi}^\dagger(x')\}. \tag{14.66}$$

Fermion creation and annihilation operators An important formula (proven below) relates the fermion field operator to the boson field operator:

$$\hat{\psi}(x) = F\left(\frac{2\pi}{L}\right)^{1/2} e^{-i\hat{\phi}^+(x)} e^{-i\hat{\phi}^-(x)}$$

$$= F\frac{1}{\sqrt{a}} e^{-i\hat{\phi}(x)}. \tag{14.67}$$

The second step follows because

$$e^{-i\hat{\phi}(x)} = e^{-i\hat{\phi}^+(x) - i\hat{\phi}^-(x)}$$

$$= e^{-i\hat{\phi}^+(x)} e^{-i\hat{\phi}^-(x)} e^{\frac{1}{2}[\hat{\phi}^+(x), \hat{\phi}^-(x)]}$$

$$= \left(\frac{2\pi a}{L}\right)^{1/2} e^{-i\hat{\phi}^+(x)} e^{-i\hat{\phi}^-(x)}, \tag{14.68}$$

where we have used the operator identity in Eq. (B.3) and the commutation relation

$$[\hat{\phi}^+(x), \hat{\phi}^-(x)] = \ln(2\pi a/L). \tag{14.69}$$

The form in the first line of Eq. (14.67), being normal ordered, is more convenient for calculations, although the form in the second line, being simpler, appears more commonly in literature. The creation operator is

$$\hat{\psi}^\dagger(x) = F^\dagger \left(\frac{2\pi}{L}\right)^{1/2} e^{i\hat{\phi}^+(x)} e^{i\hat{\phi}^-(x)} \tag{14.70}$$

$$= F^\dagger \frac{1}{\sqrt{a}} e^{i\hat{\phi}(x)}, \tag{14.71}$$

because $[\hat{\phi}^+]^\dagger = \hat{\phi}^-$ and $\hat{\phi}^\dagger = \hat{\phi}$.

The analogy between fermions and bosons in one dimension is not complete. To simulate the effect of *adding* or *removing* an electron is not possible in the boson problem, because the boson operators b_q and $b_q^\dagger$ do not change the number of electrons. Therefore, the electron creation and destruction operators cannot be written entirely in terms of bosons. For observables in which the number of electrons is conserved, this is not an issue. In general, the problem is redressed by introducing an operator F, called the Klein operator, which removes the $k = 0$ electron from the vacuum state. It can be shown that the Klein operator commutes with the boson operators, and $F^\dagger F = 1$. As a result, the Klein operators do not show up when considering an operator that contains an equal number of fermion creation and annihilation operators. We do not discuss it in any more detail, but carry it along in our formulas as a reminder.

We now proceed to show that the form in Eq. (14.67) satisfies the correct anticommutation relations. Using Eq. (B.4), we have

$$\hat{\psi}(x)\,\hat{\psi}(x') = \frac{F^2}{a} e^{-i\hat{\phi}(x)} e^{-i\hat{\phi}(x')}$$

$$= \frac{F^2}{a} e^{-i\hat{\phi}(x')} e^{-i\hat{\phi}(x)} e^{-[\hat{\phi}(x),\hat{\phi}(x')]}$$

$$= \frac{F^2}{a} e^{-i\hat{\phi}(x')} e^{-i\hat{\phi}(x)} e^{i\pi\epsilon(x-x')}$$

$$= -\frac{F^2}{a} e^{-i\hat{\phi}(x')} e^{-i\hat{\phi}(x)}$$

$$= -\hat{\psi}(x')\,\hat{\psi}(x), \tag{14.72}$$

which gives

$$\{\hat{\psi}(x), \hat{\psi}(x')\} = 0. \tag{14.73}$$

The anticommutator

$$\{\hat{\psi}^\dagger(x), \hat{\psi}^\dagger(x')\} = 0 \tag{14.74}$$

can similarly be obtained. Next, we have

$$\hat{\psi}(x)\,\hat{\psi}^\dagger(x') = \left(\frac{2\pi}{L}\right) e^{i\hat{\phi}^+(x)} e^{i\hat{\phi}^-(x)} e^{-i\hat{\phi}^+(x')} e^{-i\hat{\phi}^-(x')}$$

$$= \left(\frac{2\pi}{L}\right) e^{i\hat{\phi}^+(x)} e^{-i\hat{\phi}^+(x')} e^{i\hat{\phi}^-(x)} e^{-i\hat{\phi}^-(x')} e^{[\hat{\phi}^-(x),\hat{\phi}^+(x')]}$$

$$= -ie^{i[\hat{\phi}^+(x)-\hat{\phi}^+(x')]} e^{i[\hat{\phi}^-(x)-\hat{\phi}^+(x')]} \frac{1}{x-x'-ia}. \tag{14.75}$$

Using the analogous result,

$$\hat{\psi}^\dagger(x')\hat{\psi}(x) = -ie^{i[\hat{\phi}^+(x)-\hat{\phi}^+(x')]} e^{i[\hat{\phi}^-(x)-\hat{\phi}^+(x')]} \frac{1}{x'-x-ia}, \tag{14.76}$$

it follows that

$$\{\hat{\psi}(x),\hat{\psi}^\dagger(x')\} = -ie^{i[\hat{\phi}^+(x)-\hat{\phi}^+(x')]} e^{i[\hat{\phi}^-(x)-\hat{\phi}^-(x')]}$$

$$\times \left[\frac{1}{x-x'-ia} + \frac{1}{x'-x-ia}\right]$$

$$= 2\pi\,\delta(x-x'), \tag{14.77}$$

where we have used, for $a \to 0$,

$$\frac{1}{y \pm ia} = P\frac{1}{y} \mp i\pi\,\delta(y). \tag{14.78}$$

Any electron operator can be expressed in the boson language with the help of Eq. (14.67). Sometimes, it is easier to deduce the answer by inspecting some known commutation relations.

Kinetic energy The kinetic energy,

$$H_0 = \sum_{k=-\infty}^{\infty} k : c_k^\dagger c_k :, \tag{14.79}$$

satisfies the commutator

$$[H_0, b_q^\dagger] = \sum_{k=-\infty}^{\infty} ik\frac{1}{m_q}[c_k^\dagger c_{k-q} - c_{k+q}^\dagger c_k] \tag{14.80}$$

$$= qb_q^\dagger. \tag{14.81}$$

It implies that the application of $b_q^\dagger$ on an eigenstate increases the energy by q, which is also obvious from the definition of $b_q^\dagger$. The only bosonic operator that satisfies the above commutator is

$$H_0 = \sum_{q>0} qb_q^\dagger b_q. \tag{14.82}$$

Equivalent expressions are

$$H_0 = \int_{-L/2}^{L/2} \frac{dx}{2\pi} : \hat{\psi}^\dagger(x) i \partial_x \hat{\psi}(x) :, \tag{14.83}$$

$$H_0 = \int_{-L/2}^{L/2} \frac{dx}{2\pi} \frac{1}{2} : [\partial_x \hat{\phi}(x)]^2 : . \tag{14.84}$$

Density The normal ordered density operator is given by

$$\rho(x) = \frac{1}{2\pi} : \hat{\psi}^\dagger(x) \, \hat{\psi}(x) : \tag{14.85}$$

$$= \frac{1}{L} \sum_{k=-\infty}^{\infty} \sum_{k'=-\infty}^{\infty} e^{i(k-k')x} : c_k^\dagger c_{k'} : \tag{14.86}$$

$$= \frac{1}{L} \sum_{k=-\infty}^{\infty} \sum_{q>0} (e^{iqx} c_{k+q}^\dagger c_k + e^{-iqx} c_{k-q}^\dagger c_k)$$

$$= \frac{1}{2\pi} \partial \hat{\phi}(x), \tag{14.87}$$

where we have used Eqs. (14.52), (14.44), (14.45), and (14.43). The term with $k = k'$ (or $q = 0$) is eliminated by normal ordering.

As a check, let us evaluate the operator

$$[\rho(x), \hat{\psi}^\dagger(x')] = \frac{F^\dagger}{2\pi \sqrt{a}} [\partial_x \hat{\phi}(x), e^{i\hat{\phi}(x')}]$$

$$= \frac{F^\dagger}{2\pi \sqrt{a}} [\partial_x \hat{\phi}(x), i\hat{\phi}(x')] e^{i\hat{\phi}(x')}$$

$$= \delta(x - x') \frac{F^\dagger}{\sqrt{a}} e^{i\hat{\phi}(x)}$$

$$= \delta(x - x') \hat{\psi}^\dagger(x), \tag{14.88}$$

where we have used Eq. (B.5). This equation simply states that $\hat{\psi}^\dagger(x')$ adds precisely one electron. To see this, integrate the above equation over x to obtain

$$[\hat{N}, \hat{\psi}^\dagger(x')] = \hat{\psi}^\dagger(x'), \tag{14.89}$$

which is equivalent to

$$\hat{N} \hat{\psi}^\dagger(x')|N\rangle = (N + 1)\hat{\psi}^\dagger(x')|N\rangle. \tag{14.90}$$

Point splitting When we attempt a derivation of the bosonic forms for the kinetic energy and the density by directly substituting Eq. (14.67) in Eqs. (14.85) and (14.83), infinities arise because these quantities contain products of operators at the same point. The theory is regularized by evaluating the product at nearby points (called point splitting) and taking the desired limit at the end. The advantage is that the infinite contribution is explicitly identified and subtracted. This procedure is equivalent to normal ordering. More precisely, we define

$$: O_1(x) \, O_2(x) : = O_1(x+l) \, O_2(x) - \langle V | O_1(x+l) \, O_2(x) | V \rangle, \tag{14.91}$$

where the length l is small, but $l \gg a$.

Let us take, for example, the density:

$$\hat{\psi}^{\dagger}(x+l) \, \hat{\psi}(x) = \frac{2\pi}{L} e^{i\hat{\phi}^+(x+l)} e^{i\hat{\phi}^-(x+l)} e^{-i\hat{\phi}^+(x)} e^{-i\hat{\phi}^-(x)}$$

$$= \frac{2\pi}{L} e^{i\hat{\phi}^+(x+l)-i\hat{\phi}^+(x)} e^{i\hat{\phi}^-(x+l)-i\hat{\phi}^-(x)} e^{[\hat{\phi}^-(x+l),\hat{\phi}^+(x)]}$$

$$= \frac{2\pi}{L} [1 + il\partial_x\hat{\phi}] \exp[-\ln(i2\pi l/L)]$$

$$= -\frac{i}{l} + \partial_x\hat{\phi}(x). \tag{14.92}$$

Then,

$$: \hat{\psi}^{\dagger}(x) \, \hat{\psi}(x) : = -\frac{i}{l} + \partial_x\hat{\phi}(x) - \langle V | - \frac{i}{l} + \partial_x\hat{\phi}(x) | V \rangle$$

$$= -\frac{i}{l} + \partial_x\hat{\phi}(x) + \frac{i}{l}$$

$$= \partial_x\hat{\phi}(x). \tag{14.93}$$

Green's function for noninteracting electrons   Noninteracting electrons map into noninteracting bosons. For the latter, a normal ordered product of exponentials has ground state expectation value equal to unity, so our job is simply to normal order the product while keeping track of the additional factors obtained during the process.

$$G^{<}(x',x) = i\langle \hat{\psi}^{\dagger}(x) \, \hat{\psi}(x') \rangle$$

$$= \frac{2\pi i}{L} \langle e^{i\hat{\phi}^+(x)} e^{i\hat{\phi}^-(x)} e^{-i\hat{\phi}^+(x')} e^{-i\hat{\phi}^-(x')} \rangle$$

$$= \frac{2\pi i}{L} e^{[\hat{\phi}^-(x),\hat{\phi}^+(x')]}$$

$$= \frac{2\pi i}{L} \exp[-\ln(i(2\pi/L)(x - x' - ia))]$$

$$= \frac{1}{x - x' - ia}. \tag{14.94}$$

A similar calculation shows

$$G^{>}(x',x) = \frac{1}{x - x' + ia}. \tag{14.95}$$

We could indeed have obtained these expressions much more simply without bosonization. For example,

$$G^<(x',x) = i\langle \hat\psi^\dagger(x)\,\hat\psi(x')\rangle$$

$$= i\frac{2\pi}{L}\sum_{k,k'} e^{ikx-ik'x'}\langle c_k^\dagger c_{k'}\rangle$$

$$= i\frac{2\pi}{L}\sum_{k=-\infty}^{0} e^{ik(x-x'-ia)}$$

$$= \frac{1}{x-x'-ia}, \qquad (14.96)$$

where a regularization term e^{ka} has been introduced to eliminate the contribution from the $k=-\infty$ limit of the integral. The advantage of bosonization methods comes when one considers interacting electrons.

Unequal-time Green's functions So far, we have only calculated equal-time ($t=0$) Green's functions. For noninteracting electrons, the unequal-time Green's functions can be obtained from the equal-time ones as

$$G^>(x,t;x',t') = G^>(x-v_Ft,x'-v_Ft), \qquad (14.97)$$

$$G^<(x,t;x',t') = G^<(x-v_Ft,x'-v_Ft). \qquad (14.98)$$

This can be seen as a consequence of the relation $\hat\phi(x,t) = e^{iHt}\hat\phi(x)e^{-iHt} = \hat\phi(x-v_Ft)$, or a similar relation for the electron creation and annihilation operators.

14.4 Wen's conjecture

This section motivates a generalization of the above discussion to the FQHE edge. For the ordinary 1D system, the mapping into the boson language is rigorous, constructively derivable at the operator level. No analogous treatment is possible for the FQHE edge. We discuss here a conjecture by Wen [673], which suggests universal behavior at the edge of an incompressible FQHE state.

We would again like to construct boson operators in terms of ρ_q^- and ρ_q^+. Let us go back to Eq. (14.42):

$$[\rho_{-q}^-,\rho_q^+] = \sum_{k=-\infty}^{\infty} (\langle V|c_k^\dagger c_k|V\rangle - \langle V|c_{k+q}^\dagger c_{k+q}|V\rangle). \qquad (14.99)$$

For the ordinary 1D liquid, the right hand side is equal to m_q, which is the number of states between $k=-q$ and $k=0$. Let us now assume that in the FQHE ground state, the occupation of each k state is not unity but ν. That is certainly true deep in the bulk (which corresponds to energies much below the Fermi energy), but we assume that to be valid all the way to the edge (i.e., the Fermi level), hoping that any error close to the Fermi level does not affect the final results. In other words, a FQHE state is modeled as a 1D electron liquid for which

$$n_k = \langle V | c_k^\dagger c_k | V \rangle = v, \qquad k \le 0, \tag{14.100}$$

$$n_k = \langle V | c_k^\dagger c_k | V \rangle = 0, \qquad k > 0. \tag{14.101}$$

Then we have

$$[\rho_q^-, \rho_{q'}^+] = v m_q \delta_{q,q'}. \tag{14.102}$$

This commutator (which describes the so-called Kac–Moody algebra) differs from that in Eq. (14.43) through the factor v on the right hand side. Eq. (14.102) can also be motivated by quantizing the hydrodynamic equations of motion for edge waves [673].

To preserve the usual commutation relation $[b_q, b_{q'}^\dagger] = \delta_{q,q'}$, we now define the boson operators as

$$b_q^\dagger = \frac{i}{\sqrt{v m_q}} \sum_{k=-\infty}^{\infty} c_{k+q}^\dagger c_k, \qquad q > 0, \tag{14.103}$$

$$b_q = -\frac{i}{\sqrt{v m_q}} \sum_{k=-\infty}^{\infty} c_{k-q}^\dagger c_k, \qquad q > 0. \tag{14.104}$$

The field operators $\hat{\phi}^+(x)$ and $\hat{\phi}^-(x)$ are defined in terms of $b_q^\dagger$ and b_q as before, and all of the commutation relations remain intact. Expressing $\hat{\phi}(x)$ in terms of the density operator, Eq. (14.87) changes to

$$: \rho(x) := \frac{\sqrt{v}}{2\pi} \partial_x \hat{\phi}(x). \tag{14.105}$$

To complete the bosonization program for the FQHE edge, we need to express the fermion creation and annihilation operators in terms of bosons. An exact representation is not known; it is not known if such a representation even exists. Wen notes, however, that the following operator has many desired properties:

$$\hat{\psi}(x) \propto F e^{-\frac{i}{\sqrt{v}} \hat{\phi}^+(x)} e^{-\frac{i}{\sqrt{v}} \hat{\phi}^-(x)}$$

$$\propto F e^{-\frac{i}{\sqrt{v}} \hat{\phi}(x)}. \tag{14.106}$$

With this definition, it can be shown (Exercise 14.5) that

$$[\rho(x), \hat{\psi}^\dagger(x')] = \delta(x - x') \hat{\psi}^\dagger(x), \tag{14.107}$$

$$\hat{\psi}(x)\hat{\psi}(x') = (-1)^{1/v} \hat{\psi}(x')\hat{\psi}(x). \tag{14.108}$$

Equation (14.107) confirms that $\hat{\psi}(x)$ indeed adds one electron to the system. But, as seen in Eq. (14.108), it does not in general satisfy the desired anticommutator $\{\hat{\psi}(x), \hat{\psi}(x')\} = 0$.

The only exception is when $1/v$ is an odd integer, i.e., when $v = 1/(2p+1)$. The evaluation of $\{\hat{\psi}(x), \hat{\psi}^{\dagger}(x')\}$ is left to Exercise 14.6.

Let us specialize to the Laughlin fractions $v = 1/(2p+1)$ and calculate the edge Green's function using the expression in Eq. (14.106). With $2p + 1 \equiv m$,

$$
\begin{aligned}
G^{<}(x',x) &= \mathrm{i}\langle\hat{\psi}^{\dagger}(x)\hat{\psi}(x')\rangle \\
&\propto \mathrm{i}\langle \mathrm{e}^{\mathrm{i}\sqrt{m}\hat{\phi}^{+}(x)}\mathrm{e}^{\mathrm{i}\sqrt{m}\hat{\phi}^{-}(x)}\mathrm{e}^{-\mathrm{i}\sqrt{m}\hat{\phi}^{+}(x')}\mathrm{e}^{-\mathrm{i}\sqrt{m}\hat{\phi}^{-}(x')}\rangle \\
&= \mathrm{i}\mathrm{e}^{m[\hat{\phi}^{-}(x),\hat{\phi}^{+}(x')]} \\
&= \mathrm{i}\exp[-m\ln(\mathrm{i}(2\pi/L)(x - x' - \mathrm{i}a))] \\
&\propto \frac{1}{(x - x' - \mathrm{i}a)^{m}}.
\end{aligned}
\tag{14.109}
$$

The expression for $G^{>}(x',x)$ is given as

$$
G^{>}(x',x) \propto \frac{1}{(x - x' + \mathrm{i}a)^{m}}.
\tag{14.110}
$$

How about generalization to the other fractions of the form $v = n/(2pn \pm 1)$? This derivation made an implicit assumption of only a single bosonic field, i.e., the FQHE state maps into a 1D liquid with only one branch. This treatment can be extended to more general fractions by incorporating the possibility of several edge modes. For $v = n/(2pn + 1)$, for which all modes move in the same direction (in the simplest model), one finds [673]

$$
G(x',x) \propto \frac{1}{(x - x')^{2p+1}}, \qquad |x - x'| \gg \ell
\tag{14.111}
$$

independent of the Λ level index n or the nature of coupling between different modes. For $v = n/(2pn - 1)$, the effective description of the edge invokes multiple counterpropagating edge modes; in this case, the exponent depends on the inter-mode coupling as well as disorder.

The spectral function can be calculated using Eq. (G.18). (Here we closely follow the treatment of Giuliani and Vignale [196].) For this purpose, we need

$$
G^{<}(x,t;0,0) = G^{<}(x - v_{\mathrm{F}}t,0) \sim (x - v_{\mathrm{F}}t + \mathrm{i}a)^{-m},
\tag{14.112}
$$

which follows from $\hat{\phi}(x,t) = \mathrm{e}^{\mathrm{i}Ht}\hat{\phi}(x)\mathrm{e}^{-\mathrm{i}Ht} = \hat{\phi}(x - v_{\mathrm{F}}t)$, where v_{F} is the velocity of the free boson dispersion (remember that the expectation values are being evaluated with respect to the noninteracting bosonic ground state). Equation (G.18) produces

$$A^<(k, \omega) \propto -\frac{1}{2\pi i} \int dt \int dx \, e^{i(\omega t - kx)} \frac{1}{(x - v_F t + ia)^m}$$

$$= \frac{1}{2\pi i} 2\pi \delta(\omega - k v_F) \int d\tau \frac{e^{-i\omega \tau / v_F}}{(\tau + ia)^m}$$

$$\propto \theta(\omega) \delta(\omega - k v_F) \omega^{m-1}. \tag{14.113}$$

In the second step, we have made a change of variables from (x, t) to $(x, \tau = x - v_F t)$; and the last step makes use of (p. 532 of Ref. [196])

$$\int d\tau \frac{e^{-i\tau/v_F}}{(\tau + ia)^m} = (-i)^m \frac{2\pi a^{1-m}}{\Gamma(m)} \theta(\omega)(a\omega/v_F)^{m-1} e^{-a\omega/v_F}. \tag{14.114}$$

$A^>(k, \omega)$ can be obtained from the relation $A^>(k, \omega) = A^<(-k, -\omega)$. In the noninteracting limit $(m = 1)$, the spectral function is a delta function; for $m \neq 1$ it has a qualitatively different, power law behavior.

The edge spectral function can be measured by tunneling of an electron laterally into the edge. From Appendix G, for tunneling from an ordinary Landau Fermi liquid (say a three-dimensional Fermi liquid) into the FQHE edge, the differential tunnel conductance dI/dV is proportional to $D(eV)$, where $D(\omega)$, the tunneling density of states at the FQHE edge, is given by

$$D(\omega) = \sum_k A(k, \omega) \propto \omega^{m-1}. \tag{14.115}$$

This implies a nonlinear $I–V$ relation:

$$I \propto V^m. \tag{14.116}$$

The linear (Ohmic) resistance V/I approaches infinity in the limit $V \to 0$. This expression was calculated at $T = 0$, and is therefore valid only in the limit $k_B T \ll eV$. T and V dependence of the tunneling current for this and other geometries can also be derived using the bosonization machinery, and the resulting expressions are quoted below without proof.

The central prediction of Wen's approach is that the edge exponent for a class of FQHE states is quantized and universal (a "topological quantum number" [673]), much like the Hall resistance itself. This results, for example, in an unmodifiable exponent m in Eq. (14.116). (In contrast, the exponent of an ordinary TL liquid varies continuously with the strength of the interaction.) Many consequences of Wen's conjecture have been evaluated theoretically in the literature. We now address how it compares to experiment. Further details, both on theory and experiment, can be found in the comprehensive review by Chang [66].

14.5 Experiment

The most convincing measurements of the properties of the FQHE edge liquid come from the pioneering experiments of Chang and collaborators [63], who study tunneling from a

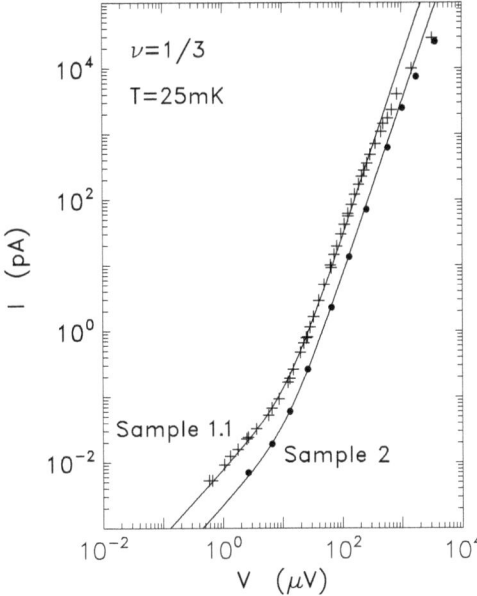

Fig. 14.2. *I–V* plot for tunneling from a three-dimensional Fermi liquid (bulk n-doped GaAs) into the edge of a FQHE state that has bulk filling factor of $v = 1/3$. The curves, for two samples, represent Eq. (14.117) with α treated as a fitting parameter. The best fits are obtained for $\alpha = 2.7$ and 2.65. Source: A. M. Chang, L. N. Pfeiffer, and K. W. West *Phys. Rev. Lett.* **77**, 2538 (1996). (Reprinted with permission.)

three-dimensional Landau Fermi liquid into the FQHE edge in samples prepared by the cleaved-edge overgrowth technique. (The sample is first grown in the 100 direction, and then, after *in-situ* cleaving, in the 011 direction perpendicular to the 2D plane.) They fit the *I–V* curves to the expression (Kane and Fisher [316, 317])

$$I \propto T^{\alpha} \left[[\Gamma((\alpha + 1)/2)]^2 x + x^{\alpha} \right], \qquad x \equiv \frac{eV}{2\pi k_{\rm B} T}, \qquad \alpha = 3. \qquad (14.117)$$

For $x \gg 1$ (but with both V and T sufficiently small), this expression reduces to $I \propto V^{\alpha}$; the exponent can thus be read directly from the slope of the *I–V* plot in the appropriate parameter region. As shown in Figure 14.2, Chang *et al.* [63] find an excellent fit over approximately three decades in current and 1.5 in voltage, with exponents $\alpha = 2.7$ and 2.65 for the two samples. The good news is that the exponent is different from $\alpha = 1$, which demonstrates Tomonaga–Luttinger liquid (rather than Landau Fermi liquid) behavior. The exponent, however, differs significantly from the predicted value $\alpha = 3$. Nonetheless, its proximity to 3 was initially taken as a confirmation of Wen's conjecture.

Other discrepancies begin to appear with further experimentation. A well-defined exponent of $\alpha \approx 1.8$ is seen also at $v = 1/2$ (Fig. 14.3; Chang, Pfeiffer, and West [64]), which is surprising because the bulk state is not a FQHE state but a compressible CF Fermi

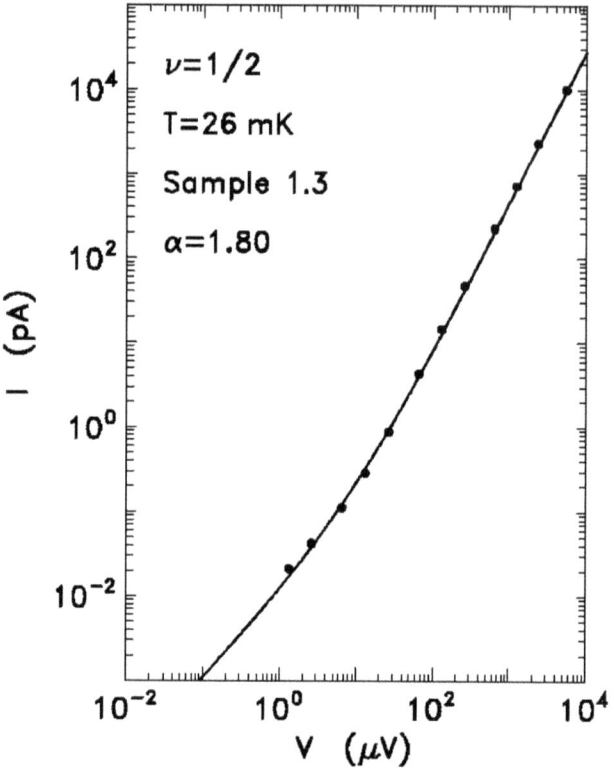

Fig. 14.3. *I–V* characteristic for tunneling from a Fermi liquid into the edge of a FQHE sample with bulk filling factor at $\nu = 1/2$. The solid line is Eq. (14.117) with $\alpha = 1.80$. Source: A. M. Chang, L. N. Pfeiffer, and K. W. West, *Physica B* **249–251**, 283 (1998). (Reprinted with permission.)

sea. Further, measurement of α as a function of the filling factor (Grayson *et al.* [212]) shows an approximately linear dependence on ν^{-1}, described by the expression

$$\alpha \approx \frac{1.16}{\nu} - 0.58 \tag{14.118}$$

in the filling factor range $\nu^{-1} > 1.4$, as displayed in Figure 14.4. It spans the range from $\alpha = 1.0$ to $\alpha = 4.0$. This is in fundamental contradiction to the effective theory prediction, according to which the edge exponent is tied to the quantized Hall conductance and not to the filling factor. The exponent is expected to be constant over any FQHE plateau, and, moreover, is predicted to be $\alpha = 3$ for all fractions of the sequence $n/(2n+1)$. In experiments on higher density samples, Hilke *et al.* [251] focus on the transition from normal Fermi liquid behavior for the IQHE regime $\nu > 1$ to TL liquid behavior for $\nu < 1$ and find

$$\alpha \approx \frac{2.0}{\nu} - 0.55. \tag{14.119}$$

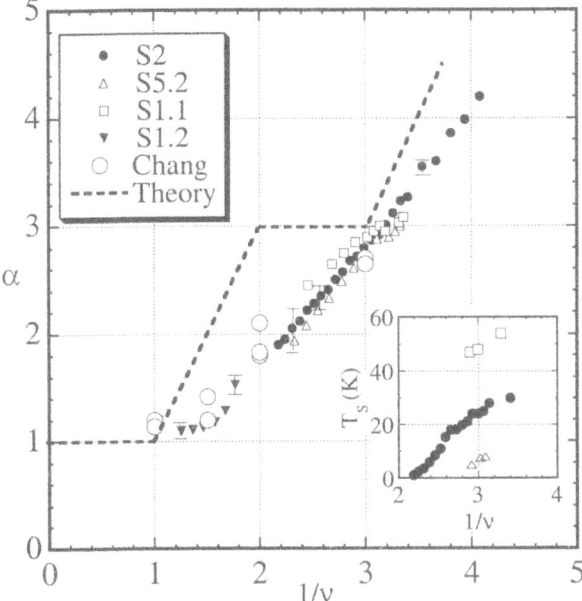

Fig. 14.4. Filling factor dependence of the measured Tomonaga–Luttinger exponent. Results are given for four samples, and also include the exponent from the previous figure. The theoretical curve is from Wen's effective approach. The inset shows the saturation temperature, T_S, such that for $eV > k_B T_S$ the I–V curve becomes linear. Source: M. Grayson, D. C. Tsui, L. N. Pfeiffer, K. W. West, and A. M. Chang, *Phys. Rev. Lett.* **80**, 1062 (1998). (Reprinted with permission.)

The difference from Eq. (14.118) demonstrates the sample-specific nature of the exponent. Extrapolated linearly, Eq. (14.119) would produce an exponent $\alpha \approx 5.45$ at $\nu = 1/3$.

Chang *et al.* [65] find, in some samples, a shoulder in the plot of the exponent (see, for example, the data for the samples 1.1 and 5.2 in Fig. 14.4). The shoulder is often shifted relative to the FQHE plateau; for example, in one case (sample 1.3 of Ref. [65]), the shoulder occurs in the range $4.12 < \nu^{-1} < 4.76$ rather than the expected $2 < \nu^{-1} < 3.33$. The shoulder does not occur in all samples (for example, sample 2 in Fig. 14.4 has no shoulder), and when it does, it does not occur at $\alpha = 3.0$ but at a smaller value of $\alpha \approx 2.7$.

Tunneling between opposite FQHE edges of the same sample has also been investigated experimentally. For resonant tunneling through a quasi-bound state on a potential hill or well, the line shape of the resonant tunneling peak between two $1/3$ edges is predicted to have a universal, non-Lorentzian shape, with peak width scaling as $T^{2/3}$ with temperature [152, 317]. Milliken, Umbach, and Webb [445] employ a point contact geometry, where resonant tunneling peaks are seen because of tunneling through unintentional impurities; their measured line shapes are in good accordance with theory. Maasilta and Goldman [409]

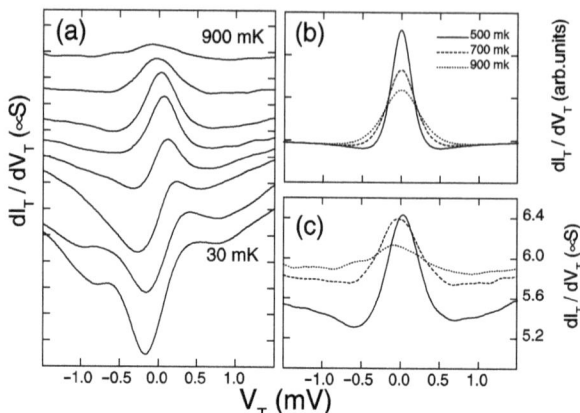

Fig. 14.5. Differential conductance for weak tunneling between two edges of a quantum point contact at $\nu = 1/3$. Panel (a) shows the conductance at 30, 100, 200, 300, 400, 500, 700, and 900 mK. Theoretical calculation is shown in panel (b) for three high temperatures, with the corresponding experimental curves shown in (c). Source: S. Roddaro, V. Pellegrini, F. Beltram, G. Biasiol, L. Sorba, R. Raimondi, and G. Vignale, *Phys. Rev. Lett.* **90**, 046805 (2003). (Reprinted with permission.)

study resonant tunneling through a quantum antidot and demonstrate that the line-shape fits equally well the Fermi liquid theory expression

$$G_T \propto T^{-1} \cosh^{-2}\left(\frac{\epsilon_0 - \mu}{2k_B T}\right), \tag{14.120}$$

which is the convolution of the intrinsic Lorentzian line-shape with the derivative of the Fermi function in the limit when the line width of the Lorentzian is small compared to the temperature (i.e., the Lorentzian effectively acts as a delta function). Here μ is the chemical potential and ϵ_0 is the energy of the resonant state. They note that the Luttinger and Fermi liquid line shapes are indistinguishable in the temperature range investigated in these experiments. Further, the temperature dependence of the width is seen to scale as $T^{1.05\pm0.07}$, which is consistent with a Fermi liquid interpretation.

Certain other discrepancies between experimental observations and the effective theory were noted in Section 9.6 in connection with shot noise measurements in point contact geometry. Roddaro *et al.* [553] study tunneling between two FQHE edges of the 1/3 state in the weak tunneling regime, where the tunneling of fractionally charged quasiparticles is expected to dominate over tunneling of electrons. In this limit, the tunneling current is predicted to behave as $I_T \sim V_T^{-1/3}$ at very low temperature ($k_B T \ll e^* V_T$), but as $I_T \sim V_T$ at high temperatures ($k_B T \gg e^* V_T$). The low-temperature differential tunnel conductance dI_T/dV_T is theoretically predicted to exhibit a zero bias peak when plotted as a function of V_T. The peak height is expected to grow (as $T^{-4/3}$) and the peak width to shrink to zero as the temperature $T \to 0$. Roddaro *et al.* find that the observed behavior is qualitatively consistent with theory in the temperatures range 600–900 mK, but deviates at lower temperatures, and for temperatures below 0.4 K the peak turns into a minimum instead (Fig. 14.5). D'Agosta, Vignale, and Raimondi [104] have attributed these features to the temperature dependence of the tunneling amplitude in the context of this experiment.

In conclusion, experiments do not support the universality of the edge exponent. The nonuniversality makes the calculation of the edge exponent necessarily more complicated, as it now becomes sensitive to various approximations and simplifications. Nonetheless, the question of how a nonquantized value can occur theoretically has motivated numerous studies. One line of approach considers edge reconstruction for smooth edges (Aleiner and Glazman [6]; Conti and Vignale [94]), which results in multiple counter-propagating edge modes, thereby modifying the edge exponent. We describe below some theories of the FQHE edge that are logically independent of Wen's approach. The origin of the nonuniversality of the edge exponent is not fully resolved as yet.

14.6 Exact diagonalization studies

Exact diagonalization studies on small systems have limited applicability for the question of the edge exponent, which deals with the *long*-distance behavior of the edge correlation functions. Goldman and Tsiper [210] obtain the edge TL liquid exponent from the exact occupations of the angular momentum orbitals at the edge. Wen [673] has demonstrated that, for the Laughlin wave function, the exponent is given by

$$\alpha = \frac{\rho(m_{\text{max}}^L - 1)}{\rho(m_{\text{max}}^L)}, \tag{14.121}$$

where $\rho(m)$ is the occupation of the angular momentum m state in the Laughlin wave function, and $m_{\text{max}}^L = 3(N - 1)$ is the angular momentum of the outermost single particle orbital with nonzero occupation. Assuming this expression to be valid for other wave functions as well, Goldman and Tsiper [210] compute the edge TL liquid exponent for several finite Coulomb systems; the results to extrapolate to $\alpha \approx 2.6$ in the thermodynamic limit.

Wan, Evers, and Rezayi [663] obtain the edge exponent from the power law behavior of the edge Green's function:

$$|G_{\text{edge}}(\boldsymbol{r} - \boldsymbol{r}')| \sim |\boldsymbol{r} - \boldsymbol{r}'|^{-\alpha} \sim |\sin(\theta/2)|^{-\alpha}, \tag{14.122}$$

where $\boldsymbol{r}$ and $\boldsymbol{r}'$ are two points at the edge at an angular separation θ. Confinement is produced by fixing the total angular momentum as well as the maximum angular momentum, m_{max}, for each electron. Wan *et al.* refer to $m_{\text{max}} = 3(N-1)$ as "hard edge," and $m_{\text{max}} = 3(N-1)+5$ as "soft edge," because incorporating larger angular momentum orbitals allows the edge to relax. For systems with up to $N = 9$ electrons, for the hard edge, oscillations occur in $G_{\text{edge}}(\theta)$ even at the largest θ, whereas, for the soft edge, a power law behavior with an exponent of $\alpha \approx 3.2 \pm 0.2$ is seen. The exponent deduced from Green's function does not agree with that from Eq. (14.121). Wan *et al.* also estimate $\alpha \approx 3.2 \pm 0.2$ for a hard edge in the presence of a confinement potential produced by a background charge at a distance of $d = 1.0\ell$. They take these results as supporting the universality of the edge exponent.

Zülicke, Palacios, and MacDonald [753] show evidence, based on finite size exact diagonalization calculations, that the electron operator, after projection into the edge states, no longer satisfies Fermi statistics. As a result, they argue, there is no reason to require antisymmetry, which is a key step in Wen's reasoning leading to a quantized exponent.

14.7 Composite fermion theories of the edge

The following points make the CF theory attractive for the edge physics:

- To the extent the physics of the edge has to do with the physics of the FQHE (which is what makes the FQHE edge interesting), it must be describable in terms of composite fermions.
- Microscopic wave functions of the CF theory have been demonstrated to be accurate for systems with an edge, specifically for electrons confined to a disk (Section 6.8).
- The edge states of the IQHE are simple Landau Fermi liquids. The CF mapping of the FQHE into the IQHE should thus provide a good starting point for an investigation of the edge. In principle, a systematic microscopic derivation of the physics of the FQHE edge may be possible within the CF theory.
- The CF theory also has the obvious potential of unifying the edges of all FQHE states.

The CF theory offers important insights into the edge physics, but a reliable deduction of its consequences for the edge dynamics has proved challenging, and our current understanding remains far from being conclusive. We briefly mention here several composite fermion based approaches developed for the edges.

14.7.1 Effective approaches

Several studies of the edge physics have used the CFCS framework. Many of these evaluate the density profile at the edge, including the possibility of edge reconstruction. Brey [37] studies the evolution of the edge states of composite fermions as a function of the steepness of the confinement potential within the standard self-consistent Hartree approximation (which assumes that the ground state wave function for the CS-transformed problem can be expressed as a product of orthogonal one-particle wave functions; notably, while the CS transformation is nonlocal, it does not affect the charge and the current densities). He finds that smooth edges contain strips of composite fermions at different filling factors. He has also generalized the Landauer–Büttiker formalism to composite fermions. This method is used by Brey and Tejedor [38] to investigate transport of composite fermions across a potential barrier, which effectively acts as a magnetic barrier due to the reduced charge density in the barrier region. Similar CF Hartree calculations have been performed by Chklovskii [76], who also considers the appearance of fractional edge channels at $\nu = 1$.

Joglekar, Nguyen, and Murty [315] find, in the Murthy–Shankar formulation of composite fermions, that the 1/3 edge is susceptible to a reconstruction as the background potential is softened, for long-range as well as short-range interactions, although the edges of 2/5, 3/7, etc. are more robust.

Kirczenow and Johnson [307,338,341] consider slowly varying edge potentials using the CF mean-field theory. They stress that when composite fermions carry an electric current, their effective chemical potential is not the same as that for electrons, because composite fermions also sense the scalar potential associated with the induced Chern–Simons electric field discussed in Section 5.16.3. As a result, the electrochemical potential sensed by composite fermions is not what would be measured by a voltmeter. Kirczenow and Johnson obtain a complex picture, with edge states in the FQHE that have no analog in the IQHE. This approach has consequences distinct from Wen's approach. The temperature dependence for the two-terminal conductance for transport through an adiabatic or a nonadiabatic constriction is predicted to be similar to a Fermi liquid [339], which is consistent with the linear temperature dependence of the widths of the resonant tunneling peaks observed experimentally [409]. Also, all carriers at the edge of $n/(2pn - 1)$ states are predicted to propagate in the same direction as noninteracting electrons, consistent with experiment [16].

Shytov, Levitov, and Halperin [390,589] derive, using the CFCS approach, an effective action for the edge of a CF state that treats incompressible and compressible states on an equal footing. It can deal with the CF Fermi sea at $\nu = 1/2$ as well as with the FQHE states with nonzero ρ_{xx}. They obtain an expression for the exponent in $I \sim V^\alpha$ in terms of the magnetoresistances. For incompressible systems, they obtain

$$\alpha = \begin{cases} 2p + 1, & \nu = \frac{n}{2pn+1} \\ 2p + 1 - \frac{2}{n}, & \nu = \frac{n}{2pn-1} \end{cases} \tag{14.123}$$

in agreement with the effective approaches of Wen [673] and Kane and Fisher [319].

Another intriguing line of inquiry seeks to make a connection between the edge excitations of the FQHE and the excitations of a certain one-dimensional model of interacting particles, known as the Calogero–Sutherland model [49,632], motivated by the feature that the ground state wave function for the latter resembles, for certain interactions, the one-dimensional limit of the Laughlin wave function. We only mention here some pertinent references (Yu, Zheng, and Zhu [733,734]; Brink, Hansson, and Vasiliev [39]).

14.7.2 Microscopic investigation

We saw in Section 6.8 that the wave functions of the CF theory work reasonably well for nonhomogeneous density FQHE droplets in the disk geometry, which is at least a necessary condition for the application of these wave functions to the edge physics. The edge Green's function has been evaluated for wave functions at various levels of approximation for $\nu = 1/3$, 2/5 and 3/7 (Mandal and Jain [423, 425]). The "unperturbed" wave functions (with no Λ level mixing) produce an exponent of $\alpha = 3$. However, when the wave functions are improved by CF diagonalization in a larger basis, allowing for Λ level mixing (but not

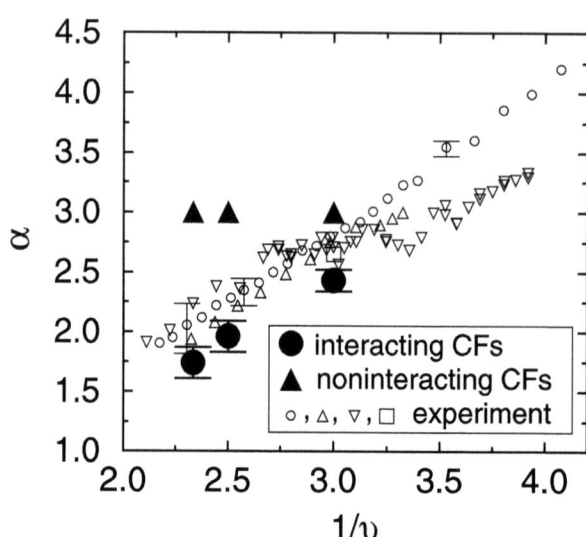

Fig. 14.6. Luttinger liquid exponents α for the 1/3, 2/5, and 3/7 edges for noninteracting (filled triangles) and interacting (filled circles) composite fermions. The experimental results (empty symbols) are taken from the following sources: square from Chang *et al.* [63]; circles and triangles from Grayson *et al.* [212] (samples M and Q); inverted triangles from Chang *et al.* [65] (samples 1 and 2). Source: S. S. Mandal and J. K. Jain, *Phys. Rev. B* **63**, 201310(R) (2001). (Reprinted with permission.)

allowing for any edge reconstruction), the exponent is reduced, as shown in Fig. 14.6. Systems with up to 40, 50, and 60 particles are studied for 1/3, 2/5, and 3/7. While the calculations are for finite N, it is noted that: (i) the maximum distance along the edge is ~30 times the magnetic length; (ii) the system is large enough to produce a well-defined exponent; (iii) the "expected" exponent is obtained for the zeroth order wave function; and finally, (iv) increasing the number of particles from 30 to 50 for 2/5 and 30 to 60 for 3/7 does not appreciably alter the exponent, while going from 30 to 40 particles at 1/3 *reduces* α slightly. These considerations suggest that the results reflect the thermodynamic limit. A comparison of the calculated exponents with the measured ones is shown in Fig. 14.6; it ought to be remembered that the theory neglects many effects present in experiments which will surely renormalize the calculated exponent. (Also, the tunneling experiment measures the frequency dependence of the equal position correlation function, whereas the theoretical calculation is for the equal time correlation function. The two are governed by the same exponent in TL liquids, however.) It appears that Wen's approach gives an effective description of the lowest-order CF theory, but Λ level mixing is a relevant perturbation. It is suggested in Ref. [425] that while the edge excitations *are* described by certain bosonic fields, the electron creation field in terms of these bosons might be a nontrivial, nonlocal operator.

Exercises

14.1 The edge can be defined as the region where the filling factor is going from its bulk value to zero, which extends over several magnetic lengths. Show that the exponent characterizing the decay of the equal-time edge Green's function at $v = 1$ is independent of the precise value of R, provided it is located at the edge.

14.2 Calculate the edge Green's function for noninteracting electrons in the second Landau level, assuming it is fully occupied.

14.3 List all excitations with $E \leq 5\bar{k}$ for the fermionic and the bosonic problems defined above and show that the degeneracies are equal.

14.4 Show that Eqs. (14.83) and (14.84) reduce to Eqs. (14.79) and (14.82).

14.5 Complete the derivations of Eqs. (14.105), (14.107), and (14.108).

14.6 For $\hat{\psi}(x)$ defined in Eq. (14.106), obtain the following expression for the anticommutator (with $v = 1/m$)

$$\{\hat{\psi}(x), \hat{\psi}^\dagger(x')\} \propto \left(\frac{\partial}{\partial x}\right)^{m-1} \delta(x - x'). \tag{E14.1}$$

(Hint: Follow the steps leading to Eq. (14.77).)

15

Composite fermion crystals

The FQHE was discovered when Tsui, Stormer, and Gossard were looking for the Wigner crystal (WC) [675], i.e. a crystal of electrons. While electrons in the lowest Landau level are now known to form a CF fluid at relatively large fillings, a solid phase is expected at sufficiently small fillings, when electrons are so far from each other that the system behaves more or less classically. This chapter is concerned with the nature of the solid phase. Theory makes a compelling case that the actual solid is not an ordinary, classical Wigner crystal of electrons, but a more complex, topological quantum crystal of composite fermions, in which quantized vortices are bound to electrons. This is called the "CF crystal," to be differentiated from the electronic Wigner crystal. Experiments have seen indications of the formation of a crystal at low fillings, but further work will be required to ascertain its quantum mechanical character. (The quantum mechanical behavior of ^{3}He and ^{4}He solids has been studied extensively [55, 101]. Quantum mechanics plays a fundamental role in the superfluid behavior of ^{4}He solid [336, 384], a subject of much current investigation.)

We also briefly mention the possibility of charge density wave states of composite fermions, for example, unidirectional stripes or bubble crystals, which have been proposed theoretically. These have also not been observed yet, but might occur in higher Λ levels (i.e., between regular FQHE states), and are analogous to similar structures believed to occur for electrons in higher Landau levels.

15.1 Wigner crystal

15.1.1 Zero magnetic field

The ground state of interacting electrons is determined by minimizing the total energy, which is a sum of the interaction and the kinetic energies. These two have opposing tendencies. The interaction between electrons prefers a crystal, which is the "classical" ground state of charged particles. For a crystal to be meaningful, the wave packet of each electron must be localized to a size smaller than the lattice spacing (determined by the density), which, due to the quantum mechanical uncertainty principle, has a kinetic energy cost associated with it. For typical electron densities encountered in ordinary metals, the kinetic energy cost is so huge that a crystal cannot be stabilized; the essential physics here is dictated by the kinetic energy, which produces the familiar Fermi sea. The Coulomb interaction leads to a renormalization of various parameters, but does not cause a qualitative change in the nature of the state. A Wigner crystal is expected, however, at sufficiently low densities, when the typical separation between electrons is large [675].

In three dimensions, the estimate for the ground state energy for the liquid in the high density limit ($r_s \to 0$) is given by (Wigner [675])

$$\frac{E}{N} = \frac{2.21}{r_s^2} - \frac{0.916}{r_s} + \cdots , \tag{15.1}$$

where $r_s = r_0/a_0$ is the dimensionless interelectron separation, with $r_0 = (3/4\pi\rho)^{1/3}$ and $a_0 = \hbar^2/m_e e^2$. The energy is quoted in units of rydberg ($e^2/2a_0$). The first term is the kinetic energy and the second is the Coulomb interaction energy (negative because it includes interaction with a positively charged uniform "jellium" background). In the dilute limit ($r_s \to \infty$), the energy per particle for an electron solid is given by (Wigner [676]; Carr [52])

$$\frac{E}{N} = -\frac{1.792}{r_s} + \frac{2.66}{r_s^{3/2}} + \frac{b}{r_s^2} \cdots \tag{15.2}$$

for a body-centered cubic lattice. The first term is the Coulomb interaction energy, the second is the kinetic energy from zero point motion, and the third comes from anharmonic corrections. If one assumes these energies to be reasonably accurate at arbitrary r_s for the respective states, then these equations imply a transition from a liquid at high densities to a solid at low densities. Such Hartree–Fock type approximations are too crude, however, for estimating where the transition from the liquid to crystal occurs, because near the transition both the liquid and the solid are likely to be quite different from the asymptotic ones.

The most reliable estimates for the energies of various states of interacting electrons come from quantum Monte Carlo calculations (Ceperley and Alder [54], Rapisarda and Senatore [535], Tanatar and Ceperley [633]).[1] These calculations predict, in two dimensions, a transition at $r_s = 37$, where the quantity $r_s = (m_e e^2/\hbar^2)/\sqrt{\pi\rho}$ ($m_e = $ electron mass) is the interelectron separation in two dimensions in units of the Bohr radius $\hbar^2/m_e e^2$.

Crandall and Williams [99] suggested that a physical realization of the Wigner crystal could occur on the surface of liquid helium, where the possibility of the formation of a 2D electron system had been predicted by Cole and Cohen [91]. Electrons at the free surface of liquid helium are trapped in the direction perpendicular to the surface (with a binding energy of 0.7 meV), in a potential well formed by the combination of an attractive potential due to interaction with the image charge and a repulsive barrier to penetration into the liquid. Electrons are free to move parallel to the surface, however. Since the helium surface is extremely smooth, electrons form an almost ideal 2D system, with very high mobilities

---

[1] The idea of stochastic simulation of the Schrödinger equation, based on its similarity to the diffusion equation, has a long history. It was discussed by Metropolis and Ulam in 1949 [443], who attributed it to Fermi. In 1975, Anderson [7] applied it to the quantum chemistry problem of determining the ground state energy of the He$_3^+$ molecular ion.

at low temperatures. The Wigner crystal phase was observed by Grimes and Adams [216], identified through its coupling with standing capillary waves on the helium surface, known as "ripplons." The electron density in these experiments was $\sim 5 \times 10^8$ cm^{-2} and the melting temperature was around 0.5 K (much higher than the Fermi energy of ~ 0.03 K). The phase transition occurs at $\Gamma = 137 \pm 15$, where $\Gamma = \sqrt{\pi \rho} e^2 / k_B T$ is a measure of the ratio of Coulomb energy to kinetic energy for a 2D system of charged classical particles governed by Maxwell–Boltzmann statistics, as appropriate for electrons much above the degeneracy temperature.

15.1.2 Lowest Landau level

Lozovik and Yudson [406] proposed that a Wigner crystal could be achieved by application of a strong magnetic field perpendicular to a 2D system. Once all electrons are in the lowest Landau level, their kinetic energy is quenched and their state is determined entirely by the interaction energy. What else can electrons do but form a Wigner crystal?

As we have seen, in a range of filling factors, electrons capture vortices to turn into composite fermions, which form a liquid state. This physics is unexpected, and, in all likelihood, would not have been conceived but for the dramatic discovery of the FQHE. From the perspective of the Wigner crystal, the lattice constant of a hypothetical crystal is comparable to the size of the maximally localizable wave packet in the lowest Landau level, and the resulting overlap between nearest neighbor electrons is so strong as to "melt" the crystal.

This argument also indicates that at sufficiently small fillings, when electrons on neighboring lattice sites are far removed, a crystal should be energetically favored. The wave packet for an electron in the lowest Landau level localized at $R = (X, Y)$ is given by [417] (Eq. E3.2)

$$\phi_R(r) = \frac{1}{\sqrt{2\pi}} \exp\left(-\frac{1}{4}(r - R)^2 + \frac{i}{2}(xY - yX)\right). \tag{15.3}$$

The Maki–Zotos (also called Hartree–Fock) wave function for the uncorrelated Wigner crystal in the lowest Landau level is constructed by placing electrons on a triangular lattice R_j, the lowest energy solution for the classical problem, and then antisymmetrizing the product [417]:

$$\Psi^{WC} = \frac{1}{\sqrt{N!}} \sum_P \epsilon_P \prod_{j=1}^N \phi_{R_j}(r_{Pj}), \tag{15.4}$$

where the sum is over all permutations P and ϵ_P is $+1$ for even permutations and -1 for odd permutations. With the lattice constant $a = (4\pi/\sqrt{3}\nu)^{1/2} l_0$, the overlap integral between nearest neighbor electron wave functions [417], $\exp(-a^2/2l_0^2) = \exp(-3.627/\nu)$, decays rapidly with decreasing ν.

A principal reason why 2D semiconductor heterostructures are more promising for the formation of a Wigner crystal is the extremely high quality of the electron system, made possible by modulation doping. In 3D doped semiconductors, electrons and the donor ions occupy the same space, and a high magnetic field localizes electrons back on to the donors.

15.1.3 Wigner crystal vs. CF liquid

Much work has been done toward estimating the filling factor where a transition from the CF liquid to the Wigner crystal takes place. An accurate determination, however, has not been possible for several reasons. (i) Most calculations compare variational energies, and are as reliable as the wave functions they use. What makes the situation somewhat murky is that the curves for the solid and liquid energies move so close to one another at small fillings that slight errors can vastly affect the theoretical prediction for the critical filling factor, the value of which has oscillated over time as one or the other variational state is subjected to further improvement. Levesque, Weiss, and MacDonald [389] compared the energy of the Laughlin wave function with an uncorrelated Hartree–Fock Wigner crystal [406, 417, 727, 729] and concluded that the transition takes place near $\nu = 1/10$. Lam and Girvin [363] (also Chui and Esfarjani [78]) improved the energy of the Wigner crystal substantially over the Hartree–Fock crystal, by including correlations that preferentially weight favorable fluctuations in particle positions, which moved the transition up to $\nu \approx 1/6$. The Laughlin wave function is not particularly accurate at low fillings, however. Better variational bounds for the energies of the liquid states at $\nu = 1/7$, 1/9, and 1/11 can be obtained by including Λ level mixing up to second order (Mandal, Peterson, and Jain [426]). While the reduction in liquid energy ($\approx 0.8\%$) is not large in absolute terms, it again shifts the phase boundary, and, with the WC energies taken from Ref. [363], FQHE down to $\nu = 1/11$ cannot be excluded. Of course, further improvements on the crystal energy would shift the phase boundary again. Yang, Haldane, and Rezayi [717] investigate the problem through exact diagonalization, with six particles on a torus, and find evidence for a charge density wave with one electron per unit cell, as expected for the Wigner crystal, at $\nu = 1/8$ but not at $\nu = 1/6$, surmising that the system becomes unstable to crystallization at around $\nu = 1/7$. An excitonic instability had earlier been found at $\nu = 1/9$ (Kamilla and Jain [313]), suggesting that FQHE does not occur here, or at smaller fillings; the instability, however, is eliminated when more accurate calculations including Λ level mixing are performed (Peterson and Jain [512]). (ii) The calculations do not include modifications in the interaction due to the nonzero thickness. (iii) They also disregard disorder, which is also known to influence the liquid–solid phase boundary [79]. (iv) Finally, because the energy of the CF liquid is not smooth as a function of the filling factor, not one but several re-entrant liquid to solid transitions may occur.

It is fair to say that only experiments will tell us precisely where the transitions take place. Our focus below is not on the critical filling factors, but on the nature of the crystal phase. For this purpose, a consideration of filling factors deep inside the crystal phase is sufficient.

15.2 Composite fermions at low v

Since we are interested in the role of crystalline correlations, we work in the disk geometry, with confinement imposed by fixing the total z-component of the angular momentum L. (The spherical geometry is biased against a crystal, because, for a general N, a 2D triangular crystal cannot be wrapped onto a sphere without creating defects.) Exact diagonalization studies of this problem [116, 189, 218, 242, 418, 708, 728] have interpreted the results in many different ways [419, 559, 560, 719, 720]. We confine our discussion below to the application of the composite fermion theory to this problem [116, 279, 281, 295, 299]. We begin by asking how the CF theory described in the earlier chapters fares at low fillings.

The CF description becomes progressively worse with decreasing filling factor. As seen in Section 6.8 the mean-field CF model, which assumes free composite fermions at L^* with an empirical effective cyclotron energy, successfully predicts cusp positions in the plot of the ground state energy versus L in the range described by ^{2}CFs (Fig. 6.10). This model becomes inadequate at larger angular momenta, however (Seki, Kuramoto, and Nishino [578]; Yannouleas and Landman [720]). As seen in Fig. 15.1, discrepancies appear between the mean-field and the actual cusp position. Similar behavior is seen for other particle numbers for which exact diagonalization has been performed.

At large L, the actual cusps occur at regular intervals in L, as seen, for example, in Fig. 6.10. This can be understood qualitatively (Maksym [419]; Manninen, Viefers, Koskinen and Reimann [427]; Ruan and collaborators [559, 560]; Yannouleas and Landman [720]) by analogy to classical crystal-like states (Bedanov and Peeters [24]), which select certain angular momenta because of their symmetry. The classical analogy is sensible at large angular momenta where electrons are farther apart (which corresponds to very small fillings).

A breakdown of the CF theory at sufficiently small fillings would not be surprising, given that the actual state here is not a CF liquid but a triangular crystal. Calculations show,

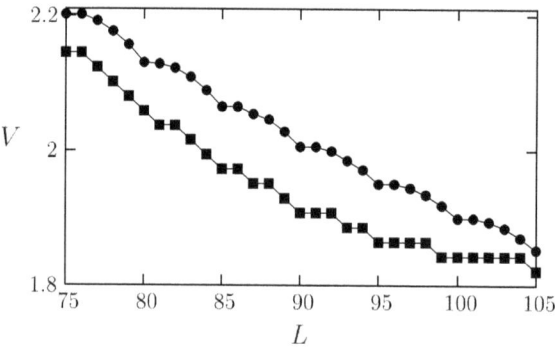

Fig. 15.1. Comparing the L dependence of the exact energy V (circles) with the prediction of the mean-field CF model (squares). Same as in Fig. 6.10, but at larger L. Source: G. S. Jeon, C.-C. Chang, and J. K. Jain, cond-mat/0611309 (2006). (Reprinted with permission.)

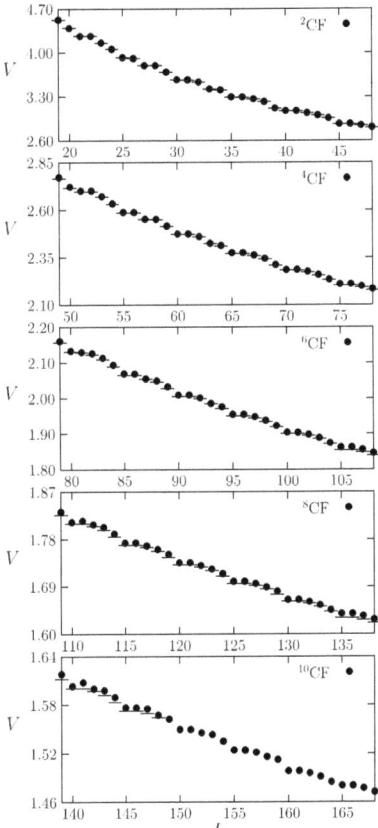

Fig. 15.2. The exact interaction energy V (dashes) as a function of the angular momentum L for $N = 6$ particles. The dots are predictions of the lowest (zeroth) order CF diagonalization. Different panels correspond to L ranges where composite fermions of different flavors are relevant. The energies are quoted in units of $e^2/\epsilon\ell$. Source: Jeon, Chang and Jain, cond-mat/0611309 (2006). (Reprinted with permission.)

however, that while the mean-field CF model fails to explain the cusp structure, the CF diagonalization gives, at the zeroth order itself, an accurate description of the ground state energy as well as of the periodic cusps. Figure 15.2 shows the exact Coulomb ground state energy up to a very large angular momentum ($L = 165$ corresponds to $\nu = 1/11$), along with the ground state energy predicted by zeroth-order CF diagonalization. Although the mean-field approximation fails, the system is well described in terms of composite fermions in the entire range studied.

The success of the CF theory at low ν (or large L) might appear puzzling, given that crystalline correlations surely play a role there. There is no inconsistency, however. The formation of composite fermions with a large number of attached vortices automatically generates crystalline correlations (Jeon, Chang, and Jain [295, 296]). Take, for example, six particles at $L = 95$, which maps into $L^* = 5$ of ^{6}CFs. The CF theory gives a unique compact state here, denoted by [4, 2]. The wave function is given by

$$\Psi = \mathcal{P}_{\text{LLL}} e^{-\sum_{l=1}^{N} |z_l|^2/4} A \left[z_1^* \cdot z_2 z_2^* \cdot \prod_{i=3}^{N} z_i^{i-3} \right] \prod_{j<k} (z_j - z_k)^6, \qquad (15.5)$$

where A denotes an antisymmetric Slater determinant, and the LLL projection amounts to replacing $z^* \rightarrow 2\partial/\partial z$. The energy of this wave function is $1.955\,35(15)e^2/\epsilon\ell$, which compares well with the exact ground state energy $1.950\,61e^2/\epsilon\ell$ obtained by a numerical diagonalization of a $69\,624 \times 69\,624$ Hamiltonian matrix. The overlap of the [4, 2] wave function with the exact ground state is 0.902. To investigate crystalline correlations it is necessary to compute the pair correlation function

$$g(r) \sim \int \prod_{j=3}^{N} d^2 r_j |\Psi(r, R, r_3, \ldots, r_N)|^2, \qquad (15.6)$$

which gives the conditional probability of finding a particle at r while one particle is held fixed at R. As seen in Fig. 15.3, the pair correlation function for the [4, 2] wave function in Eq. (15.5) compares favorably with the exact pair correlation function. Thus, the single

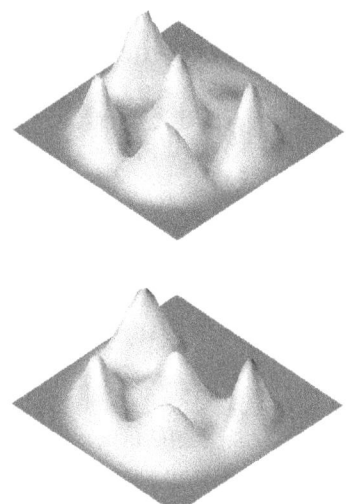

Fig. 15.3. Pair correlation function for $N = 6$ particles at $L = 95$. The upper plot is the exact result, and the lower plot displays the prediction of the CF theory. The exact ground state is a linear superposition of $69\,624$ basis states, whereas the ^{6}CF ground state is given by the unique wave function of Eq. (15.5). The "missing peak" on the outer ring indicates the location R of the fixed particle. Source: G. S. Jeon, C.-C. Chang and J. K. Jain, *J. Phys. Condens. Matter* **16**, L271 (2004). (Reprinted with permission.) Similar results were published by Yannouleas and Landman [719, 720].

wave function of the CF theory provides a decent account of the actual ground state wave function at $L = 95$. A similar level of accuracy is achieved at other L values.

So far, we have not introduced any crystalline correlations explicitly. A crystal appears spontaneously when we make composite fermions. While this brings us close to the solution, we now see that incorporating *both* crystalline and CF correlations produces almost the exact state.

15.3 Composite fermion crystal

Inspired by our understanding of the liquid state, the question naturally arises if vortices are bound to electrons even in the crystal phase. Yi and Fertig [725] consider a trial wave function

$$\Psi = \prod_{j<k}(z_j - z_k)^q \Psi^{\text{WC}}, \tag{15.7}$$

where Ψ^{WC} is the Maki–Zotos wave function for the uncorrelated triangular Wigner crystal in the lowest Landau level (Eq. 15.4), and show that it has lower energy than the Lam–Girvin wave function [363] in the filling factor range $0.1 < \nu < 0.2$. Chang, Jeon, and Jain [68] study a closely related wave function:

$$\Psi_L^{2p-\text{CFC}} = \prod_{j<k}(z_j - z_k)^{2p} \Psi_{L^*}^{\text{WC}}, \tag{15.8}$$

$$L^* = L - pN(N-1). \tag{15.9}$$

It is interpreted as a CF crystal (CFC) for $2p \neq 0$, because the Jastrow factor $\prod_{j<k}(z_j - z_k)^{2p}$ binds $2p$ quantized vortices to each electron in Ψ^{WC} to convert it into a composite fermion. The function $\Psi_{L^*}^{\text{WC}}$ is the projection of the Maki–Zotos wave function into angular momentum L^*, obtained by the method of Yannouleas and Landman [720], who give an explicit expression for it. Such projection produces a rotating Wigner crystal, and thus $\Psi_L^{2p-\text{CFC}}$ represents a rotating CF crystal. For infinite systems the Maki–Zotos crystal is a triangular crystal. For finite systems, the crystal is chosen to coincide with a stable crystal of *classical* particles [24], shown in Fig. 15.4 for some simple cases. The CFC wave function represents complex interparticle correlations in the crystal phase. The CF vorticity $2p$ is treated as a variational parameter, to be fixed by minimizing the Coulomb interaction energy. The idea, essentially, is that a few vortices dissociate from composite fermions to facilitate better crystalline order.

Chang, Jeon, and Jain [68] compare the CFC wave function with the exact wave function, the latter obtained with the help of the Lanczos method in the low-ν region of interest. For $N = 6$, the lowest energy classical configuration has one particle at the center, with the remaining five forming a ring around it [24]. Tables 15.1 and 15.2 compare the Laughlin wave function, the Wigner crystal, and the CF crystal with the exact state of six electrons at $\nu = 1/5$, 1/7, and 1/9. For the CF crystal, the optimal CF flavors at these fillings are ^{2}CF, ^{4}CF, and ^{6}CF, respectively. Results for the Wigner crystal and the Laughlin wave function

Table 15.1. *Testing electron crystal and CF crystal wave functions at low fillings*

ν	L	D	CFC	WC	Laughlin
1/3	45	1 206	—	0.394	0.964
1/5	75	19 858	0.891	0.645	0.701
1/7	105	117 788	0.994	0.723	0.504
1/9	135	436 140	0.988	0.740	0.442

Notes: Last three columns give overlaps of CF crystal (CFC), Wigner crystal (WC), and Laughlin's wave function with exact ground state wave function at several filling factors ν.
Overlap is defined as $|\langle \psi^{\text{trial}} | \psi^{\text{exact}} \rangle|^2 / \langle \psi^{\text{trial}} | \psi^{\text{trial}} \rangle \langle \psi^{\text{exact}} | \psi^{\text{exact}} \rangle$.
D is dimension of lowest Landau level basis for $N = 6$ electrons, and L is total angular momentum of the state.
Source: from Chang, Jeon, and Jain [68].

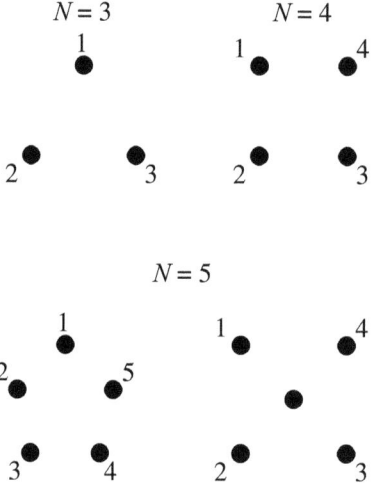

Fig. 15.4. Some stable classical crystals for small numbers of particles.

are also given for $\nu = 1/3$; the CFC wave function does not give anything new at this filling factor (it reduces to WC for $2p = 0$, and to the Laughlin wave function for $2p = 2$). As mentioned earlier, the Laughlin wave function worsens and Ψ^{WC} improves with decreasing ν, but neither is especially accurate at small fillings. In contrast, the CFC wave functions have an overlap of ~99% with the exact states at $\nu = 1/7$ and $1/9$, and the CFC energies deviate from the exact energies by 0.016% and 0.006%, respectively. The nontriviality of these agreements can be appreciated by noting that: (i) The exact state is a linear superposition

Table 15.2. *Electron crystal and CF crystal energies at low fillings*

ν	L	Exact	CFC	WC	Laughlin
1/3	45	2.8602	—	2.9163(9)	2.8643(3)
1/5	75	2.2019	2.2042(5)	2.2196	2.2093(2)
1/7	105	1.8533	1.8536(2)	1.8622	1.8617(2)
1/9	135	1.6305	1.6306(1)	1.6361	1.6388(1)

Notes: Total interaction energies for exact ground state, CF crystal (CFC), Wigner crystal (WC), and Laughlin wave function for six particles at several filling factors.

Uncertainty in last digit from Monte Carlo sampling is given in parentheses. Electron–background and background–background contributions are not included, but are expected to be the same for all states.

Source: Chang, Jeon and Jain [68]. (Total energies were mislabeled as interaction energies per particle in Ref. [68].)

of a large number of Slater determinant basis functions (see Table 15.1), involving $D - 1$ parameters. (ii) The CFC wave functions for $\nu = 1/7$ and $1/9$ are more accurate than the Laughlin wave function at $\nu = 1/3$, in spite of the much larger Fock space dimensions at $1/7$ and $1/9$. The CFC pair correlation function (not shown) is indistinguishable from exact [68]; the formation of composite fermions somewhat weakens the crystalline correlations relative to the Maki–Zotos uncorrelated Wigner crystal.

Chang *et al.* [68] consider states down to $\nu = 1/45$, using a combination of exact diagonalization and CF diagonalization for $N = 6$, and find similar accuracy at smaller ν. The optimal value of $2p$ increases as the filling factor is reduced, with $2p = 38$ producing the best energy at $\nu = 1/45$.

Finite size studies do not necessarily provide a reliable account of the thermodynamic state. For example, for $N = 6$ the CFC gives a better description of the $\nu = 1/5$ ground state than the Laughlin liquid wave function, even though the thermodynamic state here is known to be a liquid [205, 301]. (With increasing N, the Laughlin 1/5 wave function eventually begins to have lower energy than the CFC [70].) Nonetheless, an extremely precise description of a finite N state (assuming N is not too small) gives a strong indication for the nature of the state in the thermodynamic limit. Even though the finite N study cannot predict the precise ν value where a transition from liquid to crystal takes place, it makes a compelling case that whenever the thermodynamic state is a crystal, the crystal is made of composite fermions.

The energy difference per particle, $V^{CFC} - V^{WC}$, can be taken as a crude estimate of the temperature below which the quantum nature of the crystal should be robust to thermal fluctuations. From the energy differences in Ref. [68], for parameters appropriate for GaAs, the quantum crystal regime is estimated to be below 200 mK at $\nu = 1/9$ at $B = 25$ T.

The quantum character of the CF crystal ought to be observable at currently attainable temperatures.

Because every particle sees quantized vortices on every other particle, the formation of composite fermions implies a long-range quantum coherence in the crystal phase. It is a topological quantum crystal, in the same sense as the FQHE states are topological quantum fluids, although the full implications of the topological, or the quantum, nature of this crystal remain a subject for further study. It should be noted that, unlike in the ^{4}He quantum crystal, the overlap between (uncorrelated) electron wave packets at neighboring sites is negligible in the filling factor region of interest (the overlap integral is 10^{-15} for $\nu = 1/9$); the quantum nature of the CF crystal owes its origin to the long-range Coulomb interaction.

In the picture that emerges, composite fermions are relevant not only for the LLL liquid, but also for the LLL crystal. As the filling factor is decreased from unity, electrons capture the maximum possible (even) number of vortices at first (for $\nu \leq 1/5$), but then eventually capture fewer than maximum, and exploit the remaining degree of freedom to stabilize a crystal. Composite fermions with very high vorticity are predicted to occur in the crystal phase.

15.4 Experimental status

At present, a direct experimental observation of the lattice structure of the LLL crystal is lacking. The state, buried deep inside the heterostructure, is inaccessible to probes such as STM. Also, disorder turns the crystal into a "glass" with crystalline order only over a finite correlation length. Much circumstantial evidence exists, however, that a crystal is present at sufficiently low ν [156, 584]. Many transport (Goldman *et al.* [205, 391]; Jiang *et al.* [301]; Pan *et al.* [495]; Santos *et al.* [566]) and magneto-optical measurements (Andrei *et al.* [11]; Buhmann *et al.* [40]; Chen *et al.* [71, 72]; Kukushkin *et al.* [356]; Williams *et al.* [678]; Ye *et al.* [722]) have been interpreted in terms of a pinned solid phase.

Figure 15.5 shows the longitudinal resistance of a high-quality sample (Jiang *et al.* [302]). The FQHE at $\nu = 1/5$ is well developed, but the longitudinal resistance in the regions $2/9 > \nu > 1/5$ and $\nu < 1/5$ diverges exponentially with vanishing temperature. (On the scale of the figure, the resistances associated with the FQHE states $f = n/(2n \pm 1)$ are hardly visible.) Similar divergence had been seen earlier at higher fillings in worse quality samples (for example, in Ref. [681], a transition to a zero temperature insulating state was observed at $\nu \approx 1/4$), but the insulating behavior on either side of 1/5 has not gone away with significant improvements in the mobility, and is therefore very likely to be an intrinsic effect. Its vicinity to the correlated 1/5 FQHE state is seen as supporting the interpretation in terms of a pinned crystal in Ref. [301].

Transport [204, 495] and photoluminescence [40] experiments, however, report evidence for a FQHE-like structure down to $\nu = 1/9$. In very high quality samples (Fig. 15.6) the observation by Pan *et al.* [495] of a series of resistance minima at filling factors $\nu = 1/7$, 2/11, 2/13, 3/17, 3/19, 1/9, 2/15 and 2/17, which belong to the sequences $f = n/(6n \pm 1)$ and $f = n/(8n \pm 1)$, gives an indication for an incipient liquid of ^{6}CFs and ^{8}CFs in the

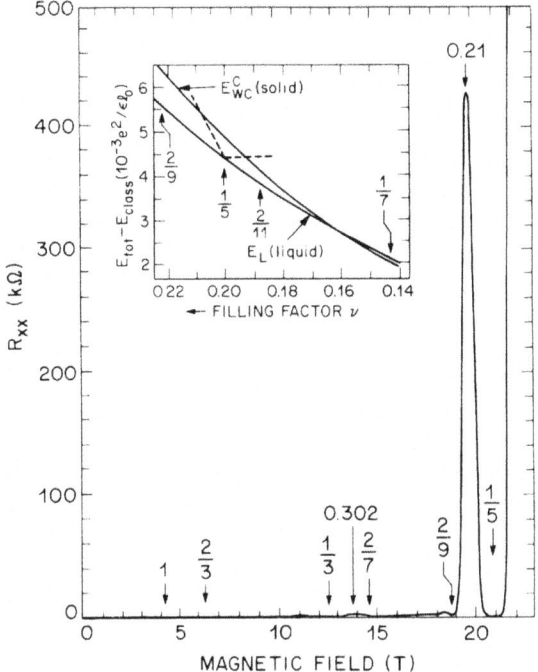

Fig. 15.5. Longitudinal resistance near $\nu = 1/5$ at 90 mK. The inset shows schematically how cusps in the liquid energy could produce re-entrant phase transitions. Source: H. W. Jiang, R. L. Willett, H. L. Stormer, D. C. Tsui, L. N. Pfeiffer, and K. W. West, *Phys. Rev. Lett.* **65**, 633636 (1990). (Reprinted with permission.)

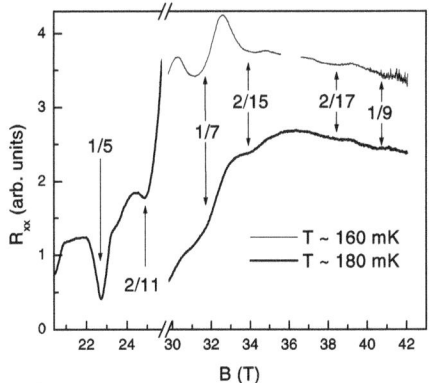

Fig. 15.6. Longitudinal resistance at low fillings. The traces are at relatively high temperatures. No minima at $\nu < 1/5$ are seen below 115 mK. Source: W. Pan, H. L. Stormer, D. C. Tsui, L. N. Pfeiffer, K. W. Baldwin, and K. W. West, *Phys. Rev. Lett.* **88**, 176802 (2002). (Reprinted with permission.)

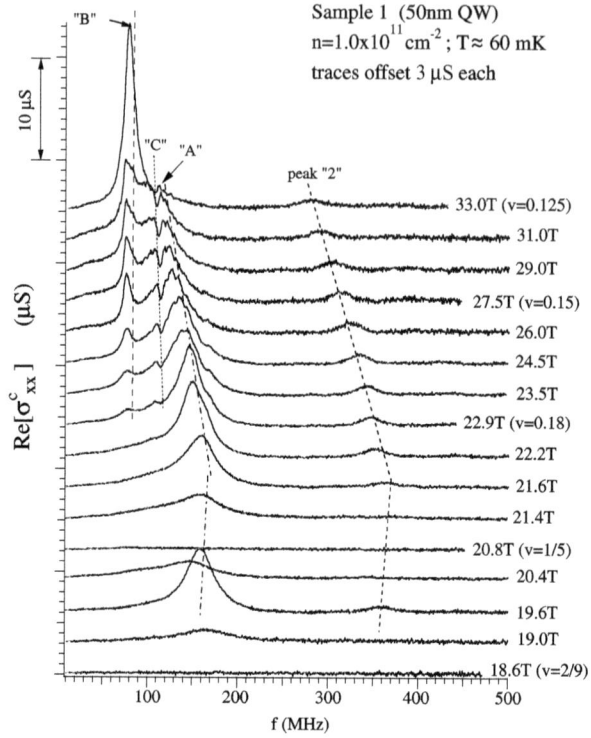

Fig. 15.7. Frequency dependence of the longitudinal conductivity at various fillings in the range $2/9 \geq \nu \geq 1/8$, obtained from microwave absorption. Source: Yong P. Chen, R. M. Lewis, L. W. Engel, D. C. Tsui, P. D. Ye, Z. H. Wang, L. N. Pfeiffer, and K. W. West. *Phys. Rev. Lett.* **93**, 206805 (2004). (Reprinted with permission.)

range $1/5 > \nu \geq 1/9$. Such minima are seen only at relatively high temperatures, however, and Pan *et al.* argue that they appear above the melting temperature of the crystal.

A sharp resonance has been observed down to filling factor $\nu = 1/25$ (see, for example, Ye *et al.* [722]) in the real part of the dynamical conductivity, $\text{Re}[\sigma_{xx}(f)]$, measured by microwave absorption experiments. This resonance has been interpreted as originating from the collective pinning mode of the crystal. (The energy of the pinning mode is equal to the gap that opens, due to disorder, in the long-wave-length limit of the magnetophonon of the solid.) Consistent with this interpretation is the disappearance of this experimental resonance for the FQHE liquids. Figure 15.7 shows the frequency dependence of the real diagonal conductivity at various fillings in the range $2/9 \geq \nu \geq 1/8$. The peak labeled A is absent at 2/9, appears between 2/9 and 1/5, disappears at 1/5, and then reappears at lower filling. Another peak labeled B appears at around $\nu = 1/6$ and dominates at smaller fillings. The observation of two peaks is interpreted as evidence for two different solids.

Given that the crystalline structure of the insulating phase is not yet conclusively established, the question of its composite fermionic character may seem premature.

Nonetheless, in view of the theoretical evidence, we venture several speculations on why certain aspects of experiments are more naturally "understandable" by thinking of the insulating phase as a pinned CF crystal rather than a pinned Wigner crystal. Such considerations may play a role in the eventual confirmation of the solid. Further work will be needed to make the connection with experiment more direct and to clarify other possible implications.

(1) Re-entrant transitions from solid to CF liquid to solid have been observed near $\nu = 1/5$. There is no theoretical reason why composite fermions must shed *all* four of their vortices in this transition, especially given that a lower energy crystal is obtained when they shed only two.

(2) Melting of the low-temperature crystal at small fillings into a CF liquid, as opposed to a featureless electron liquid [495], also becomes less baffling if the crystal is made of composite fermions rather than electrons, thus requiring a less drastic reorganization of the state at the transition.

(3) A qualitative difference between the CF crystal and the Wigner crystal descriptions is that while the latter produces a single, monolithic phase at small fillings, the former produces a sequence of crystals, composed of composite fermions with different flavors. An observation of phase transitions from one type of crystal to another would provide support to the CF crystal. The evidence for two different solids (Fig. 15.7) may be an example of that.

(4) Another result, not fully understood at present, is that the Hall resistance of the insulator, presumably a pinned crystal, has a value close to what it would have for a normal liquid [207,563]. If the current is carried by composite fermions instead, then the Hall voltage induced by the accompanying vortex current (the vortices effectively behave as magnetic flux quanta) produces, through an effective Faraday effect, a nonzero and finite value for the Hall resistance that is not too far from the "classical" value. Zheng and Fertig [746] consider a similar mechanism for transport by correlated interstitial defects.

(5) Narevich, Murthy, and Fertig [470] use the Murthy–Shankar approach for the CF crystal to estimate gaps and shear modulus on either side of the $\nu = 1/5$ quantum Hall state. They note that the measured activation energies are two orders of magnitude smaller than the ones expected from theoretical calculations based on a Hartree–Fock Wigner crystal, but are in much better agreement with the model of a CF crystal, which also explains the observed nonmonotonic filling factor dependence of the gap.

15.5 CF charge density waves

What kinds of states are possible in between the special fillings $f = n/(2pn \pm 1)$, when a Λ level is partially full? We look to electrons for guidance. When electrons partially occupy a Landau level, they can form a CF liquid (filled Λ level state, CF Fermi sea, or a paired CF state), a Wigner crystal, a stripe phase, or a bubble crystal, depending on the LL index and the filling factor within that Landau level. Let us consider the analogous states for composite fermions.

The CF crystal, which is the CF analog of the Wigner crystal, was discussed above. The FQHE of composite fermions produces new fractions in between the fractions $f = n/(2pn \pm 1)$, which was the topic of Section 7.4.1. Both CF crystal and CF FQHE (e.g., 4/11) have significant theoretical support from comparisons against exact results. The same cannot be said of the other CF states considered in the rest of this section. They either are the best known *variational* states, or occur in models in which only two-body interactions between composite fermions in the partially filled Λ level are retained. Because of the approximate nature of these calculations, the conclusions are not fully reliable, and the situation will be clarified only with the help of further theoretical and experimental work.

Fully spin-polarized ^{2}CFs at $\nu^* = n + 1/2$ have been modeled in terms of only the composite fermions in the topmost half-filled Λ level, with an effective interaction shown in Fig. 6.6 (Lee, Scarola, and Jain [382, 383]). A comparison of the variational energies of the paired state and Fermi sea of *composite fermions of composite fermions*, i.e., ^{4}CFs, and of the bubble crystal and stripes of composite fermions (^{2}CFs), shows that, with the exception of $n = 0$ ($\nu = 1/4$), the CF stripe state has the lowest energy. The period was estimated to be $10, 28$, and 34 magnetic lengths, respectively, for $n = 1, 2$, and 3. The period is rather large compared to the period of electron stripes in higher Landau levels, which is not surprising in view of weaker interaction between composite fermions and the smaller difference between the charge densities of the two neighboring FQHE states. Should the stripes at half-filled higher Λ levels of composite fermions be realized, they would show up through anisotropic transport at $\nu = 3/8, 5/12$, and $7/16$. Because of the more fragile nature of CF stripes, the conditions for their observation are expected to be more stringent than those for the electron stripes in higher Landau levels (which are themselves seen only in very high quality samples and at temperatures below $50\,\mathrm{mK}$). Away from the half-filled Λ level, the bubble crystal of CF particles or holes has lower energy than the CF-FQHE state. These calculations [382, 383] rely on two sets of approximations: the use of an effective model interaction, and the choice of variational wave functions. The variational wave functions employed for the CF Fermi sea and the FQHE are known *not* to be quite accurate in higher Landau (Λ) levels. The inadequacy of these calculations is revealed by their inability to explain FQHE at 4/11.

Shibata and Yoshioka [586, 587] have studied the physics of the lowest Landau level by the density matrix renormalization group method, from which they obtain the ground state energy, energy gap, and pair correlation function as a function of the filling factor. They find incompressible and compressible liquids, paired state, and crystal, as well as stripes. The last is found to occur only in samples with very narrow widths.

Goerbig, Lederer, and Morais Smith [198] investigate the issue using the Murthy-Shankar formulation, and argue that the CF liquid has lower energy than the competing solids at fractions like 4/11, but CF stripes produce the best energy in the vicinity of half-filled higher Λ levels. This suggests the possibility of re-entrant phase transitions in the FQHE regime $2/3 > \nu > 1/3$ of the kind observed in higher Landau levels [140, 184, 706].

CF stripes have also been proposed, by Yang, Yu, and Zhu [718], as a possible resolution for the 2/5 state with half the maximum polarization, the experimental evidence for which is presented in Section 11.6. A physical understanding of this state takes the $0\uparrow$ Λ level to be fully occupied, and $1\uparrow$ and $0\downarrow$ to be half occupied. Following the Hartree–Fock approach of Ref. [465], it was argued in Ref. [718] that Λ level mixing favors a unidirectional charge/spin density wave state of composite fermions.

The CF theory has also been applied to the re-entrant phase transitions observed in higher Landau levels (Cooper *et al.* [97]; Eisenstein *et al.* [140]; Gervais *et al.* [184]; Xia *et al.* [706]). Yang, Yu, and Su [716] compare the cohesive energies of the CF Fermi sea, obtained from the numerically obtained pair correlation function of the projected wave functions, with those for the Wigner and bubble crystal phases in higher Landau levels, and determine that the CF Fermi sea has lower energy at 1/6 filling of the third Landau level (i.e., at $\nu = 4 + 1/6$), which may be related to the re-entrant integral quantum Hall effect. Goerbig, Lederer, and Morais Smith [199] also compare the energies of various solid and liquid states in higher Landau levels through a Hartree–Fock calculation for electrons and composite fermions, and find alternation between the IQHE and the FQHE in a certain Landau level index dependent filling factor range. We stress that a charge density wave state in a higher Landau level, either a Wigner or a bubble crystal, is pinned by disorder, resulting in an integral quantum Hall effect.

Appendix A

Gaussian integral

The Gaussian integral is defined as

$$I_n(b) = \int_0^\infty dx\, x^n\, e^{-bx^2}.$$ (A.1)

For $n = 0$, it can be evaluated by noting that the square of the integral is integrable by quadrature:

$$
\begin{aligned}
[2I_0(b)]^2 &= \left(\int_{-\infty}^\infty dx\, e^{-bx^2} \right)^2 \\
&= \int_{-\infty}^\infty dx \int_{-\infty}^\infty dy\, e^{-b(x^2+y^2)} \\
&= \int_0^{2\pi} d\phi \int_0^\infty r\, dr\, e^{-br^2} \\
&= \frac{\pi}{b}.
\end{aligned}
$$ (A.2)

It then follows:

$$I_0(b) = \int_0^\infty dx\, e^{-bx^2} = \frac{1}{2}\sqrt{\frac{\pi}{b}}.$$ (A.3)

$I_1(b)$ is given by

$$I_1(b) = \int_0^\infty dx\, x\, e^{-bx^2} = \frac{1}{2b}.$$ (A.4)

For other integral values of n, $I_n(b)$ can be obtained by taking an appropriate number of derivatives of the above two equations with respect to b. The first few integrals are

$$I_2(b) = \int_0^\infty dx\, x^2\, e^{-bx^2} = \frac{1}{4b}\sqrt{\frac{\pi}{b}},$$ (A.5)

$$I_3(b) = \int_0^\infty dx\, x^3\, e^{-bx^2} = \frac{1}{2b^2},$$ (A.6)

$$I_4(b) = \int_0^\infty dx \ x^4 \ e^{-bx^2} = \frac{3}{8b^2}\sqrt{\frac{\pi}{b}}, \tag{A.7}$$

$$I_5(b) = \int_0^\infty dx \ x^5 \ e^{-bx^2} = \frac{1}{b^3}. \tag{A.8}$$

This is also a good place to note that

$$\int_{-\infty}^\infty dx \ e^{-bx^2 \pm cx} = \sqrt{\frac{\pi}{b}} \ e^{\frac{c^2}{4b}} \tag{A.9}$$

can be evaluated by "completing the square" in the argument of the exponential on the left hand side:

$$bx^2 \mp cx = b\left(x \mp \frac{c}{2b}\right)^2 - \frac{c^2}{4b}. \tag{A.10}$$

More generally, we have (for $b > 0$, $a > -1$):

$$\int_0^\infty dx \ x^a e^{-bx^2} = \frac{1}{2b^{(a+1)/2}} \Gamma\left(\frac{a+1}{2}\right). \tag{A.11}$$

The Γ function is defined as

$$\Gamma(\alpha) = \int_0^\infty e^{-t} t^{\alpha-1} \ dt, \quad (\alpha > 0). \tag{A.12}$$

Because of the property

$$\Gamma(\alpha + 1) = \alpha \Gamma(\alpha), \tag{A.13}$$

it is sufficient to know its value for the interval

$$1 \le \alpha < 2. \tag{A.14}$$

In particular, with

$$\Gamma(1) = 1, \tag{A.15}$$

$$\Gamma\left(\frac{1}{2}\right) = \sqrt{\pi}, \tag{A.16}$$

we can write explicit forms for $\Gamma(\alpha)$ for integral or half-integral values of the argument α:

$$\Gamma(n+1) = n!, \tag{A.17}$$

$$\Gamma\left(n + \frac{1}{2}\right) = \left(n - \frac{1}{2}\right) \cdot \left(n - \frac{3}{2}\right) \cdots \frac{1}{2} \cdot \sqrt{\pi}. \tag{A.18}$$

Appendix B

Useful operator identities

The celebrated Baker–Campbell–Hausdorff formula is given by

$$e^{-A} B e^{A} = B + \sum_{m=1}^{\infty} \frac{(-1)^m}{m!} [A, B]_m, \tag{B.1}$$

with

$$[A, B]_m \equiv [A, [A, \ldots [A, B]]] \qquad (m \text{ commutators}). \tag{B.2}$$

A descendent formula gives us the multiplication rule for two exponentials of linear operators A and B:

$$e^{A} e^{B} = e^{A+B} e^{\frac{1}{2}[A,B]}, \tag{B.3}$$

when both A and B commute with $[A, B]$, i.e., $[A[A, B]] = 0 = [B, [A, B]]$. Other useful expressions are

$$e^{A} e^{B} = e^{B} e^{A} e^{[A,B]}, \tag{B.4}$$

and

$$[B, e^{A}] = [B, A] e^{A}. \tag{B.5}$$

The proof of Eq. (B.3) can be found in standard textbooks [196, 416] and is reproduced here for convenience. We define

$$I(s) \equiv e^{-sA} B e^{sA}. \tag{B.6}$$

When $[A, B]$ commutes with A and B, we have

$$\frac{dI(s)}{ds} = -e^{-sA} [A, B] e^{sA} = -[A, B], \tag{B.7}$$

which can be integrated to arrive at

$$I(s) = B - s[A, B]. \tag{B.8}$$

(This is a special case of Eq. (B.1).) Next we define the operator

$$U(s) = e^{-sA} e^{s(A+B)}, \tag{B.9}$$

which satisfies the differential equation

$$\frac{dU(s)}{ds} = I(s)U(s). \tag{B.10}$$

Because $I(s)$ and $I(s')$ commute according to Eq. (B.8), the solution is

$$U(s) = \exp\left[\int_0^s I(\tau)d\tau\right] = \exp\left[sB - \frac{s^2}{2}[A, B]\right]. \tag{B.11}$$

Equating it to Eq. (B.9) for $s = 1$ yields Eq. (B.3).

Appendix C

Point flux tube

The name "flux tube" refers to a magnetic field of the form

$$\boldsymbol{B}_\alpha = \alpha\phi_0\delta^{(2)}(r)\hat{z}, \tag{C.1}$$

which represents a point flux of strength $\alpha\phi_0 = \alpha hc/e$ threading the x–y plane at the origin in the normal direction. The point flux is created by the vector potential

$$\boldsymbol{A}_\alpha = \frac{\alpha}{2\pi}\phi_0\nabla\theta$$

$$= \frac{\alpha}{2\pi}\phi_0\frac{\hat{\boldsymbol{\theta}}}{r}$$

$$= \frac{\alpha}{2\pi}\phi_0\frac{\hat{z}\times\boldsymbol{r}}{r^2}, \tag{C.2}$$

where θ is the azimuthal coordinate. To see this, we begin by noting that

$$\nabla\times\boldsymbol{A}_\alpha = 0, \quad \boldsymbol{r}\neq\boldsymbol{0}, \tag{C.3}$$

because the curl of the gradient of a regular function vanishes. The function θ is not regular at the origin, so here the curl may not vanish. To obtain its value at the origin, we calculate the flux through an area enclosing the origin:

$$\oint \nabla\times\boldsymbol{A}_\alpha\cdot\mathrm{d}\boldsymbol{s} = \oint \boldsymbol{A}_\alpha\cdot\mathrm{d}\boldsymbol{l}$$

$$= \frac{\alpha}{2\pi}\phi_0\oint\frac{1}{r}\hat{\boldsymbol{\theta}}\cdot(\mathrm{d}r\,\hat{\boldsymbol{r}}+r\,\mathrm{d}\theta\,\hat{\boldsymbol{\theta}})$$

$$= \frac{\alpha}{2\pi}\phi_0\oint\mathrm{d}\theta$$

$$= \alpha\phi_0. \tag{C.4}$$

Therefore, $\nabla\times\boldsymbol{A}_\alpha$ has the form

$$\nabla\times\boldsymbol{A}_\alpha = \alpha\phi_0\delta^{(2)}(r)\hat{z}, \tag{C.5}$$

showing that $\boldsymbol{A}_\alpha$ produces the desired flux through the origin.

Appendix D

Adiabatic insertion of a point flux

Of interest is the question of how a state evolves when we pierce the two-dimensional plane by an infinitesimally thin magnetic solenoid and change the flux through it from 0 to $\pm\phi_0$ adiabatically. This process was used by Laughlin to explain quantization of the Hall resistance [368] and to deduce the existence of fractionally charged excitations [369].

The evolution of an eigenstate of a single electron is straightforward. Consider the single particle problem in the presence of the point test flux tube $\alpha\phi_0$ through the origin:

$$H(\alpha)\Psi(\alpha) = E(\alpha)\Psi(\alpha), \tag{D.1}$$

$$H(\alpha) = \frac{1}{2m_b}\left(p + \frac{e}{c}A + \frac{e}{c}A_t\right)^2, \tag{D.2}$$

with

$$A_t = \frac{\alpha}{2\pi}\phi_0\nabla\theta = \frac{\alpha\phi_0}{2\pi r}\hat{\theta}. \tag{D.3}$$

A_t can be eliminated from the Hamiltonian through the gauge transformation

$$\Psi(\alpha) = \exp(-i\alpha\theta)\Psi'(\alpha). \tag{D.4}$$

The Schrödinger equation then becomes

$$H'(\alpha)\Psi'(\alpha) = E(\alpha)\Psi'(\alpha), \tag{D.5}$$

with

$$H' = \frac{1}{2m_b}\left(p + \frac{e}{c}A\right)^2. \tag{D.6}$$

One may verify that

$$\Psi'(\alpha) = z^{l-\alpha}\,e^{-\frac{1}{4}z\bar{z}} \tag{D.7}$$

is a solution (unnormalized) with eigenvalue $\hbar\omega_c/2$. (It is simplest to use Eq. (3.18).) The parameter l is to be fixed by demanding that

$$\Psi(\alpha) = e^{-i\alpha\theta}\,z^{l-\alpha}\,e^{-\frac{1}{4}z\bar{z}} \tag{D.8}$$

be single valued, which constrains l to be an integer. (Ψ' is not necessarily single valued; it is not a physical wave function, but merely a function that appears in the course of a calculation.)

Let us begin with $\alpha = 0$ and focus on the eigenstate $z^l e^{-z\bar{z}/4}$. As α is tuned adiabatically to $\alpha = \pm 1$, the state evolves into $e^{\mp i\theta} z^{l \mp 1} e^{-z\bar{z}/4}$. Apart from the phase factor $e^{\mp i\theta}$, the wave function is identical to the angular momentum $l \mp 1$ eigenstate $z^{l \mp 1} e^{-z\bar{z}/4}$. Thus, under the adiabatic insertion of flux from 0 to $\pm \phi_0$, each single particle orbital physically moves into the next.[1] "Gauging away" the integral flux quantum (by making a singular gauge transformation) leaves us with $z^{l \mp 1} e^{-z\bar{z}/4}$, which is the solution of the original Hamiltonian at angular momentum $l \pm 1$. (While any point flux can be gauged away, only for integral flux quanta are the boundary conditions the same as they are for $\alpha = 0$, and hence the solution is acceptable for $\alpha = 0$.)

One can thus derive a new eigenstate from a given eigenstate by first adiabatically inserting a point flux $\pm \phi_0$, and then removing it at once by a singular gauge transformation. Because we end up with the same Hamiltonian that we began with, the process is guaranteed to map an eigenstate into another eigenstate. This statement is, in principle, valid for the fully interacting many-electron system, although, in practice, it is impossible to keep track of the adiabatic evolution of such a system, given that we do not even know the exact solution for $\alpha = 0$. Nonetheless, this method has been used to deduce the charge of an excitation of a FQHE state [369].

---

[1] The state $l = 0$ cannot evolve into $l = -1$, because no such state exists in the lowest Landau level. $\Psi' = z^{-1} e^{-\frac{1}{4} z\bar{z}}$ is not a valid wave function because of its lack of normalizability. In a more complete treatment, the $l = 0$ state is seen to move into the next Landau level for $+\phi_0$ flux insertion.

Appendix E
Berry phase

The eigenfunctions and eigenvalues of a Hamiltonian depend on its parameters, denoted collectively by $\boldsymbol{R}$. How does an eigenstate (assumed to be nondegenerate) evolve when $\boldsymbol{R}$ is varied in time? The answer is in general complicated, but a simple result is obtained if the variation is very slow, or adiabatic, so the system remains in the "same" eigenstate. For a complete loop of $\boldsymbol{R}$ in the parameter space, the system returns to its original eigenstate, modulo a phase factor. The change in the phase associated with a closed loop in the parameter space is independent of the rate of traversal around the circuit (provided it is sufficiently slow) and depends only on its geometry. It is called a geometric phase, or a (Pancharatnam–) Berry phase [29, 499].

At any given instant, let $|n(\boldsymbol{R})\rangle$ be the eigenstate of $\hat{H}(\boldsymbol{R})$:

$$\hat{H}(\boldsymbol{R})|n(\boldsymbol{R})\rangle = E_n(\boldsymbol{R})|n(\boldsymbol{R})\rangle. \tag{E.1}$$

For the time-dependent Schrödinger equation,

$$\hat{H}(\boldsymbol{R}(t))|\Psi(t)\rangle = i\hbar \frac{\partial}{\partial t}|\Psi(t)\rangle, \tag{E.2}$$

we try the solution

$$|\Psi(t)\rangle = e^{i\gamma_n(t)}\, e^{-(i/\hbar)\int_0^t dt' E_n(\boldsymbol{R}(t'))}|n(\boldsymbol{R}(t))\rangle, \tag{E.3}$$

where the second phase factor on the right hand side is the familiar dynamical phase factor. A substitution into the Schrödinger equation yields

$$\dot{\gamma}_n(t) = i\langle n(\boldsymbol{R}(t))|\boldsymbol{\nabla}_{\boldsymbol{R}}|n(\boldsymbol{R}(t))\rangle \cdot \dot{\boldsymbol{R}}. \tag{E.4}$$

For a closed loop $\mathcal{C}$ it reduces to

$$\gamma_n = i \oint_{\mathcal{C}} \langle n(\boldsymbol{R}(t))|\boldsymbol{\nabla}_{\boldsymbol{R}}|n(\boldsymbol{R}(t))\rangle \cdot d\boldsymbol{R}. \tag{E.5}$$

The phase is purely real because

$$0 = \boldsymbol{\nabla}_{\boldsymbol{R}}\langle n(\boldsymbol{R}(t))|n(\boldsymbol{R}(t))\rangle$$
$$= 2\mathrm{Real}\langle n(\boldsymbol{R}(t))|\boldsymbol{\nabla}_{\boldsymbol{R}}|n(\boldsymbol{R}(t))\rangle. \tag{E.6}$$

While the overall phase of a wave function is not observable, phase *differences* are. A special case of the Berry phase is the Aharonov–Bohm phase [29], the observable consequences of which are well known. As an instructive exercise, let us see the correspondence between the two for an electron in the lowest Landau level by calculating the Berry phase associated with an electron going around in a loop enclosing an area A. The wave function for a localized wave packet centered at η is given by (Eq. E3.2)

$$\chi_\eta(z) = \frac{1}{\sqrt{2\pi}} \exp\left[\frac{1}{2}\bar{\eta}z - \frac{1}{4}|z|^2 - \frac{1}{4}|\eta|^2\right].$$

(E.7)

Let us now take $\eta \equiv Re^{-i\theta}$ in a circular loop by slowly changing $\theta = 0$ to $\theta = 2\pi$, while holding R constant. The Berry phase associated with this path is given by

$$
\begin{aligned}
\gamma &= \oint dt \left\langle \chi_\eta \left| i\frac{d}{dt} \right. \chi_\eta \right\rangle \\
&= \oint d\theta \left\langle \chi_\eta \left| i\frac{d}{d\theta} \right. \chi_\eta \right\rangle \\
&= -\frac{1}{2} \oint d\theta \int \frac{d^2r}{2\pi} z\bar{\eta} \exp\left[\frac{1}{2}\left(\bar{\eta}z + \eta\bar{z} - r^2 - R^2\right)\right] \\
&= -\frac{1}{2} \oint d\theta \int \frac{d^2r}{2\pi} z\bar{\eta} \sum_{n=0}^{\infty}\sum_{m=0}^{\infty} \frac{(\eta\bar{z})^n}{2^n n!} \frac{(\bar{\eta}z)^m}{2^m m!} \exp\left[-\frac{1}{2}\left(r^2 + R^2\right)\right] \\
&= -\frac{1}{2} \oint d\theta \int_0^\infty r\,dr \sum_{n=1}^{\infty} \frac{r^{2n} R^{2n}}{2^{2n-1} n!(n-1)!} \exp\left[-\frac{1}{2}\left(r^2 + R^2\right)\right] \\
&= -\frac{1}{2} \oint d\theta\, R^2 \\
&= -\frac{\pi R^2}{\ell^2} \\
&= -2\pi \frac{\phi}{\phi_0}.
\end{aligned}
$$

(E.8)

(In the above, the angular integral in d^2r picks out the terms with $m = n - 1$ in the double sum; the Gaussian integral is given by

$$\int_0^\infty dr\, r^{2n+1} e^{-r^2/2} = 2^n n!;$$

(E.9)

and the units have been restored in the penultimate step, with $\ell^2 = \hbar c/eB$.) This is precisely the Aharonov–Bohm phase for an electron loop enclosing a flux $\phi = \pi R^2 B$. The absence of corrections is somewhat surprising in light of the finite size of the coherent wave packet.

Appendix F

Second quantization

F.1 Fermions

"Second quantization" is a convenient method for taking care of the antisymmetry of a many-fermion wave function by introducing certain anticommuting operators. Contrary to what the name suggests, this method does not entail any additional quantization, but is merely a book-keeping device, completely equivalent to the "first quantized" Schrödinger equation. Some scientists have a tendency to use the second quantized language for every many-body physics problem, but that is not always a wise approach. It is most powerful for situations where the physics is roughly described by a single Slater determinant, that is, when the Hartree–Fock approximation is a reasonable starting point. The second quantized formulation then gives a convenient way of organizing a perturbation theory through Feynman diagrams. For the FQHE, which is a strongly correlated system, the "first quantized" wave functions play an important role, but the second quantized formulation becomes useful in the Chern–Simons theory of composite fermions. This appendix contains an introduction to second quantization. Other treatments can be found in standard textbooks (Negele and Orland [476]; Schrieffer [574]).

In general, the wave function for a system of many electrons is complicated. But when the number of occupied single particle orbitals is equal to the number of electrons, only one wave function is possible, given by the Slater determinant:

$$
\Phi_N(\alpha_i, r_j) = \frac{1}{\sqrt{N!}}
\begin{vmatrix}
\eta_{\alpha_1}(r_1) & \eta_{\alpha_1}(r_2) & \cdot & \cdot & \cdot \\
\eta_{\alpha_2}(r_1) & \eta_{\alpha_2}(r_2) & \cdot & \cdot & \cdot \\
\cdot & & \cdot & & \cdot \\
\cdot & & & \cdot & \cdot \\
\cdot & & \cdot & & \cdot
\end{vmatrix}
$$

$$
= \frac{1}{\sqrt{N!}} A[\eta_{\alpha_1}(r_1) \ldots \eta_{\alpha_N}(r_N)]
$$

$$
= \frac{1}{\sqrt{N!}} \sum_P \epsilon_P \prod_j \eta_{\alpha_j}(r_{Pj}),
\tag{F.1}
$$

where $\eta_\alpha(r)$ are the single particle orbitals (with α collectively denoting the quantum numbers), α_j are the quantum numbers of the occupied single particle orbitals, $\{Pj\}$ is a permutation of integers $\{j\}$, and $\epsilon_P = +1$ or -1 depending on whether an even or odd number of permutations bring $\{j\}$ to $\{Pj\}$. A is the antisymmetrization operator, and the sum is over all permutations $\{Pj\}$. The factor $(N!)^{-1/2}$ ensures normalization of the wave function.

At the heart of second quantization lies the fact that any antisymmetric wave function can be written as a linear superposition of Slater determinants, which, therefore, form a complete basis. A Slater determinant is completely determined, up to a sign, by specifying which orbitals are occupied. So, rather than working with the full Slater determinant, let us represent it by a vector in an abstract vector space (called the Fock space) where we simply list the occupied single particle operators:

$$\Phi_N(\alpha_i, r_j) \Leftrightarrow |\alpha_1, \dots, \alpha_N\rangle. \tag{F.2}$$

This is called the occupation number representation. The vector on the right hand side is the "image" of the Slater determinant. The image of the complex conjugate of the Slater determinant is represented by

$$\Phi_N^*(\alpha_i, r_j) \Leftrightarrow \langle\alpha_1, \dots, \alpha_N|. \tag{F.3}$$

The orthonormality of the Slater determinant basis functions then implies

$$\langle\alpha_1, \dots, \alpha_N | \beta_1, \dots, \beta_N\rangle = \epsilon_P \delta_{\{\alpha_j\}\{\beta_j\}}. \tag{F.4}$$

The right hand side is zero unless the sets $\{\alpha_j\}$ and $\{\beta_j\}$ are the same, and $\epsilon_P = +1$ or -1 depending on whether it takes an even or odd number of permutations to arrange $\{\beta_j\}$ in the same order as $\{\alpha_j\}$.

A Fock vector is conveniently represented as

$$|\alpha_1, \dots, \alpha_N\rangle = c_{\alpha_1}^\dagger \cdots c_{\alpha_N}^\dagger |0\rangle, \tag{F.5}$$

where $|0\rangle$ is the state with no particles, and each "creation operator" $c_\alpha^\dagger$ acting on the state adds a particle at the beginning of the list:

$$c_\alpha^\dagger |\alpha_1, \dots, \alpha_N\rangle = |\alpha, \alpha_1, \dots, \alpha_N\rangle. \tag{F.6}$$

The antisymmetry property, that an exchange of any two labels in $|\alpha_1, \dots, \alpha_N\rangle$ produces a negative sign, is incorporated by imposing the anticommutation relations

$$\{c_\alpha^\dagger, c_\beta^\dagger\} = 0, \tag{F.7}$$

because an exchange of two particles in $c_{\alpha_1}^\dagger \cdots c_{\alpha_N}^\dagger |0\rangle$ requires an odd number of anticommutator operations. This also implies that

$$c_\alpha^\dagger |\alpha, \alpha_1, \dots, \alpha_N\rangle = 0. \tag{F.8}$$

Because every basis vector has α either occupied or unoccupied, Eqs. (F.6) and (F.8) completely specify the operator $c_\alpha^\dagger$.

We next define a "destruction" or "annihilation" operator as

$$c_\alpha |\alpha, \alpha_1, \ldots, \alpha_N\rangle = |\alpha_1, \ldots, \alpha_N\rangle \tag{F.9}$$

and

$$c_\alpha |\alpha_1, \ldots, \alpha_N\rangle = 0, \tag{F.10}$$

where $\alpha \neq \alpha_j$. In other words, if the orbital η_α is occupied, c_α removes it (α must first be brought to the left, which may produce a negative sign), and if η_α is not occupied, c_α annihilates the state.

The anticommutators

$$\{c_\alpha, c_\beta\} = 0, \quad \{c_\alpha, c_\beta^\dagger\} = \delta_{\alpha\beta} \tag{F.11}$$

are derived in Exercise F.1. Exercise F.2 proves that the creation and annihilation operators are Hermitian conjugates:

$$c_\alpha^\dagger = (c_\alpha)^\dagger. \tag{F.12}$$

F.1.1 Field operators

Annihilation and creation "field operators" are defined as

$$\hat{\psi}(r) = \sum_\alpha \eta_\alpha(r) c_\alpha, \tag{F.13}$$

$$\hat{\psi}^\dagger(r) = \sum_\alpha \eta_\alpha^*(r) c_\alpha^\dagger, \tag{F.14}$$

where the sum is over a complete basis. They satisfy

$$\{\hat{\psi}(r), \hat{\psi}(r')\} = 0 = \{\hat{\psi}^\dagger(r), \hat{\psi}^\dagger(r')\}, \tag{F.15}$$

$$\{\hat{\psi}(r), \hat{\psi}^\dagger(r')\} = \delta^{(2)}(r - r'). \tag{F.16}$$

F.1.2 Representations of wave functions

We wish to obtain the representation of a wave function $\phi(r_1, \ldots, r_N)$ in the Fock space

$$\phi(r_1, \ldots, r_N) \Leftrightarrow |\phi\rangle. \tag{F.17}$$

Expressing it as a sum of Slater determinant basis functions

$$\phi(r_1, \ldots, r_N) = \sum_{\{\alpha_j\}} C\{\alpha_j\} \Phi_N(\alpha_j, r_k), \tag{F.18}$$

immediately yields

$$|\phi\rangle = \sum_{\{\alpha_j\}} C\{\alpha_j\}|\alpha_1,\ldots,\alpha_N\rangle. \tag{F.19}$$

An equivalent way to write the wave function is

$$|\phi\rangle = \frac{1}{\sqrt{N!}} \int d^2r_1 \ldots d^2r_N \phi(r_1,\ldots,r_N)\hat{\psi}^\dagger(r_1)\ldots\hat{\psi}^\dagger(r_N)|0\rangle. \tag{F.20}$$

It has the interpretation that

$$\hat{\psi}^\dagger(r_1)\ldots\hat{\psi}^\dagger(r_N)|0\rangle \equiv |r_1,\ldots,r_N\rangle \tag{F.21}$$

has particles at $r_1,\ldots,r_N$, and $\phi(r_1,\ldots,r_N)$ is the probability amplitude of this configuration. Equation (F.20) is established by considering a single Slater determinant:

$$\begin{aligned}
|\Phi_N\rangle &= \frac{1}{\sqrt{N!}} \int d^2r_1 \ldots d^2r_N \, \Phi_N(\alpha_i, r_j)\hat{\psi}^\dagger(r_1)\ldots\hat{\psi}^\dagger(r_N)|0\rangle \\
&= \int d^2r_1 \ldots d^2r_N \left[\prod_{j=1}^N \eta_{\alpha_j}(r_j)\right] \hat{\psi}^\dagger(r_1)\ldots\hat{\psi}^\dagger(r_N)|0\rangle \\
&= c_{\alpha_1}^\dagger \ldots c_{\alpha_N}^\dagger |0\rangle \\
&\Leftrightarrow \Phi_N(\eta_{\alpha_i}, r_j).
\end{aligned} \tag{F.22}$$

We have used that each of the $N!$ terms in the expansion of $\Phi_N(\eta_{\alpha_i}, r_j)$ makes an identical contribution to the integral. Equation (F.20) follows because a general wave function is a linear superposition of Slater determinants. The wave function in Eq. (F.20) is automatically normalized provided $\phi(r_1,\ldots,r_N)$ is

$$\begin{aligned}
\langle\phi|\phi\rangle &= \frac{1}{N!} \left[\prod_j \int d^2r_j' \int d^2r_j\right] \phi^*(\{r_j'\})\phi(\{r_j\})\langle 0|\hat{\psi}(r_N')\ldots\hat{\psi}(r_1')\hat{\psi}^\dagger(r_1)\ldots\hat{\psi}^\dagger(r_N)|0\rangle \\
&= \frac{1}{N!} \left[\prod_j \int d^2r_j' \int d^2r_j\right] \phi^*(\{r_j'\})\phi(\{r_j\}) \sum_P (-1)^P \prod_{k=1}^N \delta^{(2)}(r_{Pk}' - r_k) \\
&= \frac{1}{N!} \left[\prod_j \int d^2r_j' \int d^2r_j\right] \phi^*(\{r_j'\})\phi(\{r_j\})N! \prod_{k=1}^N \delta^{(2)}(r_k' - r_k) \\
&= \left[\prod_j \int d^2r_j\right] |\phi(\{r_j\})|^2 \\
&= 1.
\end{aligned} \tag{F.23}$$

The sum at the second step is over all permutations. The third step involves a change of variables $r'_j \to r'_{Qj}$ with $Q = P^{-1}$, and noting that

$$\prod_j d^2 r'_{Qj} = \prod_j d^2 r'_j \qquad (F.24)$$

and

$$\phi(\{r'_{Qj}\}) = (-1)^Q \phi(\{r'_j\}) = (-1)^P \phi(\{r'_j\}). \qquad (F.25)$$

F.1.3 Reduced density matrix

The one-particle reduced density matrix (Appendix H) is defined as

$$\rho_1(r',r) = \langle \hat{\psi}^\dagger(r') \hat{\psi}(r) \rangle, \qquad (F.26)$$

where the brackets on the right hand side the represent the quantum statistical mechanical average

$$\langle \mathcal{O} \rangle = \frac{\text{Tr}(\mathcal{O}\hat{\rho})}{\text{Tr}(\mathcal{O})}, \qquad (F.27)$$

where $\hat{\rho}$ is the density matrix. At zero temperature, it reduces to

$$\rho_1(r',r) = \frac{\langle \phi_0 | \hat{\psi}^\dagger(r') \hat{\psi}(r) | \phi_0 \rangle}{\langle \phi_0 | \phi_0 \rangle}, \qquad (F.28)$$

where $|\phi_0\rangle$ is the ground state. We now derive an expression for it in terms of the ground state wave function.

Assuming that ϕ_0 is normalized, we note

$$\hat{\psi}(r)|\phi_0\rangle = \frac{1}{\sqrt{N!}} \left[\prod_{j=1}^N \int d^2 r_j \right] \phi_0(r_1,\ldots,r_N) \hat{\psi}(r) \hat{\psi}^\dagger(r_1) \ldots \hat{\psi}^\dagger(r_N)|0\rangle$$

$$= \frac{N}{\sqrt{N!}} \left[\prod_{j=2}^N \int d^2 r_j \right] \phi_0(r,r_2,\ldots,r_N) \hat{\psi}^\dagger(r_2) \ldots \hat{\psi}^\dagger(r_N)|0\rangle. \qquad (F.29)$$

In other words, apart from normalization, the application of the annihilation operator $\hat{\psi}(r)$ simply replaces one of the particle coordinates by the variable r:

$$\hat{\psi}(r)|\phi_0\rangle \Leftrightarrow \sqrt{N}\phi_0(r,r_2,\ldots,r_N). \qquad (F.30)$$

It then follows that

$$
\rho_1(\boldsymbol{r}',\boldsymbol{r}) = \frac{\langle \phi_0 | \hat{\psi}^\dagger(\boldsymbol{r}') \hat{\psi}(\boldsymbol{r}) | \phi_0 \rangle}{\langle \phi_0 | \phi_0 \rangle}
$$

$$
= N \left[\prod_{j=2}^{N} \int \mathrm{d}^2 r_j \right] \phi_0^*(\boldsymbol{r}',\boldsymbol{r}_2,\ldots,\boldsymbol{r}_N) \phi_0(\boldsymbol{r},\boldsymbol{r}_2,\ldots,\boldsymbol{r}_N). \tag{F.31}
$$

F.1.4 Representation of operators

So far, we have defined the images of wave functions in the Fock space, which follows from the correspondence in Eq. (F.2). To complete the formulation of the problem in the Fock space, we also need to determine the images of operators. That can, in principle, be done for any operator, because an operator is completely defined by its matrix elements. We now show that certain useful classes of operators have simple representations in terms of the creation and annihilation operators.

We begin with a single particle operator

$$
\hat{O} = \sum_{j=1}^{N} \hat{O}(\boldsymbol{r}_j). \tag{F.32}
$$

This could be, for example, the kinetic energy. A more precise notation would have $\hat{O}^{(j)}(\boldsymbol{r}_j)$ on the right hand side, to indicate that the operator acts only on the jth particle, but we omit the superscript for simplicity; the argument of $\hat{O}(\boldsymbol{r}_j)$ specifies the relevant part of the direct product space. The action of the operator on a single particle orbital gives

$$
\hat{O}(\boldsymbol{r}_j)\eta_\alpha(\boldsymbol{r}_j) = \sum_\beta \eta_\beta(\boldsymbol{r}_j) O_{\beta\alpha}, \tag{F.33}
$$

where the matrix $O_{\beta\alpha}$ can be determined by multiplying both sides by $\eta_\beta^*(\boldsymbol{r}_j)$ and integrating over $\boldsymbol{r}_j$:

$$
O_{\beta\alpha} = \int \mathrm{d}^2 r \, \eta_\beta^*(\boldsymbol{r}) \hat{O}(\boldsymbol{r}) \eta_\alpha(\boldsymbol{r}). \tag{F.34}
$$

To obtain the image of $\hat{O}$ in the Fock space, we act with it on a Slater determinant basis state:

$$
\hat{O}|\alpha_1,\ldots,\alpha_N\rangle \Leftrightarrow \sum_{j=1}^{N} \hat{O}(\boldsymbol{r}_j) \frac{1}{\sqrt{N!}} A[\eta_{\alpha_1}(\boldsymbol{r}_1)\ldots\eta_{\alpha_N}(\boldsymbol{r}_N)]
$$

$$
= \sum_{j=1}^{N} \frac{1}{\sqrt{N!}} A\left[\left(\sum_\beta \eta_\beta(\boldsymbol{r}_j) O_{\beta\alpha_j} \right) \prod_{k}' \eta_{\alpha_k}(\boldsymbol{r}_k) \right]
$$

$$= \sum_{j=1}^{N} \sum_{\beta} O_{\beta\alpha_j} \frac{1}{\sqrt{N!}} A \left[\eta_\beta(\boldsymbol{r}_j) \prod_k{}' \eta_{\alpha_k}(\boldsymbol{r}_k) \right]$$

$$\Leftrightarrow \sum_{j=1}^{N} \sum_{\beta} O_{\beta\alpha_j} c_\beta^\dagger c_{\alpha_j} [c_{\alpha_1}^\dagger \dots c_{\alpha_N}^\dagger |0\rangle]$$

$$= \sum_{\gamma} \sum_{\beta} O_{\beta\gamma} c_\beta^\dagger c_\gamma [c_{\alpha_1}^\dagger \dots c_{\alpha_N}^\dagger |0\rangle]$$

$$= \sum_{\gamma} \sum_{\beta} O_{\beta\gamma} c_\beta^\dagger c_\gamma |\alpha_1, \dots, \alpha_N\rangle. \tag{F.35}$$

The second step follows because the operator commutes with antisymmetrization, being symmetric under an exchange of two coordinates. The prime denotes the condition $k \neq j$. The Slater determinant in the third line differs from the one we started with by the replacement $\eta_{\alpha_j} \leftrightarrow \eta_\beta$, which is accomplished in the Fock space by the operator $c_\beta^\dagger c_{\alpha_j}$. (Reader: please check that no extra signs are introduced.) In the penultimate line, we replace the sum over α_j by the sum over the complete set of single particle orbitals γ, because the terms with $\gamma \neq \alpha_j$ automatically vanish. This step is crucial, because now the operator is written in a form that does not depend on which Fock state it is being applied to. Because the operators are defined by their action on states, we have established the correspondence

$$\sum_{j=1}^{N} \hat{O}(\boldsymbol{r}_j) \Leftrightarrow \sum_{\alpha,\beta} c_\beta^\dagger O_{\beta\alpha} c_\alpha. \tag{F.36}$$

The quantity on the right has exactly the same effect in the Fock space as the quantity on the left does in the real space. In terms of field operators, we have

$$\sum_{j=1}^{N} \hat{O}(\boldsymbol{r}_j) \Leftrightarrow \int d^2r \, \hat{\psi}^\dagger(\boldsymbol{r}) \hat{O}(\boldsymbol{r}) \hat{\psi}(\boldsymbol{r}). \tag{F.37}$$

Next consider a two-particle operator, defined as

$$\hat{V} = \sum_{i \neq j} \hat{V}(\boldsymbol{r}_i, \boldsymbol{r}_j), \tag{F.38}$$

$$\hat{V}(\boldsymbol{r}_i, \boldsymbol{r}_j) \eta_\alpha(\boldsymbol{r}_i) \eta_\beta(\boldsymbol{r}_j) = \sum_{\gamma\delta} \eta_\gamma(\boldsymbol{r}_i) \eta_\delta(\boldsymbol{r}_j) V_{\gamma\delta,\alpha\beta}, \tag{F.39}$$

$$V_{\gamma\delta,\alpha\beta} = \int d^2r \, d^2r' \eta_\gamma^*(\boldsymbol{r}) \eta_\delta^*(\boldsymbol{r}') \hat{V}(\boldsymbol{r}, \boldsymbol{r}') \eta_\alpha(\boldsymbol{r}) \eta_\beta(\boldsymbol{r}'). \tag{F.40}$$

An example of such an operator is the Coulomb interaction between particles.

Proceeding as before, we have

$$
\hat{V}|\alpha_1,\ldots,\alpha_N\rangle = \sum_{i\neq j=1}^{N} \hat{V}(r_i,r_j)\frac{1}{\sqrt{N!}}A\left[\prod_l \eta_{\alpha_l}(r_l)\right]
$$

$$
= \sum_{i\neq j=1}^{N}\frac{1}{\sqrt{N!}}A\left[\left(\sum_{\beta\gamma}\eta_\beta(r_i)\eta_\gamma(r_j)V_{\beta\gamma;\alpha_i\alpha_j}\right)\prod_l{}'\eta_{\alpha_l}(r_l)\right]
$$

$$
= \sum_{i\neq j=1}^{N}\sum_{\beta\gamma}V_{\beta\gamma;\alpha_i\alpha_j}\frac{1}{\sqrt{N!}}A\left[\eta_\beta(r_i)\eta_\gamma(r_j)\prod_l{}'\eta_{\alpha_l}(r_l)\right]
$$

$$
\Leftrightarrow \sum_{i\neq j=1}^{N}\sum_{\beta,\gamma}V_{\beta\gamma;\alpha_i\alpha_j}c_\beta^\dagger c_\gamma^\dagger c_{\alpha_j}c_{\alpha_i}[c_{\alpha_1}^\dagger\ldots c_{\alpha_N}^\dagger|0\rangle]
$$

$$
= \sum_{\beta\gamma\delta\sigma}V_{\beta\gamma;\delta\sigma}c_\beta^\dagger c_\gamma^\dagger c_\sigma c_\delta[c_{\alpha_1}^\dagger\ldots c_{\alpha_N}^\dagger|0\rangle]
$$

$$
= \sum_{\beta\gamma\delta\sigma}V_{\beta\gamma;\delta\sigma}c_\beta^\dagger c_\gamma^\dagger c_\sigma c_\delta|\alpha_1,\ldots,\alpha_N\rangle. \tag{F.41}
$$

The prime denotes the condition $l \neq i,j$. The ordering of operators in the fourth line ought to be noted; it ensures that no extra sign is introduced. We therefore have

$$
\sum_{i\neq j}\hat{V}(r_i,r_j) \Leftrightarrow \sum_{\beta\gamma\delta\sigma}c_\beta^\dagger c_\gamma^\dagger V_{\beta\gamma;\delta\sigma}c_\sigma c_\delta \tag{F.42}
$$

$$
= \int d^2r\int d^2r'\,\hat{\psi}^\dagger(r)\hat{\psi}^\dagger(r')V(r,r')\hat{\psi}(r')\hat{\psi}(r).
$$

F.2 Bosons

Second quantization for bosons can be accomplished similarly, keeping in mind that now the wave function is symmetric under an exchange of any two particles. The (unnormalized) basis functions for bosons are defined as

$$
\Phi_N(\alpha_i,r_j) = \frac{1}{\sqrt{N!}}\sum_P\prod_j\eta_{\alpha_j}(r_{Pj})
$$

$$
\Leftrightarrow ||\alpha_1,\ldots,\alpha_N\rangle\rangle
$$

$$
\equiv c_{\alpha_1}^\dagger\ldots c_{\alpha_N}^\dagger|0\rangle. \tag{F.43}
$$

Symmetrization of a product of single particle states produces a sum over all permutations of N particles in the given N single particle states. Now, of course, many of the α_j may be identical. The double bracket notation reminds us that the basis function defined above is not normalized. The creation operator is defined by

$$c_\alpha^\dagger ||\alpha_1, \alpha_2, \ldots\rangle\rangle = ||\alpha, \alpha_1, \alpha_2, \ldots\rangle\rangle. \tag{F.44}$$

The action of the annihilation operator, on the other hand, depends on whether the single particle state α is occupied or not:

$$c_\alpha ||\alpha, \alpha_1, \alpha_2, \ldots\rangle\rangle = ||\alpha_1, \alpha_2, \ldots\rangle\rangle, \tag{F.45}$$

$$c_\alpha ||\alpha_1, \alpha_2, \ldots\rangle\rangle = 0 \tag{F.46}$$

where in the last $||\ldots\rangle\rangle$ the state α is assumed to be not occupied. The symmetry under exchange is reflected in the commutation relations:

$$[c_\alpha^\dagger, c_\beta^\dagger] = 0, \qquad [c_\alpha, c_\beta] = 0, \qquad [c_\alpha, c_\beta^\dagger] = \delta_{\alpha\beta}. \tag{F.47}$$

With these definitions, the discussion of the previous section goes through, and the representations for operators are identical to those in Eqs. (F.36) and (F.40). It is customary to label the basis states by the number of bosons in any given single particle state:

$$||\alpha_1, \ldots, \alpha_N\rangle\rangle \Leftrightarrow ||\{n_{\alpha_j}\}\rangle\rangle \tag{F.48}$$

where n_α is the number of times the state α appears in the set $\{\alpha_j\}$. We leave it to the reader to show that the normalized basis functions are given by

$$|\{n_{\alpha_j}\}\rangle = \left(\prod_j n_{\alpha_j}!\right)^{-1/2} ||\{n_{\alpha_j}\}\rangle\rangle$$

$$= \left(\prod_j \frac{\left(c_{\alpha_j}^\dagger\right)^{n_{\alpha_j}}}{\sqrt{n_{\alpha_j}!}}\right) |0\rangle. \tag{F.49}$$

The action of the creation and annihilation operators on the normalized wave functions can be readily seen to be

$$c_1^\dagger |n_1, \ldots\rangle = \sqrt{n_1 + 1} |n_1 + 1, \ldots\rangle, \tag{F.50}$$

and

$$c_1 |n_1, \ldots\rangle = \sqrt{n_1} |n_1 - 1, \ldots\rangle. \tag{F.51}$$

Exercises

F.1 Prove the anticommutators in Eq. (F.11) by showing that the equality holds for application on an arbitrary Fock state. (Treat the cases $\alpha \neq \beta$ and $\alpha = \beta$ separately.)

F.2 This exercise concerns a proof of the statement $c_\alpha^\dagger = (c_\alpha)^\dagger$. Consider

$$\langle \alpha_1, \alpha_2, \dots | (c_\alpha)^\dagger = [c_\alpha |\alpha_1, \alpha_2, \dots \rangle]^\dagger$$

and

$$\langle \alpha_1, \alpha_2, \dots | c_\alpha^\dagger.$$

Show that the two are equal by application on an arbitrary Fock vector (which may or may not contain α).

F.3 Determine the Fock space representation of the N body operator $\prod_{j=1}^{N} \hat{T}(r_j)$.

Appendix G
Green's functions, spectral function, tunneling

This appendix lists some standard definitions and results from conventional many-body physics based on diagrammatic perturbation theory. For simplicity, the spin quantum numbers are suppressed and zero temperature is assumed. Extensive treatments can be found in the excellent textbooks by Mahan [416] and Giuliani and Vignale [196].

G.1 Green's functions

The time ordered Green's function is given by

$$
\begin{aligned}
G(\mathbf{r}, t; \mathbf{r}', t') &= -i\langle 0 | T \hat{\psi}(\mathbf{r}, t) \hat{\psi}^\dagger(\mathbf{r}', t') | 0 \rangle \\
&= -i\theta(t - t')\langle 0 | \hat{\psi}(\mathbf{r}, t) \hat{\psi}^\dagger(\mathbf{r}', t') | 0 \rangle \\
&\quad + i\theta(t' - t)\langle 0 | \hat{\psi}^\dagger(\mathbf{r}', t') \hat{\psi}(\mathbf{r}, t) | 0 \rangle \\
&= \theta(t) G^>(\mathbf{r}, t; \mathbf{r}', t') + \theta(-t) G^<(\mathbf{r}, t; \mathbf{r}', t').
\end{aligned}
\tag{G.1}
$$

The second equality defines time ordering (denoted by T) for fermions, and the third defines $G^>(\mathbf{r}, t; \mathbf{r}', t')$ and $G^<(\mathbf{r}, t; \mathbf{r}', t')$. The expectation value is with respect to the fully interacting ground state $|0\rangle$ (assumed to be normalized) in the Heisenberg representation, and the time evolution of operators is governed by the full Hamiltonian. For a time-independent Hamiltonian and a translationally invariant problem, the Green's function depends only on $\mathbf{r} - \mathbf{r}'$ and $t - t'$. The retarded and advanced Green's functions are defined as

$$
G^{\mathrm{ret}}(\mathbf{r}, t; \mathbf{r}', t') = -i\theta(t - t')\langle 0 | \{ \hat{\psi}(\mathbf{r}, t), \hat{\psi}^\dagger(\mathbf{r}', t') \} | 0 \rangle
\tag{G.2}
$$

and

$$
G^{\mathrm{adv}}(\mathbf{r}, t; \mathbf{r}', t') = -i\theta(t' - t)\langle 0 | \{ \hat{\psi}(\mathbf{r}, t), \hat{\psi}^\dagger(\mathbf{r}', t') \} | 0 \rangle,
\tag{G.3}
$$

where the curly brackets denote the anticommutator.

Many exact relations between the Green's functions with different superscripts, as well as between Green's functions and the spectral function (defined in the next section) can be established by expressing them in what is know as the Lehmann representation [385], obtained by inserting a complete set of states between the two fermionic operators. Let us consider, for example, $G^>$ (setting $\hbar = 1$):

$$G^>(r, t; r', t') = -i\langle 0, N | e^{iHt} \hat{\psi}(r) e^{-iH(t-t')} \hat{\psi}^\dagger(r') e^{-iHt'} | 0, N \rangle$$

$$= -i \sum_{\alpha, \beta} \eta_\alpha(r) \eta_\beta^*(r') \sum_m e^{-i(E_m^{N+1} - E_0^N)(t-t')}$$

$$\times \langle 0, N | c_\alpha | m, N+1 \rangle \langle m, N+1 | c_\beta^\dagger | 0, N \rangle, \tag{G.4}$$

where $|m, N+1\rangle$ denote the exact eigenstates of the $N+1$ particle system with eigenenergy E_m^{N+1} (with $m = 0$ reserved for the ground state), and we have used the definition Eq. (F.13) for the field operator. The "Fourier transform" (in d spatial dimensions) is given by

$$G^>(\alpha, \beta, E) = \int d^d r \int d^d r' \int d(t - t') e^{iE(t-t')} \eta_\alpha^*(r) \eta_\beta(r') G^>(r, t; r', t')$$

$$= -2\pi i \sum_m \langle 0, N | c_\alpha | m, N+1 \rangle \langle m, N+1 | c_\beta^\dagger | 0, N \rangle$$

$$\times \delta(E - E_m^{N+1} + E_0^N). \tag{G.5}$$

The quantum numbers α and β often label conserved quantities (for examples, the wave vector or the angular momentum), in which case $G^>(\alpha, \beta, E) = \delta_{\alpha, \beta} G^>(\alpha, E)$, and

$$G^>(\alpha, E) = -2\pi i \sum_m |\langle m, N+1 | c_\alpha^\dagger | 0, N \rangle|^2 \delta(E - E_m^{N+1} + E_0^N). \tag{G.6}$$

The noninteracting time ordered Green's function plays a central role in diagrammatic perturbation theory, because Wick's theorem expresses time ordered products of operators in terms of products of time ordered Green's functions. The poles of the (fully interacting) Green's functions identify exact excitation energies of the system (which becomes explicit in the Lehmann representation). The Green's functions also yield the experimentally measurable spectral function; the spectral function is useful for the question of whether electron-like quasiparticles are well defined (as assumed in Landau's Fermi liquid theory), and, if so, what is their lifetime. The one-particle reduced density matrix can also be obtained from the Green's function:

$$\rho_1(r, r') = \lim_{t'-t \to 0^-} -iG(r', t'; r, t) = -iG^<(r', 0; r, 0). \tag{G.7}$$

G.2 Spectral function

The electron spectral function is an important quantity in the standard many-body physics. It is defined (at zero temperature) as

$$A(\alpha, E) = A^>(\alpha, E) + A^<(\alpha, E), \tag{G.8}$$

where $A^>$ and $A^<$ are the positive and negative energy parts, defined as

$$A^>(\alpha, E) = \sum_m |\langle m, N+1|c_\alpha^\dagger|0, N\rangle|^2 \delta(E - E_m^{N+1} + E_0^N) \qquad (G.9)$$

and

$$A^<(\alpha, E) = \sum_m |\langle m, N-1|c_\alpha|0, N\rangle|^2 \delta(E + E_m^{N-1} - E_0^N). \qquad (G.10)$$

Here $|0, N\rangle$ is the ground state of the N particle interacting system, $c_\alpha^\dagger$ creates an electron with quantum numbers α (for the ordinary Landau Fermi liquid, α usually denotes the momentum and spin quantum numbers; for the quantum Hall effect, α could denote the angular momentum), c_α destroys an electron in the state α, and $|m, N \pm 1\rangle$ denotes the exact eigenstates of the fully interacting $N \pm 1$ particle system. Because the right hand side of $A(\alpha, E)$ contains only non-negative factors, we have $A(\alpha, E) \geq 0$. In addition, it satisfies the property

$$\int dE\, A(\alpha, E) = 1, \qquad (G.11)$$

because the anticommutator $\{c_\alpha, c_\alpha^\dagger\} = 1$. The spectral function is therefore interpreted as a probability function: It is the probability that an electron or a hole added in the state α has energy E. The positive (negative) energy part tells us how an electron (hole) added in a state with definite quantum numbers α distributes itself over the exact eigenstates of the fully interacting $N + 1$ $(N - 1)$ particle system.

One often writes

$$E - E_m^{N+1} + E_0^N = E - \mu^N - \Delta E_m^{N+1} \qquad (G.12)$$

and

$$E + E_m^{N-1} - E_0^N = E - \mu^{N-1} + \Delta E_m^{N-1}, \qquad (G.13)$$

where the chemical potential $\mu^N = E_0^{N+1} - E_0^N$ is usually N independent in the thermodynamic limit, and $\Delta E_m^{N\pm1} = E_m^{N\pm1} - E_0^{N\pm1}$ is a non-negative excitation energy. It then follows, for zero temperature, that $A^>(\alpha, E)$ vanishes for $E < \mu$, whereas $A^<(\alpha, E)$ vanishes for $E > \mu$ (hence the name positive and negative energy parts).

The spectral function can be obtained from the Green's functions. For example, the results in the previous section imply that [196]

$$A^>(\alpha, E) = -\frac{1}{2\pi i} G^>(\alpha, E). \qquad (G.14)$$

$A^<$ is similarly related to $G^<$. (Another relation is derived in Exercise 8.2.)

For concreteness, consider momentum eigenstates $\eta_k(r) = L^{-d/2}e^{ik\cdot r}$. Then, assuming translational invariance, Eq. (G.4) reduces to

$$G^>(r, t; r', t') = -i \sum_k \sum_m e^{-i(E_m^{N+1} - E_0^N)(t-t')} \frac{e^{ik\cdot(r-r')}}{L^d} |\langle m, N+1|c_k^\dagger|0, N\rangle|^2. \qquad (G.15)$$

The Fourier transform, defined as

$$G^>(\boldsymbol{p}, E) = \int dt \int d^d r \, e^{iEt - i\boldsymbol{p}\cdot\boldsymbol{r}} \, G^>(\boldsymbol{r}, t; \boldsymbol{0}, 0), \tag{G.16}$$

is given by

$$G^>(\boldsymbol{p}, E) = -2\pi i \sum_m |\langle m, N+1 | c_p^\dagger | 0, N \rangle|^2 \delta(E - E_m^{N+1} + E_0^N). \tag{G.17}$$

In particular, we have

$$A^>(\boldsymbol{p}, E) = -\frac{1}{2\pi i} \int dt \int d^d r \, e^{iEt - i\boldsymbol{p}\cdot\boldsymbol{r}} \, G^>(\boldsymbol{r}, t; \boldsymbol{0}, 0). \tag{G.18}$$

G.3 Tunneling

The electron spectral function can be measured in tunneling experiments where an electron is injected from the outside. The current between two weakly coupled systems (labeled by L and R), held at a chemical potential difference of eV, is given by [196, 416]

$$I(eV) \propto \sum_{\alpha,\beta} |T_{\alpha,\beta}|^2 \int_0^{eV} dE \, A_L(\alpha, E) \, A_R(\beta, eV - E), \tag{G.19}$$

where α and β label the quantum numbers of the quasiparticles on the two sides, all energies are measured from the chemical potential, and $T_{\alpha,\beta}$ is the matrix element connecting the two states. Only $A_L^>$ and $A_R^<$ contribute to the integral. If $T_{\alpha,\beta}$ is independent of α and β, then

$$I(eV) \propto |T|^2 \int_0^{eV} dE \, D_L(E) \, D_R(eV - E), \tag{G.20}$$

where the tunneling density of states is defined as

$$D(E) = \sum_\alpha A(\alpha, E). \tag{G.21}$$

If one of the two systems is an ordinary metal, for which the tunneling density of states can be taken as a constant, then

$$\frac{dI}{dV} \propto D(eV), \tag{G.22}$$

where the right hand side is the density of states for the other system. The differential conductance for tunnel transport can thus be used as a direct measure of the tunneling density of states.

Exercises

G.1 By going to the Lehmann representation, establish that

$$[G^{\text{ret}}(\boldsymbol{k}, E)]^* = G^{\text{adv}}(\boldsymbol{k}, E).$$

Here, the following representation of the step function may be useful:

$$\theta(t - t') = -\int_{-\infty}^{\infty} \frac{dE}{2\pi i} \frac{e^{-iE(t-t')}}{E + i\eta},$$

where $\eta \to 0^+$; this can be readily established by closing the contour in the upper or lower half of the complex E plane, depending on whether $t - t' < 0$ or $t - t' > 0$, respectively.

G.2 Assuming translational invariance, show that

$$A(\boldsymbol{k}, E) = -\frac{1}{\pi} \text{Im}\, G^{\text{ret}}(\boldsymbol{k}, E).$$

Appendix H
Off-diagonal long-range order

H.1 ODLRO: definition

Bose–Einstein condensation (BEC) refers to a macroscopic occupation of the zero momentum state. BEC is associated with an off-diagonal long-range order (ODLRO) in real space (Penrose [510, 511]; Yang [712]). We give here a brief account of the general concept of ODLRO. BEC has an intimate connection with superfluidity.[1]

We consider a system of N bosons in a periodic box of volume V. The reduced one-particle density matrix $\hat{\rho}_1$ is defined through its matrix element

$$\rho_1(r, r') \equiv \langle \hat{\psi}^\dagger(r) \hat{\psi}(r') \rangle. \tag{H.1}$$

$\hat{\psi}(r)$ is the field operator for bosons:

$$\hat{\psi}(r) = \sum_k \frac{e^{ik \cdot r}}{\sqrt{V}} b_k, \tag{H.2}$$

where b_k annihilates a boson in the state k. The average of an operator is defined in the usual manner as

$$\langle \mathcal{O} \rangle = \text{Tr}(\mathcal{O}\hat{\rho}), \tag{H.3}$$

where $\hat{\rho}$ is the density matrix. The trace is independent of the basis. At zero temperature, $\langle \cdots \rangle$ is equal to the expectation value with respect to the ground state. For $r = r'$, the quantity $\langle \hat{\psi}^\dagger(r) \hat{\psi}(r) \rangle$ is equal to the density. At zero temperature, $\rho_1(r, r')$ can be expressed in terms of the ground state wave function, as shown in Eq. (F.31).

Definition: The term ODLRO refers to the property [510, 511]

$$\lim_{|r-r'| \to \infty} \rho_1(r, r') = \lim_{|r-r'| \to \infty} \langle \hat{\psi}^\dagger(r) \hat{\psi}(r') \rangle \neq 0, \tag{H.4}$$

which signifies a long-range order in the off-diagonal matrix element of the one-particle reduced density matrix in the coordinate representation.

---

[1] There *are* examples where one occurs without the other. BEC in a *non*interacting Bose gas is not a superfluid, and a Kosterlitz–Thouless superfluidity occurs in two dimensions without BEC.

The connection with BEC is established as follows:

$$\langle \hat{\psi}^\dagger(\mathbf{r})\,\hat{\psi}(\mathbf{r}') \rangle = \frac{1}{V} \sum_{\mathbf{k},\mathbf{k}'} e^{-i\mathbf{k}\cdot\mathbf{r}}\,e^{i\mathbf{k}'\cdot\mathbf{r}'}\,\langle b_{\mathbf{k}}^\dagger b_{\mathbf{k}'} \rangle$$

$$= \frac{1}{V} \sum_{\mathbf{k}} e^{-i\mathbf{k}\cdot(\mathbf{r}-\mathbf{r}')} n_{\mathbf{k}}$$

$$\rightarrow \frac{n_0}{V}, \tag{H.5}$$

where $n_{\mathbf{k}}$ is the Bose occupation factor. We have assumed translational invariance, which implies that $\langle b_{\mathbf{k}}^\dagger b_{\mathbf{k}'} \rangle = \delta_{\mathbf{k}\mathbf{k}'} n_{\mathbf{k}}$. In the last step, we have taken the limit $|\mathbf{r}-\mathbf{r}'| \rightarrow \infty$, where only the term $\mathbf{k} \rightarrow 0$ survives (the other terms vanish upon integration because of strong fluctuations in the phase factor). ODLRO thus implies a nonzero value for n_0/V, hence Bose–Einstein condensation.

A necessary and sufficient condition for ODLRO is that $\rho_1(\mathbf{r},\mathbf{r}')$ have a large eigenvalue (Yang [712]), where large means $O(N)$ or $O(V)$. The eigenvalue equation is given by

$$\int d\mathbf{r}'\rho_1(\mathbf{r},\mathbf{r}')\Phi_n(\mathbf{r}') = \lambda_n\Phi_n(\mathbf{r}), \tag{H.6}$$

where λ_n are the eigenvalues and $\Phi_n(\mathbf{r})$ the eigenfunctions. To show that Eq. (H.4) implies a large eigenvalue, we substitute $\Phi(\mathbf{r}) = 1/\sqrt{V}$. It is an eigenfunction, with eigenvalue given by

$$\lambda = \int d\mathbf{r}'\rho_1(\mathbf{r},\mathbf{r}') = O(V). \tag{H.7}$$

The second equality follows from Eq. (H.4). To prove the converse, that an $O(N)$ eigenvalue implies Eq. (H.4), we express $\rho_1(\mathbf{r},\mathbf{r}')$ in the spectral representation:

$$\rho_1(\mathbf{r},\mathbf{r}') = \sum_n \lambda_n\Phi_n(\mathbf{r})\Phi_n^*(\mathbf{r}')$$

$$= \lambda\Phi(\mathbf{r})\Phi^*(\mathbf{r}') + \rho_1'(\mathbf{r},\mathbf{r}'), \tag{H.8}$$

where the term corresponding to the large eigenvalue has been separated in the last step. The eigenvalue $\Phi(\mathbf{r})$ contributes a factor $1/\sqrt{V}$, so if λ is $O(V)$, then the right hand side is nonzero in the limit $V \rightarrow \infty$ (with $\rho_1' \rightarrow 0$). The presence of ODLRO thus implies that the one-particle reduced density matrix factorizes as

$$\rho_1(\mathbf{r},\mathbf{r}') \rightarrow \lambda\Phi(\mathbf{r})\Phi^*(\mathbf{r}'), \quad |\mathbf{r}-\mathbf{r}'| \rightarrow \infty, \tag{H.9}$$

where λ is the large eigenvalue and $\Phi(\mathbf{r})$ the corresponding eigenfunction.

The particle number is assumed to be fixed in the above discussion. By going to the grand canonical ensemble, one can write wave functions for which the condition for ODLRO simplifies to

$$\langle \hat{\psi}(\mathbf{r}) \rangle \neq 0, \tag{H.10}$$

where now the expectation value is with respect to ground states with different numbers of bosons (N bosons in the ket and $N - 1$ in the bra). In the Fourier space, we have

$$\langle \hat{\psi}(r) \rangle = \frac{1}{\sqrt{V}} \sum_k e^{ik \cdot r} \langle b_k \rangle. \tag{H.11}$$

For a macroscopic occupation of the zero momentum state, we have

$$\langle b_k \rangle = \sqrt{n_0}, \tag{H.12}$$

producing a nonvanishing value for $\langle \hat{\psi}(r) \rangle$.

H.2 ODLRO for fermions

It follows from the above discussion that ODLRO in the one-particle reduced density matrix cannot occur for fermions, because the occupation factor for any given single particle state is no greater than unity. Superconductivity is characterized instead by ODLRO in the coordinate representation of the *two*-particle reduced density matrix [712]:

$$\lim_{|r-r'| \to \infty} \langle \hat{\psi}_\uparrow^\dagger(r) \hat{\psi}_\downarrow^\dagger(r) \hat{\psi}_\downarrow(r') \hat{\psi}_\uparrow(r') \rangle \neq 0. \tag{H.13}$$

$\hat{\psi}_\uparrow(r)$ and $\hat{\psi}_\downarrow(r)$ are now the annihilation field operators for spin-up and spin-down electrons, with the subscript denoting the direction of the electron spin. An ODLRO implies that if a pair of electrons is destroyed at r' in the ground state and another is created at a far away point r, then the resulting state has a nonzero overlap with the ground state. By analogy to the previous subsection, such an ODLRO is interpreted as a Bose–Einstein condensation of Cooper pairs.

ODLRO can be demonstrated in the BCS wave function

$$|\Psi_{\text{BCS}}\rangle = \prod_k (u_k + v_k c_{-k\downarrow}^\dagger c_{k\uparrow}^\dagger)|0\rangle, \tag{H.14}$$

where, following the standard notation, u_k and v_k are the coherence factors (not to be confused with the spherical coordinates). The two-particle reduced density matrix is given by

$$\langle \hat{\psi}_\uparrow^\dagger(r) \hat{\psi}_\downarrow^\dagger(r) \hat{\psi}_\downarrow(r') \hat{\psi}_\uparrow(r') \rangle$$

$$= \frac{1}{V^2} \sum_{k,k',q} e^{-iq \cdot (r-r')} \langle c_{k'\uparrow}^\dagger c_{q-k'\downarrow}^\dagger c_{q-k\downarrow} c_{k\uparrow} \rangle$$

$$\to \sum_{k,k'} \left\langle \left(\frac{1}{V} \sum_{k'} c_{k'\uparrow}^\dagger c_{-k'\downarrow}^\dagger \right) \left(\frac{1}{V} \sum_k c_{-k\downarrow} c_{k'\uparrow} \right) \right\rangle$$

$$= \left(\frac{1}{V} \sum_k u_k v_k \right)^2$$

$$= \left(\frac{1}{V} \sum_k \frac{\Delta_k}{2E_k} \right)^2$$

$$\approx \left(\frac{\Delta}{W} \right)^2. \tag{H.15}$$

In the second step the limit $|r - r'| \to \infty$ has been taken, which picks out $q = 0$. In the third step, the expectation value has been evaluated with respect to the BCS wave function, which requires a little algebra. In the subsequent steps, we have used the following results from the BCS theory:

$$u_k v_k = \frac{\Delta_k}{2E_k}, \tag{H.16}$$

$$E_k = +\sqrt{\Delta_k^2 + \epsilon_k^2}, \tag{H.17}$$

$$\Delta_k = -\frac{1}{V} \sum_{k'} W_{kk'} \frac{\Delta_{k'}}{2E_{k'}}, \tag{H.18}$$

where ϵ_k is the kinetic energy measured from the chemical potential and $W_{kk'}$ is the interaction. The last step in Eq. (H.15) results after making the simplifying assumption for the form of the interaction

$$W_{kk'} = -W; \qquad |\epsilon_k|, |\epsilon_{k'}| < \hbar \omega_D, \tag{H.19}$$

where $\hbar \omega_D$ is the Debye energy, and $W_{kk'} = 0$ otherwise. With this form, Δ_k becomes k independent for k satisfying $|\epsilon_k| < \hbar \omega_D$ and zero otherwise, the sums are restricted to the same energy shell near the Fermi energy, and the gap equation (Eq. H.18) reduces to

$$1 = \frac{W}{V} \sum_k \frac{1}{2E_k}. \tag{H.20}$$

The BCS wave function is expressed in the grand canonical ensemble, i.e., is a superposition of states with different numbers of particles. We leave it to the reader to confirm that the "pair amplitude" $\langle \hat{\psi}_\uparrow^\dagger(r) \hat{\psi}_\downarrow^\dagger(r) \rangle$, which is the standard choice for the order parameter of a superconductor, has a nonzero expectation value

$$\langle \hat{\psi}_\uparrow^\dagger(r) \hat{\psi}_\downarrow^\dagger(r) \rangle = \frac{1}{V} \sum_k \frac{\Delta_k}{2E_k} \neq 0. \tag{H.21}$$

For a fixed N, $\langle \hat{\psi}_\uparrow^\dagger(r) \hat{\psi}_\downarrow^\dagger(r) \rangle$ must vanish and one must go back to Eq. (H.13).

Appendix I
Total energies and energy gaps

This appendix deals with the definitions of the total energy and energy differences in the spherical geometry.

I.1 Incorporating the neutralizing background

The total energy is given by

$$E = E_{\text{el}-\text{el}} + E_{\text{el}-\text{bg}} + E_{\text{bg}-\text{bg}}, \tag{I.1}$$

where "el" refers to electron and "bg" to background. The positive charge of the background is assumed to be uniformly distributed. None of the individual energies on the right hand side has a well-defined thermodynamic limit. Each term is proportional to N^2, but the sum is properly extensive, proportional to N.

The last two terms are determined straightforwardly in the spherical geometry, which has N electrons on the surface, and a background charge $+Ne$ distributed uniformly on the surface. The energy of the interaction between electrons and the background,

$$E_{\text{el}-\text{bg}} = -\frac{N^2}{\sqrt{Q}} \frac{e^2}{\epsilon \ell}, \tag{I.2}$$

follows by noting that all of the background charge $+Ne$ can be taken to be located at the origin, and the radius of the sphere is $R = \sqrt{Q}\ell$. The energy of the background charge interacting with itself is similarly given by

$$E_{\text{bg}-\text{bg}} = \frac{N^2}{2\sqrt{Q}} \frac{e^2}{\epsilon \ell}, \tag{I.3}$$

with the factor of $1/2$ correcting for double counting. The total energy is thus

$$E = E_{\text{el}-\text{el}} - \frac{N^2}{2\sqrt{Q}} \frac{e^2}{\epsilon \ell}. \tag{I.4}$$

The energies quoted in this book generally include the contributions $E_{\text{el}-\text{bg}}$ and $E_{\text{bg}-\text{bg}}$.

I.2 Density correction

The incompressible state at $\nu = n/(2pn + 1)$ occurs at Q given by Eq. (5.58). The density of electrons is N dependent, given by

$$\rho_N = \frac{N}{4\pi R^2} = \frac{N}{4\pi Q\ell^2},$$
(I.5)

which differs slightly from the thermodynamic density

$$\rho_\infty = \frac{\nu}{2\pi \ell^2}.$$
(I.6)

Because the energy is proportional to the inverse interparticle separation, i.e., to $\sqrt{\rho_N}$, some of the N dependence can be eliminated [456] by defining

$$E_N' = \left(\frac{\rho_\infty}{\rho_N}\right)^{1/2} E_N = \left(\frac{2Q\nu}{N}\right)^{1/2} E_N.$$
(I.7)

E_N and E_N' are identical in the thermodynamic limit, but the latter has a weaker dependence on N.

I.3 Neutral excitation

A neutral excitation (i.e., a CF-exciton) of an incompressible state is obtained by exciting a composite fermion from an occupied Λ level to an empty Λ level. Such an excitation does not involve any change in the value of either N or Q. Therefore, E_{el-bg} and E_{bg-bg}, being the same for the ground and excited states, do not contribute to the energy gaps. Density correction for both the ground and the excited states gives better extrapolations to the thermodynamic limit.

I.4 Charged excitations

The energy gap to creating a far-separated CF-quasiparticle–CF-quasihole pair, identified with the gap measured in transport experiments, can be determined in two ways.

(i) It can be obtained by evaluating the thermodynamic limit of the energy of the CF-exciton on the sphere with the CF-quasiparticle and CF-quasihole at the farthest separation, i.e., at the opposite poles. (This corresponds to the large wave vector limit of the CF-exciton dispersion.) A shortcoming of this method is that, for finite systems, the energy also includes the interaction between the CF-quasiparticle and the CF-quasihole, which does not have a smooth dependence on N, and a determination of the thermodynamic limit often requires a study of rather large systems. The interaction energy for two point particles of charge $e^* = +e/(2pn + 1)$ and $-e^*$ at the two poles is given by

$$V_{\text{int}} = -\frac{e^{*2}}{\epsilon(2R)},$$
(I.8)

which is usually subtracted from the CF-exciton energy before taking the thermodynamic limit.

(ii) The second method evaluates the energy of a CF-quasiparticle and a CF-quasihole separately, the sum of which gives the transport gap. Considering again the filling factor $\nu = n/(2pn + 1)$, a single CF-quasiparticle or CF-quasihole is obtained at (Eq. E8.15)

$$Q' = \left(p + \frac{1}{2n}\right)N' - \left(p + \frac{n}{2} \pm \frac{1}{n}\right). \tag{I.9}$$

(The CF-quasiparticle or the CF-quasihole is the *ground* state of the system of N' electrons at monopole strength Q'.) We consider, for simplicity, the CF-quasiparticle below (the CF-quasihole being completely analogous), denoting various quantities for the CF-quasiparticle state by primed symbols (N', Q', R', ℓ').

The charge density of the state containing the CF-quasiparticle is uniform everywhere except in the vicinity of the CF-quasiparticle, where an excess charge $qe = -e/(2pn + 1)$ is present. (Of course, the state contains an integral number of electrons.) To make sure that the positively charged background cancels the electron charge density in the regions where the latter is uniform, but not in the vicinity of the CF-quasiparticle [222], we take the total background charge to be $(N' + q)e$. Now we have, in units of $e^2/\epsilon\ell'$,

$$E_{\text{bg-bg}} + E_{\text{el-bg}} = \frac{(N' + q)^2}{2\sqrt{Q'}} \frac{e^2}{\epsilon\ell'} - \frac{N'(N' + q)}{\sqrt{Q'}} \frac{e^2}{\epsilon\ell'}$$

$$= -\frac{N'^2 - q^2}{2\sqrt{Q'}} \frac{e^2}{\epsilon\ell'}. \tag{I.10}$$

A complication is that the CF-quasiparticle state occurs at different N' or Q' values than the uniform density incompressible state. Let us first take the simplest case of $n = 1$, where the two states occur at the same N but for different monopole strengths. It is customary to define the "proper" energy [151, 230, 455] of the CF-quasiparticle as the energy difference keeping the *radius* of the sphere constant:

$$R = \sqrt{Q}\ell = \sqrt{Q'}\ell', \tag{I.11}$$

which requires different magnetic fields B and B', related by

$$\frac{Q}{B} = \frac{Q'}{B'}. \tag{I.12}$$

The energy of the state containing the CF-quasiparticle is naturally obtained in units of $e^2/\epsilon\ell'$, which is converted into the units $e^2/\epsilon\ell$ as

$$\frac{e^2}{\epsilon\ell'} = \left(\frac{Q'}{Q}\right)^{1/2} \frac{e^2}{\epsilon\ell}. \tag{I.13}$$

Subtracting the energy of the uniform density incompressible state gives the proper energy for the CF-quasiparticle. The proper energy of the CF-quasihole can be obtained similarly.[1]

When both Q' and N' are different from Q and N, it is convenient to take the energies of the uniform state for the allowed values of N, and then interpolate to determine what its energy would have been at N'. One then proceeds in the same manner as above.

Exercise

I.1 As a concrete example, consider the state with $N = 8$ at $\nu = 1/3$. From exact diagonalization, we have the interaction energies $6.362\,649e^2/\epsilon\ell$ and $6.593\,175e^2/\epsilon\ell'$ for the ground and the CF-quasiparticle states. Determine Q, Q', and the background corrections for the two states. Calculate the energy per particle for the uniform state, and the proper energy of the CF-quasiparticle. (Do not worry about the density correction for this problem.) (Fano, Ortolani, and Colombo [151].)

---

[1] Another contribution to the energy comes from the kinetic energy difference

$$N\frac{\hbar eB'}{2m_b c} - N\frac{\hbar eB}{2m_b c} = \frac{\Delta Q}{Q}N\frac{\hbar eB}{2m_b c},\tag{EI.14}$$

where $\Delta Q = Q' - Q$. The net contribution to the kinetic energy for the CF-quasiparticle and the CF-quasihole drops out, because the ΔQ's for them are equal in magnitude but differ in sign. Therefore, it is common to drop the kinetic energy contribution even when quoting the energy of a single CF-quasiparticle or a single CF-quasihole. (Only the sum of the CF-quasiparticle and CF-quasihole energies is physically meaningful.)

Appendix J

Lowest Landau level projection

Section 5.14.4 shows that unprojected wave functions of the type $\Phi\Phi_1^{2p}$ can be projected into the lowest Landau level by replacing the single particle wave functions in Φ by certain operators. This appendix contains the explicit forms for these operators in the disk and the spherical geometries. Further details can be found in Jain and Kamilla [280, 281], Möller and Simon [451], and Ruuska and Manninen [561].

J.1 Disk geometry

Theorem J.1 For arbitrary k, we can write

$$\mathcal{P}_{\text{LLL}}\eta_{n,m}(z,\bar{z})z^k = \hat{\eta}_{n,m}(z,\bar{z})z^k, \tag{J.1}$$

where

$$\hat{\eta}_{n,m}(\mathbf{r}_j) = N_{n,m}\frac{(-1)^n}{n!}\,e^{-\frac{1}{4}|z|^2}z^m\frac{\partial^n}{\partial z^n}. \tag{J.2}$$

Proof

In the disk geometry the single electron eigenstates are given by

$$\eta_{n,m}(z,\bar{z}) = N_{n,m}\,e^{-\frac{1}{4}|z|^2}z^m L_n^m\left(\frac{z\bar{z}}{2}\right), \tag{J.3}$$

where the magnetic length is set equal to unity, the normalization coefficient is

$$N_{n,m} = (-1)^n\sqrt{\frac{n!}{2\pi\,2^m(n+m)!}}, \tag{J.4}$$

$n = 0, 1, 2, \ldots$ is the LL index, and the angular momentum is given by $m = -n, -n+1, \ldots$ Using the explicit form for L_n^m we write

$$\eta_{n,m}(z) = N_{n,m}\,e^{-\frac{1}{4}|z|^2}\sum_{k=k_0}^{n}(-1)^k\binom{n+m}{n-k}\frac{1}{2^k k!}\bar{z}^k z^{k+m}, \tag{J.5}$$

where $k_0 = \max(0, -m)$.

490

We proved in Section 5.14

$$\mathcal{P}_{\mathrm{LLL}} \bar{z}^n z^k \, e^{-\frac{1}{4}z\bar{z}} = e^{-\frac{1}{4}z\bar{z}} \left(2\frac{\partial}{\partial z} \right)^n z^k. \tag{J.6}$$

That is, the LLL projection of any wave function can be obtained by normal ordering the wave function (bringing all the $\bar{z}$'s to the left) followed by the substitution $\bar{z} \rightarrow 2\partial/\partial z$, with the understanding that the derivatives do not act upon the Gaussian part of the wave function.

Making the replacement $\bar{z} \rightarrow 2\partial/\partial z \equiv 2\partial$ in Eq. (J.5) produces for $\hat{\eta}$:

$$\hat{\eta}_{n,m}(\mathbf{r}) = N_{n,m} \, e^{-\frac{1}{4}|z|^2} \sum_{k=0}^{n} (-1)^k \binom{n+m}{n-k} \frac{1}{k!} \partial^k z^{k+m}. \tag{J.7}$$

The expression can be further simplified. Substituting

$$\partial^k z^{k+m} = \sum_{\alpha=k_0}^{k} \frac{k!}{\alpha!} \binom{k+m}{k-\alpha} z^{m+\alpha} \partial^\alpha,$$

and using

$$\sum_{k=k_0}^{n} \sum_{\alpha=k_0}^{k} = \sum_{\alpha=k_0}^{n} \sum_{k=\alpha}^{n},$$

the sum over k is seen to be proportional to $(1-1)^{n-\alpha}$ which vanishes unless $\alpha = n$. Thus, only the term with $k = \alpha = n$ survives, and Eq. (J.7) reduces to the simpler expression in Eq. (J.2). That the derivatives can be so moved to the right becomes less surprising in view of the requirement that the projection of $\eta_{n,m}$ must vanish for $n \neq 0$, i.e.,

$$\mathcal{P}_{\mathrm{LLL}} \eta_{n,m} = \hat{\eta}_{n,m} \cdot 1 = 0 \ (n \neq 0). \tag{J.8}$$

□

J.2 Spherical geometry

We adopt a slightly different notation in this section, and work with $Y_{q,n,m}$ (instead of $Y_{q,l,m}$), where $n = l - |q| = 0, 1, \ldots$ is the LL index. We are interested in the LLL projection of $Y_{q,n,m}(\mathbf{\Omega}) Y_{q',0,m'}(\mathbf{\Omega})$, where $q' > 0$ and $Q = q + q' > 0$, but q can be either positive or negative. We assume $q > 0$ except in Section J.2.1, which outlines the generalization to $q < 0$. The single particle orbitals (for $q > 0$) are given by (Eq. 3.156)

$$Y_{|q|,n,m}(\mathbf{\Omega}) = N_{qnm} (-1)^{q+n-m} u^{|q|+m} v^{|q|-m}$$

$$\times \sum_{s=0}^{n} (-1)^s \binom{n}{s} \binom{2|q|+n}{|q|+m+s} (v^*v)^{n-s} (u^*u)^s, \tag{J.9}$$

where θ and ϕ are the usual angular coordinates of the spherical geometry, the normalization coefficient is given by

$$N_{qnm} = \left(\frac{(2|q| + 2n + 1)}{4\pi} \frac{(|q| + n - m)!(|q| + n + m)!}{n!(2|q| + n)!} \right)^{1/2}, \qquad (\text{J.10})$$

and

$$u = \cos(\theta/2) \exp(i\phi/2), \qquad (\text{J.11})$$

$$v = \sin(\theta/2) \exp(-i\phi/2). \qquad (\text{J.12})$$

The binomial coefficient $\binom{\gamma}{\beta}$ is to be set equal to zero if either $\beta > \gamma$ or $\beta < 0$.

Theorem J.2 *There exists an operator* $\hat{Y}^{q'}_{|q|,n,m}$ *satisfying the property that*

$$\mathcal{P}_{\mathrm{LLL}} Y_{|q|,n,m}(\mathbf{\Omega}) Y_{q',0,m'}(\mathbf{\Omega}) = \hat{Y}^{q'}_{|q|,n,m}(\mathbf{\Omega}) Y_{q',0,m'}(\mathbf{\Omega}), \qquad (\text{J.13})$$

where $Y_{q',0,m'}(\mathbf{\Omega}) \sim u_j^{q'+m'} v_j^{q'-m'}$ *is a LLL wave function at monopole strength* q'. *It is given by*

$$\hat{Y}^{q'}_{|q|,n,m}(\mathbf{\Omega}) = \frac{(2Q + 1)}{(2Q + n + 1)} :: Y_{|q|,n,m}\left(u^* \to \frac{\partial}{\partial u}; v^* \to \frac{\partial}{\partial v} \right) :: \qquad (\text{J.14})$$

where $Q = q' + q$, $Y_{|q|,n,m}$ *are given by Eqs. (J.9) and (J.23), and the normal ordering* $::$ Y $::$ *implies that* u^* *and* v^* *are moved to the extreme* right *before replacing them by the derivatives. In other words:*

$$\hat{Y}^{q'}_{|q|,n,m}(\mathbf{\Omega}) = N_{qnm}(-1)^{q+n-m} \frac{(2Q + 1)!}{(2Q + n + 1)!} u^{|q|+m} v^{|q|-m}$$

$$\times \sum_{s=0}^{n} (-1)^s \binom{n}{s} \binom{2|q| + n}{|q| + n - m - s} u^s v^{n-s} \left(\frac{\partial}{\partial u} \right)^s \left(\frac{\partial}{\partial v} \right)^{n-s}. \qquad (\text{J.15})$$

Proof   To construct the operator $\hat{Y}^{q'}_{q,n,m}$, we multiply one of the terms on the right hand side of Eq. (J.9) by the LLL wave function $Y_{q',0,m'}$ and write (with $Q \equiv q + q'$, $M \equiv m + m'$, and the subscript j suppressed):

$$(v^*v)^{n-s}(u^*u)^s u^{Q+M} v^{Q-M} = a_0 u^{Q+M} v^{Q-M} + \text{higher LL states.} \qquad (\text{J.16})$$

For $|M| > Q$, a_0 must vanish, since only $|M| \le Q$ is possible in the lowest Landau level. Let us first consider the case $|M| \le Q$. Multiplying both sides by $u^{*Q+M}v^{*Q-M}$ and integrating over the angular coordinates gives[1]

$$a_0 = \frac{(Q - M + n - s)!(Q + M + s)!(2Q + 1)!}{(Q + M)!(Q - M)!(2Q + n + 1)!}.$$ (J.17)

This shows that, apart from an m'-independent multiplicative constant $(2Q + 1)!/(2Q + n + 1)!$, the LLL projection of the left hand side of Eq. (J.16) can be accomplished by first bringing all u^* and v^* to the left and then making the replacement

$$u^* \to \frac{\partial}{\partial u}, \qquad v^* \to \frac{\partial}{\partial v}.$$ (J.18)

While true in general for $|M| \le Q$, this prescription can be shown [280, 281] also to produce the correct result (i.e., zero) for $|M| > Q$ for the LLL projection of states of the form $Y_{q,n,m}Y_{q',0,m'}$.

Thus, in Eq. (J.13),

$$\hat{Y}_{q,n,m}^{q'} = \frac{(2Q + 1)!}{(2Q + n + 1)!} N_{qnm}(-1)^{q+n-m} e^{iq\phi_j}$$

$$\times \sum_{s=0}^{n}(-1)^s \binom{n}{s}\binom{2q + n}{q + n - m - s}\left(\frac{\partial}{\partial u}\right)^s u^{q+m+s}\left(\frac{\partial}{\partial v}\right)^{n-s} v^{q-m+n-s}.$$

(J.19)

Further simplification occurs when one brings all the derivatives to the right in Eq. (J.19) using

$$\left(\frac{\partial}{\partial v}\right)^\beta v^\gamma = \sum_{\alpha=0}^{\beta} \frac{\beta!}{\alpha!}\binom{\gamma}{\beta - \alpha} v^{\gamma-\beta+\alpha}\left(\frac{\partial}{\partial v}\right)^\alpha,$$ (J.20)

and a similar equation for the derivative with respect to u (with the summation index α'). The sum over s in Eq. (J.19) then takes the form

$$\sum_{s=\alpha'}^{n-\alpha}(-1)^s\binom{n - \alpha - \alpha'}{s - \alpha'} = \sum_{s'=0}^{n-\alpha-\alpha'}(-1)^{\alpha'+s'}\binom{n - \alpha - \alpha'}{s'},$$ (J.21)

---

[1] The expression for the beta function is useful:

$$B(p,q) = 2\int_0^{\pi/2} \cos^{2p-1}x\sin^{2q-1}x\,dx = \frac{\Gamma(p)\Gamma(q)}{\Gamma(p + q)}; \; p,q > 0.$$

which is equal to $(-1)^{\alpha'}(1-1)^{n-\alpha-\alpha'}$ and vanishes unless $n = \alpha + \alpha'$. The only term satisfying this condition is one with $\alpha = n - s$ and $\alpha' = s$. Consequently, the derivatives in Eq. (J.19) can be moved to the extreme right and act only on the following LLL wave function. This yields Eq. (J.14), completing the proof. $\qquad\square$

$\hat{Y}^{q'}_{q,n,m}$ is independent of m' (although it depends on q'). Apart from a multiplicative constant, $\hat{Y}$ is obtained from Y by the replacement

$$u^* \to \frac{\partial}{\partial u}, \qquad v^* \to \frac{\partial}{\partial v}. \tag{J.22}$$

The multiplicative constant only contributes to the overall normalization factor when Φ is a single Slater determinant, but must be carefully accounted for when Φ is a superposition of several Slater determinants with different occupations of various Landau levels.

With the help of the identity $u^*u + v^*v = 1$, the Eq. (J.9) can be cast into different equivalent forms; for example, the sum over s can be expressed as a power series entirely in v^*v or u^*u. While the projection can be carried out with any form, simplification is achieved when the expression Eq. (J.9) is used. Otherwise, the replacement of u^* or v^* by corresponding derivatives is not equivalent to LLL projection, and neither are the cancellations indicated above obtained.

J.2.1 Negative B^*

The above considerations were generalized to negative values of q by Möller and Simon [451]. The expression for monopole harmonics for negative q is obtained by complex conjugation of $Y_{|q|,n,m}$ in Eq. (J.9):

$$Y_{-|q|,n,m}(\mathbf{\Omega}) = N_{qnm}(-1)^{q+n-m}u^{*|q|+m}v^{*|q|-m}$$
$$\times \sum_{s=0}^{n}(-1)^s\binom{n}{s}\binom{2|q|+n}{|q|+m+s}(v^*v)^{n-s}(u^*u)^s. \tag{J.23}$$

Proceeding as before, with $Q = q' - |q|$ and $M = m' - m$, we get

$$\hat{Y}^{q'}_{-|q|,n,m}(\mathbf{\Omega}) = N_{qnm}(-1)^{q+n-m}\frac{(2Q+1)!}{(2q'+n+1)!}$$
$$\times \sum_{s=0}^{n}(-1)^s\binom{n}{s}\binom{2|q|+n}{|q|+n-m-s}u^sv^{n-s}\left(\frac{\partial}{\partial u}\right)^{|q|+m+s}\left(\frac{\partial}{\partial v}\right)^{|q|-m+n-s}.$$
$$\tag{J.24}$$

The trick of replacing u^* and v^* by derivatives is valid for negative q as well. The multiplicative factors are different for positive and negative values of q, however.

J.2.2 Derivatives

As seen in Eq. (5.104), the projected wave functions are determinants of a matrix with elements of the form

$$
\hat{Y}^{q'}_{q,n,m}(\Omega_j)J^p_j = N_{qnm}(-1)^{q+n-m}\frac{(2Q+1)!}{(2Q+n+1)!}u^{q+m}_j v^{q-m}_j
$$

$$
\times \sum_{s=0}^{n}(-1)^s\binom{n}{s}\binom{2q+n}{q+n-m-s}u^s_j v^{n-s}_j\left[\left(\frac{\partial}{\partial u_j}\right)^s\left(\frac{\partial}{\partial v_j}\right)^{n-s}J^p_j\right].
$$

$$(J.25)$$

It in general complicated and must be evaluated numerically. For this purpose, one conveniently writes the derivatives as

$$
\left(\frac{\partial}{\partial u_j}\right)^s\left(\frac{\partial}{\partial v_j}\right)^{n-s}J^p_j = J^p_j\left[\hat{U}^s_j\hat{V}^{n-s}_j 1\right],
$$

$$(J.26)$$

where

$$
\hat{U}_j = J^{-p}_j\frac{\partial}{\partial u_j}J^p_j = p\sum_k{}'\frac{v_k}{u_j v_k - v_j u_k} + \frac{\partial}{\partial u_j},
$$

$$(J.27)$$

$$
\hat{V}_j = J^{-p}_j\frac{\partial}{\partial v_j}J^p_j = p\sum_k{}'\frac{-u_k}{u_j v_k - v_j u_k} + \frac{\partial}{\partial v_j}.
$$

$$(J.28)$$

For a given n, the explicit analytical form of the derivatives is used in the evaluation of the wave function. The advantage of this form is that the J_j's can be factored out of the Slater determinants to give back the usual Jastrow factor

$$
\prod_j J^p_j = \Phi^{2p}_1.
$$

$$(J.29)$$

For the projected wave functions, we need an explicit expression for

$$
P_j(s, \bar{n} - s) \equiv \left[\hat{U}^s_j\hat{V}^{\bar{n}-s}_j 1\right],
$$

$$(J.30)$$

which we evaluate by the following method (Park [503]). First define

$$
f_j(\alpha, \beta) \equiv \sum_k{}'\left(\frac{v_k}{u_j v_k - v_j u_k}\right)^\alpha\left(\frac{-u_k}{u_j v_k - v_j u_k}\right)^\beta.
$$

$$(J.31)$$

Eqs. (J.27) and (J.28) can be rewritten as

$$\hat{U}_j = pf_j(1,0) + \frac{\partial}{\partial u_j},$$
(J.32)

$$\hat{V}_j = pf_j(0,1) + \frac{\partial}{\partial v_j}.$$
(J.33)

Also, the derivatives of $f_j(\alpha, \beta)$ with respect to u_j and v_j have the simple form

$$\frac{\partial}{\partial u_j} f_j(\alpha, \beta) = -(\alpha + \beta)f_j(\alpha + 1, \beta),$$
(J.34)

$$\frac{\partial}{\partial v_j} f_j(\alpha, \beta) = -(\alpha + \beta)f_j(\alpha, \beta + 1).$$
(J.35)

Explicit expressions for $P_j(s, \bar{n} - s)$ in terms of $f(\alpha, \beta)$ for various values of arguments are given below for the lowest six Landau levels in Φ. (For yet higher Landau levels, the expressions become too long to reproduce here, but can be evaluated following the same method using Mathematica.) $P_j(\bar{n} - s, s)$ can be obtained from $P_j(s, \bar{n} - s)$ by swapping the arguments of all $f_j(\alpha, \beta)$ on the right hand side. Therefore, we list only $P_j(s, \bar{n} - s)$ with $s \geq \bar{n} - s$. In the following, the subscript j on the quantities P and f is suppressed.

$\bar{n} = 0$

$P(0,0) = 1$

$\bar{n} = 1$

$P(1,0) = pf(1,0)$

$\bar{n} = 2$

$P(2,0) = p^2 f(1,0)^2 - pf(2,0)$

$P(1,1) = p^2 f(0,1)f(1,0) - pf(1,1)$

$\bar{n} = 3$

$P(3,0) = p^3 f(1,0)^3 - 3p^2 f(1,0)f(2,0) + 2pf(3,0)$

$P(2,1) = p^3 f(0,1)f(1,0)^2 - 2p^2 f(1,0)f(1,1)$

$\qquad - p^2 f(0,1)f(2,0) + 2pf(2,1)$

$\bar{n} = 4$

$$P(4,0) = p^4 f(1,0)^4 - 6 p^3 f(1,0)^2 f(2,0)$$
$$+ 3 p^2 f(2,0)^2 + 8 p^2 f(1,0) f(3,0) - 6 p f(4,0)$$
$$P(3,1) = p^4 f(0,1) f(1,0)^3 - 3 p^3 f(1,0)^2 f(1,1)$$
$$- 3 p^3 f(0,1) f(1,0) f(2,0) + 3 p^2 f(1,1) f(2,0)$$
$$+ 6 p^2 f(1,0) f(2,1) + 2 p^2 f(0,1) f(3,0)$$
$$- 6 p f(3,1)$$
$$P(2,2) = p^4 f(0,1)^2 f(1,0)^2 - p^3 f(0,2) f(1,0)^2$$
$$- 4 p^3 f(0,1) f(1,0) f(1,1) + 2 p^2 f(1,1)^2$$
$$+ 4 p^2 f(1,0) f(1,2) - p^3 f(0,1)^2 f(2,0)$$
$$+ p^2 f(0,2) f(2,0) + 4 p^2 f(0,1) f(2,1)$$
$$- 6 p f(2,2)$$

$\bar{n} = 5$

$$P(5,0) = p^5 f(1,0)^5 - 10 p^4 f(1,0)^3 f(2,0)$$
$$+ 15 p^3 f(1,0) f(2,0)^2 + 20 p^3 f(1,0)^2 f(3,0)$$
$$- 20 p^2 f(2,0) f(3,0) - 30 p^2 f(1,0) f(4,0)$$
$$+ 24 p f(5,0)$$
$$P(4,1) = p^5 f(0,1) f(1,0)^4 - 4 p^4 f(1,0)^3 f(1,1)$$
$$- 6 p^4 f(0,1) f(1,0)^2 f(2,0)$$
$$+ 12 p^3 f(1,0) f(1,1) f(2,0)$$
$$+ 3 p^3 f(0,1) f(2,0)^2 + 12 p^3 f(1,0)^2 f(2,1)$$
$$- 12 p^2 f(2,0) f(2,1) + 8 p^3 f(0,1) f(1,0) f(3,0)$$
$$- 8 p^2 f(1,1) f(3,0) - 24 p^2 f(1,0) f(3,1)$$
$$- 6 p^2 f(0,1) f(4,0) + 24 p f(4,1)$$
$$P(3,2) = p^5 f(0,1)^2 f(1,0)^3 - p^4 f(0,2) f(1,0)^3$$
$$- 6 p^4 f(0,1) f(1,0)^2 f(1,1) + 6 p^3 f(1,0) f(1,1)^2$$
$$+ 6 p^3 f(1,0)^2 f(1,2) - 3 p^4 f(0,1)^2 f(1,0) f(2,0)$$
$$+ 3 p^3 f(0,2) f(1,0) f(2,0)$$

$$+ 6 p^3 f(0, 1) f(1, 1) f(2, 0)$$

$$- 6 p^2 f(1, 2) f(2, 0) + 12 p^3 f(0, 1) f(1, 0) f(2, 1)$$

$$- 12 p^2 f(1, 1) f(2, 1) - 18 p^2 f(1, 0) f(2, 2)$$

$$+ 2 p^3 f(0, 1)^2 f(3, 0) - 2 p^2 f(0, 2) f(3, 0)$$

$$- 12 p^2 f(0, 1) f(3, 1) + 24 p f(3, 2)$$

Appendix K

Metropolis Monte Carlo

We encounter multi-dimensional integrals of the form

$$I = \frac{\int d^2r_1 \ldots d^2r_N |\Psi(r_1,\ldots,r_N)|^2 \mathcal{O}(r_1,\ldots,r_N)}{\int d^2r_1 \ldots d^2r_N |\Psi(r_1,\ldots,r_N)|^2} \, , \tag{K.1}$$

where $\mathcal{O}$ represents the interaction energy or some other variable. Such integrals must be evaluated approximately. The standard trapezoidal rule is impractical except for very small N. For example, for $N = 10$, if we take 20 points in each direction, we have a grid with a total of $20^{20} \approx 10^{26}$ points. Even a fast supercomputer would take a prohibitively long time to evaluate the function on all these points.

A powerful method for computing such multi-dimensional integrals uses the Monte Carlo strategy developed by Metropolis *et al.* [444]. The basic idea is to approximate the integral by the sum

$$I \approx \frac{1}{M} \sum_{n=1}^{M} \mathcal{O}(R^{(n)}) \, , \tag{K.2}$$

where the vectors $\{R^{(n)}\}$ (R collectively denotes all coordinates $\{r_j\}$) are distributed according to the probability $|\Psi(R^{(n)})|^2$. Calculating the above sum many times yields different values. The "central limit theorem" tells us that, independent of the original distribution, these values are normally distributed about a mean (which is the actual value of the integral I) with variance σ^2/M, where σ^2 is the variance of $\mathcal{O}(R^{(n)})$.

The Metropolis "importance sampling" algorithm generates a sequence of points $R^{(0)}, R^{(1)}, R^{(3)}, \ldots$ according to a specified distribution (with probability $|\Psi(R)|^2$ in our case) through a random walk in the multi-dimensional space. The rules for the random walk are as follows. Suppose the system is at $R^{(n)}$ at the nth step. We make a "trial" step to a nearby point R_t by moving one or many coordinates randomly, and define

$$r = \frac{|\Psi(R_t)|^2}{|\Psi(R^{(n)})|^2} \, . \tag{K.3}$$

If $r > \eta$, where η is a random number in the range $[0, 1]$, then we accept the step, setting $R^{(n+1)} = R_t$. If $r < \eta$, then the trial step is rejected, and we set $R^{(n+1)} = R^{(n)}$.

(A more probable trial point is always accepted, but even a less probable one is accepted with a nonzero probability.) An iteration of this process generates, in the long run, points $\{R^{(n)}\}$ distributed with probability $|\Psi(R_n)|^2$. (The proof, which relies on the principle of detailed balance, can be found in standard books on numerical methods [349].)

Equation (K.2) gives an increasingly more accurate value for the integral the longer the Monte Carlo is run. The following considerations help. (i) A proper choice of the sampling function $|\Psi|^2$ is crucial. In the case of the expectation value of the Coulomb interaction, the choice is obvious. For the evaluation of an off-diagonal element, e.g.,

$$\frac{\int dR\,\Psi_1^*(R)V(R)\Psi_2(R)}{\left(\int dR\,|\Psi_1(R)|^2 \int dR'\,|\Psi_2(R')|^2\right)^{1/2}},\tag{K.4}$$

(where Ψ_1 and Ψ_2 may be nonorthogonal) we may choose $|\Psi_2(R)|^2$ as the sampling function to write the integral as

$$\frac{\dfrac{\int dR\,V(R)\,[\Psi_1^*(R)/\Psi_2^*(R)]|\Psi_2(R)|^2}{\int dR\,|\Psi_2(R)|^2}}{\left(\dfrac{\int dR\,|\Psi_1(R)/\Psi_2(R)]|^2|\Psi_2(R)|^2}{\int dR\,|\Psi_2(R)|^2}\right)^{1/2}},\tag{K.5}$$

for which both the numerator and the denominator can be evaluated in a single Monte Carlo run. (ii) The probability with which trial steps are accepted is called the acceptance ratio, the value of which depends on the step size. If the step size is too large, then the random walker will be stuck near the maximum probability. If it is too small, the random walker will not move far from its initial position. The rule of thumb is to choose the step size to make the acceptance ratio approximately equal to 0.5. (iii) We run the Monte Carlo for a while initially (i.e., let the system "thermalize") before evaluating the integral.

Monte Carlo for composite fermions The wave functions of the form $\Psi = \mathcal{P}_{LLL}\Phi\Phi_1^{2p}$, projected into the lowest Landau level by method II, are amenable to the Metropolis Monte Carlo method. The evaluation of a Slater determinant takes $\mathcal{O}(N^3)$ operations. For a free fermion Slater determinant, if a single particle is moved, only one row (or column) of the density matrix is altered; in that case, N operations are needed to evaluate the new determinant in terms of the old inverse matrix, and $\mathcal{O}(N^2)$ operations to update the inverse matrix if the move is accepted [53]. For composite fermions, however, each term in the Slater determinant depends on all the particle coordinates, so even the movement of a single composite fermion alters all matrix elements. Therefore, an update of the trial $N \times N$ density matrix of the CF system proceeds through a full evaluation of the determinant at every Metropolis step, requiring $\mathcal{O}(N^3)$ operations. Nonetheless, systems with as many as 100 composite fermions can be handled without much difficulty, especially for the states

Ψ derived from a single Slater determinant Φ. For the neutral exciton of the n filled Λ level state, the wave function is a linear superposition of $\sim N/n$ determinants, making the computation significantly more time consuming. The energies of the ground and excited states must be evaluated with high accuracy to obtain reasonable values for the energy gaps, especially when the gaps are small; up to 4×10^8 Monte Carlo steps have been performed in the evaluation of the exciton energy. Energy differences as small as $0.001 e^2/\epsilon \ell$ have been computed, which are approximately two orders of magnitude smaller than the transport gap of the 1/3 state.

Appendix L

Composite fermion diagonalization

The CF basis functions of LLL projected wave functions Ψ_α must, in general, be orthogonalized via the well-known Gram–Schmid procedure [423, 424].[1] Defining the normalized states as

$$|\eta_\alpha\rangle \equiv \frac{|\Psi_\alpha\rangle}{\sqrt{\langle\Psi_\alpha|\Psi_\alpha\rangle}}, \tag{L.1}$$

the scalar products

$$\mathcal{O}_{\alpha\beta} \equiv \langle\eta_\alpha|\eta_\beta\rangle \tag{L.2}$$

and the Coulomb matrix elements

$$\langle\eta_\alpha|V|\eta_\beta\rangle \tag{L.3}$$

can be evaluated conveniently by the Monte Carlo method (Metropolis importance sampling). We outline how the Hamiltonian matrix in a Gram–Schmid orthogonalized basis can be expressed in terms of these quantities.

The orthogonal basis states $|\xi\rangle$ are expressed as

$$|\xi_\alpha\rangle = |\eta_\alpha\rangle - \sum_{\gamma=1} \frac{\langle\xi_\gamma|\eta_\alpha\rangle}{\langle\xi_\gamma|\xi_\gamma\rangle}|\xi_\gamma\rangle. \tag{L.4}$$

The relation in Eq. (L.4) implies the following recursion relation for the transformation matrix $U_{\alpha\beta}$, defined by $|\xi_\alpha\rangle \equiv \sum_\beta U_{\alpha\beta}|\eta_\beta\rangle$:

$$U_{\alpha\beta} = \begin{cases} -\sum\limits_{\gamma=1}^{\alpha-1} \dfrac{\sum_{\delta=1}^\gamma U_{\gamma\delta}^* \mathcal{O}_{\delta\alpha}}{\sum_{\delta,\epsilon=1}^\gamma U_{\gamma\delta}^* U_{\gamma\epsilon} \mathcal{O}_{\delta\epsilon}} U_{\gamma\beta} & \text{for } \beta < \alpha, \\ 1 & \text{for } \beta = \alpha, \\ 0 & \text{for } \beta > \alpha. \end{cases} \tag{L.5}$$

The computation of $U_{\alpha\beta}$ enables us to calculate the Coulomb Hamiltonian matrix elements $V_{\alpha\beta}$ in the orthonormal basis:

$$V_{\alpha\beta} = \frac{\langle\xi_\alpha|V|\xi_\beta\rangle}{\sqrt{\langle\xi_\alpha|\xi_\alpha\rangle\langle\xi_\beta|\xi_\beta\rangle}}, \tag{L.6}$$

---

[1] The treatment in this appendix closely follows Jeon et al. [299]

where

$$\langle \xi_\alpha | V | \xi_\beta \rangle \;=\; \sum_{\gamma,\delta} U^*_{\alpha\gamma} U_{\beta\delta} \langle \eta_\gamma | V | \eta_\delta \rangle, \tag{L.7}$$

$$\langle \xi_\alpha | \xi_\alpha \rangle \;=\; \sum_{\gamma,\delta} U^*_{\alpha\gamma} U_{\beta\delta} \mathcal{O}_{\gamma\delta}. \tag{L.8}$$

The Coulomb Hamiltonian matrix in Eq. (L.6) is diagonalized to obtain the CF predictions for the energies and the wave functions for the ground state and low-energy excitations. The statistical error can be estimated by calculating the spectra in several independent runs.

References

[1] E. Abrahams, P. W. Anderson, D. C. Licciardello, and T. V. Ramakrishnan, Scaling theory of localization: Absence of quantum diffusion in two dimensions. *Phys. Rev. Lett.* **42**, 673 (1979).

[2] *Handbook of Mathematical Functions*, edited by M. Abramowitz and I. A. Stegun (Washington, DC. US GPO, 1964).

[3] A. A. Abrikosov, On the magnetic properties of superconductors of the second group. *Sov. Phys. JETP* **5**, 1174 (1957).

[4] E. H. Aifer, B. B. Goldberg, and D. A. Broido, Evidence of skyrmion excitations about $\nu = 1$ in n-modulation-doped single quantum wells by interband optical transmission. *Phys. Rev. Lett.* **76**, 680 (1996).

[5] C. Albrecht, J. H. Smet, K. von Klitzing, *et al.*, Evidence of Hofstadter's fractal energy spectrum in the quantized Hall conductance. *Physica E* **20**, 133 (2003).

[6] I. L. Aleiner and L. I. Glazman, Novel edge excitations of two-dimensional electron liquid in a magnetic field. *Phys. Rev. Lett.* **72**, 2935 (1994).

[7] J. B. Anderson, Random-walk simulation of Schrödinger equation: H_3^+. *J. Chem. Phys.* **63**, 1499 (1975).

[8] P. W. Anderson, More is different. *Science* **177**, 393 (1972).

[9] T. Ando, Theory of quantum transport in a two-dimensional electron system under magnetic fields: Oscillatory conductivity. *J. Phys. Soc. Jpn.* **37**, 1233 (1974).

[10] T. Ando, A. B. Fowler, and F. Stern, Electronic properties of two-dimensional systems. *Rev. Mod. Phys.* **54**, 437 (1982).

[11] E. Y. Andrei, G. Deville, D. C. Glattli, *et al.*, Observation of a magnetically induced Wigner solid. *Phys. Rev. Lett.* **60**, 2765 (1988).

[12] A. G. Aronov, E. Altshuler, A. D. Mirlin, and P. Wölfle, Theory of Shubnikov–de Haas oscillations around the $\nu = 1/2$ filling factor of the Landau level: Effect of gauge-field fluctuations. *Phys. Rev. B* **52**, 4708 (1995).

[13] D. P. Arovas, J. R. Schrieffer, and F. Wilczek, Fractional statistics and the quantum Hall effect. *Phys. Rev. Lett.* **53**, 722 (1984).

[14] N. W. Ashcroft and N. D. Mermin, *Solid State Physics* (New York: Holt, Rinehart and Winston, 1976).

[15] R. C. Ashoori, J. A. Lebens, N. P. Bigelow, and R. H. Silsbee, Equilibrium tunneling from the two-dimensional electron gas in GaAs: Evidence for a magnetic-field-induced energy gap. *Phys. Rev. Lett.* **64**, 681 (1990).

[16] R. C. Ashoori, H. L. Stormer, L. N. Pfeiffer, K. W. Baldwin, and K. West, Edge magnetoplasmons in the time domain. *Phys. Rev. B* **45**, 3894 (1992).

[17] H. Bachmair, E. O. Gobel, H. Hein, The von Klitzing resistance standard. *Physica E* **20**, 14 (2003).

[18] N. Q. Balaban, U. Meirav, and I. Bar-Joseph, Absence of scaling in the integer quantum Hall effect. *Phys. Rev. Lett.* **81**, 4967 (1998).

[19] L. Balents and L. Radzihovsky, Interlayer tunneling in double-layer quantum Hall pseudoferromagnets. *Phys. Rev. Lett.* **86**, 1825 (2001).

[20] E. Balthes, D. Schweitzer, and P. Wyder, Low integer Landau level filling factors v and indications for the fractional $v = 1/2$ in the 2D organic metal κ-(BEDT-TTF)$_2$I$_3$. *Solid State Commun.* **136**, 238 (2005).

[21] S. E. Barrett, G. Dabbagh, L. N. Pfeiffer, K. W. West, and R. Tycko, Optically pumped NMR evidence for finite-size skyrmions in GaAs quantum wells near Landau level filling $v = 1$. *Phys. Rev. Lett.* **74**, 5112 (1995).

[22] G. Baym and L. P. Kadanoff, Conservation laws and correlation functions. *Phys. Rev.* **124**, 287 (1961).

[23] V. Bayot, E. Grivei, H. C. Manoharan, X. Ying, and M. Shayegan, Thermopower of composite fermions. *Phys. Rev. B* **52**, R8621 (1995).

[24] V. M. Bedanov and F. M. Peeters, Ordering and phase transitions of charged particles in a classical finite two-dimensional system. *Phys. Rev. B* **49**, 2667 (1994).

[25] C. W. J. Beenakker, Guiding-center-drift resonance in a periodically modulated two-dimensional electron gas. *Phys. Rev. Lett.* **62**, 2020 (1989).

[26] C. W. J. Beenakker and B. Rejaei, Single electron tunneling in the fractional quantum Hall effect regime. *Physica B* **189**, 147 (1993).

[27] E. J. Bergholtz and A. Karlhede, Half-filled lowest Landau level on a thin torus. *Phys. Rev. Lett.* **92**, 026802 (2005).

[28] E. J. Bergholtz and A. Karlhede, One-dimensional theory of the quantum Hall system. *J. Stat. Mech. Theory Exp.*, L04001 (2006).

[29] M. V. Berry, Quantal phase factors accompanying adiabatic changes. *Proc. R. Soc. London A* **392**, 45 (1984).

[30] R. N. Bhatt and Y. Huo, Current carrying states in the lowest Landau level. *Phys. Rev. Lett.* **68**, 1375 (1992).

[31] A. Bijl, The lowest wave function of the symmetrical many particles system. *Physica* **7**, 869 (1940).

[32] D. Bohm and D. Pines, A collective description of electron interactions: III. Coulomb interactions in a degenerate electron gas. *Phys. Rev.* **92**, 609 (1953).

[33] N. E. Bonesteel, Composite fermions and the energy gap in the fractional quantum Hall effect. *Phys. Rev. B* **51**, 9917 (1995).

[34] N. E. Bonesteel, I. A. McDonald, and C. Nayak, Gauge fields and pairing in double-layer composite fermion metals. *Phys. Rev. Lett.* **77**, 3009 (1996).

[35] L. Bonsall and A. A. Maradudin, Some static and dynamical properties of a two-dimensional Wigner crystal. *Phys. Rev. B* **15**, 1959 (1977).

[36] J. P. Bouchaud, A. Georges, and C. Lhuillier, Pair wave functions for strongly correlated fermions and their determinantal representation. *J. Physique* **49**, 553 (1988).

[37] L. Brey, Edge states of composite fermions. *Phys. Rev. B* **50**, 11861 (1994).

[38] L. Brey and C. Tejedor, Composite fermions traversing a potential barrier. *Phys. Rev. B* **51**, 17259 (1995).

[39] L. Brink, T. H. Hansson, and M. A. Vasiliev, Explicit solution to the N-body Calogero problem. *Phys. Lett. B* **286**, 109 (1992).

[40] H. Buhmann, W. Joss, K. von Klitzing, Magneto-optical evidence for fractional quantum Hall states down to filling factor 1/9. *Phys. Rev. Lett.* **65**, 1056 (1990).

[41] H. Buhmann, W. Joss, K. von Klitzing, *et al.*, Novel magneto-optical behavior in the Wigner-solid regime. *Phys. Rev. Lett.* **66**, 926 (1991).

[42] M. Büttiker, Four-terminal phase-coherent conductance. *Phys. Rev. Lett.* **57**, 1761 (1986).

[43] M. Büttiker, Absence of backscattering in the quantum Hall effect in multiprobe conductors. *Phys. Rev. B* **38**, 9375 (1988).

[44] M. Büttiker, Transmission probabilities and the quantum Hall effect. *Phys. Rev. Lett.* **62**, 229 (1989).

[45] Yu. A. Bychkov, S. V. Iordanskii, and G. M. Eliashberg, Two-dimensional electrons in a strong magnetic field. *Pis'ma Zh. Eksp. Teor. Fiz.* **33**, 152 (1981) [*JETP Lett.* **33**, 143 (1981)].

[46] N. Byers and C. N. Yang, Theoretical considerations concerning quantized magnetic flux in superconducting cylinders. *Phys. Rev. Lett.* **7**, 46 (1961).

[47] M. Byszewski, B. Chwalisz, D. K. Maude, Optical probing of composite fermions in a two-dimensional electron gas. *Nature Physics* **2**, 239 (2006).

[48] J. M. Caillol, D. Levesque, J. J. Weis, and J. P. Hansen, A Monte Carlo study of the classical two-dimensional one-component plasma. *J. Stat. Phys.* **28** 325 (1982).

[49] F. Calogero, Ground state of a one-dimensional N-body system. *J. Math. Phys.* **10**, 2197 (1967).

[50] A. Cappelli, C. Mendez, J. Simonin, and G. R. Zemba, Numerical study of hierarchical quantum Hall edge states in the disk geometry. *Phys. Rev. B* **58**, 16291 (1998).

[51] H. A. Carmona, A. K. Geim, A. Nogaret, *et al.*, Two dimensional electrons in a lateral magnetic superlattice. *Phys. Rev. Lett.* **74**, 3009 (1995).

[52] W. J. Carr, Energy, specific heat, and magnetic properties of the low-density electron gas. *Phys. Rev.* **122**, 1437 (1961).

[53] D. Ceperley, G. V. Chester, and M. H. Kalos, Monte Carlo simulation of a many-fermion study. *Phys. Rev. B* **16**, 3081 (1977).

[54] D. M. Ceperley and B. J. Alder, Ground state of the electron gas by a stochastic method. *Phys. Rev. Lett.* **45**, 566 (1980).

[55] D. M. Ceperley, Path integrals in the theory of condensed helium. *Rev. Mod. Phys.* **67**, 279 (1995).

[56] T. Chakraborty, P. Pietiläinen, and F. C. Zhang, Elementary excitations in the fractional quantum Hall effect and the spin-reversed quasiparticles. *Phys. Rev. Lett.* **57**, 130 (1986).

[57] T. Chakraborty and P. Pietiläinen, Fractional quantum Hall effect at half-filled Landau level in a multiple-layer electron system. *Phys. Rev. Lett.* **59**, 2784 (1987).

[58] J. T. Chalker and P. D. Coddington, Percolation, quantum tunnelling and the integer Hall effect. *J. Phys. C* **21**, 2665 (1988).

[59] C. de C. Chamon, D. E. Freed, and X. G. Wen, Tunneling and quantum noise in one-dimensional Luttinger liquids. *Phys. Rev. B* **51**, 2363 (1995).

[60] A. M. Chang, M. A. Paalanen, D. C. Tsui, H. L. Stormer, and J. C. M. Hwang, Fractional quantum Hall effect at low temperatures. *Phys. Rev. B* **28**, 6133 (1983).

[61] A. M. Chang, M. A. Paalanen, D. C. Tsui, H. L. Stormer, and J. C. M. Hwang, Higher-order states in the multiple-series, fractional, quantum Hall effect. *Phys. Rev. Lett.* **53**, 997 (1984).

[62] A. M. Chang and D. C. Tsui, Experimental observation of a striking similarity between quantum Hall transport coefficients. *Solid State Commun.* **56**, 153 (1985).

[63] A. M. Chang, L. N. Pfeiffer, and K. W. West, Observation of chiral Luttinger behavior in electron tunneling into fractional quantum Hall edges. *Phys. Rev. Lett.* **77**, 2538 (1996).

[64] A. M. Chang, L. N. Pfeiffer, and K. W. West, An apparent $g = 1/2$ chiral Luttinger liquid at the edge of the $\nu = 1/2$ compressible fractional quantum Hall liquid. *Physica B* **249–251**, 283 (1998).

[65] A. M. Chang, M. K. Wu, J. C. C. Chi, L. N. Pfeiffer, and K. W. West, Plateau behavior in the chiral Luttinger liquid exponent. *Phys. Rev. Lett.* **86**, 143 (2001).

[66] A. M. Chang, Chiral Luttinger liquids at the fractional quantum Hall edge. *Rev. Mod. Phys.* **75**, 1449 (2003).

[67] C.-C. Chang and J. K. Jain, Microscopic origin of the next-generation fractional quantum Hall effect. *Phys. Rev. Lett.* **92**, 196806 (2004).

[68] C.-C. Chang, G. S. Jeon, and J. K. Jain, Microscopic verification of topological electron-vortex binding in the lowest-Landau-level crystal state. *Phys. Rev. Lett.* **94**, 016809 (2005).

[69] C.-C. Chang, N. Regnault, Th. Jolicoeur, and J. K. Jain, Composite fermionization of bosons in rapidly rotating atomic traps. *Phys. Rev. A* **72**, 013611 (2005).

[70] C.-C. Chang, C. Tőke, G. S. Jeon, and J. K. Jain, Competition between composite-fermion-crystal and liquid orders at $\nu = 1/5$. *Phys. Rev. B* **73**, 155323 (2006).

[71] Yong Chen, R. M. Lewis, L. W. Engel, *et al.*, Microwave resonance of the 2D Wigner crystal around integer Landau fillings. *Phys. Rev. Lett.* **91**, 016801 (2003).

[72] Yong P. Chen, R. M. Lewis, L. W. Engel, *et al.*, Evidence for two different solid phases of two-dimensional electrons in high magnetic fields. *Phys. Rev. Lett.* **93**, 206805 (2004).

[73] Yu-Ming Cheng, Tsai-Yu Huang, C.-T. Liang, *et al.*, Experimental studies of composite fermion conductivity: Dependence on carrier density. *Physica E* **12**, 105 (2002).

[74] G. V. Chester, Speculations on Bose-Einstein condensation and quantum crystals. *Phys. Rev. A* **2**, 256 (1970).

[75] D. B. Chklovskii and P. A. Lee, Transport properties between quantum Hall plateaus. *Phys. Rev. B* **48**, 18 060 (1993).

[76] D. B. Chklovskii, Structure of fractional edge states: A composite-fermion approach. *Phys. Rev. B* **51**, 9895 (1995).

[77] R. Chughtai, V. Zhitomirsky, R. J. Nicholas, and M. Henini, Measurements of composite fermion masses from the spin polarization of two-dimensional electrons in the region $1 < \nu < 2$. *Phys. Rev. B* **65**, 161305(R) (2002).

[78] S. T. Chui and K. Esfarjani, Solidification of the two-dimensional electron gas in high magnetic fields. *Phys. Rev. B* **42**, 10758 (1990).

[79] S. T. Chui and B. Tanatar, Phase diagram of the two-dimensional quantum electron freezing with external impurities. *Phys. Rev. B* **55**, 9330 (1997).

[80] Y. C. Chung, M. Heiblum and V. Umansky, Scattering of bunched fractionally charged quasiparticles. *Phys. Rev. Lett.* **91**, 216804 (2003).

[81] Y. C. Chung, M. Heiblum, Yuval Oreg, V. Umansky, and D. Mahalu, Anomalous chiral Luttinger liquid behavior of diluted fractionally charged quasiparticles. *Phys. Rev. B* **67**, 201104(R) (2003).

[82] O. Ciftja and S. Fantoni, Application of Fermi-hypernetted-chain theory to composite-fermion quantum Hall states. *Phys. Rev. B* **56**, 13290 (1997).

[83] O. Ciftja and S. Fantoni, Fermi-hypernetted-chain study of unprojected wave functions to describe the half-filled state of the fractional quantum Hall effect. *Phys. Rev. B* **58**, 7898 (1998).

[84] O. Ciftja, Exact results for a composite-fermion wave function. *Phys. Rev. B* **59**, 8132 (1999).

[85] O. Ciftja, The Fermi-sea-like limit of the composite fermion wave function. *Eur. Phys. J. B* **13**, 671 (2000).

[86] O. Ciftja and C. Wexler, Energy gaps for fractional quantum Hall states described by a Chern–Simons composite fermion wavefunction. *Eur. Phys. J. B* **23**, 437 (2001).

[87] O. Ciftja, Theoretical estimates for the correlation energy of the unprojected composite fermion wave function. *Physica E* **9**, 226 (2001).

[88] O. Ciftja and C. Wexler, Coulomb energy of quasiparticle excitations in Chern–Simons composite fermion states. *Solid State Commun.* **122** 401 (2002).

[89] R. G. Clark, R. J. Nicholas, A. Usher, C. T. Foxon, and J. J. Harris, Odd and even fractionally quantized states in GaAs-GaAlAs heterojunctions. *Surface Science* **170**, 141 (1986).

[90] R. G. Clark, S. R. Haynes, A. M. Suckling, *et al.*, Spin configurations and quasiparticle fractional charge of fractional-quantum-Hall-effect ground states in the $N = 0$ Landau level. *Phys. Rev. Lett.* **62**, 1536 (1989).

[91] M. W. Cole and M. H. Cohen, Image-potential-induced surface bands in insulators. *Phys. Rev. Lett.* **23**, 1238 (1969).

[92] P. T. Coleridge, Z. W. Wasilewski, P. Zawadzki, A. S. Sachrajda, and H. A. Carmona, Composite-fermion effective masses. *Phys. Rev. B* **52**, R11603 (1995).

[93] E. Comforti, Y. C. Chung, M. Heiblum, V. Umansky, and D. Mahalu, Bunching of fractionally-charged quasiparticles tunnelling through high potential barriers. *Nature (London)* **416**, 515 (2002).

[94] S. Conti and G. Vignale, Collective modes and electronic spectral function in smooth edges of quantum Hall systems. *Phys. Rev. B* **54**, R14309 (1996).

[95] S. Conti and G. Vignale, Dynamics of the two-dimensional electron gas in the lowest Landau level: A continuum elasticity approach. *J. Phys. Condens. Matter* **10**, L779 (1998).

[96] N. R. Cooper and N. K. Wilkin, Composite fermion description of rotating Bose–Einstein condensates. *Phys. Rev. B* **60**, R16279 (1999).

[97] K. B. Cooper, M. P. Lilly, J. P. Eisenstein, L. N. Pfeiffer, and K. W. West, Insulating phases of two-dimensional electrons in high Landau levels: Observation of sharp thresholds to conduction. *Phys. Rev. B* **60**, R11285 (1999).

[98] N. R. Cooper, N. K. Wilkin, and J. M. F. Gunn, Quantum phases of vortices in rotating Bose–Einstein condensates. *Phys. Rev. Lett.* **87**, 120405 (2001).

[99] R. S. Crandall and R. Williams, Crystallization of electrons on surface of liquid helium. *Phys. Lett.* **34A**, 404 (1971).

[100] C. Cristofano, G. Maiella, R. Musto, and F. Nicodemi, Coulomb gas approach to quantum Hall effect. *Phys. Lett.* **B262**, 88 (1991).

[101] M. C. Cross and D. S. Fisher, Magnetism in solid ^{3}He: Confrontation between theory and experiment. *Rev. Mod. Phys.* **57**, 881 (1985).

[102] P. A. Crump, B. Tieke, R. J. Barraclough, *et al.*, Evidence for composite fermions from the magnetothermopower of 2D holes. *Surf. Sci.* **362**, 50 (1996).

[103] S. Curone and P. C. E. Stamp, Singular behaviour of electrons and of composite fermions in a finite effective field. *J. Phys., Condens. Matter* **8**, 6073 (1996).

[104] R. D'Agosta, G. Vignale, and R. Raimondi, Temperature dependence of the tunneling amplitude between quantum Hall edges. *Phys. Rev. Lett.* **94**, 086801 (2005).

[105] N. d'Ambrumenil and A. M. Reynolds, Fractional quantum Hall states in higher Landau levels. *J. Phys. C: Solid State Phys.* **21**, 119 (1988).

[106] C. G. Darwin, *Proc. Cambridge Philos. Soc.* **27**, 86 (1930).

[107] S. Das Sarma, Localization, metal-insulator transitions, and quantum Hall effect. In *Perspectives in Quantum Hall Effects*, eds. S. Das Sarma and A. Pinczuk (New York: Wiley, 1997).

[108] S. Das Sarma and Daw-Wei Wang, Resonant raman scattering by elementary electronic excitations in semiconductor structures. *Phys. Rev. Lett.* **83**, 816 (1999).

[109] H. D. M. Davies, J. C. Harris, J. F. Ryan, and A. J. Turberfield, Spin and charge density excitations and the collapse of the fractional quantum Hall state at $\nu = 1/3$. *Phys. Rev. Lett.* **78**, 4095 (1997).

[110] P. G. deGennes, *Superconductivity of Metals and Alloys* (Reading, MA: Addison-Wesley, 1989).

[111] F. Delahaye, T. J. Witt, R. E. Elmquist, and R. F. Dziuba, Comparison of quantum Hall resistance standards of the NIST and the BIPM. *Metrologia* **37**, 173 (2000).

[112] F. Delahaye and B. Jackelmann, Revised technical guidelines for reliable dc measurements of the quantized Hall resistance. *Metrologia* **40**, 217 (2003).

[113] A. E. Dementyev, N. N. Kuzma, P. Khandelwal, *et al.*, Optically pumped NMR studies of electron spin polarization and dynamics: New constraints on the composite fermion description of $\nu = 1/2$. *Phys. Rev. Lett.* **83**, 5074 (1999).

[114] R. de Picciotto, M. Rezhnikov, M. Heiblum, *et al.*, Direct observation of a fractional charge. *Nature* **389**, 162 (1997).

[115] G. Dev and J. K. Jain, Band structure of the fractional quantum Hall effect. *Phys. Rev. Lett.* **69**, 2843 (1992).

[116] G. Dev and J. K. Jain, Jastrow–Slater trial wave functions for the fractional quantum Hall effect: Results for few-particle systems. *Phys. Rev. B* **45**, 1223 (1992).

[117] A. M. Devitt, S. H. Roshko, U. Zeitler, *et al.*, Ballistic phonon studies in the lowest Landau level. *Physica E* **5**, 47 (2000).

[118] P. Di Francesco, P. Mathieu, D. Senechal, *Conformal Field Theory* (New York, Springer-Verlag: 1997).

[119] R. Dingle, H. L. Stormer, A. C. Gossard, and W. Wiegmann, Electron mobilities in modulation-doped semiconductor heterojunction superlattices. *Appl. Phys. Lett.* **33** (7), 665 (1978).

[120] R. L. Doretto, M. O. Goerbig, P. Lederer, A. O. Caldeira, and C. Morais Smith, Spin-excitations of the quantum Hall ferromagnet of composite fermions. *Phys. Rev. B* **72**, 035341 (2005).

[121] S. I. Dorozhkin, J. H. Smet, K. von Klitzing, *et al.*, Comparison between the compressibilities of the zero field and composite-fermion metallic states of the 2D electron system. *Phys. Rev. B* **63**, 121301(R) (2001).

[122] S. I. Dorozhkin, J. H. Smet, K. von Klitzing, *et al.*, Measurements of the compressibility of the composite fermion metallic state in a 2D electron system. *Physica E: Low-dimensional Systems and Nanostructures* **12**, 97 (2002).

[123] R. R. Du, H. L. Stormer, D. C. Tsui, L. N. Pfeiffer, and K. W. West, Experimental evidence for new particles in the fractional quantum Hall effect. *Phys. Rev. Lett.* **70**, 2944 (1993).

[124] R. R. Du, H. L. Stormer, D. C. Tsui, L. N. Pfeiffer, and K. W. West, Shubnikov-de Haas oscillations around $\nu = 1/2$ Landau level filling factor. *Solid State Commun.* **90**, 71 (1994).

[125] R. R. Du, H. L. Stormer, D. C. Tsui, *et al.*, Drastic enhancement of composite fermion mass near Landau level filling $\nu = 1/2$. *Phys. Rev. Lett.* **73**, 3274 (1994).

[126] R. R. Du, A. S. Yeh, H. L. Stormer, *et al.*, Fractional quantum Hall effect around $\nu = 3/2$: Composite fermions with a spin. *Phys. Rev. Lett.* **75**, 3926 (1995).

[127] R. R. Du, A. S. Yeh, H. L. Stormer, *et al.*, g factor of composite fermions around $\nu = 3/2$ from angular-dependent activation-energy measurements. *Phys. Rev. B* **55**, R7351 (1997).

[128] R. R. Du, D. C. Tsui, H. L. Stormer, *et al.*, Strongly anisotropic transport in higher two-dimensional Landau levels. *Solid State Commun.* **109**, 389 (1999).

[129] I. Dujovne, A. Pinczuk, M. Kang, *et al.*, Evidence of Landau levels and interactions in low-lying excitations of composite fermions at $1/3 \leq \nu \leq 2/5$. *Phys. Rev. Lett.* **90**, 036803 (2003).

[130] I. Dujovne, A. Pinczuk, M. Kang, *et al.*, Light scattering observations of spin reversal excitations in the fractional quantum Hall regime. *Solid State Commun.* **127**, 109 (2003).

[131] I. Dujovne, A. Pinczuk, M. Kang, *et al.*, Composite-fermion spin excitations as ν approaches 1/2: Interactions in the Fermi sea. *Phys. Rev. Lett.* **95**, 056808 (2005).

[132] G. Ebert, K. von. Klitzing, J. C. Maan, *et al.*, Fractional quantum Hall effect at filling factors up to $\nu = 3$. *J. Phys. C: Solid State Phys.* **17**, L775 (1984).

[133] A. L. Efros and B. I. Shklovskii, Coulomb gap and low temperature conductivity of disordered systems. *J. Phys. C* **8**, L49 (1975).

[134] A. L. Efros and F. G. Pikus, Classical approach to the gap in the tunneling density of states of a two-dimensional electron liquid in a strong magnetic field. *Phys. Rev. B* **48**, 14694 (1993).

[135] J. P. Eisenstein, R. L. Willett, H. L. Stormer, *et al.*, Collapse of the even-denominator fractional quantum Hall effect in tilted fields. *Phys. Rev. Lett.* **61**, 997 (1988).

[136] J. P. Eisenstein, H. L. Stormer, L. Pfeiffer, and K. W. West, Evidence for a phase transition in the fractional quantum Hall effect. *Phys. Rev. Lett.* **62**, 1540 (1989).

[137] J. P. Eisenstein, G. S. Boebinger, L. N. Pfeiffer, K. W. West, and S. He, New fractional quantum Hall state in double-layer two-dimensional electron systems. *Phys. Rev. Lett.* **68**, 1383 (1992).

[138] J. P. Eisenstein, L. N. Pfeiffer, and K. W. West, Coulomb barrier to tunneling between parallel two-dimensional electron systems. *Phys. Rev. Lett.* **69** 3804 (1992).

[139] J. P. Eisenstein, L. N. Pfeiffer, and K. W. West, Compressibility of the two-dimensional electron gas: Measurements of the zero-field exchange energy and fractional quantum Hall gap. *Phys. Rev. B.* **50** 1760 (1994).

[140] J. P. Eisenstein, K. B. Cooper, L. N. Pfeiffer, and K. W. West, Insulating and fractional quantum Hall states in the first excited Landau level. *Phys. Rev. Lett.* **88**, 076801 (2002).

[141] A. Endo, M. Kawamura, S. Katsumoto, and Y. Iye, Magnetotransport of $\nu = 3/2$ composite fermions under periodic effective magnetic-field modulation. *Phys. Rev. B* **63**, 113310 (2001).

[142] L. W. Engel, H. P. Wei, D. C. Tsui, and M. Shayegan, Critical exponent in the fractional quantum Hall effect. *Surf. Sci.* **229**, 13 (1990).

[143] L. W. Engel, S. W. Hwang, T. Sajoto, D. C. Tsui, and M. Shayegan, New results on scaling in the integral quantum Hall effect. *Phys. Rev. B* **45**, 3418 (1992).

[144] F. Evers, A. D. Mirlin, D. G. Polyakov, and P. Wölfle, Nonadiabatic scattering of a classical particle in an inhomogeneous magnetic field. *Phys. Rev. B* **58**, 15321 (1998).

[145] F. Evers, A. D. Mirlin, D. G. Polyakov, and P. Wölfle, Semiclassical theory of transport in a random magnetic field. *Phys. Rev. B* **60**, 8951 (1999).

[146] F. Evers, A. D. Mirlin, D. G. Polyakov, and P. Wölfle, Finite-frequency transport of composite fermions. *Physica B* **298**, 187 (2001).

[147] Z. F. Ezawa and A. Iwazaki, Chern–Simons gauge theory for dobule-layer electron system. *Int. J. Mod. Phys. B* **6**, 3205 (1992).

[148] Z. F. Ezawa and A. Iwazaki, Lowest Landau level constraint, Goldstone mode, and Josephson effect in a double-layer quantum Hall system. *Phys. Rev. B* **48**, 15189 (1993).

[149] V. I. Fal'ko and S. V. Iordanskii, Electron-phonon drag effect at 2D Landau levels. *J. Phys. Condens. Matt.* **4**, 9201 (1992).

[150] F. F. Fang and W. E. Howard, Negative field-effect mobility on (100) Si surfaces. *Phys. Rev. Lett.* **16**, 797 (1966).

[151] G. Fano, F. Ortolani, and E. Colombo, Configuration-interaction calculations on the fractional quantum Hall effect. *Phys. Rev. B* **34**, 2670 (1986).

[152] P. Fendley W. W. Ludwig, and H. Saleur, Exact conductance through point contacts in the $\nu = 1/3$ fractional quantum Hall effect. *Phys. Rev. Lett.* **74**, 3005 (1995).

[153] P. Fendley W. W. Ludwig, and H. Saleur, Exact nonequilibrium dc shot noise in Luttinger liquids and fractional quantum Hall devices. *Phys. Rev. B* **75**, 2196 (1995).

[154] H. A. Fertig, Energy spectrum of a layered system in a strong magnetic field. *Phys. Rev. B* **40**, 1087 (1989).

[155] H. A. Fertig, L. Brey, R. Cote, and A. H. MacDonald, Charged spin-texture excitations and the Hartree–Fock approximation in the quantum Hall effect. *Phys. Rev. B* **50**, 11018 (1994).

[156] H. A. Fertig, Properties of the electron solid. In *Perspectives in Quantum Hall Effects*, eds. S. Das Sarma and A. Pinczuk (New York: Wiley, 1997).

[157] H. A. Fertig and J. P. Straley, Deconfinement and dissipation in quantum Hall Josephson tunneling. *Phys. Rev. Lett.* **91**, 046806 (2003).

[158] H. A. Fertig and G. Murthy, Coherence network in the quantum Hall bilayer. *Phys. Rev. Lett.* **95**, 156802 (2005).

[159] A. L. Fetter, C. B. Hanna, and R. B. Laughlin, Random-phase approximation in the fractional-statistics gas. *Phys. Rev. B* **39**, 9679 (1989).

[160] R. P. Feynman, Atomic theory of the lambda transition in helium. *Phys. Rev.* **91**, 1291 (1953).

[161] R. P. Feynman. In *Progress in Low Temperature Physics*, vol. 1, Chapter 2, ed. C. G. Gorter (Amsterdam: North Holland, 1955).

[162] R. P. Feynman and M. Cohen, Energy spectrum of the excitations in liquid helium. *Phys. Rev.* **102**, 1189 (1956).

[163] R. P. Feynman and A. R. Hibbs, *Quantum Mechanics and Path Integrals* (New York: McGraw-Hill, 1965).

[164] M. P. A. Fisher, G. Grinstein, and S. M. Girvin, Presence of quantum diffusion in two dimensions: Universal resistance at the superconductor-insulator transition. *Phys. Rev. Lett.* **64**, 587 (1990).

[165] R. Fleischmann, T. Geisel and R. Ketzmerick, Magnetoresistance due to chaos and nonlinear resonances in lateral surface superlattices. *Phys. Rev. Lett.* **68**, 1367 (1992).

[166] R. Fleischmann, T. Geisel, C. Holzknecht, and R. Ketzmerick, Nonlinear dynamics of composite fermions in nanostructures. *Europhys. Lett.* **36**, 167 (1996).

[167] M. Flohr, Fusion and tensoring of conformal field theory and composite fermion picture of fractional quantum Hall effect. *Mod. Phys. Lett.* **A11**, 55 (1996).

[168] M. Flohr and K. Osterloh, Conformal field theory approach to bulk wave functions in the fractional quantum Hall effect. *Phys. Rev. B* **67**, 235316 (2003).

[169] V. Fock, *Z. Phys.* **47**, 446 (1928).

[170] M. M. Fogler and A. A. Koulakov, Laughlin liquid to charge-density-wave transition at high Landau levels. *Phys. Rev. B* **55**, 9326 (1997).

[171] K. C. Foster, N. E. Bonesteel, and S. H. Simon, Conductivity of paired composite fermions. *Phys. Rev. Lett.* **91**, 046804 (2003).

[172] J. D. F. Franklin, I. Zailer, C. J. B. Ford, *et al.*, The Aharonov–Bohm effect in the fractional quantum Hall regime. *Surf. Sci.* **361**, 17 (1996).

[173] N. Freytag, Y. Tokunaga, M. Horvatić, *et al.*, New phase transition between partially and fully polarized quantum Hall states with charge and spin gaps at $\nu = 2/3$. *Phys. Rev. Lett.* **87**, 136801 (2001).

[174] N. Freytag, M. Horvatić, C. Berthier, M. Shayegan, and L. P. Lévy, NMR investigation of how free composite fermions are at $\nu = 1/2$. *Phys. Rev. Lett.* **89**, 246804 (2002).

[175] J. E. F. Frost, C.-T. Liang, D. R. Mace, *et al.*, Ballistic composite fermions in semiconductor nanostructures. *Phys. Rev. B* **53**, 9602 (1996).

[176] S. Fubini, Vertex operators and quantum Hall effect. *Mod. Phys. Lett. A* **6**, 347 (1991).

[177] Y. Gallais, T. H. Kirschenmann, C. F. Hirjibehedin, *et al.*, Spin excitations in the fractional quantum Hall regime at $\nu \leq 1/3$. *Physica E: Low-dimensional Systems and Nanostructures* **34**, 144 (2006).

[178] Y. Gallais, T. H. Kirschenmann, I. Dujovne, *et al.*, Transition from free to interacting composite fermions away from $\nu = 1/3$. *Phys. Rev. Lett.* **97**, 036804 (2006).

[179] P. J. Gee, J. Singleton, S. Uji, *et al.*, On the dependence on the magnetic field orientation of the composite fermion effective mass. *J. Phys. Condens. Matter* **8**, 10407 (1996).

[180] P. J. Gee, F. M. Peeters, J. Singleton, *et al.*, Composite fermions in tilted magnetic fields and the effect of the confining potential width on the composite-fermion effective mass. *Phys. Rev. B* **54**, 14313 (1996).

[181] M. C. Geisler, J. H. Smet, V. Umansky, *et al.*, Detection of a Landau band-coupling-induced rearrangement of the Hofstadter butterfly. *Phys. Rev. Lett.* **92**, 256801 (2004).

[182] R. R. Gerhardts, D. Weiss, and K. von Klitzing, Novel magnetoresistance oscillations in a periodically modulated two-dimensional electron gas. *Phys. Rev. Lett.* **62**, 1173 (1989).

[183] R. R. Gerhardts, Quasiclassical calculation of magnetoresistance oscillations of a two-dimensional electron gas in spatially periodic magnetic and electrostatic fields. *Phys. Rev. B* **53**, 11064 (1996).

[184] G. Gervais, L. W. Engel, H. L. Stormer, *et al.*, Competition between a fractional quantum Hall liquid and bubble and Wigner crystal phases in the third Landau level. *Phys. Rev. Lett.* **93**, 266804 (2004).

[185] T. K. Ghosh and G. Baskaran, Modeling two-roton bound state formation in the fractional quantum Hall system. *Phys. Rev. Lett.* **87**, 186803 (2001).

[186] I. Giaever, Energy gap in superconductors measured by electron tunneling. *Phys. Rev. Lett.* **5**, 147 (1960).

[187] M. Girardeau, Relationship between systems of impenetrable bosons and fermions in one dimension. *J. Math. Phys.* **1**, 516 (1960).

[188] U. Girlich and M. Hellmund, Comparison of fractional-quantum-Hall-effect quasielectron trial wave functions on a sphere. *Phys. Rev. B* **49**, 17488 (1994).

[189] S. M. Girvin and T. Jach, Interacting electrons in two-dimensional Landau levels: Results for small clusters. *Phys. Rev. B* **28**, 4506 (1983).

[190] S. M. Girvin and T. Jach, Formalism for the quantum Hall effect: Hilbert space of analytic functions. *Phys. Rev. B* **29**, 5617 (1984).

[191] S. M. Girvin, A. H. MacDonald, and P. M. Platzman, Collective-excitation gap in the fractional quantum Hall effect. *Phys. Rev. Lett.* **54**, 581 (1985).

[192] S. M. Girvin, A. H. MacDonald, and P. M. Platzman, Magneto-roton theory of collective excitations in the fractional quantum Hall effect. *Phys. Rev. B* **33**, 2481 (1986).

[193] S. M. Girvin and A. H. MacDonald, Off-diagonal long-range order, oblique confinement, and the fractional quantum Hall effect. *Phys. Rev. Lett.* **58**, 1252 (1987).

[194] S. M. Girvin and A. H. MacDonald, Multicomponent quantum Hall systems: The sum of their parts and more. In *Perspectives in Quantum Hall Effects*, eds. S. Das Sarma and A. Pinczuk (New York: Wiley, 1997).

[195] G. F. Giuliani and J. J. Quinn, Lifetime of a quasiparticle in a two-dimensional electron gas. *Phys. Rev. B* **26**, 4421 (1982).

[196] G. F. Giuliani and G. Vignale, *Quantum Theory of the Electron Liquid* (Cambridge University Press, 2005).

[197] M. O. Goerbig, P. Lederer, and C. Morais Smith, Second generation of composite fermions in the Hamiltonian theory. *Phys. Rev. B* **69**, 155324 (2004). [Erratum: *Phys. Rev. B* **70**, 249903 (2004).]

[198] M. O. Goerbig, P. Lederer, and C. Morais Smith, Possible reentrance of the fractional quantum Hall effect in the lowest Landau level. *Phys. Rev. Lett.* **93**, 216802 (2004).

[199] M. O. Goerbig, P. Lederer, and C. M. Smith, Competition between quantum-liquid and electron-solid phases in intermediate Landau levels. *Phys. Rev. B* **69**, 115327 (2004).

[200] N. Goldenfeld. *Lectures on Phase Transitions and the Renormalization Group* (Reading, MA: Addison–Wesley, 1992).

[201] A. S. Goldhaber and J. K. Jain, Characterization of fractional-quantum-Hall-effect quasiparticles. *Phys. Lett. A* **199**, 267 (1995).

[202] G. A. Goldin, R. Menikoff, and D. H. Sharp, Particle statistics from induced representations of a local current group. *J. Math. Phys.* **21**, 650 (1980).

[203] G. A. Goldin, R. Menikoff, and D. H. Sharp, Representations of a local current algebra in nonsimply connected space and the Aharonov–Bohm effect. *J. Math. Phys.* **22**, 1664 (1981).

[204] V. J. Goldman, M. Shayegan, and D. C. Tsui, Evidence for the fractional quantum Hall state at $\nu = 1/7$. *Phys. Rev. Lett.* **61**, 881 (1988).

[205] V. J. Goldman, M. Santos, M. Shayegan, and J. E. Cunningham, Evidence for two-dimensional quantum Wigner crystal. *Phys. Rev. Lett.* **65**, 2189 (1990).

[206] V. J. Goldman and M. Shayegan, Fractional quantum Hall states at $\nu = 7/11$ and 9/13. *Surf. Sci.* **229**, 10 (1990).

[207] V. J. Goldman, J. K. Wang, B. Su, and M. Shayegan, Universality of the Hall effect in a magnetic-field-localized two-dimensional electron system. *Phys. Rev. Lett.* **70**, 647 (1993).

[208] V. J. Goldman, B. Su, and J. K. Jain, Detection of composite fermions by magnetic focusing. *Phys. Rev. Lett.* **72**, 2065 (1994).

[209] V. J. Goldman and B. Su, Resonant tunneling in the quantum Hall regime: Measurement of fractional charge. *Science* **267**, 1010 (1995).

[210] V. J. Goldman and E. V. Tsiper, Dependence of the fractional quantum Hall edge critical exponent on the range of interaction. *Phys. Rev. Lett.* **86**, 5841 (2001).

[211] E. Gornik, R. Lassnig, G. Strasser, *et al.*, Specific heat of two-dimensional electrons in GaAs-GaAlAs multilayers. *Phys. Rev. Lett.* **54**, 1820 (1985).

[212] M. Grayson, D. C. Tsui, L. N. Pfeiffer, K. W. West, and A. M. Chang, Continuum of chiral Luttinger liquids at the fractional quantum Hall edge. *Phys. Rev. Lett.* **80**, 1062 (1998).

[213] M. Greiter, X. G. Wen, and F. Wilczek, Paired Hall state at half filling. *Phys. Rev. Lett.* **66**, 3205 (1991).

[214] M. Greiter, X. G. Wen, and F. Wilczek, Paired Hall states in double-layer electron systems. *Phys. Rev. B* **46**, 9586 (1992).

[215] M. Greiter, X. G. Wen, and F. Wilczek, Paired Hall states. *Nucl. Phys. B* **374**, 567 (1992).

[216] C. C. Grimes and G. Adams, Evidence for a liquid-to-crystal phase transition in a classical, two-dimensional sheet of electrons. *Phys. Rev. Lett.* **42**, 795 (1979).

[217] C. Gros and A. H. MacDonald, Conjecture concerning the fractional Hall hierarchy. *Phys. Rev. B* **42**, 9514 (1990).

[218] A. D. Güçlü and C. J. Umrigar, Maximum-density droplet to lower-density droplet transition in quantum dots. *Phys. Rev. B* **72**, 045309 (2005).

[219] A. D. Güçlü, G. S. Jeon, C. J. Umrigar, and J. K. Jain, Quantum Monte Carlo study of composite fermions in quantum dots: The effect of Landau level mixing. *Phys. Rev. B* **72**, 205327 (2005).

[220] G. Hackenbroich and F. von Oppen, Periodic-orbit theory of quantum transport in antidot lattices. *Europhys. Lett.* **29**, 151 (1995).

[221] F. D. M. Haldane, Fractional quantization of the Hall effect: A hierarchy of incompressible quantum fluid states. *Phys. Rev. Lett.* **51**, 605 (1983).

[222] F. D. M. Haldane and E. H. Rezayi, Finite-size studies of the incompressible state of the fractionally quantized Hall effect and its excitations. *Phys. Rev. Lett.* **54**, 237 (1985).

[223] F. D. M. Haldane and E. H. Rezayi, Periodic Laughlin–Jastrow wave functions for the fractional quantized Hall effect. *Phys. Rev. B* **31**, 2529 (1985).

[224] F. D. M. Haldane and E. H. Rezayi, Spin-singlet wave function for the half-integral quantum Hall effect. *Phys. Rev. Lett.* **60**, 956 (1988).

[225] F. D. M. Haldane and E. H. Rezayi, Erratum: Spin-singlet wave function for the half-integral quantum Hall effect [*Phys. Rev. Lett.* **60**, 956 (1988)]. *Phys. Rev. Lett.* **60**, 1886 (1988).

[226] F. D. M. Haldane, The hierarchy of fractional states and numerical studies. In *The Quantum Hall Effect*, ed. R. E. Prange and S. M. Girvin (New York: Springer, 1990).

[227] E. H. Hall, On a new action of the magnet on electric currents. *Am. J. Math.* **2**, 287 (1879).

[228] B. I. Halperin, Quantized Hall conductance, current-carrying edge states, and the existence of extended states in a two-dimensional disordered potential. *Phys. Rev. B* **25**, 2185 (1982).

[229] B. I. Halperin, Theory of the quantized Hall conductance. *Helv. Phys. Acta* **56**, 75 (1983).

[230] B. I. Halperin, Statistics of quasiparticles and the hierarchy of fractional quantized Hall states. *Phys. Rev. Lett.* **52**, 1583 (1984).

[231] B. I. Halperin, P. A. Lee, and N. Read, Theory of the half-filled Landau level. *Phys. Rev. B* **47**, 7312 (1993).

[232] B. I. Halperin, Fermion Chern-Simons theory and the unquantized quantum Hall effect. In *Perspectives in Quantum Hall Effects*, eds. S. Das Sarma and A. Pinczuk (New York: Wiley, 1996).

[233] B. I. Halperin, Composite fermions and the Fermion–Chern–Simons theory. *Physica E* **20**, 71 (2003).

[234] M. Hamermesh, *Group Theory and Its Application to Physical Problems* (New York: Dover, 1962).

[235] J. H. Han and S.-R. Eric Yang, Ground-state wave functions of general filling factors in the lowest Landau level. *Phys. Rev. B* **58**, R10163 (1998).

[236] J. P. Hansen and I. R. MacDonald, *Theory of Simple Liquids* (London: Academic Press, 1976).

[237] T. H. Hansson, A. Kerlhede, and J. M. Leinaas, Nonlinear sigma model for partially polarized quantum Hall states. *Phys. Rev. B* **54**, 11110 (1996).

[238] T. H. Hansson, A. Karlhede, J. M. Leinaas and U. Nilsson, Field theory for partially polarized quantum Hall states. *Phys. Rev. B* **60**, 4866 (1999).

[239] T. H. Hansson, C.-C. Chang, J. K. Jain, and S. Viefers, Conformal field theory of composite fermions. cond-mat/0603125 (2006).

[240] Y. Hatsugai, P.-A. Bares, and X. G. Wen, Electron spectral function of an interacting two dimensional electron gas in a strong magnetic field. *Phys. Rev. Lett.* **71**, 424 (1993).

[241] R. J. Haug, A. H. MacDonald, P. Streda, and K. von Klitzing, Quantized multichannel magnetotransport through a barrier in two dimension. *Phys. Rev. Lett.* **61**, 2797 (1988).

[242] P. Hawrylak, Single-electron capacitance spectroscopy of few-electron artificial atoms in a magnetic field: Theory and experiment. *Phys. Rev. Lett.* **71**, 3347 (1993).

[243] S. He, X. C. Xie, S. Das Sarma and F. C. Zhang, Quantum Hall effect in double-quantum-well systems. *Phys. Rev. B* **43**, 9339 (1991).

[244] S. He, X. C. Xie, and F. C. Zhang, Anyons, boundary constraint, and hierarchy in fractional quantum Hall effect. *Phys. Rev. Lett.* **68**, 3460 (1992).

[245] S. He, S. Das Sarma, and X. C. Xie, Quantized Hall effect and quantum phase transitions in coupled two-layer electron systems. *Phys. Rev. B* **47**, 4394 (1993).

[246] S. He, P. M. Platzman, and B. I. Halperin, Tunneling into a two-dimensional electron system in a strong magnetic field. *Phys. Rev. Lett.* **71**, 777 (1993).

[247] S. He, S. H. Simon, and B. I. Halperin, Response function of the fractional quantized Hall state on a sphere. II. Exact diagonalization. *Phys. Rev. B* **50**, 1823 (1994).

[248] Song He, Stabilizing the Richardson eigenvector algorithm by controlling chaos. *Computers in Physics* **11**, 194 (1997).

[249] O. Heinonen (Ed.) *Composite Fermions* (New York: World Scientific, 1998).

[250] C. Hermann and C. Weisbuch, $k \cdot p$ perturbation theory in III-V compounds and alloys: A reexamination. *Phys. Rev. B* **15**, 823 (1977).

[251] M. Hilke, D. C. Tsui, M. Grayson, L. N. Pfeiffer, and K. W. West, Fermi liquid to Luttinger liquid transition at the edge of a two-dimensional electron gas. *Phys. Rev. Lett.* **87**, 186806 (2001).

[252] H. Hirai, S. Komiyama, S. Sasa, and T. Fujii, Linear magnetoresistance due to an inhomogeneous electron-density distribution in a two-dimensional electron gas. *Solid State Commun.* **72**, 1033 (1989).

[253] C. F. Hirjibehedin, A. Pinczuk, B. S. Dennis, L. N. Pfeiffer, and K. W. West, Crossover and coexistence of quasiparticle excitations in the fractional quantum Hall regime at $\nu \leq 1/3$. *Phys. Rev. Lett.* **91**, 186802 (2003).

[254] C. F. Hirjibehedin, I. Dujovne, A. Pinczuk, *et al.*, Splitting of long-wavelength modes of the fractional quantum Hall liquid at $\nu = 1/3$. *Phys. Rev. Lett.* **95**, 066803 (2005).

[255] T.-L. Ho, Broken symmetry of two-component $\nu = 1/2$ quantum Hall states. *Phys. Rev. Lett.* **75**, 1186 (1995).

[256] C. Hodges, H. Smith, and J. W. Wilkins, Effect of fermi surface geometry on electron-electron scattering. *Phys. Rev. B* **4**, 302 (1971).

[257] S. Holmes, D. K. Maude, M. L. Williams, *et al.*, Experimental determination of the transport properties of composite fermions with reduced Lande g-factor. *Semicond. Sci. Technol.* **9**, 1549 (1994).

[258] S. Holmes, D. K. Maude, M. L. Williams, *et al.*, Determination of the transport properties of composite fermions under high hydrostatic pressure. *J. Phys. Chem. Solids* **56**, 459 (1995).

[259] D. R. Hofstadter, Energy levels and wave functions of Bloch electrons in rational and irrational magnetic fields. *Phys. Rev. B* **14**, 2239 (1976).

[260] B. Huckestein and B. Kramer, One-parameter scaing in the lowest Landau band: Precise determination of the critical behavior of the localization length. *Phys. Rev. Lett.* **64**, 1437 (1990).

[261] B. Huckestein, Scaling theory of the integer quantum Hall effect. *Rev. Mod. Phys.* **67**, 357 (1995).

[262] Ikuo Ichinose, Tetsuo Matsui, Masaru Onoda, Particle-flux separation in Chern–Simons theory and the Landau-level gap of composite fermions. *Phys. Rev. B* **52**, 10547 (1995).

[263] I. Ichinose and T. Matsui, Mechanism of particle-flux separation and generation of the dynamical gauge field in the Chern–Simons theory with fermions. *Nucl. Phys. B* **468**, 487 (1996).

[264] I. Ichinose and T. Matsui, Particle-flux separation of electrons in the half-filled Landau level: Chargeon-fluxon approach. *Nucl. Phys. B* **483**, 681 (1997).

[265] I. Ichinose and T. Matsui, Gauge theory of composite fermions: Particle-flux separation in quantum Hall systems. *Phys. Rev. B* **68**, 085322 (2003).

[266] S. Ilani, J. Martin, E. Teitelbaum, *et al.*, The microscopic nature of localization in the quantum Hall effect. *Nature* **427** 328 (2004).

[267] S. V. Iordansky, On the conductivity of two-dimensional electrons in a strong magnetic field. *Solid State Commun.* **43**, 1 (1982).

[268] K. Ismail, Electron transport in the quantum Hall regime in strained Si/SiGe. *Physica (Amsterdam) B* **227**, 310 (1996).

[269] J. K. Jain and S. A. Kivelson, Landauer-type formulation of quantum-Hall transport: Critical currents and narrow channels. *Phys. Rev. B* **37**, 4276 (1988).

[270] J. K. Jain and S. A. Kivelson, Quantum Hall effect in quasi one-dimensional systems: Resistance fluctuations and breakdown. *Phys. Rev. Lett.* **60**, 1542 (1988).

[271] J. K. Jain, Prediction of Aharonov–Bohm oscillations on the quantum Hall plateaus of small and narrow rings. *Phys. Rev. Lett.* **60**, 2074 (1988).

[272] J. K. Jain, Composite-fermion approach for the fractional quantum Hall effect. *Phys. Rev. Lett.* **63**, 199 (1989).

[273] J. K. Jain and S. A. Kivelson, Jain and Kivelson reply. *Phys. Rev. Lett.* **62**, 231 (1989).

[274] J. K. Jain, Incompressible quantum Hall states. *Phys. Rev. B* **40**, 8079 (1989).

[275] J. K. Jain, Theory of the fractional quantum Hall effect. *Phys. Rev. B* **41**, 7653 (1990).

[276] J. K. Jain, S. A. Kivelson, and N. Trivedi, Scaling theory of the fractional quantum Hall effect. *Phys. Rev. Lett.* **64**, 1297 (1990).

[277] J. K. Jain, S. A. Kivelson, and N. Trivedi, Erratum: Scaling theory of the fractional quantum Hall effect. *Phys. Rev. Lett.* **64**, 1993 (E) (1990).

[278] J. K. Jain and X. G. Wu, Hund's rule for composite fermions. *Phys. Rev. B* **49**, 5085 (1994).

[279] J. K. Jain and T. Kawamura, Composite fermions in quantum dots. *Europhys. Lett.* **29**, 321 (1995).

[280] J. K. Jain and R. K. Kamilla, Quantitative study of large composite-fermion systems. *Phys. Rev. B* **55**, R4895 (1997).

[281] J. K. Jain and R. K. Kamilla, Composite fermions in the Hilbert space of the lowest electronic Landau level. *Int. J. Mod. Phys. B* **11**, 2621 (1997).

[282] J. K. Jain and R. K. Kamilla, Composite fermions: Particles of the lowest Landau level. Chapter 1 In *Composite Fermions*, ed. O. Heinonen (New York: World Scientific, 1998).

[283] J. K. Jain, The composite fermion: A quantum particle and its quantum fluids. *Physics Today* **53** (4), 39 (2000).

[284] J. K. Jain, The role of analogy in unraveling the fractional-quantum-Hall-effect mystery. *Physica E* **20**, 79 (2003).

[285] J. K. Jain, C.-C. Chang, G. S. Jeon, and M. R. Peterson, Composite fermions in the neighborhood of $\nu = 1/3$. *Solid State Commun.* **127**, 805 (2003).

[286] J. K. Jain, K. Park, M. R. Peterson, and V. W. Scarola, Composite fermion theory of excitations in the fractional quantum Hall effect. *Solid State Commun.* **135**, 602 (2005).

[287] J. K. Jain and M. R. Peterson, Reconstructing the electron in a fractionalized quantum fluid. *Phys. Rev. Lett.* **94**, 186808 (2005).

[288] J. K. Jain and C. Shi, Resonant tunneling in fractional quantum Hall effect: Superperiods and braiding statistics. *Phys. Rev. Lett.* **96**, 136802 (2006).

[289] R. Jastrow, Many-body problem with strong forces. *Phys. Rev.* **98**, 1479 (1955).

[290] B. Jeckelmann, A. D. Inglis, B. Jeanneret, Material, device, and step independence of the quantized Hall resistance. *IEEE Trans. Instrum. Meas.* **44**, 269–272 (1995).

[291] A. Jeffery, R. E. Elmquist, J. Q. Shields, *et al.*, Determination of the von Klitzing constant and the fine-structure constant through a comparison of the quantized Hall resistance and the ohm derived from the NIST calculable capacitor. *Metrologia* **35**, 83 (1998).

[292] G.-S. Jeon and J. K. Jain, Nature of quasiparticle excitations in the fractional quantum Hall effect. *Phys. Rev. B* **68**, 165346 (2003).

[293] G. S. Jeon, K. L. Graham, J. K. Jain, Fractional statistics in the fractional quantum Hall effect. *Phys. Rev. Lett.* **91**, 036801 (2003).

[294] G. S. Jeon, K. L. Graham, J. K. Jain, Berry phases for composite fermions: Effective magnetic field and fractional statistics. *Phys. Rev. B* **70**, 125316 (2004).

[295] G. S. Jeon, C.-C. Chang and J. K. Jain, Composite-fermion crystallites in quantum dots. *J. Phys.: Condens. Matter* **16**, L271–L277 (2004).

[296] G. S. Jeon, C.-C. Chang and J. K. Jain, Composite fermion theory of correlated electrons in semiconductor quantum dots in high magnetic fields. *Phys. Rev. B* **69**, 241304(R) (2004).

[297] G. S. Jeon and J. Ye, Trial wave function approach to bilayer quantum Hall systems. *Phys. Rev. B* **71**, 035348 (2005).

[298] G. S. Jeon, A. D. Güçlü, C. J. Umrigar, and J. K. Jain, Composite-fermion antiparticle description of the hole excitation in the maximum-density droplet with a small number of electrons. *Phys. Rev. B* **72**, 245312 (2005).

[299] G. S. Jeon, C.-C. Chang and J. K. Jain, Semiconductor quantum dots in high magnetic fields: The composite fermion view. cond-mat/0611309 (2006).

[300] H. W. Jiang, H. L. Stormer, D. C. Tsui, L. N. Pfeiffer, and K. W. West, Transport anomalies in the lowest Landau level of two-dimensional electrons at half-filling. *Phys. Rev. B* **40**, 12013 (1989).

[301] H. W. Jiang, R. L. Willett, H. L. Stormer, *et al.*, Quantum liquid versus electron solid around $v = 1/5$ Landau-level filling. *Phys. Rev. Lett.* **65**, 633 (1990).

[302] H. W. Jiang, H. L. Stormer, D. C. Tsui, L. N. Pfeiffer, and K. W. West, Magnetotransport studies of the insulating phase around $v = 1/5$ Landau-level filling. *Phys. Rev. B* **44**, 8107 (1991).

[303] H. W. Jiang, C. E. Johnson, K. L. Wang, and S. T. Hannahs, Observation of magnetic-field-induced delocalization: Transition from Anderson insulator to quantum Hall conductor. *Phys. Rev. Lett.* **71**, 1439 (1993).

[304] Y. N. Joglekar, H. K. Nguyen, and G. Murthy, Edge reconstructions in fractional quantum Hall systems. *Phys. Rev. B* **68**, 035332 (2003).

[305] P. Johansson and J. M. Kinaret, Magnetophonon shakeup in a Wigner crystal: Applications to tunneling spectroscopy in the quantum Hall regime. *Phys. Rev. Lett.* **71**, 1435 (1993).

[306] M. H. Johnson and B. A. Lippmann, Motion in a constant magnetic field. *Phys. Rev.* **76**, 828 (1949).

[307] B. L. Johnson and G. Kirczenow, Composite fermions in the quantum Hall effect. *Rep. Prog. Phys.* **60**, 889 (1997).

[308] C. Jörger, W. Dietsche, W. Wegscheider, and K. von Klitzing, Drag effect between 2D electron gases: Composite fermion and screening effects. *Physica E* **6**, 586 (2000).

[309] V. Kalmeyer and S. C. Zhang, Metallic phase of the quantum Hall system at even-denominator filling fractions. *Phys. Rev. B* **46**, 9889 (1992).

[310] R. K. Kamilla, X. G. Wu, and J. K. Jain, Composite fermion theory of collective excitations in fractional quantum Hall effect. *Phys. Rev. Lett.* **76**, 1332 (1996).

[311] R. K. Kamilla, X. G. Wu, and J. K. Jain, Skyrmions in the fractional quantum Hall effect. *Solid State Commun.* **99**, 289 (1996).

[312] R. K. Kamilla, X. G. Wu, and J. K. Jain, Excitons of composite fermions. *Phys. Rev. B* **54**, 4873 (1996).

[313] R. K. Kamilla and J. K. Jain, Excitonic instability and termination of fractional quantum Hall effect. *Phys. Rev. B* **55**, R13417 (1997).

[314] R. K. Kamilla and J. K. Jain, Variational study of the vortex structure of composite fermions. *Phys. Rev. B* **55**, 9824 (1997).

[315] R. K. Kamilla, J. K. Jain, and S. M. Girvin, Fermi-sea-like correlations in a partially filled Landau level. *Phys. Rev. B* **56**, 12411 (1997).

[316] C. L. Kane and M. P. A. Fisher, Resonant tunneling in an interacting one-dimensional electron gas. *Phys. Rev. B* **46**, 7268 (1992).

[317] C. L. Kane and M. P. A. Fisher, Transmission through barriers and resonant tunneling in an interacting one-dimensional electron gas. *Phys. Rev. B* **46**, 15233 (1992).

[318] C. L. Kane and M. P. A. Fisher, Nonequilibrium noise and fractional charge in the quantum Hall effect. *Phys. Rev. Lett.* **72**, 724 (1994).

[319] C. L. Kane and M. P. A. Fisher, Impurity scattering and transport of fractional quantum Hall edge states. *Phys. Rev. B* **51**, 13449 (1995).

[320] W. Kang, H. L. Stormer, L. N. Pfeiffer, K. W. Baldwin, and K. W. West, How real are composite fermions? *Phys. Rev. Lett.* **71**, 3850 (1993).

[321] W. Kang, H. L. Stormer, L. N. Pfeiffer, K. W. Baldwin and K. W. West, How real are composite fermions? *Physica B* **211**, 396 (1995).

[322] W. Kang, Song He, H. L. Stormer, *et al.*, Temperature dependent scattering of composite fermions. *Phys. Rev. Lett.* **75**, 4106 (1995).

[323] M. Kang, A. Pinczuk, B. S. Dennis, *et al.*, Inelastic light scattering by gap excitations of fractional quantum Hall states at $1/3 \leq \nu \leq 2/3$. *Phys. Rev. Lett.* **84**, 546 (2000).

[324] M. Kang, A. Pinczuk, B. S. Dennis, L. N. Pfeiffer, and K. W. West, Observation of multiple magnetorotons in the fractional quantum Hall effect. *Phys. Rev. Lett.* **86**, 2637 (2001).

[325] V. C. Karavolas, G. P. Triberis, and F. M. Peeters, Electrical and thermal transport of composite fermions. *Phys. Rev. B* **56**, 15289 (1997).

[326] V. C. Karavolas and G. P. Triberis, Thermoelectric transport of composite fermions at $\nu = 1/2$ and $\nu = 3/2$: A simple way of evaluating p. *Phys. Rev. B* **65**, 033307 (2001).

[327] V. C. Karavolas and G. P. Triberis, Systematic study of thermal transport of composite fermions around filling factors $\nu = 1 \pm 1/2m$. *Phys. Rev. B* **66**, 155315 (2002).

[328] M. Kasner and W. Apel, Comparison of fractional-quantum-Hall-effect quasielectron trial wave functions on the disk. *Phys. Rev. B* **48**, 11435 (1993).

[329] R. F. Kazarinov and S. Luryi, Quantum percolation and quantization of Hall resistance in two-dimensional electron gas. *Phys. Rev. B* **25**, 7626 (1982).

[330] M. Kellogg, J. P. Eisenstein, L. N. Pfeiffer, and K. W. West, Vanishing Hall resistance at high magnetic field in a double-layer two-dimensional electron system. *Phys. Rev. Lett.* **93**, 036801 (2004).

[331] P. Khandelwal, N. N. Kuzma, S. E. Barrett, L. N. Pfeiffer, and K. W. West, Optically pumped nuclear magnetic resonance measurements of the electron spin polarization in GaAs quantum wells near Landau level filling factor $\nu = 1/3$. *Phys. Rev. Lett.* **81**, 673 (1998).

[332] P. Khandelwal, A. E. Dementyev, N. N. Kuzma, *et al.*, Spectroscopic evidence for the localization of skyrmions near $\nu = 1$ as $T \to 0$. *Phys. Rev. Lett.* **86**, 5353 (2001).

[333] D. V. Khveshchenko, Logarithmic temperature dependence of conductivity at half filled Landau level. *Phys. Rev. Lett.* **77**, 362 (1996).

[334] Y. B. Kim, P. A. Lee and X. G. Wen, Quantum Boltzmann equation of composite fermions interacting with a gauge field. *Phys. Rev. B* **52**, 17275 (1995).

[335] Y. B. Kim and P. A. Lee, Specific heat and validity of the quasiparticle approximation in the half-filled Landau level. *Phys. Rev. B* **54**, 2715 (1996).

[336] E. Kim and M. H. W. Chan, Observation of superflow in solid helium. *Science* **305**, 1941 (2004).

[337] E.-A. Kim, M. Lawler, S. Vishveshwara, and E. Fradkin, Signatures of fractional statistics in noise experiments in quantum Hall fluids. *Phys. Rev. Lett.* **95**, 176402 (2005).

[338] G. Kirczenow and B. L. Johnson, Composite fermions, edge currents, and the fractional quantum Hall effect. *Phys. Rev. B* **51**, 17579 (1995).

[339] G. Kirczenow, Composite-fermion edge states and transport through nanostructures in the fractional quantum Hall regime. *Phys. Rev. B* **53**, 15767 (1996).

[340] G. Kirczenow, Composite-fermion edge states, fractional charge, and current noise. *Phys. Rev. B* **58**, 1457 (1998).

[341] G. Kirczenow, Composite fermion approach to the edge state transport. In *Composite Fermions*, ed. O. Heinonen (New York: World Scientific, 1998).

[342] S. Kivelson, C. Kallin, D. P. Arovas, and J. R. Schrieffer, Cooperative ring exchange and the fractional quantum Hall effect. *Phys. Rev. B* **36**, 1620 (1987).

[343] S. A. Kivelson, Semiclassical theory of localized many-anyon states. *Phys. Rev. Lett.* **65**, 3369 (1990).

[344] S. A. Kivelson, D.-H. Lee, Y. Krotov, and J. Gan, Composite-fermion Hall conductance at $\nu = 1/2$. *Phys. Rev. B* **55**, 15552 (1997).

[345] H. Kjønsberg and J. Myrheim, Numerical study of charge and statistics of Laughlin quasiparticles. *Int. J. Mod. Phys. A* **14**, 537 (1999).

[346] H. Kjønsberg and J. M. Leinaas, Charge and statistics of quantum Hall quasi-particles – a numerical study of mean values and fluctuations. *Nucl. Phys. B* **559**, 705 (1999).

[347] S. R. Koch, J. Haug, K. von Klitzing, and K. Ploog, Size-dependent analysis of the metal-insulator transition in the integral quantum Hall effect. *Phys. Rev. Lett.* **67**, 883 (1991).

[348] S. Koch, R. J. Haug, K. von Klitzing, and K. Ploog, Experiments on scaling in $Al_xGa_{1-x}As/GaAs$ heterostructures under quantum Hall conditions. *Phys. Rev. B* **43**, 6828 (1991).

[349] S. E. Koonin. *Computational Physics* (New York: Addison-Wesley Publishing, 1986).

[350] A. A. Koulakov, M. M. Fogler, and B. I. Shklovskii, Charge density wave in two-dimensional electron liquid in weak magnetic field. *Phys. Rev. Lett.* **76**, 499 (1996).

[351] W. Kohn, Cyclotron resonance and de Haas–van Alphen oscillations of an interacting electron gas. *Phys. Rev.* **123**, 1242 (1961).

[352] W. Kohn and J. M. Luttinger, New mechanism for superconductivity. *Phys. Rev. Lett.* **15**, 524 (1965).

[353] S. V. Kravchenko and M. P. Sarachik, Metal-insulator transition in two-dimensional electron systems. *Rep. Prog. Phys.* **67**, 1 (2004).

[354] N. Kumada, D. Terasawa, Y. Shimoda, *et al.*, Phase diagram of interacting composite fermions in the bilayer $\nu = 2/3$ quantum Hall effect. *Phys. Rev. Lett.* **89**, 116802 (2002).

[355] T. S. Kuhn, *The Structure of Scientific Revolutions*, 3rd edn. (Chicago University Press, 1996).

[356] I. V. Kukushkin, V. I. Falko, *et al.*, Evidence of the triangular lattice of crystallized electrons from time resolved luminescence. *Phys. Rev. Lett.* **72**, 3594 (1994).

[357] I. V. Kukushkin, K. von Klitzing, and K. Eberl, Spin polarization of composite fermions: Measurements of the Fermi energy. *Phys. Rev. Lett.* **82**, 3665 (1999).

[358] I. V. Kukushkin, K. von Klitzing, K. G. Levchenko, and Yu. E. Lozovik, Temperature dependence of the spin polarization of composite fermions. *JETP Lett.* **70**, 730 (1999).

[359] I. V. Kukushkin, J. H. Smet, K. von Klitzing, and K. Eberl, Optical investigation of spin-wave excitations in fractional quantum Hall states and of interaction between composite fermions. *Phys. Rev. Lett.* **85**, 3688 (2000).

[360] I. V. Kukushkin, J. H. Smet, K. von Klitzing, and W. Wegscheider, Cyclotron resonance of composite fermions. *Nature* **415**, 409 (2002).

[361] I. V. Kukushkin, J. H. Smet, K. von Klitzing and W. Wegscheider, Cyclotron resonance of composite fermions with two and four flux quanta. *Physica E* **20**, 96 (2003).

[362] K. Lai, W. Pan, D. C. Tsui, *et al.*, Two-flux composite fermion series of the fractional quantum Hall states in strained Si. *Phys. Rev. Lett.* **93**, 156805 (2004).

[363] P. K. Lam and S. M. Girvin, Liquid-solid transition and the fractional quantum-Hall effect. *Phys. Rev. B* **30**, 473 (1984).

[364] L. D. Landau, *Z. Phys.* **64**, 623 (1930).

[365] L. D. Landau, *J. Phys. USSR* **5**, 71 (1940).

[366] R. Landauer, Electrical resistance of disordered one-dimensional lattices. *Philos. Mag.* **21**, 863 (1970).

[367] G. Landwehr, 25 years of quantum Hall effect: How it all came about. *Physica E* **20**, 1 (2003).

[368] R. B. Laughlin, Quantized Hall conductivity in two dimensions. *Phys. Rev. B* **23**, 5632 (1981).

[369] R. B. Laughlin, Anomalous quantum Hall effect: An incompressible quantum fluid with fractionally charged excitations. *Phys. Rev. Lett.* **50**, 1395 (1983).

[370] R. B. Laughlin, Elementary theory: The incompressible quantum fluid. In *The Quantum Hall Effect*, 1st edn. R. E. Prange and S. M. Girvin (New York: Springer-Verlag, 1987).

[371] R. B. Laughlin, Superconducting ground state of noninteracting particles obeying fractional statistics. *Phys. Rev. Lett.* **60**, 2677 (1988).

[372] R. B. Laughlin, Nobel Lecture: Fractional quantization. *Rev. Mod. Phys.* **71**, 863 (1999).

[373] R. B. Laughlin, *A Different Universe: Reinventing Physics from the Bottom Down* (New York: Basic Books, 2005).

[374] D. R. Leadley, R. J. Nicholas, C. T. Foxon, and J. J. Harris, Measurements of the effective mass and scattering times of composite fermions from magnetotransport analysis. *Phys. Rev. Lett.* **72**, 1906 (1994).

[375] D. R. Leadley, M. van der Burgt, R. J. Nicholas, C. T. Foxon, and J. J. Harris, Pulsed-magnetic-field measurements of the composite-fermion effective mass. *Phys. Rev. B* **53**, 2057 (1996).

[376] D. R. Leadley, M. van der Burgt, R. J. Nicholas, *et al.*, The unifying role of effective field in the composite fermion model of the fractional quantum Hall effect. *Surf. Sci.* **362**, 22 (1996).

[377] D. R. Leadley, R. J. Nicholas, D. K. Maude, *et al.*, Fractional quantum Hall effect measurements at zero g factor. *Phys. Rev. Lett.* **79**, 4246 (1997).

[378] P. A. Lee and T. V. Ramakrishnan, Disordered electronic systems. *Rev. Mod. Phys.* **57**, 287 (1985).

[379] D.-H. Lee and C. L. Kane, Boson-vortex-skyrmion duality, spin-singlet fractional quantum Hall effect, and spin-1/2 anyon superconductivity. *Phys. Rev. Lett.* **64**, 1313 (1990).

[380] P. A. Lee and N. Nagaosa, Gauge theory of the normal state of high-T_c superconductors. *Phys. Rev. B* **46**, 5621 (1992).

[381] D. K. K. Lee and J. T. Chalker, Unified model for two localization problems: Electron states in spin-degenerate Landau levels and in a random magnetic field. *Phys. Rev. Lett.* **72**, 1510 (1994).

[382] S.-Y. Lee, V. W. Scarola, and J. K. Jain, Stripe formation in the fractional quantum Hall regime. *Phys. Rev. Lett.* **87**, 256803 (2001).

[383] S.-Y. Lee, V. W. Scarola and J. K. Jain, Structures for interacting composite fermions: Stripes, bubbles, and fractional quantum Hall effect. *Phys. Rev. B* **66**, 085336 (2002).

[384] A. J. Leggett, Can a solid be superfluid? *Phys. Rev. Lett.* **25**, 1543 (1970).

[385] H. Lehmann, Über Eigenschaften von Ausbreitungsfunktionen und Renormierungskonstanten Quantisierter Felder. *Nouvo Cimento* **11**, 342 (1954).

[386] J. M. Leinaas and J. Myrheim, On the theory of identical particles. *Nuovo Cimento B* **37**, 1 (1977).

[387] D. J. T. Leonard and N. F. Johnson, Dynamical properties of a two-dimensional electron gas in a magnetic field within the composite fermion model. *Phys. Rev. B* **58**, 15468 (1998).

[388] D. J. T. Leonard, T. Portengen, V. N. Nicopoulos, and N. F. Johnson, Orthogonality catastrophe in a composite fermion liquid. *J. Phys. Condens. Matter.* **10**, L453 (1998).

[389] D. Levesque, J. J. Weis, and A. H. MacDonald, Crystallization of the incompressible quantum-fluid state of a two-dimensional electron gas in a strong magnetic field. *Phys. Rev. B* **30**, 1056 (1984).

[390] L. S. Levitov, A. V. Shytov, and B. I. Halperin, Effective action of a compressible quantum Hall state edge: Application to tunneling. *Phys. Rev. B* **64**, 075322 (2001).

[391] Y. P. Li, T. Sajoto, L. W. Engel, D. C. Tsui, and M. Shayegan, Low-frequency noise in the reentrant insulating phase around the 1/5 fractional quantum Hall liquid. *Phys. Rev. Lett.* **67**, 1630 (1991).

[392] W. Li, G. A. Csáthy, D. C. Tsui, L. N. Pfeiffer, and K. W. West, Scaling and universality of integer quantum Hall plateau-to-plateau transitions. *Phys. Rev. Lett.* **94**, 206807 (2005).

[393] C.-T. Liang, C. G. Smith, D. R. Mace, *et al.*, Measurements of a composite fermion split-gate device. *Phys. Rev. B* **53**, R7596 (1996).

[394] C. -T. Liang, J. E. F. Frost, M. Y. Simmons, D. A. Ritchie and M. Pepper, Experimental evidence of a metal-insulator transition in a half-filled Landau level. *Solid State Commun.* **102**, 327 (1997).

[395] C. -T. Liang, J. E. F. Frost, M. Y. Simmons, D. A. Ritchie and M. Pepper, Experimental evidence for a metal-insulator transition and geometric effect in a half-filled Landau level. *Physica E* **2**, 78 (1998).

[396] C. T. Liang, M. Y. Simmons, D. A. Ritchie, and M. Pepper, Measurements of composite fermion conductivity dependence on carrier density. *J. Phys. Condens. Matter* **16**, 1095 (2004).

[397] M. P. Lilly, J. P. Eisenstein, L. N. Pfeiffer, and K. W. West, Coulomb drag in the extreme quantum limit. *Phys. Rev. Lett.* **80**, 1714 (1998).

[398] M. P. Lilly, K. B. Cooper, J. P. Eisenstein, L. N. Pfeiffer, and K. W. West, Evidence for an anisotropic state of two-dimensional electrons in high Landau levels. *Phys. Rev. Lett.* **82**, 394 (1999).

[399] M. P. Lilly, K. B. Cooper, J. P. Eisenstein, L. N. Pfeiffer, and K. W. West, Anisotropic states of two-dimensional electron systems in high Landau levels: Effect of an in-plane magnetic field. *Phys. Rev. Lett.* **83**, 824 (1999).

[400] D. Z. Liu, X. C. Xie, S. Das Sarma, and S. C. Zhang, Electron localization in a two-dimensional system with random magnetic flux. *Phys. Rev. B* **52**, 5858 (1995).

[401] A. Lopez and E. Fradkin, Fractional quantum Hall effect and Chern–Simons gauge theories. *Phys. Rev. B* **44**, 5246 (1991).

[402] A. Lopez and E. Fradkin, Response functions and spectrum of collective excitations of fractional-quantum-Hall-effect systems. *Phys. Rev. B* **47**, 7080 (1993).

[403] A. Lopez and E. Fradkin, Fermionic Chern–Simons theory for the fractional quantum Hall effect in bilayers. *Phys. Rev. B* **51**, 4347 (1995).

[404] A. Lopez and E. Fradkin, Fermionic Chern–Simons field theory for the fractional quantum Hall effect. In *Composite Fermions*, ed. O. Heinonen (New York: World Scientific, 1998).

[405] A. Lopez and E. Fradkin, Fermion Chern–Simons theory of hierarchical fractional quantum Hall states. *Phys. Rev. B* **69**, 155322 (2004).

[406] Y. E. Lozovik and V. I. Yudson, Feasibility of superfluidity of paired spatially separated electrons and holes - new superconductivity mechanism. *JETP Lett.* **22**, 274 (1975).

[407] Y. E. Lozovik and I. V. Ovchinnikov, BCS instability of a two-layer system of composite fermions. *JETP Lett.* **79** 76 (2004).

[408] J. M. Luttinger, An exactly soluble model of a many fermion system. *J. Math. Phys. N. Y.* **4**, 1154 (1963).

[409] I. J. Maasilta and V. J. Goldman, Line shape of resonant tunneling between fractional quantum Hall edges. *Phys. Rev. B* **55**, 4081 (1997).

[410] I. G. MacDonald, *Symmetric Functions and Hall Polynomials* (Clarendon, Oxford, 1979).

[411] A. H. MacDonald and S. M. Girvin, Collective excitations of fractional Hall states and Wigner crystallization in higher Landau levels. *Phys. Rev. B* **33**, 4009 (1986).

[412] A. H. MacDonald, Introduction to the physics of the quantum Hall regime. cond-mat/9410047 (1994).

[413] A. H. MacDonald, H. A. Fertig, and L. Brey, Skyrmions without sigma models in quantum Hall ferromagnets. *Phys. Rev. Lett.* **76**, 2153 (1996).

[414] A. H. MacDonald and J. J. Palacios, Magnons and skyrmions in fractional Hall ferromagnets. *Phys. Rev. B* **58**, R10171 (1998).

[415] G. D. Mahan, Excitons in metals: Infinite hole mass. *Phys. Rev.* **163**, 612 (1967).

[416] G. D. Mahan, *Many Particle Physics*, 3rd edn. (New York: Plenum, 2000).

[417] K. Maki and X. Zotos, Static and dynamic properties of a two-dimensional Wigner crystal in a strong magnetic field. *Phys. Rev. B* **28**, 4349 (1983).

[418] P. A. Maksym and T. Chakraborty, Quantum dots in a magnetic field: Role of electron-electron interactions. *Phys. Rev. Lett.* **65**, 108 (1990).

[419] P. A. Maksym, Eckardt frame theory of interacting electrons in quantum dots. *Phys. Rev. B* **53**, 10871 (1996).

[420] S. S. Mandal and V. Ravishankar, Theory of arbitrarily polarized quantum Hall states: Filling fractions and wave functions. *Phys. Rev. B* **54**, 8688 (1996).

[421] S. S. Mandal and V. Ravishankar, Direct test of the composite-fermion model in quantum Hall systems. *Phys. Rev. B* **54**, 8699 (1996).

[422] S. S. Mandal and J. K. Jain, Low-energy spin rotons in the fractional quantum Hall effect. *Phys. Rev. B* **63**, 201310(R) (2001).

[423] S. S. Mandal and J. K. Jain, How universal is the fractional-quantum-Hall edge Luttinger liquid? *Solid State Commun.* **118**, 503 (2001).

[424] S. S. Mandal and J. K. Jain, Theoretical search for the nested quantum Hall effect of composite fermions. *Phys. Rev. B* **66**, 155302 (2002).

[425] S. S. Mandal and J. K. Jain, Relevance of inter-composite-fermion interaction to the edge Tomonaga–Luttinger liquid. *Phys. Rev. Lett.* **89**, 096801 (2002).

[426] S. S. Mandal, M. R. Peterson, and J. K. Jain, Two-dimensional electron system in high magnetic fields: Wigner crystal versus composite-fermion liquid. *Phys. Rev. Lett.* **90**, 106403 (2003).

[427] M. Manninen, S. Viefers, M. Koskinen, and S. M. Reimann, Many-body spectrum and particle localization in quantum dots and finite rotating Bose condensates. *Phys. Rev. B* **64**, 245322 (2001).

[428] H. C. Manoharan, M. Shayegan, and S. J. Klepper, Signatures of a novel Fermi liquid in a two-dimensional composite particle metal. *Phys. Rev. Lett.* **73**, 3270 (1994).

[429] E. Manousakis and V. R. Pandharipande, Theoretical studies of elementary excitations in liquid ^{4}He. *Phys. Rev. B* **30**, 5062 (1984).

[430] E. Mariani, N. Magnoli, F. Napoli, M. Sassetti, and B. Kramer, Spin-pairing instabilities at the coincidence of two Landau levels. *Phys. Rev. B* **66**, 241303(R) (2002).

[431] E. Mariani, R. Mazzarello, M. Sassetti, and B. Kramer, Spin polarization transitions in the FQHE. *Ann. Phys.* **11**, 926 (2002).

[432] J. Martin, S. Ilani, B. Verdene, *et al.*, Localization of fractionally charged quasiparticles. *Science* **305**, 980 (2004).

[433] D. K. Maude, M. Potemski, J. C. Portal, *et al.*, Spin excitations of a two-dimensional electron gas in the limit of vanishing Landé g factor. *Phys. Rev. Lett.* **77**, 4604 (1996).

[434] V. Melik-Alaverdian and N. E. Bonesteel, Composite fermions and Landau-level mixing in the fractional quantum Hall effect. *Phys. Rev. B* **52**, R17032 (1995).

[435] V. Melik-Alaverdian and N. E. Bonesteel, Quantum Hall fluids on the Haldane sphere: A diffusion Monte Carlo study. *Phys. Rev. Lett.* **79**, 5286 (1997).

[436] V. Melik-Alaverdian and N. E. Bonesteel, Monte Carlo comparison of quasielectron wave functions. *Phys. Rev. B* **58**, 1451 (1998).

[437] S. Melinte, N. Freytag, M. Horvatić, *et al.*, NMR determination of 2D Electron spin polarization at $\nu = 1/2$. *Phys. Rev. Lett.* **84**, 354 (2000).

[438] S. Melinte, M. Berciu, C. Zhou, *et al.*, Laterally modulated 2D electron system in the extreme quantum limit. *Phys. Rev. Lett.* **92**, 036802 (2004).

[439] C. J. Mellor, R. H. Eyles, J. E. Digby, *et al.*, Phonon absorption at the magnetoroton minimum in the fractional quantum Hall effect. *Phys. Rev. Lett.* **74**, 2339 (1995).

[440] E. E. Mendez, M. Heiblum, L. L. Chang, and L. Esaki, High-magnetic-field transport in a dilute two-dimensional electron gas. *Phys. Rev. B* **28**, 4886 (1983).

[441] E. E. Mendez, W. I. Wang, L. L. Chang, and L. Esaki, Fractional quantum Hall effect in a two-dimensional hole system. *Phys. Rev. B* **30**, 1087 (1984).

[442] M. Merlo, N. Magnoli, M. Sassetti, and B. Kramer, Electron conductivity and second generation composite fermions. *Phys. Rev. B* **72**, 153308 (2005).

[443] M. Metropolis and S. Ulam, The Monte Carlo method. *J. Am. Stat. Assoc.* **44**, 335 (1949).

[444] N. Metropolis, A. W. Rosenbluth, M. N. Rosenbluth, A. H. Teller, and E. Teller, Equation of state calculations by fast computing machines. *J. Chem. Phys.* **21**, 1087 (1953).

[445] F. P. Milliken, C. P. Umbach, and R. A. Webb, Indications of a Luttinger liquid in the fractional quantum Hall regime. *Solid State Commun.* **97**, 309 (1996).

[446] A. D. Mirlin, E. Altshuler, and P. Wölfle, Quasiclassical approach to impurity effect on magnetooscillations in 2D metals. *Ann. Phys.* **5**, 281 (1996).

[447] A. D. Mirlin and P. Wölfle, Composite fermions in the fractional quantum Hall effect: Transport at finite wave vector. *Phys. Rev. Lett.* **78** (1997).

[448] A. D. Mirlin, P. Wölfle, Y. Levinson, and O. Entin-Wohlman, Velocity shift of surface acoustic waves due to interaction with composite fermions in a modulated structure. *Phys. Rev. Lett.* **81**, 1070 (1998).

[449] A. D. Mirlin, D. G. Polyakov, and P. Wölfle, Composite fermions in a long-range random magnetic field: Quantum Hall effect versus Shubnikov–de Haas oscillations. *Phys. Rev. Lett.* **80**, 2429 (1998).

[450] B. Mieck, Nonuniversal delocalization in a strong magnetic. *Europhys. Lett.* **13**, 453 (1990).

[451] G. Möller and S. H. Simon, Composite fermions in a negative effective magnetic field: A Monte Carlo study. *Phys. Rev. B* **72**, 045344 (2005).

[452] D. Monroe, Y. H. Xie, E. A. Fitzgerald, and P. J. Silverman, Quantized Hall effects in high-electron-mobility Si/Ge structures. *Phys. Rev. B* **46**, 7935 (1992).

[453] K. Moon, H. Mori, K. Yang, *et al.*, Spontaneous interlayer coherence in double-layer quantum Hall systems: Charged vortices and Kosterlitz–Thouless phase transitions. *Phys. Rev. B* **51**, 5138 (1995).

[454] G. Moore and N. Read, Nonabelions in the fractional quantum Hall effect. *Nucl. Phys. B* **360**, 362 (1991).

[455] R. H. Morf and B. I. Halperin, Monte Carlo evaluation of trial wave functions for the fractional quantized Hall effect: Disk geometry. *Phys. Rev. B* **33**, 2221 (1986).

[456] R. H. Morf, N. d'Ambrumenil, and B. I. Halperin, Microscopic wave functions for the fractional quantized Hall states at $\nu = 2/5$ and $2/7$. *Phys. Rev. B* **34**, 3037 (1986).

[457] R. H. Morf, Transition from quantum Hall to compressible states in the second Landau level: New light on the $\nu = 5/2$ enigma. *Phys. Rev. Lett.* **80**, 1505 (1998).

[458] R. H. Morf, Comment on 'Activation gaps and mass enhancement of composite fermions'. *Phys. Rev. Lett.* **83**, 1485 (1999).

[459] T. Morinari, Composite-fermion pairing in bilayer quantum Hall systems. *Phys. Rev. B* **59**, 7320 (1999).

[460] S. Q. Murphy, J. P. Eisenstein, G. S. Boebinger, L. N. Pfeiffer, and K. W. West, Many-body integer quantum Hall effect: Evidence for new phase transitions. *Phys. Rev. Lett.* **72**, 728 (1994).

[461] S. Q. Murphy, J. P. Eisenstein, L. N. Pfeiffer, and K. W. West, Lifetime of two-dimensional electrons measured by tunneling spectroscopy. *Phys. Rev. B* **52**, 14825 (1995).

[462] G. Murthy and R. Shankar, Field theory of the fractional quantum Hall effect. In *Composite Fermions*, ed. O. Heinonen (New York: World Scientific, 1998).

[463] G. Murthy and R. Shankar, Hamiltonian description of composite fermions: Calculation of gaps. *Phys. Rev. B* **59**, 12260 (1999).

[464] G. Murthy, Hamiltonian description of composite fermions: Magnetoexciton dispersions. *Phys. Rev. B* **60**, 13702 (1999).

[465] G. Murthy, Composite fermion Hofstadter problem: Partially polarized density wave states in the $\nu = 2/5$ fractional quantum Hall effect. *Phys. Rev. Lett.* **84**, 350 (2000).

[466] G. Murthy, Finite-temperature magnetism in fractional quantum Hall systems: The composite-fermion Hartree–Fock approximation and beyond. *J. Phys. Condens. Matter* **12** 10543 (2000).

[467] G. Murthy, Effects of disorder on the $\nu = 1$ quantum Hall state. *Phys. Rev. B* **64**, 241309 (2001).

[468] G. Murthy and R. Shankar, Hamiltonian theories of the fractional quantum Hall effect. *Rev. Mod. Phys.* **75**, 1101 (2003).

[469] Y. Nagaoka, Ferromagnetism in a narrow, almost half-filled s band. *Phys. Rev.* **147**, 392 (1966).

[470] R. Narevich, G. Murthy, and H. A. Fertig, Hamiltonian theory of the composite-fermion Wigner crystal. *Phys. Rev. B* **64**, 245326 (2001).

[471] T. Nakajima and H. Aoki, Composite-fermion picture for the spin-wave excitation in the fractional quantum Hall system. *Phys. Rev. Lett.* **73**, 3568 (1994).

[472] T. Nakajima and H. Aoki, Composite-fermion analysis of the double-layer fractional quantum Hall system. *Phys. Rev. B* **52** (1995).

[473] T. Nakajima and H. Aoki, Composite-fermion picture for the double-layer fractional quantum Hall system. *Surf. Sci.* **362**, 83 (1996).

[474] C. Nayak and F. Wilczek, Renormalization group approach to low temperature properties of a non-Fermi liquid metal. *Nucl. Phys. B* **430**, 534 (1994).

[475] C. Nayak and F. Wilczek, $2n$-quasihole states realize $2(n-1)$-dimensional spinor braiding statistics in paired quantum Hall states. *Nucl. Phys.* **479**, 529 (1996).

[476] J. W. Negele and H. Orland, *Quantum Many-Particle Systems* (Boulder, CO: Westview Press, 1988).

[477] R. J. Nicholas, D. R. Leadley, M. S. Daly, *et al.*, The dependence of the composite fermion effective mass on carrier density and Zeeman energy. *Semicond. Sci. Technol.* **11**, 1477 (1996).

[478] M. P. Nightingale and C. J. Umrigar. In *Recent Advances in Quantum Monte Carlo Methods*, ed. W. A. Lester (Singapore: World Scientific Publishing, 1997).

[479] F. Nihey, K. Nakamura, T. Takamasu, *et al.*, Orbital quantization of composite fermions in antidot lattices. *Phys. Rev. B* **59** 14872 (1999).

[480] K. S. Novoselov, A. K. Geim, S. V. Morozov, *et al.*, Two-dimensional gas of massless Dirac fermions in graphene. *Nature* **438**, 197 (2005).

[481] P. Nozières and C. T. DeDominicis, Singularities in the x-ray absorption and emission of metals III. *Phys. Rev.* **178**, 1097 (1969).

[482] P. Nozières, *J. Low Temp. Phys.* **137**, 45 (2004).

[483] J. H. Oaknin, L. Martin-Moreno, J. J. Palacios, and C. Tejedor, Low-lying excitations of quantum Hall droplets. *Phys. Rev. Lett.* **74**, 5120 (1995).

[484] M. Onoda, T. Mizusaki, T. Otsuka, and H. Aoki, Excitation spectrum and effective mass of the even-fraction quantum Hall liquid. *Phys. Rev. Lett.* **84**, 3942 (2000).

[485] M. Onoda, T. Mizusaki, and H. Aoki, Effective-mass staircase and the Fermi-liquid parameters for the fractional quantum Hall composite fermions. *Phys. Rev. B* **64**, 235315 (2001).

[486] M. Onoda, T. Mizusaki, T. Otsuka, and H. Aoki, Composite fermion picture and the spin states in the fractional quantum Hall system – a numerical study. *Physica B* **298**, 173 (2001).

[487] M. Onoda, T. Mizusaki, and H. Aoki, How heavy and how strongly interacting are composite fermions? *Physica E* **12**, 101 (2002).

[488] M. W. Ortalano, S. He, and S. Das Sarma, Realistic calculations of correlated incompressible electronic states in GaAs-$Al_xGa_{1-x}As$ heterostructures and quantum wells. *Phys. Rev. B* **55**, 7702 (1997).

[489] W. Pan, J.-S. Xia, V. Shvarts, *et al.*, Exact quantization of the even-denominator fractional quantum Hall state at $\nu = 5/2$ Landau level filling factor. *Phys. Rev. Lett.* **83**, 3530 (1999).

[490] W. Pan, R. R. Du, H. L. Stormer, *et al.*, Strongly anisotropic electronic transport at Landau level filling factor $\nu = 9/2$ and $\nu = 5/2$ under a tilted magnetic field. *Phys. Rev. Lett.* **83**, 820 (1999).

[491] W. Pan, H. L. Stormer, D. C. Tsui, *et al.*, Effective mass of the four-flux composite fermion at $\nu = 1/4$. *Phys. Rev. B* **61**, R5101 (2000).

[492] W. Pan, A. S. Yeh, J. S. Xia, *et al.*, The other even-denominator fractions. *Physica E* **9**, 9 (2001).

[493] W. Pan, J. S. Xia, E. D. Adams, *et al.*, New results at half-fillings in the second and third Landau levels. *Physica B* **298**, 113 (2001).

[494] W. Pan, H. L. Stormer, D. C. Tsui, *et al.*, Some fractions are more special than others: News from the fractional quantum Hall zone. *Int. J. Mod. Phys.* **16**, 2940 (2002).

[495] W. Pan, H. L. Stormer, D. C. Tsui, *et al.*, Transition from an electron solid to the sequence of fractional quantum Hall states at very low Landau level filling factor. *Phys. Rev. Lett.* **88**, 176802 (2002).

[496] W. Pan, H. L. Stormer, D. C. Tsui, *et al.*, Fractional quantum Hall effect of composite fermions. *Phys. Rev. Lett.* **90**, 016801 (2003).

[497] W. Pan, J. S. Xia, H. L. Stormer, *et al.*, Quantization of the diagonal resistance: Density gradients and the empirical resistance rule in a 2D system. *Phys. Rev. Lett.* **95**, 066808 (2005).

[498] W. Pan, H. L. Stormer, D. C. Tsui, *et al.*, Resistance scaling for composite fermions in the presence of a density gradient. cond-mat/0601627 (2006).

[499] S. Pancharatnam, *Proc. Ind. Acad. Sci.* **A 44**, 247 (1956).

[500] K. Park and J. K. Jain, Phase diagram of the spin polarization of composite fermions and a new effective mass. *Phys. Rev. Lett.* **80**, 4237 (1998).

[501] K. Park, V. Melik-Alaverdian, N. E. Bonesteel, and J. K. Jain, Possibility of p-wave pairing of composite fermions at $\nu = 1/2$. *Phys. Rev. B* **58**, R10167 (1998).

[502] K. Park, N. Meskini, and J. K. Jain, Activation gaps for fractional quantum Hall effect: Realistic treatment of transverse thickness. *J. Phys. Condens. Matter* **11**, 7283 (1999).

[503] K. Park, *Composite Fermions*, Ph.D. thesis (Stony Brook: State University of New York, 2000).

[504] K. Park and J. K. Jain, Two-roton bound state in the fractional quantum Hall effect. *Phys. Rev. Lett.* **84**, 5576 (2000).

[505] K. Park and J. K. Jain, Girvin-Macdonald-Platzman collective mode for general fractional Hall states: Magneto roton minimum at half-filled Landau level. *Solid State Commun.* **115**, 353 (2000).

[506] K. Park and J. K. Jain, The spin physics of composite fermions. *Solid State Commun.* **119**, 291 (2001).

[507] K. Park, Charged excitons of composite fermions in the fractional quantum Hall effect. *Solid State Commun.* **121**, 19 (2001).

[508] E. Peled, D. Shahar, Y. Chen, D. L. Sivco, and A. Y. Cho, Observation of a quantized Hall resistivity in the presence of mesoscopic fluctuations. *Phys. Rev. Lett.* **90**, 246802 (2003).

[509] E. Peled, Y. Chen, E. Diez, *et al.*, Symmetries of the resistance of mesoscopic samples in the quantum Hall regime. *Phys. Rev. B.* **69**, 241305 (2004).

[510] O. Penrose, On the quantum mechanics of He-2. *Philos. Mag.* **42**, 1373 (1951).

[511] O. Penrose and L. Onsager, Bose–Einstein condensation and liquid helium. *Phys. Rev.* **104**, 576 (1956).

[512] M. R. Peterson and J. K. Jain, Possible persistence of fractional quantum Hall effect down to ultralow fillings. *Phys. Rev. B* **68**, 195310 (2003).

[513] M. R. Peterson and J. K. Jain, Flavor altering excitations of composite fermions. *Phys. Rev. Lett.* **93**, 046402 (2004).

[514] L. N. Pfeiffer and K. W. West, The role of MBE in recent quantum Hall effect physics discoveries. *Physica E* **20**, 57 (2003).

[515] A. Pinczuk, B. S. Dennis, L. N. Pfeiffer, and K.W. West, Observation of collective excitations in the fractional quantum Hall effect. *Phys. Rev. Lett.* **70** 3983 (1993).

[516] A. Pinczuk, B. S. Dennis, L. N. Pfeiffer, and K.W. West, Inelastic light scattering in the regimes of the integer and fractional quantum Hall effect. *Semicond. Sci. Technol.* **9**, 1865 (1994).

[517] A. Pinczuk, B. S. Dennis, L. N. Pfeiffer, and K. W. West, Light scattering by low-energy collective excitations of quantum Hall states. *Proc. 12th Int. Conf. on High Magnetic Fields in Physics of Semiconductors* (Singapore: World Scientific, 1997) p. 83.

[518] A. Pinczuk, Resonant inelastic light scattering from quantum Hall systems. In *Perspectives in Quantum Hall Effects*, eds. S. Das Sarma and A. Pinczuk (New York: Wiley, 1997).

[519] A. Pinczuk, B. S. Dennis, L. N. Pfeiffer, and K. W. West, Light scattering by collective excitations in the fractional quantum Hall regime. *Physica B* **249**, 40 (1998).

[520] P. M. Platzman and R. Price, Quantum freezing of the fractional quantum Hall liquid. *Phys. Rev. Lett.* **70**, 3487 (1993).

[521] P. M. Platzman and S. He, Resonant Raman scattering from mobile electrons in the fractional quantum Hall regime. *Phys. Rev. B* **49**, 13674 (1994).

[522] P. M. Platzman and S. He, Resonant Raman scattering from magnetorotons in the fractional quantum Hall liquid . *Physica Scripta* **T66**, 167 (1996).

[523] D. G. Polyakov and B. I. Shklovskii, Variable range hopping as the mechanism of the conductivity peak broadening in the quantum Hall regime. *Phys. Rev. Lett.* **70**, 3796 (1993).

[524] D. G. Polyakov, F. Evers, A. D. Mirlin, and P. Wölfle, Kinetics and localization of composite fermions in a random magnetic field. *Physica B* **258**, 441 (1998).

[525] R. E. Prange and L. P. Kadanoff, Transport theory for electron-phonon interactions in metals. *Phys. Rev.* **134**, A566 (1964).

[526] R. E. Prange, Quantized Hall resistance and the measurement of the fine-structure constant. *Phys. Rev. B* **23**, 4802 (1981).

[527] R. E. Prange and R. Joynt, Conduction in a strong field in two dimensions: The quantum Hall effect. *Phys. Rev. B* **25**, 2943 (1982).

[528] P. J. Price, Low temperature two-dimensional mobility of a GaAs heterolayer. *Surf. Sci.* **143**, 145 (1984).

[529] R. Price and S. Das Sarma, Conduction in a strong field in two dimensions: The quantum Hall effect. *Phys. Rev. B* **54**, 8033 (1996).

[530] J. J. Quinn and A. Wójs, Composite fermions in fractional quantum Hall systems. *J. Phys. Condens. Matter* **12**, 265(R) (2000).

[531] J. J. Quinn and A. Wójs, Composite fermions and the fractional quantum Hall effect: Essential role of the pseudopotential. *Physica E* **6**, 1 (2000).

[532] J. J. Quinn, A. Wójs, and K. S. Yi, Novel families of fractional quantum Hall states: Pairing of composite fermions. *Phys. Lett. A* **318**, 152 (2003).

[533] J. J. Quinn, A. Wójs, and K. S. Yi, Residual interactions and pairing of composite fermion quasiparticles and novel condensed quantum fluid states. *J. Korean Phys. Soc.* **45**, S491 (2004).

[534] R. Rajaraman, Generalized Chern–Simons theory of composite fermions in bilayer Hall systems. *Phys. Rev. B* **56** 6788 (1997).

[535] F. Rapisarda and G. Senatore, Diffusion Monte Carlo study of electrons in two-dimensional layers. *Austrian J. Phys.* **49**, 161 (1996).

[536] M. Rasolt, B. I. Halperin, and D. Vanderbilt, Dissipation due to a valley wave channel in the quantum Hall effect of a multivalley semiconductor. *Phys. Rev. Lett.* **57**, 126 (1986).

[537] N. Read, Order parameter and Ginzburg–Landau theory for the fractional quantum Hall effect. *Phys. Rev. Lett.* **62**, 86 (1989).

[538] N. Read, Excitation structure of the hierarchy scheme in the fractional quantum Hall effect. *Phys. Rev. Lett.* **65**, 1502 (1990).

[539] N. Read and E. H. Rezayi, Quasiholes and fermionic zero modes of paired fractional quantum Hall states: The mechanism for non-Abelian statistics. *Phys. Rev. B* **54**, 16864 (1996).

[540] L. Reatto, Bose–Einstein condensation for a class of wave functions. *Phys. Rev.* **183**, 334 (1969).

[541] N. Regnault and Th. Jolicoeur, Quantum Hall fractions in rotating Bose-Einstein condensate. *Phys. Rev. Lett.* **91**, 030402 (2003).

[542] N. Regnault and Th. Jolicoeur, Quantum Hall fractions for spinless bosons. *Phys. Rev. B* **69**, 235309 (2004).

[543] N. Regnault, C.-C. Chang, Th. Jolicoeur, and J. K. Jain, Composite fermion theory of rapidly rotating two-dimensional bosons. *J. Phys. B* **39**, S89 (2006).

[544] E. H. Rezayi, Reversed-spin excitations of the fractionally quantized Hall effect from finite-size calculations. *Phys. Rev. B* **36**, 5454 (1987).

[545] E. H. Rezayi, Electron and quasiparticle density of states in the fractionally quantized Hall effect. *Phys. Rev. B* **35**, 3032 (1987).

[546] E. H. Rezayi and F.D. M. Haldane, Off-diagonal long-range order in fractional quantum Hall effect states. *Phys. Rev. Lett.* **61**, 1985 (1988).

[547] E. H. Rezayi, Wave functions and other properties of spin-reversed quasiparticles at $\nu = 1/m$ Landau-level occupation. *Phys. Rev. B* **43**, 5944 (1991).

[548] E. H. Rezayi and A. H. MacDonald, Origin of the $\nu = 2/5$ fractional quantum Hall effect. *Phys. Rev. B* **44**, 8395 (1991).

[549] E. H. Rezayi and N. Read, Fermi-liquid-like state in a half-filled Landau level. *Phys. Rev. Lett.* **72**, 900 (1994).

[550] E. H. Rezayi and F.D. M. Haldane, Incompressible paired Hall state, stripe order, and the composite fermion liquid phase in half-filled Landau levels. *Phys. Rev. Lett.* **84**, 4685 (2000).

[551] M. Rezhnikov, R. de Picciotto, T. G. Griffiths, *et al.*, Observation of quasiparticles with one-fifth of an electron's charge. *Nature* **399**, 238 (1999).

[552] K. Richter, Phase coherence effects in antidot lattices – a semiclassical approach to bulk conductivity. *Europhys. Lett.* **29**, 7 (1995).

[553] S. Roddaro, V. Pellegrini, F. Beltram, *et al.*, Nonlinear quasiparticle tunneling between fractional quantum Hall edges. *Phys. Rev. Lett.* **90**, 046805 (2003).

[554] L. P. Rokhinson, B. Su, and V. J. Goldman, Logarithmic temperature dependence of conductivity at half-integer filling factors: Evidence for interaction between composite fermions. *Phys. Rev. B* **52**, R11588 (1995).

[555] L. P. Rokhinson and V. J. Goldman, Magnetoresistance of composite fermions at $\nu = 1/2$. *Phys. Rev. B* **56**, R1672 (1997).

[556] T. Rötger, G. J. C. L. Bruls, J. C. Maan, *et al.*, Relation between low-temperature quantum and high-temperature classical magnetotransport in a two-dimensional electron gas. *Phys. Rev. Lett.* **62**, 9093 (1989).

[557] L. M. Roth, B. Lax, and S. Swerdling, Theory of optical magneto-absorption effects in semiconductors. *Phys. Rev.* **114**, 90 (1959).

[558] M. L. Roukes, A. Scherer, S. J. Allen, Jr., *et al.*, Quenching of the Hall effect in a one-dimensional wire. *Phys. Rev. Lett.* **59**, 3011 (1987).

[559] W. Y. Ruan, Y. Y. Liu, C. G. Bao, and Z. Q. Zhang, Origin of magic angular momenta in few-electron quantum dots. *Phys. Rev. B* **51**, 7942 (1995).

[560] W. Y. Ruan and H.-F. Cheung, Correlations and ground-state transitions of few-electron quantum dots in a strong magnetic field. *J. Phys. Condens. Mat.* **11**, 435 (1999).

[561] V. Ruuska and M. Manninen, Algebraic analysis of composite fermion wavefunctions. *Phys. Rev. B* **72**, 153309 (2005).

[562] S. Sachdev, *Quantum Phase Transitions* (Cambridge University Press, 1999).

[563] T. Sajoto, Y. P. Li, L. W. Engel, D. C. Tsui, and M. Shayegan, Hall resistance of the reentrant insulating phase around the 1/5 fractional quantum Hall liquid. *Phys. Rev. Lett.* **70**, 2321 (1993).

[564] S. Sakhi, Coulomb drag in double-layer electron systems at even-denominator filling factors. *Phys. Rev. B* **56**, 4098 (1997).

[565] L. Saminadayar, R. V. Glattli, Y. Jin, and B. Etienne, Observation of the $e/3$ fractionally charged Laughlin quasiparticle. *Phys. Rev. Lett.* **79**, 2526 (1997).

[566] M. B. Santos, Y. W. Suen, M. Shayegan, *et al.*, Observation of a reentrant insulating phase near the 1/3 fractional quantum Hall liquid in a two-dimensional hole system. *Phys. Rev. Lett.* **68**, 1188 (1992).

[567] V. W. Scarola, K. Park, and J. K. Jain, Rotons of composite fermions: Comparison between theory and experiment. *Phys. Rev. B* **61**, 13064 (2000).

[568] V. W. Scarola, K. Park, and J. K. Jain, Cooper instability of composite fermions. *Nature* **406**, 863 (2000).

[569] V. W. Scarola and J. K. Jain, Phase diagram of bilayer composite fermion states. *Phys. Rev. B* **64**, 085313 (2001).

[570] V. W. Scarola, J. K. Jain, and E. H. Rezayi, Possible pairing-induced even-denominator fractional quantum Hall effect in the lowest Landau level. *Phys. Rev. Lett.* **88**, 216804 (2002).

[571] V. W. Scarola, S. Lee and J. K. Jain, Excitation gaps of incompressible composite fermion states: Approach to the Fermi sea. *Phys. Rev. B* **66**, 155320 (2002).

[572] A. Schmeller, J. P. Eisenstein, L. N. Pfeiffer, and K. W. West, Evidence for skyrmions and single spin flips in the integer quantized Hall effect. *Phys. Rev. Lett.* **75**, 4290 (1995).

[573] D. Schoenberg, *Magnetic Oscillations in Metals* (Cambridge University Press, 1984).

[574] J. R. Schrieffer, *Theory of Superconductivity* (Boulde, CO: Westview Press, 1964).

[575] L. S. Schulman, *Techniques and Applications of Path Integration* (New York: Wiley, 1981).

[576] F. Schulze-Wischeler, E. Mariani, F. Hohls, and R. J. Haug, Direct measurement of the g factor of compose fermions. *Phys. Rev. Lett.* **92**, 156401 (2004).

[577] F. Schulze-Wischeler, F. Hohls, U. Zeitler, *et al.*, Phonon excitations of composite-fermion Landau levels. *Int. J. Mod. Phys. B* **18**, 3857 (2004).

[578] T. Seki, Y. Kuramoto, and T. Nishino, Origin of magic angular momentum in a quantum dot under strong magnetic field. *J. Phys. Soc. Jpn.* **65**, 3945 (1996).

[579] D. Shahar, M. Hilke, C. C. Li, *et al.*, A new transport regime in the quantum Hall effect. *Solid State Commun.* **107**, 19 (1998).

[580] R. Shankar, Renormalization-group approach to interacting fermions. *Rev. Mod. Phys.* **66**, 129 (1994).

[581] R. Shankar and G. Murthy, Towards a field theory of fractional quantum Hall states. *Phys. Rev. Lett.* **79**, 4437 (1997).

[582] R. Shankar, Magnetic phenomena at and near $v = 1/2$ and $1/4$: Theory, experiment, and interpretation. *Phys. Rev. Lett.* **84**, 3946 (2000).

[583] R. Shankar, Hamiltonian theory of gaps, masses, and polarization in quantum Hall states. *Phys. Rev. B* **63**, 085322 (2001).

[584] M. Shayegan, Case for the magnetic-field-induced two-dimensional Wigner crystal. In *Perspectives in Quantum Hall Effects*, eds. S. Das Sarma and A. Pinczuk (New York: Wiley, 1997).

[585] D. Sheng, L. Balents, and Z. Wang, Phase diagram for quantum Hall bilayers at $v = 1$. *Phys. Rev. Lett.* **91**, 116802 (2003).

[586] N. Shibata and D. Yoshioka, Ground state phase diagram of 2D electrons in high magnetic field. *J. Phys. Soc. Jpn.* **72**, 664 (2003).

[587] N. Shibata and D. Yoshioka, Stripe state in the lowest Landau level. *J. Phys. Soc. Jpn.* **73**, 1 (2004).

[588] N. Shibata and D. Yoshioka, Ground state of $v = 1$ bilayer quantum Hall systems. *J. Phys. Soc. Jpn.* **75**, 043712 (2006).

[589] A. V. Shytov, L. S. Levitov, and B. I. Halperin, Tunneling into the edge of a compressible quantum Hall state. *Phys. Rev. Lett.* **80**, 141 (1998).

[590] J. A. Simmons, H. P. Wei, L. W. Engel, D. C. Tsui, and M. Shayegan, Resistance fluctuations in narrow AlGaAs/GaAs heterostructures: Direct evidence of fractional charge in the fractional quantum Hall effect. *Phys. Rev. Lett.* **63**, 1731 (1989).

[591] S. H. Simon and B. I. Halperin, Finite-wave-vector electromagnetic response of fractional quantized Hall states. *Phys. Rev. B* **48**, 17368 (1993).

[592] S. H. Simon and B. I. Halperin, Response function of the fractional quantized Hall state on a sphere. I. Fermion Chern–Simons theory. *Phys. Rev. B* **50**, 1807 (1994).

[593] S. H. Simon, A. Stern and B. I. Halperin, Composite fermions with orbital magnetization. *Phys. Rev. B* **54**, R11114 (1996).

[594] S. H. Simon, The Chern–Simons Fermi liquid description of fractional quantum Hall states. In *Composite Fermions*, ed. O. Heinonen (New York: World Scientific, 1998).

[595] P. Sitko and L. Jacak, Hartree–Fock ground-state of the composite fermion metal. *Mod. Phys. Lett. B* **9**, 889 (1995).

[596] P. Sitko, S. N. Yi, K. S. Yi, and J. J. Quinn, Fermi liquid shell model approach to composite fermion excitation spectra in fractional quantum Hall states. *Phys. Rev. Lett.* **76**, 3396 (1996).

[597] P. Sitko, On the interaction energy of two-dimensional electron FQHE systems within the Chern–Simons approach. *Int. J. Mod. Phys. B* **13**, 2263 (1999).

[598] P. Sitko, Charge and statistics of quasiholes in Pfaffian states of composite fermion excitations. *Phys. Lett. A* **324**, 120 (2004).

[599] J. H. Smet, D. Weiss, R. H. Blick, *et al.*, Magnetic focusing of composite fermions through arrays of cavities. *Phys. Rev. Lett.* **77**, 2272 (1996).

[600] J. H. Smet, R. Fleischmann, D. Weiss, *et al.*, Evidence for quasi-classical transport of composite fermions in an inhomogeneous effective magnetic field. *Semicond. Sci. Technol.* **11**, 1482 (1996).

[601] J. H. Smet, D. Weiss, K. von Klitzing, *et al.*, Composite fermions in periodic and random antidot lattices. *Phys. Rev. B* **56**, 3598 (1997).

[602] J. H. Smet, K. von Klitzing, D. Weiss, and W. Wegscheider, dc transport of composite fermions in weak periodic potential. *Phys. Rev. Lett.* **80**, 4538 (1998).

[603] J. H. Smet, Ballistic transport of composite fermions in semiconductor nanostructures. In *Composite Fermions*, ed. O. Heinonen (New York: World Scientific, 1998).

[604] J. H. Smet, S. Jobst, K. von Klitzing, *et al.*, Commensurate composite fermions in weak periodic electrostatic potentials: Direct evidence of a periodic effective magnetic field. *Phys. Rev. Lett.* **83**, 2620 (1999).

[605] J. H. Smet, Wheels within wheels. *Nature* **422**, 391 (2003).

[606] M. J. Snelling, G. P. Flinn, A. S. Plaut, *et al.*, Magnetic g factor of electrons in GaAs/Al$_x$Ga$_{1-x}$As quantum wells. *Phys. Rev. B* **44**, 11345 (1991).

[607] S. L. Sondhi, A. Karlhede, S. A. Kivelson, and E. H. Rezayi, Skyrmions and the crossover from the integer to fractional quantum Hall effect at small Zeeman energies. *Phys. Rev. B* **47**, 16419 (1993).

[608] S. L. Sondhi, S. M. Girvin, J. P. Carini, and D. Shahar, Continuous quantum phase transitions. *Rev. Mod. Phys.* **69**, 315 (1997).

[609] I. B. Spielman, J. P. Eisenstein, L. N. Pfeiffer, and K. W. West, Resonantly enhanced tunneling in a double layer quantum Hall ferromagnet. *Phys. Rev. Lett.* **84**, 5808 (2000).

[610] I. B. Spielman, L. A. Tracy, J. P. Eisenstein, L. N. Pfeiffer, and K. W. West, Spin transition in strongly correlated bilayer two-dimensional electron systems. *Phys. Rev. Lett.* **94**, 076803 (2005).

[611] F. Stern, Self-consistent results for n-type Si inversion layers. *Phys. Rev. B* **5**, 4891 (1972).

[612] F. Stern and S. Das Sarma, Electron energy levels in GaAs-Ga$_{1-x}$Al$_x$As heterojunctions. *Phys. Rev. B* **30**, 840 (1984).

[613] A. Stern and B. I. Halperin, Singularities in the Fermi-liquid description of a partially filled Landau level and the energy gaps of fractional quantum Hall states. *Phys. Rev. B* **52**, 5890 (1995).

[614] A. Stern, S. M. Girvin, A. H. MacDonald, and Ning Ma, Theory of interlayer tunneling in bilayer quantum Hall ferromagnets. *Phys. Rev. Lett.* **86**, 1829 (2001).

[615] M. Stone, Schur functions, chiral bosons, and the quantum-Hall-effect edge states. *Phys. Rev. B* **42**, 8399 (1990).

[616] M. Stone, Edge waves in the quantum Hall effect. *Annal. Phys.* **207**, 38 (1991).

[617] M. Stone, H. W. Wyld, and R. L. Schultz, Edge waves in the quantum Hall effect and quantum dots. *Phys. Rev. B* **45**, 14156 (1992).

[618] H. L. Stormer, R. Dingle, A. C. Gossard, and W. Wiegmann, *IOP Conf. Proc.* No. 43 (London: Institute of Physics), 557 (1978).

[619] H. L. Stormer, R. Dingle, A. C. Gossard, W. Wiegmann, and M. D. Sturge, Two-dimensional electron-gas at a semiconductor-semiconductor interface. *Solid State Commun.* **24**, 705 (1979).

[620] H. L. Stormer, A. M. Chang, D. C. Tsui, *et al.*, Fractional quantization of the Hall effect. *Phys. Rev. Lett.* **50**, 1953 (1983).

[621] H. L. Stormer, K. W. Baldwin, L. N. Pfeiffer, and K. W. West, Strikingly linear magnetic-field dependence of the magnetoresistivity in high-quality two-dimensional electron systems. *Solid State Commun.* **84**, 95 (1992).

[622] H. L. Stormer, L. N. Pfeiffer, K. W. Baldwin, and K. W. West, Observation of a Bloch–Grüneisen regime in two-dimensional electron transport. *Phys. Rev. B* **41**, 1278 (1990).

[623] H. L. Stormer, quoted in Half-filled Landau level yields intriguing data and theory. *Physics Today* **46**(7), 17 (1993).

[624] H. L. Stormer, D. C. Tsui, A. C. Gossard, and J. C. M. Hwang, *Physica* **117B & 118B**, 688 (1983).

[625] H. L. Stormer and D. C. Tsui, Composite fermions in the fractional quantum Hall effect. In *Perspectives in Quantum Hall Effects*, eds. S. Das Sarma and A. Pinczuk (New York: Wiley, 1997).

[626] H. L. Stormer, W. Kang, S. He, *et al.*, Composite fermions: Their scattering and their spin. *Physica B* **227**, 164 (1996).

[627] H. L. Stormer, Nobel Lecture: The fractional quantum Hall effect. *Rev. Mod. Phys.* **71**, 875 (1999).

[628] P. Streda, J. Kucera, and A. H. MacDonald, Edge states, transmission matrices, and the Hall resistance. *Phys. Rev. Lett.* **59**, 1973 (1987).

[629] W. P. Su, Statistics of the fractionally charged excitations in the quantum Hall effect. *Phys. Rev. B* **34**, 1031 (1986).

[630] Y. W. Suen, L. W. Engel, M. B. Santos, M. Shayegan, and D. C. Tsui, Observation of a $v = 1/2$ fractional quantum Hall state in a double-layer electron system. *Phys. Rev. Lett.* **68**, 1379 (1992).

[631] T. Sugiyama and N. Nagaosa, Localization in a random magnetic field in 2D. *Phys. Rev. Lett.* **70**, 1980 (1993).

[632] B. Sutherland, Quantum many-body problem in one dimension: Ground state. *J. Math. Phys.* **12**, 246 (1971).

[633] B. Tanatar and D. M. Ceperley, Ground state of the two-dimensional electron gas. *Phys. Rev. B* **39**, 5005 (1989).

[634] D. J. Thouless, M. Kohmoto, M. P. Nightingale, and M. den Nijs, Quantized Hall conductance in a two-dimensional periodic potential. *Phys. Rev. Lett.* **49**, 405 (1982).

[635] D. J. Thouless, Topological considerations. In *The Quantum Hall Effect*, 1st edn. R. E. Prange and S. M. Girvin (New York: Springer-Verlag, 1987).

[636] B. Tieke, U. Zeitler, R. Fletcher, *et al.*, Even-denominator filling factors in the thermoelectric power of a two-dimensional electron gas. *Phys. Rev. Lett.* **76**, 3630 (1996).

[637] G. Timp, P. M. Mankiewich, P. deVegvar, *et al.*, Suppression of the Aharonov–Bohm effect in the quantized Hall regime. *Phys. Rev. B* **39**, 6227 (1989).

[638] I. V. Tokatly, Magneto-elasticity theory of incompressible quantum Hall liquids. *Phys. Rev. B* **73**, 205340 (2006).

[639] I. V. Tokatly, Unified hydrodynamics of the lowest Landau level. *Phys. Rev. B* **74**, 035333 (2006).

[640] I. V. Tokatly and G. Vignale, A new collective mode in the fractional quantum Hall liquid. cond-mat/0607705 (2006).

[641] C. Tőke and J. K. Jain, Understanding the 5/2 fractional quantum Hall effect without the Pfaffian wave function. *Phys. Rev. Lett* **96**, 246805 (2006).

[642] C. Tőke, N. Regnault, and J. K. Jain, Nature of excitations of the 5/2 fractional quantum Hall effect. cond-mat/0609177 (2006).

[643] S. Tomonaga, Remarks on Bloch's method of sound waves applied to many-fermion problems. *Prog. Theor. Phys. (Kyoto)* **5**, 544 (1950).

[644] N. Trivedi and J. K. Jain, Numerical study of Jastrow-Slater trial states for the fractional quantum Hall effect. *Mod. Phys. Lett. B* **5**, 503 (1991).

[645] S. A. Trugman, Localization, percolation, and the quantum Hall effect. *Phys. Rev. B* **27**, 7539 (1983).

[646] S. A. Trugman and S. A. Kivelson, Exact results for the fractional quantum Hall effect with general interactions. *Phys. Rev. B* **31**, 5280 (1985).

[647] M. Tsaousidou, P. N. Butcher, and S. S. Kubakaddi, Quantitative interpretation of thermopower data for composite fermions in a half-filled Landau level. *Phys. Rev. Lett.* **83**, 4820 (1999).

[648] Y. Tserkovnyak and S. H. Simon, Monte Carlo evaluation of non-Abelian statistics. *Phys. Rev. Lett.* **90**, 016802 (2003).

[649] E. V. Tsiper, Analytic Coulomb matrix elements in the lowest Landau level in disk geometry. *J. Math. Phys.* **43**, 1664 (2002).

[650] D. C. Tsui, H. L. Stormer, and A. C. Gossard, Two-dimensional magnetotransport in the extreme quantum limit. *Phys. Rev. Lett.* **48**, 1559 (1982).

[651] D. C. Tsui, Nobel Lecture: Interplay of disorder and interaction in two-dimensional electron gas in intense magnetic fields. *Rev. Mod. Phys.* **71**, 891 (1999).

[652] E. Tutuc, M. Shayegan, and D. A. Huse, Counterflow measurements in strongly correlated GaAs hole bilayers: Evidence for electron-hole pairing. *Phys. Rev. Lett.* **93**, 036802 (2004).

[653] I. Ussishkin and A. Stern, Coulomb drag in compressible quantum Hall states. *Phys. Rev. B* **56**, 4013 (1997).

[654] I. Ussishkin and A. Stern, Coulomb drag at $\nu = 1/2$: Composite fermion pairing fluctuations. *Phys. Rev. Lett.* **81**, 3932 (1998).

[655] H. van Houten, C.W. J. Beenakker, J. G. Williamson, *et al.*, Coherent electron focusing with quantum point contacts in a two-dimensional electron gas. *Phys. Rev. B* **39**, 8556 (1989).

[656] S. Viefers, T. H. Hansson, and S. M. Reimann, Bose condensates at high angular momenta. *Phys. Rev. A* **62**, 053604 (2000).

[657] G. Vignale, Integral charge quasiparticles in a fractional quantum Hall liquid. *Phys. Rev. B* **73**, 073306 (2006).

[658] J. von Delft and H. Schoeller, Bosonization for beginners – Refermionization for experts. *Annal. Phys.* **7**, 225 (1998).

[659] K. von Klitzing, G. Dorda, and M. Pepper, New method for high-accuracy determination of the fine-structure constant based on quantized Hall resistance. *Phys. Rev. Lett.* **45**, 494 (1980).

[660] K. von Klitzing, The quantized Hall effect. *Rev. Mod. Phys.* **58**, 519 (1986).

[661] K. von Klitzing, Developments in the quantum Hall effect. *Philos. Trans. R. Soc. London A* **363** (1834), 2203 (2005).

[662] F. von Oppen, A. Stern, and B. I. Halperin, Composite fermions in modulated structures: Transport and surface acoustic waves. *Phys. Rev. Lett.* **80**, 4494 (1998).

[663] X. Wan, F. Evers, and E. H. Rezayi, Universality of the edge-tunneling exponent of fractional quantum Hall liquids. *Phys. Rev. Lett.* **94**, 166804 (2005).

[664] H. S. Wang, Y. H. Chiu, G. H. Kim, C. T. Liang, *et al.*, Upshift of the fractional quantum Hall plateaux: Evidence for repulsive scattering for composite fermions. *Physica E* **22**, 135 (2004).

[665] Z. Wang, Tunneling, dissipation, and superfluid transition in quantum Hall bilayers. *Phys. Rev. Lett.* **92**, 136803 (2004).

[666] J. Wakabayashi, S. Kawaji, T. Goto, T. Fukase, and Y. Koike, Localization in Landau subbands with the Landau quantum number 0 and 1 in Si-MOS inversion layers. *J. Phys. Soc. Jpn.* **61**, 1691 (1992).

[667] J. K. Wang, D. C. Tsui, M. Santos, and M. Shayegan, Heat-capacity study of two-dimensional electrons in $GaAs/Al_xGa_{1-x}As$ multiple-quantum-well structures in high magnetic fields: Spin-split Landau levels. *Phys. Rev. B* **45**, 4384 (1992).

[668] S. Washburn, A. B. Fowler, H. Schmid, and D. Kern, Quantized Hall effect in the presence of backscattering. *Phys. Rev. Lett.* **61**, 2801 (1988).

[669] H. P. Wei, D. C. Tsui, M. A. Paalanen, and A. M. M. Pruisken, Experiments on delocalization and university in the integral quantum Hall effect. *Phys. Rev. Lett.* **61**, 1294 (1988).

[670] H. P. Wei, L. W. Engel, and D. C. Tsui, Current scaling in the integer quantum Hall effect. *Phys. Rev. B* **50**, 14609 (1994).

[671] D. Weiss, M. L. Roukes, A. Menschig, *et al.*, Electron pinball and commensurate orbits in a periodic array of scatterers. *Phys. Rev. Lett.* **66**, 2790 (1991).

[672] X.-G. Wen, Non-Abelian statistics in the fractional quantum Hall states. *Phys. Rev. Lett.* **66**, 802 (1991).

[673] X.-G. Wen, Theory of the edge states in fractional quantum Hall effects. *Int. J. Mod. Phys. B* **6**, 1711 (1992).

[674] X.-G. Wen and A. Zee, Neutral superfluid modes and magnetic monopoles in multilayer quantum Hall systems. *Phys. Rev. Lett.* **69**, 1811 (1992).

[675] E. P. Wigner, On the interaction of electrons in metals. *Phys. Rev.* **46**, 1002 (1934).

[676] E. P. Wigner, Effects of electron interaction on the energy levels of electrons in metals. *Trans. Faraday Soc.* **34**, 678 (1938).

[677] F. Wilczek, Quantum mechanics of fractional-spin particles. *Phys. Rev. Lett.* **49**, 957 (1982).

[678] F. I. B. Williams, P. A. Wright, R. G. Clark, *et al.*, Conduction threshold and pinning frequency of magnetically induced Wigner solid. *Phys. Rev. Lett.* **66**, 3285 (1991).

[679] N. K. Wilkin and J.M. F. Gunn, Condensation of composite bosons in a rotating BEC. *Phys. Rev. Lett.* **84**, 6 (2000).

[680] R. L. Willett, J. P. Eisenstein, H. L. Stormer, *et al.*, Observation of an even-denominator quantum number in the fractional quantum Hall effect. *Phys. Rev. Lett.* **59**, 1776 (1987).

[681] R. L. Willett, H. L. Stormer, D. C. Tsui, *et al.*, Termination of the series of fractional quantum Hall states at small filling factors. *Phys. Rev. B* **38**, 7881 (1988).

[682] R. L. Willett, M. A. Paalanen, R. R. Ruel, *et al.*, Anomalous sound propagation at $\nu = 1/2$ in a 2D electron gas: Observation of a spontaneously broken translational symmetry? *Phys. Rev. Lett.* **65**, 112 (1990).

[683] R. L. Willett, R. R. Ruel, K. W. West, and L.N. Pfeiffer, Experimental demonstration of a Fermi surface at one-half filling of the lowest Landau level. *Phys. Rev. Lett.* **71**, 3846 (1993).

[684] R. L. Willett and L. N. Pfeiffer, Composite fermions examined with surface acoustic waves. *Surf. Sci.* **362**, 38 (1996).

[685] R. L. Willett, K. W. West, and L. N. Pfeiffer, Transition in the correlated 2D electron system induced by a periodic density modulation. *Phys. Rev. Lett.* **78**, 4478 (1997).

[686] R. L. Willett, K. W. West, and L. N. Pfeiffer, Geometric resonance of composite fermion cyclotron orbits with a fictitious magnetic field modulation. *Phys. Rev. Lett.* **83**, 2624 (1999).

[687] R. L. Willett, K. W. West, and L. N. Pfeiffer, Experimental demonstration of Fermi surface effects at filling factor 5/2. *Phys. Rev. Lett.* **88**, 066801 (2002).

[688] R. W. Winkler, J. P. Kotthaus, and K. Ploog, Landau band conductivity in a two-dimensional electron system modulated by an artificial one-dimensional superlattice potential. *Phys. Rev. Lett.* **62**, 1177 (1989).

[689] A. Wixforth, J. P. Kotthaus, and G. Weimann, Quantum oscillations in the surface-acoustic-wave attenuation caused by a two-dimensional electron system. *Phys. Rev. Lett.* **56**, 2104 (1986).

[690] A. Wójs and J. J. Quinn, Composite fermion approach to the quantum Hall hierarchy: When it works and why. *Solid State Commun.* **108**, 493 (1998).

[691] A. Wójs and J. J. Quinn, Hund's rule for monopole harmonics, or why the composite fermion picture works. *Solid State Commun.* **110**, 45 (1999).

[692] A. Wójs and J. J. Quinn, Composite fermions and the fractional quantum Hall effect. *Acta Physica Polonica A* **96**, 593 (1999).

[693] A. Wójs and J. J. Quinn, Quasiparticle interactions in fractional quantum Hall systems: Justification of different hierarchy schemes. *Phys. Rev. B* **61**, 2846 (2000).

[694] A. Wójs and J. J. Quinn, Composite fermions and quantum Hall systems: Role of the Coulomb pseudopotential. *Philos. Mag. B* **80**, 1405 (2000).

[695] A. Wójs, Electron correlations in partially filled lowest and excited Landau levels. *Phys. Rev. B* **63**, 125312 (2001).

[696] A. Wójs and J. J. Quinn, Spin excitation spectra of integral and fractional quantum Hall systems. *Phys. Rev. B* **66**, 045323 (2002).

[697] A. Wójs, K.-S. Yi, and J. J. Quinn, Fractional quantum Hall states of clustered composite fermions. *Phys. Rev. B* **69**, 205322 (2004).

[698] A. Wójs, D. Wodziński, and J. J. Quinn, Pair-distribution functions of correlated composite fermions. *Phys. Rev. B* **71**, 245331 (2005).

[699] A. Wójs, A. Gladysiewicz, and J. J. Quinn, Quasiexcitons in incompressible quantum liquids. *Phys. Rev. B* **73**, 235338 (2006).

[700] T. T. Wu and C. N. Yang, Concept of nonintegrable phase factors and global formulation of gauge fields. *Phys. Rev. D* **12**, 3845 (1975).

[701] T. T. Wu and C. N. Yang, Dirac monopoles without strings – Monopole harmonics. *Nucl. Phys. B* **107**, 365 (1976).

[702] T. T. Wu and C. N. Yang, Some properties of monopole harmonics. *Phys. Rev. D* **16**, 1018 (1977).

[703] X. G. Wu, G. Dev, and J. K. Jain, Mixed-spin incompressible states in the fractional quantum Hall effect. *Phys. Rev. Lett.* **71**, 153 (1993).

[704] X. G. Wu and J. K. Jain, Fractional quantum Hall states in the low-Zeeman-energy limit. *Phys. Rev. B* **49**, 7515 (1994).

[705] X. G. Wu and J. K. Jain, Excitation spectrum and collective modes of composite fermions. *Phys. Rev. B* **51**, 1752 (1995).

[706] J.-S. Xia, W. Pan, C. L. Vicente, *et al.*, Electron correlation in the second Landau level: A competition between many nearly degenerate quantum phases. *Phys. Rev. Lett.* **93**, 176809 (2004).

[707] X. C. Xie, Y. Guo, and F. C. Zhang, Fractional quantum Hall effect with spin reversal. *Phys. Rev. B* **40**, 3487 (1989).

[708] X. C. Xie, S. Das Sarma, and S. He, Elementary excitations in a finite fractional quantum Hall droplet. *Phys. Rev. B* **47**, 15942 (1993).

[709] X. C. Xie, Theoretical study of energy gaps and collective excitations in the fractional quantum Hall effect. *Phys. Rev. B* **49**, 16833 (1994).

[710] X. C. Xie and S. He, Skyrmion excitations in quantum Hall systems. *Phys. Rev. B* **53**, 1046 (1996).

[711] X. C. Xie, X. R. Wang, and D. Z. Liu, Kosterlitz–Thouless-type metal-insulator transition of a 2D electron gas in a random magnetic field. *Phys. Rev. Lett.* **80**, 3563 (1998).

[712] C. N. Yang, Concept of off-diagonal long-range order and the quantum phases of liquid He and of superconductors. *Rev. Mod. Phys.* **34**, 694 (1962).

[713] S.-R. Eric Yang, A. H. MacDonald, and M. D. Johnson, Addition spectra of quantum dots in strong magnetic fields. *Phys. Rev. Lett.* **71**, 3194 (1993).

[714] J. Yang, Decomposition approach to a two-dimensional electron system with an even-denominator filling factor. *Phys. Rev. B* **50**, 8028 (1994).

[715] K. Yang, K. Moon, L. Zheng, *et al.*, Quantum ferromagnetism and phase transitions in double-layer quantum Hall systems. *Phys. Rev. Lett.* **72**, 732 (1994).

[716] S.-J. Yang, Y. Yu, and Z.-B. Su, Possible composite-fermion liquid as a crossover from Wigner crystal to bubble phase in higher Landau levels. *Phys. Rev. B* **62** (2000).

[717] K. Yang, F. D. M. Haldane and E. H. Rezayi, Wigner crystals in the lowest Landau level at low filling factors. *Phys. Rev. B* **64**, 081301(R) (2001).

[718] S.-J. Yang, Y. Yu, and B.-G. Zhu, Anisotropic transport for the $\nu = 2/5$ fractional quantum Hall state at intermediate magnetic fields. *J. Phys. Condens. Matters* **14**, 9615 (2002).

[719] E. Yannouleas and U. Landman, Trial wave functions with long-range Coulomb correlations for two-dimensional N-electron systems in high magnetic fields. *Phys. Rev. B* **66**, 115315 (2002).

[720] C. Yannouleas and U. Landman, Two-dimensional quantum dots in high magnetic fields: Rotating-electron-molecule versus composite-fermion approach. *Phys. Rev. B* **68**, 035326 (2003).

[721] P. D. Ye, D. Weiss, R. R. Gerhardts, *et al.*, Electrons in a periodic magnetic field induced by a regular array of micromagnets. *Phys. Rev. Lett.* **74**, 3013 (1995).

[722] P. D. Ye, L. W. Engel, D. C. Tsui, *et al.*, Correlation lengths of the Wigner-crystal order in a two-dimensional electron system at high magnetic fields. *Phys. Rev. Lett.* **89**, 176802 (2002).

[723] A. S. Yeh, H. L. Stormer, D. C. Tsui, *et al.*, Effective mass and g factor of four-flux-quanta composite fermions. *Phys. Rev. Lett.* **82**, 592 (1999).

[724] S. N. Yi, X. M. Chen, and J. J. Quinn, Composite-fermion excitations and the electronic spectra of fractional quantum Hall systems. *Phys. Rev. B* **53**, 9599 (1996).

[725] H. Yi and H. A. Fertig, Laughlin–Jastrow-correlated Wigner crystal in a strong magnetic field. *Phys. Rev. B* **58**, 4019 (1998).

[726] X. Ying, V. Bayot, M. B. Santos, and M. Shayegan, Observation of composite-fermion thermopower at half-filled Landau levels. *Phys. Rev. B* **50**, 4969 (1994).

[727] D. Yoshioka and H. Fukuyama, Charge density wave state of two-dimensional electrons in strong magnetic fields. *J. Phys. Soc. Jpn.* **47**, 394 (1979).

[728] D. Yoshioka, B. I. Halperin, and P. A. Lee, Ground state of two-dimensional electrons in strong magnetic fields and 1/3 quantized Hall effect. *Phys. Rev. Lett.* **50**, 1219 (1983).

[729] D. Yoshioka and P. A. Lee, Ground-state energy of a two-dimensional charge-density-wave state in a strong magnetic field. *Phys. Rev. B* **27**, 4986 (1983).

[730] D. Yoshioka, Excitation energies of the fractional quantum Hall effect. *J. Phys. Soc. Jpn.* **55**, 885 (1986).

[731] D. Yoshioka, A. H. MacDonald and S. M. Girvin, Fractional quantum Hall effect in two-layered systems. *Phys. Rev. B* **39**, 1932 (1989).

[732] D. Yoshioka, Composite fermions with a spin freedom. *Physica E* **6**, 83 (2000).

[733] Y. Yu, W. Zheng, and Z. Zhu, Microscopic picture of a chiral Luttinger liquid: Composite fermion theory of edge states. *Phys. Rev. B* **56**, 13279 (1997).

[734] Y. Yu, From composite fermions to the Calogero–Sutherland model: Edge of the fractional quantum Hall liquid and the dimension reduction. *Phys. Rev. B* **61**, 4465 (2000).

[735] J. Zak, Magnetic translation group. *Phys. Rev. A* **134**, 1602 (1964).

[736] S. Zelakiewicz, H. Noh, T. J. Gramila, L. N. Pfeiffer, and K. W. West, Missing $2k_F$ response for composite fermions in phonon drag. *Phys. Rev. Lett.* **85**, 1942 (2000).

[737] W. Zawadski and R. Lassnig, Magnetization, specific heat, magneto-thermal effect and thermoelectric power of two-dimensional electron gas in a quantizing magnetic field. *Surf. Sci.* **142**, 225 (1984).

[738] U. Zeitler, J. C. Maan, P. Wyder, *et al.*, Investigation of the electron-phonon interaction in the fractional quantum Hall regime using the thermoelectric effect. *Phys. Rev. B* **47**, 16008 (1993).

[739] U. Zeitler, B. Tieke, S. A. J. Wiegers, *et al.*, The quantum limit thermopower in two-dimensional electron systems. In *High Magnetic Fields in the Physics of Semiconductors,* ed. D. Heiman (World Scientific, Singapore, 1995) p. 38.

[740] U. Zeitler, A. M. Devitt, J. E. Digby, *et al.*, Ballistic heating of a two-dimensional electron system by phonon excitation of the magnetoroton minimum at $\nu = 1/3$. *Phys. Rev. Lett.* **82**, 5333 (1999).

[741] F. C. Zhang and T. Chakraborty, Ground state of two-dimensional electrons and the reversed spins in the fractional quantum Hall effect. *Phys. Rev. B* **30**, 7320 (1984).

[742] F. C. Zhang and S. Das Sarma, Excitation gap in the fractional quantum Hall effect: Finite layer thickness corrections. *Phys. Rev. B* **33**, 2903 (1986).

[743] S. C. Zhang, H. Hansson, and S. Kivelson, Effective-field-theory model for the fractional quantum Hall effect. *Phys. Rev. Lett.* **62**, 82 (1989).

[744] S. C. Zhang, The Chern–Simons–Landau–Ginzburg theory of the fractional quantum Hall efffect. *Int. J. Mod. Phys. B* **6**, 25 (1992).

[745] Yuanbo Zhang, Yan-Wen Tan, H. Horst, L. Stormer, P. Kim, Experimental observation of quantum Hall effect and Berry's phase in graphene. *Nature* **438**, 201 (2005).

[746] L. Zheng and H. A. Fertig, Quantum correlated interstitials and the Hall resistivity of the magnetically induced Wigner crystal. *Phys. Rev. Lett.* **73**, 878 (1994).

[747] F. Zhou and Y. B. Kim, Coulomb drag as a signature of the paired quantum Hall state. *Phys. Rev. B* **59**, R7825 (1999).

[748] C. Zhou and M. Berciu, Resistance fluctuations near integer quantum Hall transitions in mesoscopic samples. *Europhys. Lett.* **69**, 602 (2005).

[749] C. Zhou and M. Berciu, Correlated mesoscopic fluctuations in integer quantum Hall transitions. *Phys. Rev. B* **72**, 085306 (2005).

[750] C. Zhou, M. Berciu, and R. N. Bhatt, Effects of large disorder on the Hofstadter butterfly. *Phys. Rev. B* **71**, 125310 (2005).

[751] J. Zhu, W. Pan, H. L. Stormer, L. N. Pfeiffer, and K. W. West, Density-induced interchange of anisotropy axes at half-filled high Landau levels. *Phys. Rev. Lett.* **88**, 116803 (2002).

[752] J. M. Ziman, *Electrons and Phonons* (Oxford University Press, 1960).

[753] U. Zülicke, J. J. Palacios, and A. H. MacDonald, Fractional-quantum-Hall edge electrons and Fermi statistics. *Phys. Rev. B* **67**, 045303 (2003).

[754] S. D. M. Zwerschke and R. R. Gerhardts, Positive magnetoresistance of composite fermion systems with a weak one-dimensional density modulation. *Phys. Rev. Lett.* **83**, 2616 (1999).

Index

For EU product safety concerns, contact us at Calle de José Abascal, 56–1°,
28003 Madrid, Spain or eugpsr@cambridge.org.

www.ingramcontent.com/pod-product-compliance
Ingram Content Group UK Ltd.
Pitfield, Milton Keynes, MK11 3LW, UK
UKHW051007240426
470322UK00018B/546